D1520280

Generalized Method of Eigenoscillations in Diffraction Theory

Generalized Method of Eigenoscillations in Diffraction Theory

M. S. Agranovich
B. Z. Katsenelenbaum
A. N. Sivov
N. N. Voitovich

Berlin · Weinheim · New York · Chichester · Brisbane · Singapore · Toronto

Authors:

Prof. Dr. M. S. Agranovich, Moscow State Institute of Electronics and Mathematics, Moscow, Russia
Prof. Dr. B. Z. Katsenelenbaum, Institute of Radio Engineering and Electronics of the Russian Academy of Sciences, Moscow, Russia
Prof. Dr. A. N. Sivov, Institute of Radio Engineering and Electronics of the Russian Academy of Sciences, Frjasino, Russia
Prof. Dr. N. N. Voitovich, Institute of Applied Problems of Mechanics and Mathematics of the National Academy of Sciences of the Ukraine, Lviv, Ukraine

Translator:

Dr. V. Nazaikinskii, Moscow State University, Moscow, Russia

1st edition
With 24 figures and 11 tables

Die Deutsche Bibliothek – CIP-Einheitsaufnahme

Generalized method of eigenoscillations in diffraction theory / M. S. Agranovich ...
[Transl.: V. Nazaikinskii]. – Berlin ; Weinheim ; New York ; Chichester ;
Brisbane ; Singapore ; Toronto : WILEY-VCH, 1999
ISBN 3-527-40092-3

Printed on non-acid paper.
The paper used corresponds to both the U. S. standard ANSI Z.39.48 – 1984
and the European standard ISO TC 46.

Printing: GAM Media GmbH, Berlin

Bookbinding: Druckhaus „Thomas Müntzer“, Bad Langensalza

Printed in the Federal Republic of Germany

WILEY-VCH Verlag Berlin GmbH
Bühringstraße 10
D-13086 Berlin
Federal Republic of Germany

Preface

The method of eigenoscillations is usually viewed as a self-contained method for solving diffraction problems, which has little in common with any other method. In this book, the authors show that this method can be viewed as a special case of some more general method, in which the spectral parameter of the auxiliary homogeneous problem is not necessarily the frequency. In many diffraction problems, primarily in the study of open resonators, other versions of this general method prove to be natural and most efficient. In these versions, the homogeneous problem is stated in a way such that other physical parameters play the role of the spectral parameter.

The method can be applied to a wide range of diffraction and scattering problems in acoustics, electromagnetic theory, and quantum mechanics (the Schrödinger equation). It is particularly efficient in the analysis of resonant systems and permits one to represent the solution in an infinite domain by a series (the spectrum is discrete), carry out partial summation of the nonresonant background, widely use the variational technique, etc. A number of new problems are solved by this method.

The main part of the book contains the exposition of the formal technique of various versions of the generalized method of eigenoscillations (Chapters 1–2), the construction of stationary functonals (Chapter 3), and examples of applications to specific problems (Chapter 4).

Chapter 5 deals with mathematically rigorous treatment and justification of the technique constructed in Chapters 1–2 and connected with the use of nonselfadjoint operators. To this end, the results of some parts of modern analysis are used. This contact of classical mathematical physics with modern analysis proves to be useful to both parties; we hope that it will be interesting for physicists as well as mathematicians.

Most of the results of the first four chapters were obtained jointly by B. Z. Katsenelenbaum, A. N. Sivov, and N. N. Voitovich. Sections 0.1–1.6 were written by B. Z. Katsenelenbaum. Sections 2.1–2.6, 4.5–4.7, 4.9, and 4.10 were written by A. N. Sivov. Sections 3.1–3.6 and 4.1–4.3 were written by N. N. Voitovich. Sections 4.4, 4.8, and 4.11 were written jointly by the above three authors. Chapter 5 was written by M. S. Agranovich. The authors are grateful to V. A. Dikarev, A. D. Shatrov, and V. V. Shevchenko, who kindly permitted to include their results in the book (these results are contained in Section 3.7, Subsections 1.3.6 and 2.1.4, and Subsection 4.3.5, respectively).

This book is partly based on an earlier book by the same authors published in Russian by Nauka Publishing House, Moscow in 1977. Since then, considerable progress has been made both in physical and mathematical aspects of the subject and lots of new results have been obtained, which has made writing a new book virtually inevitable. The authors are deeply grateful to Prof. Reinhard Mennicken, who put forward the idea of publishing this new book in English and was an eager supporter of this project.

The authors are also indebted to Dr. Vladimir Nazaikinskii, who carefully translated the manuscript into English and made many useful mathematical remarks. Special thanks are due to Gesine Reiher for editing the book thoroughly and for her permanent attention to the project at all stages.

Moscow, December 1998

M. S. Agranovich, B. Z. Katsenelenbaum,
A. N. Sivov, and N. N. Voitovich

Contents

Introduction

0.1 Essentials of the Method. Outline of the Book

1. The method of eigenoscillations is widely used for solving interior problems in diffraction theory. In this method, the field due to the excitation of a closed volume (that is, the solution of a nonhomogeneous problem) is sought in the form of a series expansion in some auxiliary functions, depending on the domain in which the problem is considered. These functions are the eigenfunctions of an auxiliary homogeneous problem corresponding to various eigenfrequencies and form a complete orthogonal system. The method is especially efficient for resonators with low losses and at frequencies close to one of the eigenfrequencies.

The generalized eigenoscillation method, whose foundations are presented in this book, is also based on the *representaton of the solution of a stationary diffraction problem by a series* with respect to some orthogonal system of functions and is also efficient primarily near resonances. The method also applies to open resonators and even to arbitrary problems of diffraction on bounded bodies. The main idea of the method is as follows. As eigenfunctions, one introduces the solutions of the corresponding homogeneous problem, in which the spectral parameter is no longer the frequency (as in the ordinary method); instead, *the spectral parameter is some electrodynamic parameter*, say, the permittivity of some auxiliary body occupying the same volume as the body on which the diffraction occurs. The choice of the specific electrodynamic variable used as the spectral parameter depends of the form of the diffraction problem; the book presents several versions of the method. In all these versions, *the eigenfunctions correspond to a real frequency*; in particular, for open systems they satisfy the radiation condition, and the diffracted field (that is, the difference between the total and the incident field) can be expanded in a series in these eigenfunctions without an additional integral over the continuous spectrum.

In the ordinary method of eigenoscillations, which we treat as a special case of the generalized method, the amplitudes of the terms of the series representing the field are inversely proportional to the difference between the eigenfrequency and the frequency of the sources (the latter is just the frequency entering the diffraction problem), and the resonance occurs when these frequencies are close to each other and hence the amplitude of one of the terms becomes much larger than the amplitudes of all the other terms. In any other version of the method, the amplitudes are inversely proportional to the difference between the eigenvalue and the actual value in the diffraction problem of the electrodynamic parameter used as the spectral parameter in that version. The eigenvalue is a function of the frequency. For example, for the method in which the spectral parameter is the impedance of the surface of the body, the resonance occurs near the frequency for which the eigenvalue is close to the actual impedance, and then the amplitude of the corresponding term of the series is large.

For specific problems, some version of the generalized method usually proves to be simpler than the eigenoscillation method. If the problem is reduced to a trancendental equation, then this equation is the same in all methods, but it is usually easier to solve it for some parameter other than frequency. In some methods, it suffices to calculate the left-hand side of the equation; if the equation is written out appropriately, then the left-hand side itself is the desired eigenvalue and hence completely determines the resonance curve. For systems with losses, one often manages to avoid computations in the complex domain. For example, if the permittivity of the body is complex, then it is advisable to apply a method in which the spectral parameter is just the permittivity; then the eigenvalues are real (provided that there are no other losses in the problem other than the dielectric ones) and can be found from a real equation. Similar relations hold for the variational technique. In all versions of the generalized method, there exist stationary functionals for the eigenvalues, and, for example, an application of the Ritz method results in the same equations, which are usually easier to solve for the eigenvalues of the generalized method than for the eigenfrequencies.

For problems reducible to integral equations, there are versions of the generalized method for which the kernel has a particularly simple form and contains the spectral parameter as a factor (rather than as an argument of special functons, like the frequency). If the diffraction problem can be reduced to a nonhomogeneous integral equation, then the corresponding homogeneous integral equation of the second kind can usually be treated as an equation for the eigenfunctions of one of the generalized methods. The eigenvalues of these equations have a simple physical meaning. Once they are known, one can comprehensively study the vicinity of the resonance frequencies.

The present book suggests a unified approach to various possibilities occuring in the generalized eigenoscillation method.

2. Let us illustrate some features of the generalized method using the version in which the spectral parameter is the impedance of the surface as an example. (A detailed exposition of this technique is given in Section 2.1.) Consider the simplest two-dimensional scalar diffraction problem (interior or exterior) for a circular cylinder of radius a. The unknown field $U(r,\varphi)$ must satisfy the boundary condition $U(a,\varphi) = 0$, the wave equation with nonzero right-hand side (exciting currents), and the radiation condition if the exterior problem is to be solved. The eigenfunctions and eigenvalues of the cited version of the generalized problem can be written out explicitly for this problem.

For the interior problem, the eigenfunctions are given by the formulas (cf. the eigenfunctions (0.23) for the same body in the eigenoscillation method)

$$u_n(r,\varphi) = J_n(kr)\cos n\varphi, \quad n = 0, 1, 2, \ldots. \tag{0.1}$$

(To simplify the notation, we restrict ourselves to even functions of φ.)

The dependence on time is assumed in the form $\exp(i\omega t)$. The frequency will be characterized by the wave number $k = \omega/c$; from now on, *k will be referred to as the frequency.*

On the surface of the body ($r = a$), the functions (0.1) are orthogonal to each other. Each of them satisfies the impedance condition (the boundary condition of the third kind)

$$u_n + w_n \frac{\partial u_n}{\partial r}\bigg|_{r=a} = 0. \tag{0.2}$$

The numbers w_n are treated as the eigenvalues with the corresponding eigenfunctions (0.1). In this simple example, it is easy to find these eigenvalues:

$$w_n = \frac{-J_n(ka)}{kJ_n'(ka)}. \tag{0.3}$$

The solution of the diffraction problem can be represented in the form

$$U(r,\varphi) = U^0(r,\varphi) + \sum_{n=0}^{\infty} A_n u_n(r,\varphi). \tag{0.4}$$

Here U^0 is the field of the same sources in the absence of the obstacle (the cylinder), that is, in vacuum. Here, as well as everywhere in the following, the difference $U - U^0$ rather than U is expanded in the series.

The series (0.4) satisfies the nonhomogeneous wave equation, but it does not satisfy the boundary condition for U at $r = a$ termwise. The coefficients A_n can be found from the requirement that the sum of the series must satisfy the boundary condition. Since the functions u_n are orthogonal at $r = a$, we obtain closed-form expressions for the A_n. We can represent these expressions in the form

$$A_n = -\frac{\int_0^{2\pi} U^0(a,\varphi)\cos n\varphi\, d\varphi}{\pi(1+\delta_{0n})kw_n J_n'(ka)}, \tag{0.5}$$

where the dependence of A_n on w_n is shown explicitly. The coefficients A_n are inversely proportional to w_n and actually contain $J_n(ka)$ in the denominator (see (0.3); the resonance occurs as the frequency approaches the roots of the denominator.

The same solution (0.4) of the diffraction problem with the same u_n (0.1) and w_n (0.3) remains valid if the desired field must satisfy the impedance condition

$$\left(U + w\frac{\partial U}{\partial r}\right)\bigg|_{r=a} = 0 \tag{0.6}$$

at $r = a$, where w is a given number. Then A_n contains the difference $w - w_n$ in the denominator, and the resonance corresponds to small values of this difference.

The technique is by no means more complicated for exterior problems. *The eigenfunctions*, as well as the desired solution, *must satisfy the radiation condition*; for the time dependence accepted in the book, they are given by

$$u_n(r, \varphi) = H_n^{(2)}(kr) \cos n\varphi. \tag{0.7}$$

The formulas for w_n are modified accordingly. The eigenvalues w_n are now complex,

$$w_n = \frac{H_n^{(2)}(ka)}{kH_n^{(2)\prime}(ka)}. \tag{0.8}$$

This is due to the fact that there are radiation losses in the system, and so w_n never vanishes and A_n never becomes infinite, whatever the frequency. The signs in (0.3) and (0.8) are opposite, since so are the senses of the normal in the two problems.

The functional

$$W(u) = \frac{\int_S u^2 \, ds}{\int_V [-(\nabla u)^2 + k^2 u^2] \, dV} \tag{0.9}$$

(where V and S are the domain in question and its boundary, respectively) is stationary on the functions (0.1) and is equal to $-w_n$ on these functions according to (0.3); the integration is carried out over the interior of the disk. For the exterior problem, the functional is stationary on the functions (0.7) and is equal to (0.8) on these functions; the integral is taken over the exterior of the disk, and one must set $\operatorname{Im} k = +0$ in the integral. The stationary state property of (0.9) permits one to use direct methods for calculating w_n and hence for finding the resonance curve, which is described by the dependence of $1/|w_n(k) - w|$ on k.

In this problem, needless to say, all the results are trivial and can be obtained merely by separation of variables. In the applications of this method to diffraction on a cylinder or the excitation of a cylindrical resonator, one normally uses series of the form (0.4). In this version (Section 2.1) of the generalized method of eigenoscillations, such series are also used for the case in which each u_n is not a product of functions of one coordinate. The main fact is that u_n satisfies the wave equaton and the boundary condition (0.2).

3. In all versions of the method described in the book, the solution of the diffraction problem (that is, the solution of the nonhomogeneous equation) is sought in the form of a series similar to (0.4), and various versions differ in how the eigenfunctions u_n are introduced, that is, what homogeneous problems they satisfy. Both the equation and the boundary conditions for u_n must be homogeneous. Should the equation for u_n differ from that for U only by the fact that the right-hand side is zero, with the boundary conditions for u_n being the same as for U, the u_n would be identically zero in general. To obtain a nontrivial system of functions u_n, one has to modify either the equation or the boundary conditions by introducing some free parameter, which will play the role of the spectral parameter.

In the second section of the introduction and in the first chapter, we present a technique in which the left-hand side of the equation undergoes some modification and the boundary conditions for u_n coincide with those for U. In the second chapter, the u_n satisfy the equation with the original left-hand side, but the boundary conditions for u_n differ from those for U.

Accordingly, *either the equation or the boundary condition for U will not be satisfied termwise.* The requirement that the sum of the series nevertheless satisfies the corresponding condition permits one to find the coeficients A_n. It is important that if the A_n are sought from a condition on S, then the u_n are orthogonal (or orthogonal in the real sense) on S (just as in the previous example); if the A_n must be found from a condition in V, then the u_n are orthogonal in V.

4. In Section 0.2 we give an exposition of the usual eigenfrequency method. Of all requirements imposed on the solution of the diffraction problem, only one is violated by the functions u_n in this method; specifically, the requirement that the frequency must be equal to the frequency of the sources. The frequency is here the spectral parameter. This method is well-known and is presented to ensure a complete coverage of the technique developed in the book.

We always name a method after the parameter that is used as an eigenvalue in the method; since we characterize the frequency by the wave number k, the eigenfrequency method is referred to as the *k-method* in our book.

In Chapter 1 we present a method in which *permittivity is the spectral parameter.* The method applies to the problem of diffraction on a dielectric body. The functions u_n satisfy the homogeneous wave equaton in which the permittivity ε of the body is replaced by the eigenvalue ε_n. The functions u_n are orthogonal with respect to integration over the body, and the coefficients A_n contain the difference $\varepsilon - \varepsilon_n$ in the denominator. If there are no losses at all in the system, or at least only dielectric losses, that is, losses due to the fact that $\operatorname{Im} \varepsilon \neq 0$, are present, then the ε_n are real. For open resonators, and, more generally, for any diffraction problems with radiation losses, one has $\operatorname{Im} \varepsilon_n > 0$, that is, ε_n is the permittivity of some *active body* (a body *releasing energy as the field is applied*). The ε-method can be generalized to diffraction problems for inhomogeneous dielectric bodies. In particular, the method applies to the quantum-mechanical problem of scattering by a potential field, which is briefly considered in Sections 1.5 and 4.2.

5. In Chapter 2 we introduce *functions u_n which satisfy the "correct" equation* (that is, the homogeneous equation with the same ε and k as in the diffraction problem) *and an "incorrect" boundary condition* (that is, a condition different from the boundary condition for U). For various diffraction problems, one has various convenient boundary conditions determining the u_n and containing the eigenvalues. In Section 2.1 we present a version outlined in Subsection 2: instead of condition (0.6), which is valid for U, condition (0.2) is used for u_n, where w_n is the eigenvalue. The functions u_n are orthogonal on S, and the coefficients A_n contain the difference $w - w_n$ in the denominator. If there are no losses in the system, or if only losses due to a nonzero imaginary part of w are present, then the w_n are real, just as in (0.3); in open systems one has

$\operatorname{Im} w_n > 0$, just as in (0.8).

In one of the two versions of the ρ-method (Section 2.2), the condition requiring that the jump of the normal derivative of u_n must be proportional to u_n on the surface S of the metal body on which the diffraction occurs is imposed; the function u_n itself is continuous on S. Here the eigenvalue is the proportionality coefficient ρ_n. The function u_n satisfies the following remarkably simple integral equation on S:

$$-\rho_n u_n = \int_S G u_n \, dS, \tag{0.10}$$

where the kernel G is the Green function of vacuum. The method also applies to problems of diffraction on dielectric bodies.

In the second version of the ρ-method (Section 2.3), the normal derivative of u_n is continuous on S, but u_n itself undergoes a jump on S. The jump is proportional to the normal derivative, and the proportionality coefficient $\tilde{\rho}_n$ is an eigenvalue of the problem. The method given in Section 2.4 generalizes both the w- and the ρ-method; linear relations are imposed in this method on the values of u_n and its normal derivative on both sides of the surface. This method applies to diffraction problems under quite general boundary conditons for u on S, conditions like those descibing a nonisotropic semitransparent grating.

In the method given in Section 2.5, the spectral parameter is introduced into the conditions at infinity for the eigenfunctions. This method is convenient in diffraction problem in which only radiation losses are present.

The specific feature of all methods of Chapter 2, as opposed to the methods given in Chapter 1, is as follows. The series like (0.4) in the case of separation of variables have one summation index less; for example, for a two-dimensional problem the sum (0.4) is one-dimensional, and for a one-dimensional problem this sum is reduced to a single term. This "dimension reduction" dramatically simplifies specific computations even for the general case of a body of an arbitrary shape.

6. In Chapter 3 we derive *stationary functionals for the eigenvalues*, which have nearly the same form as (0.9) or the well-known Rayleigh type functional for the square of the eigenfrequency in an empty resonator:

$$K(u) = \frac{\int_V (\nabla u)^2 \, dV}{\int_V u^2 \, dV}. \tag{0.11}$$

Each of these functionals is stationary on the eigenfunctions of the corresponding homogeneous problems and and assumes the stationary values ε_n, ρ_n, or $\tilde{\rho}_n$, just in the same way as (0.11) produces k_n^2 and (0.9) produces w_n.

It turns out that there is a simple relationship between all these functionals. For example, let us show how one can obtain (0.9) from (0.11). Consider the simplest case of an interior problem in which condition (0.6) with given w must be satisfied on a closed surface. Let us write out a functional of the type (0.11) which will be stationary in the class of functions that need not satisfy the boundary condition (0.6). To this end, it suffices to supplement the numerator in (0.11) by an appropriately chosen surface integral, that is, replace (0.11) by

$$K(u) = \frac{\int_V (\nabla u)^2 \, dV - \frac{1}{w} \int_S u^2 \, dS}{\int_V u^2 \, dV}. \tag{0.12}$$

Now if in (0.12) we replace $K(u)$ by k^2 and w by $W(u)$ and solve the resultant relation for $W(u)$, then we obtain (0.9). This formal procedure, justified in Chapter 3, corresponds to the relationship between the homogeneous problems in the k-method (the eigenvalue is k_n^2, and w is a parameter of the problem) and in the w-method (k^2 is a parameter of the problem, and w_n is the eigenvalue). In all versions of the generalized method, stationary functionals can be obtained from some equation containing the parameters k^2, ε, w, ρ, and $\tilde{\rho}$ by solving this equation for the corresponding parameter and by taking the right-hand side of the resulting equation as the desired functional.

7. Chapter 4 contains applications of the technique developed in the preceding chapters to problems which cannot be solved by other methods at all or at least without considerable effort. Only the first section and partly the second contain illustrative examples.

The first two subsections of the last section in Chapter 4 contain bibliographical remarks. Although the book comprises the authors' work (partly unpublished), many of the ideas, often in a very similar form, were known previously or published independently. In these sections we review the corresponding papers and compare the results with those presented in the book. The list of general papers on mathematics and electromagnetic theory used in the book is rather brief. In the third subsection we present some ideas concerning further development of the method.

8. The mathematical justification of the technique developed in Chapters 1 and 2 is based on some parts of modern functional analysis. Chapter 5 contains a concise exposition of necessary information from these parts of functional analysis. Further, this information is used to analyze the properties of the operators associated with the most important problems considered in this book. These operators are nonselfadjoint (which is related to the essence of the problems considered), and the specific feature of the technique developed in the book is the use of series in the eigenfuctions of these nonselfadjoint operators. However, in Chapter 5 these operators are shown to be very close to selfadjoint operators. This permits one to prove that the diffracted field admits the desired series expansions, and moreover, under an appropriate choice of the summation method, these series rapidly converge and can be differentiated term by term. Chapter 5 also contains the asymptotics of the eigenvalues and *a priori* estimates for the solutions of the problems considered in the book. In more detail, the contents of Chapter 5 is revealed in the first, introductory section of that chapter.

To simplify the formal technique, throughout Chapters 1–4 we assume that the eigenvalues are simple. As is shown in Chapter 5, the general case is not much more complicated.

0.2 Method of Eigenfrequencies (the k-Method)

In this section we use the simplest scalar problem to expose a *well-known method for solving the problem on the excitation of closed resonators.* The methods described in the subsequent chapters can be viewed as generalizations of the method given in this section, which is emphasized in the title of the book.

1. Suppose that we seek a function U that is a solution of the Helmholtz equation

$$\Delta U + k^2 U = f \tag{0.13}$$

in a bounded domain V and satisfies the condition

$$U|_S = 0 \tag{0.14a}$$

on the boundary S of the domain.

Let us introduce a system of functions u_n satisfying the homogeneous equation

$$\Delta u_n + k_n^2 u_n = 0 \tag{0.15}$$

and the boundary condition

$$u_n|_S = 0. \tag{0.16a}$$

In the nonhomogeneous problem (0.13), (0.14a), k is a given number, whereas in the homogeneous problem (0.15), (0.16a), k_n is an eigenvalue. There are countably many such eigenvalues (eigenfrequencies) for which the problem has nonzero solutions, known as eigenfunctions.

The eigenvalues of problem (0.15), (0.16a) are real. This follows from the first Green formula

$$\int_V (u\Delta v + \nabla u \nabla v)\, dV = \int_S u \frac{\partial v}{\partial N}\, dS. \tag{0.17}$$

In (0.17) and throughout the following, N is the outward (with respect to V) normal on S. In we apply (0.17) to $u = u_n^*$ and $v = u_n$, substitute Δu_n according to (0.15), and take into account the fact that the surface integral is zero by virtue of (0.16a), then we obtain a real representation of k_n^2 via the eigenfunction, namely,

$$k_n^2 = \frac{\int |\nabla u_n|^2\, dV}{\int |u_n^2|\, dV}. \tag{0.18}$$

The functions corresponding to distinct values of k_n are orthogonal to each other in the real sense. In other words, integration over the volume yields

$$\int u_n u_m\, dV = 0, \quad n \neq m. \tag{0.19}$$

This readily follows from the second Green formula

$$\int_V (u\Delta v - v\Delta u)\, dV = \int_S \left(u\frac{\partial v}{\partial N} - v\frac{\partial u}{\partial N}\right) dS \tag{0.20}$$

applied to $u = u_n$ and $v = u_m$. By (0.15), the left-hand side is equal to

$$(k_n^2 - k_m^2) \int u_n u_m \, dV, \tag{0.21}$$

whereas the right-hand side is zero by virtue of (0.16a).

Under the boundary condition (0.16a), we can always ensure that the functions u_n will be real-valued. Hence, Eq. (0.19) obviously implies the Hermitian orthogonality

$$\int u_n u_m^* \, dV = 0, \quad n \neq m. \tag{0.22}$$

The subscript n in all these formulas is a multi-index; it contains three components in three-dimensional problems and two components on two-dimensional problems (obtained by separation of variables). If V is a disk of radius a (a two-dimensional problem), then the eigenfunctions have the form (we restrict ourselves to even functions of φ)

$$u_n = J_m(k_{mq}r)\cos m\varphi, \quad m = 0, 1, \ldots, \; q = 1, 2, \ldots, \tag{0.23}$$

and the eigenvalues k_{mq} are found from the equation

$$J_m(k_{mq}a) = 0. \tag{0.24a}$$

For the other values of k_{mq}, there are no nonzero functions that simultaneously satisfy Eq. (0.15) (and, in particular, are finite for all $r \leq a$) and condition (0.16a). In the construction of the systems of functions u_n, one does not take into account the form of the function f, that is, the character of the excitation in the diffraction problem (0.13), (0.14a). *This system of eigenfunctions is prepared as if in advance for solving an arbitrary problem on the excitation of the domain bounded by the surface S with the boundary condition* (0.14a). Once we know the functions u_n and the numbers k_n, we can represent the solution of problem (0.13), (0.14a) in the form

$$U = \sum_n A_n u_n. \tag{0.25}$$

In three-dimensional problems, threefold summation is implied in (0.25). The series (0.25) satisfies the boundary condition (0.14a) term by term, and the validity of Eq. (0.13) is ensured by choosing the coefficients A_n. By substituting (0.25) into (0.13) and by applying the operator Δ term by term, we obtain, with regard to the orthogonality condition (0.19),

$$A_n = \frac{1}{k^2 - k_n^2} \frac{\int f u_n \, dV}{\int u_n^2 \, dV}. \tag{0.26}$$

This is the main formula of the method of eigenoscillations. Together with (0.25), it provides the solution of problem (0.13), (0.14a) for an arbitrary excitation f.

2. The technique remains the same if the diffraction field must satisfy the boundary condition

$$\left.\frac{\partial U}{\partial N}\right|_S = 0 \tag{0.14b}$$

instead of (0.14a). Then the eigenfunctions u_n must also be different: they must satisfy the same boundary condition

$$\left.\frac{\partial u_n}{\partial N}\right|_S = 0. \tag{0.16b}$$

Hence we obtain new eigenvalues k_n, but the surface integrals in (0.17) and (0.20) are again zero, and hence all the other formulas remain valid. For example, the eigenfunctions for the disk are still the Bessel functions (0.23), but the eigenvalues k_{mq} are now determined from the equation

$$J'_m(k_{mq}a) = 0 \tag{0.24b}$$

rather than from (0.24a). Some modifications occur if U must satisfy the boundary condition of the third kind on S,

$$\left.\left(U + w\frac{\partial U}{\partial N}\right)\right|_S = 0, \tag{0.14c}$$

where w is independent of the frequency. The eigenfunctions must satisfy the same condition

$$\left.\left(u_n + w\frac{\partial u_n}{\partial N}\right)\right|_S = 0 \tag{0.16c}$$

with the same impedance w, and the eigenfunctions corresponding to distinct k_n are still orthogonal in the sense of (0.19); this follows from the fact that the right-hand side of (0.20) vanishes after the substitution of (0.16c). However, the orthogonality (0.22) no longer takes place in general. It is preserved only if $\operatorname{Im} w = 0$.

If the wave equation for U differs from (0.13) and has, say, the form

$$\Delta U + k^2 \varepsilon U = f \tag{0.27}$$

(where ε may be a function of the coordinates), then the eigenfunctions u_n must satisfy the homogeneous equation with the same ε. These functions will be orthogonal, but not in the sense of (0.19). Instead, we have the formula

$$\int \varepsilon u_n u_m \, dV = 0, \tag{0.28}$$

which can readily be obtained from (0.20). Accordingly, we have the new formula

$$A_n = \frac{1}{k^2 - k_n^2} \frac{\int f u_n \, dV}{\int \varepsilon u_n^2 \, dV} \tag{0.29}$$

for the coefficients, as well as the following expression for k_n^2:

$$k_n^2 = \frac{\int |\nabla u_n|^2 \, dV}{\int \varepsilon |u_n|^2 \, dV}. \tag{0.30}$$

It follows from the last formula that in general the k_n^2 are real only if $\mathrm{Im}\,\varepsilon = 0$.

There is nothing essentially new for electromagnetic problems. Here we do not present the formulas for the coefficients of the expansions of the fields $\mathbf{E}$ and $\mathbf{H}$ in the vector eigenfunctions (these formulas can be found in textbooks); let us only point out that these coefficients also contain the typical denominator $k^2 - k_n^2$.

3. Although the above technique can be applied for arbitrary frequencies and arbitrary complex ε and w, it is most efficient when some term in the series (0.25) for the total field is much larger than all the other terms, so that the field U in the diffraction problem is close to the field of one of the eigenoscillations. This is the case for systems with small losses ($\mathrm{Im}\,k_n$ is small) provided that the frequency of k of the exciting field is close to $\mathrm{Re}\,k_n$ (one of the denominators $k^2 - k_n^2$ is small in modulus). Under these conditions, that is, near a resonance in a high-Q system, to reveal the structure of the diffraction field, it suffices to find the field of a single eigenoscillation, and the frequency characteristic is almost completely determined by one complex number, namely, the eigenfrequency of this oscillation. In other words, the method of eigenoscillations is especially convenient if the solution of problem (0.13) (or (0.27)), (0.14) is given by the approximate one-term formula

$$U \sim \frac{c_n}{k^2 - k_n^2} u_n. \tag{0.31}$$

The exciting currents f occur only in the proportionality coefficient c_n in this formula.

Other versions of the generalized method of eigenoscillations also possess this property of the method of eigenfrequencies.

4. The method of eigenfrequencies cannot be transferred directly to systems like open resonators, i.e., to exterior diffraction problems. One can readily show that *the eigenfunctions provided by this method necessarily grow at infinity.* Indeed, let us apply (0.17) to the domain bounded by a sphere of radius r_0 ($kr_0 \gg 1$) and the surface S, on which we assume condition (0.14a) or (0.14b) for simplicity. On the sphere, the field u_n and its normal derivative must have the form

$$u_n \simeq \frac{e^{-ik_n r}}{r} \Phi_n(\theta, \varphi), \quad \frac{\partial u_n}{\partial r} \simeq -ik_n \frac{e^{-ik_n r}}{r} \Phi_n(\theta, \varphi) \tag{0.32}$$

(in the two-dimensional problem, r is replaced by $\sqrt{r}$ in the denominator, and $\Phi_n(\theta,\varphi)$ is replaced by $\Phi_n(\varphi)$). By setting $k_n = k'_n + ik''_n$ and by taking the imaginary part of Eq. (0.17) with $u = u^*_n$ and $v = u_n$, we obtain

$$k''_n \int |u_n|^2 \, dV = e^{2k''_n r_0} \int |\Phi_n|^2 \, d\Omega, \tag{0.33}$$

where Ω is the solid angle, $d\Omega = \sin\theta \, d\varphi \, d\theta$.

For $k''_n = 0$, the left-hand side would be zero, and hence in the asymptotic formula (0.32) one would have $\Phi_n(\theta,\varphi) = 0$ and hence $u_n \equiv 0$ (the latter implication is well known). Obviously, $k''_n < 0$ is also impossible, because the right-hand side of (0.33) is nonnegative. Thus, $\operatorname{Im} k_n > 0$ for all eigenfunctions, and hence they increase unboundedly remote from the body. For example, for the exterior of a cylinder of radius a with the boundary condition (0.14a), the eigenfrequencies k_{mq} are the roots of the equation

$$H^{(2)}_m(k_{mq}a) = 0, \tag{0.34}$$

and the eigenfunctions grow at infinity as

$$\exp(k''_{mq} r)/\sqrt{r}, \quad k''_{mq} > 0.$$

In diffraction problems, the diffracted field must satisfy the radiation condition, that is, have the form

$$u \simeq \frac{e^{-ikr}}{r}\Phi(\theta,\varphi) \tag{0.35}$$

as $r \to \infty$ (with $1/r$ replaced by $1/\sqrt{r}$ in two-dimensional problems). It follows that the total field cannot grow at infinity and hence cannot be expanded in the series (0.25) in the growing eigenfunctions of the homogeneous problem (0.15), (0.16a). This has been well known for a long time. To preserve the series of the type (0.25) together with their main advantage, Eq. (0.31), one must add an integral over the continuous spectrum to these series. Then for nearly resonant frequencies in high-Q systems, Eq. (0.31) will be valid near the body, and the integral over the continuous spectrum will compensate for the exponential growth of all the terms in (0.25) remote from the body.

In the generalized method, the discrete representation of the field is preserved for exterior diffraction problems as well.

Chapter 1

Spectral Parameter in the Equation (the ε-Method)

The idea of the method developed in this chapter is to take the *permittivity* to be the spectral parameter of the homogeneous problems used to define the system of eigenfunctions. The diffracted field is represented by a series in these functions. The eigenvalue ε_n is the permittivity of an *auxiliary body* that occupies the same domain as the diffracting body. *The actual permittivity does not occur in the homogeneous problem.* That is why, in particular, the fact that the actual ε is complex does not affect the eigenvalues. The eigenvalues are real if there are no losses other than dielectric ones in the problem. However, if, for example, radiation is present, then the method remains the same, the diffracted field can still be represented by a series in the eigenfunctions, but the eigenvalues are complex. The imaginary parts of the eigenvalues are positive; this corresponds to the fact that in the auxiliary homogeneous problem the body is active, and there is energy release in the body, which compensates for the losses. This chapter also contains generalizations to the case of diffraction on an inhomogeneous body. In Section 1.5 this technique is applied to the quantum-mechanical problem on elastic scattering by a potential field. The technique is first presented for scalar problems. The generalization to vector problems described by the Maxwell equations is given in Section 1.6; it requires no additional physical or mathematical ideas. The mathematical aspects of the ε-method for the "first polarization" will be considered in Chapter 5.

1.1 Dielectric Body in a Closed Resonator with Ideal Walls

1. We seek a function U that satisfies the equation

$$\Delta U + k^2 \varepsilon U = f \quad \text{in} \quad V^+ \tag{1.1a}$$

in a bounded domain V^+, the equation

$$\Delta U + k^2 U = f \quad \text{in} \quad V^- \tag{1.1b}$$

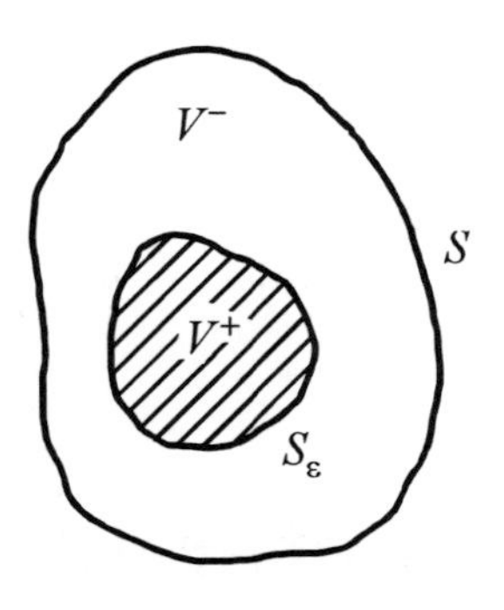

Figure 1.1

outside V^+ (that is, in V^-) (see Fig. 1.1), the conditions

$$(U^+ - U^-)|_{S_\varepsilon} = 0, \quad \left(\frac{\partial U^+}{\partial N} - \frac{\partial U^-}{\partial N}\right)\Big|_{S_\varepsilon} = 0 \tag{1.1c}$$

on the boundary S_ε of the dielectric (here N is the outward normal to S_ε and U^+ and U^- are the values of U on the corresponding sides of S_ε), and the condition

$$U|_S = 0 \tag{1.1d}$$

on the boundary S of the resonator.

Here the permittivity ε of the body is allowed to be a complex number in general.

Let U^0 be the field that satisfies the equation

$$\Delta U^0 + k^2 U^0 = f \tag{1.2a}$$

in the entire resonator and the boundary condition

$$U^0|_S = 0 \tag{1.2b}$$

on S. In other words, U^0 *is the field generated by the same sources in the absence of the dielectric body.*

We seek the solution of the diffraction problem (1.1a)–(1.1d) in the form of the series

$$U = U^0 + \sum_n A_n u_n, \tag{1.3}$$

where the u_n are the eigenfunctions of the auxiliary homogeneous problem

$$\Delta u_n + k^2 \varepsilon_n u_n = 0 \quad \text{in} \quad V^+, \tag{1.4a}$$

$$\Delta u_n + k^2 u_n = 0 \quad \text{in} \quad V^-, \tag{1.4b}$$

$$(u_n^+ - u_n^-)|_{S_\varepsilon} = 0, \quad \left(\frac{\partial u_n^+}{\partial N} - \frac{\partial u_n^-}{\partial N}\right)\Big|_{S_\varepsilon} = 0, \tag{1.4c}$$

$$u_n|_S = 0. \tag{1.4d}$$

The number ε_n is an eigenvalue of the homogeneous problem.

The auxiliary problem (1.4a)–(1.4d) has also an independent physical meaning. It describes an eigenoscillation, that is, a field existing in the absence of sources, *which occurs at the frequency of the diffraction problem* in some auxiliary system of dielectric and metal bodies. The comparison of (1.4) with (1.1) shows that this system consists of the same resonator and a dielectric body of the same shape. In contrast with the system for which the diffraction problem is to be solved, the permittivity ε_n of the body in the homogeneous problem is not equal to ε. Hence, for some values of ε_n the homogeneous problem has a nontrivial solution for a given k.

This method for introducing the eigenfunctions has the specific feature that *outside the dielectric body, the eigenfunctions satisfy the same equation as the diffracted field* $U - U^0$.

Let us compare problem (1.4) with the corresponding homogeneous problem in the k-method. For (1.1), the field U in this method would be expanded in the eigenfunctions satisfying the equations

$$\begin{aligned} \Delta u_n + k_n^2 \varepsilon u_n &= 0 \quad \text{in} \quad V^+, && (1.5a) \\ \Delta u_n + k_n^2 u_n &= 0 \quad \text{in} \quad V^- && (1.5b) \end{aligned}$$

and the same boundary conditions (1.4c) and (1.4d). Here the eigenvalue is k_n^2.

The eigenfunctions of problems (1.4) and (1.5) are different, since (1.5b), in contrast with (1.4b), contains the eigenvalue k_n^2 rather than the square of the actual frequency k. In other words, in the k-method the auxiliary field u_n satisfies an equation different from the equation for $U - U^0$ (more precisely, having different coefficients) not only inside, but also outside the body. It is this feature that does not permit one to transfer the k-method to problems for open resonators without considering the continuous spectrum.

2. Let us show that the eigenfunctions of problem (1.4) are orthogonal in the sense that

$$\int_{V^+} u_n u_m \, dV = 0, \quad n \neq m. \tag{1.6}$$

Here the integral is taken only over the volume occupied by the dielectric body rather than over the entire volume of the resonator, as in the orthogonality conditions of the k-method.

To prove (1.6), we multiply (1.4a) and (1.4b) by u_m, multiply the same equations for u_m by u_n, and integrate the difference of the resulting equations over the entire volume. By applying the Green formula (0.20), we obtain

$$\begin{aligned} (\varepsilon_n - \varepsilon_m) \int_{V^+} u_n u_m \, dV \; &+ \int_{S_\varepsilon} \left(u_n^+ \frac{\partial u_m^+}{\partial N} - u_m^+ \frac{\partial u_n^+}{\partial N} \right) dS \\ &+ \int_S \left(u_n \frac{\partial u_m}{\partial N} - u_m \frac{\partial u_n}{\partial N} \right) dS \\ &- \int_{S_\varepsilon} \left(u_n^- \frac{\partial u_m^-}{\partial N} - u_m^- \frac{\partial u_n^-}{\partial N} \right) dS = 0. \end{aligned} \tag{1.7}$$

Taking account of the boundary conditions on S and S_ε, we readily obtain (1.6). We have assumed that the eigenvalues are simple; a more general case will be considered in Chapter 5.

The eigenvalues ε_n are real. To prove this, it suffices to write out the expression similar to (1.7) with u_m replaced by u_n^*. The eigenfunctions u_n are real-valued (up to a constant factor), and problem (1.4) enjoys the Hermitian orthogonality

$$\int_{V+} u_n u_m^* \, dV = 0, \quad n \neq m \tag{1.8}$$

along with (1.6). However, since later on in this book we shall consider problems for which (1.8) does not hold, we use only the orthogonality conditions in the form (1.6) even in this section.

3. The series (1.3) with arbitrary A_n satisfies Eq. (1.1b) and conditions (1.1c) and (1.1d) term by term. The coefficients A_n can be found from the requirement that the expansion (1.3) must satisfy Eq. (1.1a) inside the dielectric. We substitute (1.3) into (1.1a), perform termwise differentiation, and use Eq. (1.4a) for u_n and Eq. (1.2a) for U^0, thus obtaining

$$\sum_n A_n(\varepsilon - \varepsilon_n)u_n = (1-\varepsilon)U^0 \quad \text{in} \quad V^+. \tag{1.9}$$

On multiplying (1.9) by u_m and integrating over V^+, with regard to (1.6) we obtain the desired expression

$$A_n = \frac{1-\varepsilon}{\varepsilon - \varepsilon_n} \frac{\int_{V+} U^0 u_n \, dV}{\int_{V+} u_n^2 \, dV} \tag{1.10a}$$

for the coefficients of the expansion.

It is possible to obtain a formula expressing A_n directly via the given sources f rather then the field U^0 generated by these sources in the absence of the dielectric, as in (1.10a). To this end, one must use the identity

$$\int_{V+} U^0 u_n \, dV = \frac{1}{k^2(1-\varepsilon_n)} \int_V u_n f \, dV. \tag{1.11}$$

To prove this identity, we use the formula

$$U^0 \Delta u_n - u_n \Delta U^0 = -u_n f + k^2(1-\varepsilon_n)U^0 u_n \quad \text{in} \quad V^+, \tag{1.12a}$$

which follows from (1.2a) and (1.4a), and the equation

$$U^0 \Delta u_n - u_n \Delta U^0 = -u_n f \quad \text{in} \quad V^-, \tag{1.12b}$$

which follows from (1.2a) and (1.4b). Let us integrate (1.12a) over V^+ and (1.12b) over V^-, apply the Green formula to the result (just as in (1.7)), and use conditions (1.2b) and (1.4d) on the metal together with the continuity of the fields U^0 and u_n and their normal derivatives on the boundary of the dielectric. Then we obtain (1.11). The substitution of (1.11) into (1.10a) yields

$$A_n = \frac{1}{\varepsilon - \varepsilon_n} \frac{1-\varepsilon}{k^2(1-\varepsilon_n)} \frac{\int_V u_n f \, dV}{\int_{V+} u_n^2 \, dV}. \tag{1.10b}$$

Note that the integration in the numerator in (1.10b), in contrast with (1.10a), is over the entire volume V.

4. Formulas (1.3) and (1.10), together with (1.2) and (1.4), provide a formal solution of the diffraction problem (1.1). This solution has nearly the same structure as that given by the k-method, but the term U^0 is not singled out in the k-method. In the ε-method, we have to single out the term U^0 in V^-, that is, outside the body, because the set of functions u_n in V^- is not sufficient to provide the expansion of an arbitrary function. However, inside the body, in V^+, the field U^0 can also be expanded in a series with respect to u_n according to (1.9). Consequently, *inside the body* instead of the series (1.3), which contains the isolated term U^0, we can use the series

$$U = \sum_n B_n u_n. \tag{1.13}$$

The expression of B_n via A_n can readily be found from (1.3) and (1.9):

$$B_n = \frac{1-\varepsilon_n}{1-\varepsilon} A_n. \tag{1.14}$$

The simplest way to study the field inside the dielectric is to use formula (1.13) with the coefficients B_n evaluated by substituting (1.10b) into (1.14), that is, by using the formula

$$B_n = \frac{1}{1-\varepsilon_n} \frac{1}{k^2} \frac{\int_V u_n f \, dV}{\int_{V+} u_n^2 \, dV}. \tag{1.15}$$

The solution in the form (1.13), (1.15) does not contain the field U^0. For the special case in which all the sources are contained inside the dielectric, that is, in V^+, it is possible not to single out the term U^0 in V^- as well.

5. Let us compare the above solution with the solution obtained by the k-method. In the k-method, the coefficients of the expansion contain the factor

$$(k^2 - k_n^2)^{-1}, \tag{1.16a}$$

which determines the dependence of various terms of the series on the frequency. This factor describes the resonant properties of the system. In the above simplest version of the ε-method, the resonance properties are described by the factor

$$(\varepsilon - \varepsilon_n)^{-1} \tag{1.16b}$$

in (1.10) and (1.15). The eigenvalues ε_n of problem (1.4) depend on the frequency k. If the frequency is such that one of the ε_n is close to the actual ε, then the corresponding term in (1.3) and (1.13) becomes dominant, and the field U is close to the eigenfunction u_n. Near the source, the field may have singularities, which are described in (1.3) by the term U^0. Thus, for a nearly resonant frequency, the field in the entire space is well approximated by the sum $U^0 + A_n u_n$.

Formulas (1.16a) and (1.16b) describe the frequency characteristic of the fields near the resonance equally well, and in this sense both graphs deserve the name of resonance curves. We shall return to this question in the end of the next section. However, being treated as functions of k, the factors (1.16a) and (1.16b) are different in general. The difference is noticeable far from the resonance, where the frequency characteristic of the total field is not described by the simple function (1.16a) or (1.16b).

The eigenvalues ε_n and k_n of the two methods are closely related to each other. If the diffracting bodies do not produce losses (that is, $\mathrm{Im}\,\varepsilon = 0$), then the k_n are real, and each k_n is a root of one of the equations

$$\varepsilon - \varepsilon_n(k) = 0. \tag{1.17a}$$

By differentiating the left-hand sides of Eqs. (1.4a) and (1.4b) with respect to k^2 and by combining the resulting equations with (1.4a) and (1.4b), one can readily find that

$$\frac{d\varepsilon_n}{dk^2} = -\frac{1}{k^2}\left(\varepsilon_n + \frac{\int_{V-} u_n^2\, dV}{\int_{V+} u_n^2\, dV}\right), \tag{1.18}$$

so that $\varepsilon_n(k)$ for positive ε_n is a monotone decreasing function of k. Since $\varepsilon_n \to \infty$ as $k \to 0$ and $\varepsilon_n \to 0$ as $k \to \infty$, it follows that for $\varepsilon > 0$ Eq. (1.17a) has exactly one root, to which we have assigned the same index n.

If we extend $\varepsilon_n(k)$ to the plane of the complex variable k, then this relationship between ε_n and k_n remains valid for complex ε. Formula (1.17a) is convenient in that it permits one to replace $\varepsilon_n - \varepsilon$ by the first term of the Taylor series near the resonance,

$$\varepsilon_n - \varepsilon = (k - k_n) \left.\frac{d\varepsilon_n}{dk}\right|_{k=k_n}, \tag{1.19}$$

and then the dependence of the denominator of A_n on the frequency will be determined by the factor $1/(k - k_n)$, just as in the k-method.

6. The homogeneous problem (1.4) does not contain the actual ε. Thus, by solving (1.4) and, in particular, by finding $\varepsilon_n(k)$, we simultaneously find the resonance curves for a body of a given shape with all possible values of permittivity. In particular, the solution of the homogeneous problem is by no means more complicated if ε is complex in the diffraction problem, that is, if the diffracting dielectric body is absorbing. (This is very important for numerical calculations.) In other words, *despite the presence of losses, the most intricate part of the calculations, namely, determining the eigenvalue,* *deals with real variables.* The presence of losses manifests itself only in the fact that ε will be complex in (1.16b) and the denominator will be nonzero for any real frequency. The maximum of the resonant term (1.16b) will be approximately equal to $1/\mathrm{Im}\,\varepsilon$ and will be attained at a value of k close to the root of the real equation

$$\varepsilon_n - \mathrm{Re}\,\varepsilon = 0. \tag{1.17b}$$

In the k-method, the finiteness of the Q-factor results in complex values of k_n, which must be found from a problem with complex ε, requiring more complicated computations.

In problems pertaining to closed systems, the k-method has a certain advantage over the ε-method. Specifically, once we have found a single number k_n, we can use (1.16a) to determine the entire *resonance curve in frequencies* for k close to k_n. In the ε-method, to this end we must find the function $\varepsilon_n(k)$. However, just for the resonance conditions, this complication of the ε-method is not too severe, since for using (1.19) it suffices to calculate $\varepsilon_n(k)$ for two values of k, eliminate $d\varepsilon_n/dk$, find k_n, and use the same formula (1.16a).

On the other hand, in the ε-method, once we have found a single ε_n, we can readily reconstruct the entire dependence of the field on ε, that is, the *resonance curve in permittivity* for ε close to ε_n. In the k-method this would require finding k_n as a function of ε; by analogy with the preceding, this can be accomplished by calculating k_n for two distinct values of ε.

If the given k coincides with an eigenfrequency of the empty resonator, then we encounter an additional difficulty. Indeed, in this case U^0 fails to exist (becomes infinite). As k approaches this eigenfrequency, one of the eigenfunctions u_n becomes proportional to U^0, and the corresponding eigenvalue ε_n tends to 1. In this situation, the coefficient A_n in (1.3) tends to infinity, as shown by (1.10b). The series (1.3) becomes an indeterminate of the form $\infty - \infty$. However, if we use the representation (1.13), then, according to (1.15), the indeterminacy does not occur in V^+. To avoid this irrelevant resonance in V^- as well, one can, say, introduce U^0 in a different way. Then it is necessary to modify the statement of the homogeneous problem so as to preserve the possibility of constructing the expansion (1.3). We can define U^0 as a solution of the equation

$$\Delta U^0 + k^2 \varepsilon U^0 = f \quad \text{in} \quad V, \tag{1.20}$$

that is, the field produced by the same sources f in the completely filled resonator. Then u_n must satisfy the equations

$$\Delta u_n + k^2 \varepsilon u_n = 0 \quad \text{in} \quad V^+, \tag{1.21a}$$
$$\Delta u_n + k^2 \varepsilon_n u_n = 0 \quad \text{in} \quad V^- \tag{1.21b}$$

and the previous conditions on S_ε and on the metal. These functions will be orthogonal with respect to integration over V^-, and the formulas for A_n and B_n will be slightly different from (1.10) and (1.15). Other definitions of U^0 and u_n are also possible.

7. The spectral methods are most often applied under resonance conditions, and then the series (0.25), (1.3), (1.13), and other series of the same type are reduced to a few first terms. However, in the numerical implementation of spectral methods for bodies with low Q-factor or for high-Q resonators far from the resonance, the problem concerning the *convergence* of the series and the *rate* at which the terms of the series *tend to zero* becomes essential. The convergence rate depends in particular on the smoothness of the source function $f(\mathbf{r})$ in (1.1). Let us consider the worst situation (as concerns the convergence) in which the source is a delta function; thus, we must study the convergence of various spectral series for the Green function [20]. We shall use the asymptotic estimates obtained in Chapter 5 for the eigenvalues ε_n for large n.

In this item we compare two series, (1.3) and (1.13), for the problem considered in this section. For the Green function, the coefficients B_n in (1.13) and A_n in (1.3) have the following form according to (1.10b) and (1.15):

$$B_n = \frac{1}{\varepsilon_n - \varepsilon} u_n(0), \tag{1.22a}$$

$$A_n = \frac{1}{(\varepsilon_n - \varepsilon)(\varepsilon_n - 1)} u_n(0). \tag{1.22b}$$

Here we have omitted factors unessential to our question (independent of n); the δ-source is situated at the origin, and the eigenfunctions are normalized by the condition

$$\int_{V+} u_n^2 \, dV = 1. \tag{1.23}$$

Thus, we study the convergence of the series

$$\sum \frac{1}{\varepsilon_n - \varepsilon} u_n(0) u_n(\mathbf{r}), \tag{1.24a}$$

$$\sum \frac{1}{(\varepsilon_n - \varepsilon)(\varepsilon_n - 1)} u_n(0) u_n(\mathbf{r}). \tag{1.24b}$$

Recall that the first series is a series for $U(\mathbf{r})$, and the second, for $U(\mathbf{r}) - U^0(\mathbf{r})$. We use a numbering in which the series are one-dimensional regardless of the dimension of the problem and $|\varepsilon_n|$ is a nondecreasing function of n.

Let us start from two-dimensional problems. According to the results of Chapter 5, in this case we have $\varepsilon_n \sim n$, so that the series (1.24) are equivalent to the series

$$\sum \frac{1}{n} u_n(0) u_n(\mathbf{r}), \tag{1.25a}$$

$$\sum \frac{1}{n^2} u_n(0) u_n(\mathbf{r}). \tag{1.25b}$$

Let us assume that the $|u_n(\mathbf{r})|$ are bounded for all n. Then the second series absolutely converges everywhere, whereas the first series diverges at $r = 0$. This is essentially due to the fact that it describes the total field, which has a δ-singularity at $r = 0$.

If in the function $U(\mathbf{r})$ we single out the term $U^0(\mathbf{r})$, then, as shown below in several examples in Chapter 4, in many cases the field in the entire space can be described by the two-term formula $U = U^0 + A_1 u_1$. As follows from the comparison of the two series in (1.22), the isolation of U^0 can be treated as an analog of the ordinary method for improving the convergence of ill-convergent series. We consider a possible generalization of this method in application to the series occurring in our problem in the next item; now let us qualitatively estimate the convergence of the series (1.25a). Namely, let us estimate the number of essential terms in this series, the dependence of this number on r, and the behavior of the sum as $r \to 0$.

For $r \neq 0$, the series (1.25a) is convergent, since the product $u_n(0)u_n(\mathbf{r})$, which is positive for all n if $r = 0$, changes its sign as n increases. For a given small r, let us find the value of n starting from which $u_n(\mathbf{r})$ becomes substantially different from $u_n(0)$; it is only for these n that the product $u_n(0)u_n(\mathbf{r})$ may be negative. The first three terms of the Taylor series for $u_n(\mathbf{r})$ have the form (for an appropriately chosen direction of the Cartesian coordinate axes)

$$u_n(\mathbf{r}) = u_n(0) + \mathbf{r}\nabla u_n + \frac{1}{2}\left(\frac{\partial^2 u_n}{\partial x^2}x^2 + \frac{\partial^2 u_n}{\partial y^2}y^2\right). \tag{1.26}$$

All the components of the gradient can be arbitrarily small, but the second derivatives grow with n. They are of the order of Δu_n, that is, according to (1.4), of the order of $k^2\varepsilon_n u_n(0)$, so that $u_n(\mathbf{r})$ differs from $u_n(0)$ by $O(k^2\varepsilon_n u_n(0)r^2)$. If n is so large that $k^2\varepsilon_n r^2$ is of the order of unity, then $u_n(\mathbf{r})$ becomes to differ significantly from $u_n(0)$. Consequently, the terms of the series (1.25a) start changing signs for n such that $\varepsilon_n \sim 1/(kr)^2$. The series (1.25a) is of the order of the finite sum $\sum_{n=1}^{N} 1/n$, since the tail of the series is alternating and remains finite as $N \to \infty$. The number N is of the order of $1/(kr)^2$. This sum for large N is approximately equal to $\ln N$, so that the overall series is of the order of $\ln[1/(kr)]$. Thus, this coarse estimate gives a correct expression for the singularity of the two-dimensional Green function as $r \to 0$.

Let us briefly repeat our considerations for the three-dimensional problem. Here, according to the results of Chapter 5, we have $\varepsilon_n \sim n^{2/3}$, so that the series for $U(\mathbf{r})$ and $U(\mathbf{r}) - U^0(\mathbf{r})$ are equivalent to the series

$$\sum \frac{1}{n^{2/3}} u_n(0)u_n(\mathbf{r}), \tag{1.27a}$$

$$\sum \frac{1}{n^{4/3}} u_n(0)u_n(\mathbf{r}), \tag{1.27b}$$

respectively. Only the second series converges absolutely for all $\mathbf{r}$. Let us estimate the series (1.27a). Just as in the two-dimensional case, $u_n(\mathbf{r})$ differs from $u_n(0)$ by $O(k^2\varepsilon_n u_n(0)r^2)$. The terms of this series start changing signs at $\varepsilon_n \sim 1/(kr)^2$, that is, $n \sim 1/(kr)^3$. The series is of the order of the finite sum $\sum_{n=1}^{N} 1/n^{2/3}$, where $N \sim 1/(kr)^3$. This sum is approximately equal to $N^{1/3}$, that is, $1/kr$. This estimate also agrees with the character of the singularity as $r \to 0$ of the Green function for three-dimensional space.

8. In this item we shall show that one can *infinitely improve the convergence* by successively replacing $U(\mathbf{r})$ by new functions $W^M(\mathbf{r})$, $M = 0, 1, \ldots$, for each of which the right-hand side of the wave equation is more differentiable than that for a function with a smaller number M. First, along with $U^0(\mathbf{r})$, we introduce another M functions $U^m(\mathbf{r})$ by the recurrence relations

$$\Delta U^m + k^2 U^m = k^2(1-\varepsilon)U^{m-1}, \quad m = 1, 2, \ldots, M. \tag{1.28}$$

Let us denote $f/k^2(1-\varepsilon)$ by U^{-1}; then the same formula applies for $m = 0$, that is, for U^0. If the Green function G of the diffraction problem for $\varepsilon = 1$ is known, then the evaluation of each new function U^m from the known function U^{m-1} is reduced to the quadrature (see (1.91 below)

$$U^m = k^2 \int (1-\varepsilon)U^{m-1} G \, dV. \tag{1.29}$$

Then we seek the solution of the diffraction problem in the form

$$U = \sum_{m=0}^{M} U^m + W^M. \tag{1.30}$$

According to (1.29), the sum in (1.30) is similar to a segment of the Neumann series. The functions $W^M(\mathbf{r})$ introduced by this formula satisfy the wave equation whose right-hand side contains U^M alone:

$$\Delta W^M + k^2 \varepsilon W^M = k^2(1-\varepsilon)U^M. \tag{1.31}$$

Since the integration (1.29) is a smoothing procedure, it follows that the smoothness of the right-hand side increases with M. This manifests itself in the rate of decay of the coefficients of the series for W^M in the eigenfunctions u_n. We denote these coefficients by A_n^M, that is, we set

$$W^M(\mathbf{r}) = \sum_n A_n^M u_n(\mathbf{r}). \tag{1.32}$$

In this notation, the coefficients A_n in (1.3) must be denoted by A_n^0. The coefficients B_n in (1.13) can be denoted by A_n^{-1} with regard to the fact that for $M = -1$ the sum in (1.30) is lacking, so that $W^{-1} = U$.

The main advantage gained by introducing the function $W^M(\mathbf{r})$ instead of $U(\mathbf{r})$ is as follows. It is easy to show according to (1.28) and (1.30) that the coefficients A_n^M are obtained from A_n^{M-1} by the formula

$$A_n^M = \frac{\varepsilon - 1}{\varepsilon_n - 1} A_n^{M-1}; \tag{1.33}$$

since $\varepsilon_n \to \infty$ as $n \to \infty$, it follows that the convergence of the series (1.32) is faster for larger M. Formula (1.33) is a generalization of (1.14), which can be obtained from the former by setting $M = 0$.

The trick described in this item allows one to pass to series whose terms decrease as rapidly as an arbitrary power of ε_n, that is, improve the convergence of spectral series. However, at the same time we increase the number of quadratures needed to obtain $U(\mathbf{r})$. This trick can be used equally well in the problems concerning inhomogeneous dielectric bodies, considered in the subsequent sections, that is, in diffraction problems in which $\varepsilon = \varepsilon(\mathbf{r})$.

In conclusion, let us make two remarks pertaining to other spectral methods. In the k-method, the analog of ε_n is given by the squared eigenfrequency k_n^2. For large n, it grows as ε_n (see Chapter 5), that is, as n or $n^{2/3}$, depending on the dimension of the problem. The Fourier coefficients of the spectral series (0.25) for the Green function decay, according to (0.29), as $1/k_n^2$, that is, in the same way as the coefficients B_n in (1.22a). The series (0.25) is convergent in the same way as (1.13), that is, slowly, and only for $r \neq 0$. Since in the k-method we cannot single out the term U^0 (and all the more, the terms U^m), we see that the above procedure for improving the convergence cannot be applied directly to the k-method.

In the "surface" methods considered below in Chapter 2, the coefficients of the spectral series decay very rapidly as $n \to \infty$, so that there is no need to improve the convergence. This is due to the fact that these coefficients are proportional to integrals of the product of $u_n(\mathbf{r})$ by functions that are infinitely differentiable for virtually all excitations.

1.2 Dielectric Body in a Resonator with Absorbing Walls, in an Open Resonator, or in Vacuum

1. Now suppose that there are some losses in the resonator other than those due to the fact that ε is complex-valued. In this subsection we consider the modifications that must be introduced in the ε-method if we wish to take account of the losses in the walls of a closed resonator. For our purposes, the losses in the walls can be characterized by a complex impedance w; that is, we replace condition (1.1d) or the condition $\partial U/\partial N|_S = 0$ by the condition

$$\left(U + w\frac{\partial U}{\partial N}\right)\bigg|_S = 0 \tag{1.34}$$

(the normal is directed inside the wall), where $\operatorname{Im} w < 0$. The same condition must hold for the field U^0,

$$\left(U^0 + w\frac{\partial U^0}{\partial N}\right)\bigg|_S = 0, \tag{1.35}$$

and for the eigenfunctions of the homogeneus problem (1.4) (instead of (1.4d)):

$$\left(u_n + w\frac{\partial u_n}{\partial N}\right)\bigg|_S = 0. \tag{1.36}$$

Once we introduce U^0 and u_n this way, all the technique of the preceding section can be applied. However, in contrast with the diffraction problem without losses in the walls, the new homogeneous problem is not selfadjoint. Hence the eigenvalues are complex and the ortogonality of the eigenfunctions holds only in the form (1.6). This difference is very important. The situation is just the same as it is in the k-method in the presence of any losses. Note that if the only losses are those in the resonator walls, then it is more convenient to use another version of the generalized method (see Section 2.1), for which the homogeneous problem is selfadjoint and the eigenvalues can be without leaving the real domain.

One can easily relate the imaginary part of ε_n to that of the impedance w. Let us consider Eqs. (1.4a) and (1.4b) for u_n and u_n^* together with the boundary conditions on the surface of the dielectric. Transformations similar to those used in the derivation of (1.7) give

$$k^2(\varepsilon_n - \varepsilon_n^*) \int_{V+} |u_n|^2 \, dV = 2i \,\mathrm{Im} \int_S u_n \frac{\partial u_n^*}{\partial N} \, dS. \tag{1.37}$$

Using (1.36), we obtain

$$\mathrm{Im}\,\varepsilon_n = -\mathrm{Im}\, w \frac{1}{k^2|w|^2} \frac{\int_S |u_n|^2 \, dS}{\int_{V+} |u_n|^2 \, dV}. \tag{1.38}$$

If there are losses in the resonator walls ($\mathrm{Im}\, w < 0$), then

$$\mathrm{Im}\,\varepsilon_n > 0. \tag{1.39}$$

Thus, the imaginary part of ε_n and that of the permittivity for bodies exhibiting dielectric losses have opposite signs. This result has a simple physical meaning. Just as in Section 1.1, the homogeneous problem for u_n in a resonator with losses describes undamped eigenoscillations in some auxiliary system (a dielectric body in a resonator). Since, according to (1.36), there are losses in this system, we see that the eigenoscillations with real k are only possible if there is energy release in the dielectric. This just means that the imaginary part of ε_n must be positive. Needless to say, the energy balance holds in this case: the energy released in the body (the left-hand side of (1.37)) is equal to the energy absorbrd by the walls.

There exists only a discrete set of complex numbers ε_n for which the eigenoscillations occur at a given frequency k and the losses in the walls at a given w are compensated for by the energy release in the active dielectric (see Chapter 5).

Suppose that $\mathrm{Im}\,\varepsilon_n(k)$ is a slowly varying function of the frequency. Then the resonance factor (1.16b) attains its maximum at a frequency close to the root of the equation

$$\mathrm{Re}\,\varepsilon_n(k) - \mathrm{Re}\,\varepsilon = 0, \tag{1.40}$$

where, for generality, we allow ε to be complex. The maximum is approximately equal to

$$\frac{1}{\operatorname{Im}\varepsilon_n - \operatorname{Im}\varepsilon}. \tag{1.41}$$

The two terms in the denominator have opposite signs, namely, $\operatorname{Im}\varepsilon_n > 0$ and $\operatorname{Im}\varepsilon \leq 0$, so that the denominator is nonzero.

In Subsections 1.1.6 and 1.2.1 we have studied two cases of application of the ε-method to diffraction problems. In the first case, the losses occur in the dielectric body, that is, ε is complex. Then the homogeneous problem is selfadjoint and ε_n is real. In the second case, the losses are not in the dielectric; accordingly, the homogeneous problem is nonselfadjoint and ε_n is complex. Likewise, in all other versions of the method we shall deal with one of the following two cases: either the losses are solely due to the fact that the diffraction problem parameter that is the spectral parameter in the corresponding homogeneous problem is complex, or there are also some other losses. In the first case, the homogeneous problem can always be made selfadjoint. In the second case, the problem is nonselfadjoint, and the eigenvalues are complex. From the viewpoint of physics, this means that the auxiliary homogeneous problem involves an active domain in which there is energy release that compensates for the losses.

2. Now suppose that we deal with diffraction on a dielectric body in vacuum or in an open resonator. *Then the diffracted field $U - U^0$ must satisfy the radiation condition.* As we have already mentioned in Section 0.2, an attempt to apply the eigenfrequency method to this problem encounters some difficulties. Specifically, these difficulties are due to the fact that the diffracted field satisfies the equation with a real frequency k, whereas the eigenfunctions of the k-method satisfy the condition with a complex frequency k_n, $\operatorname{Im} k_n > 0$, in the entire infinite domain outside the body.

In that respect, the forthcoming versions of the generalized method are advantageous in that they permit simply transferring the formal technique to open problems with the discrete representation of the field preserved everywhere. This is due to the fact that in all homogeneous problems of these versions the frequency of oscillations is just the frequency of the actual source in the diffraction problem, which occurs as a real parameter in the eigenelement problems.

The modification required to adapt the ε-method to open problems essentially amounts to *introducing the radiation condition into the homogeneous problem.* Let us use this method to formally construct the solution of the problem on the diffraction on a dielectric body in vacuum.

The mathematical statement of the diffraction problem is as follows. We seek a function U that satisfies the equation

$$\Delta U + k^2 \varepsilon U = f \tag{1.42}$$

in the domain V^+ (the interior of the dielectric), the equation

$$\Delta U + k^2 U = f \tag{1.43}$$

in the infinite domain V^-, the conditions

$$(U^+ - U^-)|_{S_\varepsilon} = 0, \quad \left(\frac{\partial U^+}{\partial N} - \frac{\partial U^-}{\partial N}\right)\bigg|_{S_\varepsilon} = 0 \tag{1.44}$$

on the boundary S_ε of the dielectric, and the radiation condition (0.35) at infinity provided that the sources are all located at finite distances. If a plane incident wave is considered, then the radiation condition is imposed on the diffracted field.

The solution of this problem is again sought in the form of the series (1.3), where U^0 is the field of the same sources f in vacuum (the primary, or the incident field); thus U^0 satisfies the equation

$$\Delta U^0 + k^2 U^0 = f \tag{1.45}$$

in the entire volume. The functions u_n in which the diffracted field $U - U^0$ is to be expanded must satisfy Eqs. (1.4a) and (1.4b) as well as the boundary conditions (1.4c), which state that the field and its normal derivative are continuous on S_ε.

Finally, we require that each function u_n must satisfy the radiation conditions at infinity. The numbers ε_n are the eigenvalues of the above-stated homogeneous problem.

Since outside the body any eigenfunction satisfies the same equation as the diffracted field, it follows that the series (1.3) satisfies the radiation condition termwise. It will be shown in Chapter 5 that the sum of the series satisfies the same condition.

The physical meaning of the homogeneous problem is as follows. It describes undamped, though accompanied by radiation, eigenoscillations with true frequency k. They exist in the absence of sources (the right-hand side of the corresponding wave equation is zero) in a system with the same geometry as in the diffraction problem but with permittivity $\varepsilon_n \neq \varepsilon$. Hence for a given frequency k the homogeneous problem has nontrivial solutions only for certain values of ε, $\varepsilon = \varepsilon_n$.

One can readily show by using the equations for u_n and u_n^*, the boundary conditions on S_ε, and the asymptotic representation of the remote field u_n that the imaginary part of the eigenvalue must be positive, that is, (1.39) holds. Indeed, according to the radiation conditions, u_n must have a form similar to (0.32) remote from the body, but with k_n replaced by k:

$$u_n \simeq \frac{e^{-ikr}}{r}\Phi_n(\varphi, \theta).$$

By carrying out integration over the sphere at infinity in (1.37), we obtain $2ik \int |\Phi_n|^2 \, d\Omega$ on the right-hand side. Hence $\operatorname{Im} \varepsilon_n$ is positive.

This property of ε_n has a physical explanation as simple as it has for a resonator with impedance walls: undamped eigenoscillations can exist in the presence of radiation only if they are maintained by energy released in an auxiliary body and proportional to the square of the field in that body.

Let us return to the formal representation of the solution of the diffraction problem. It follows from the statement of the homogeneous problem that the series (1.3) with arbitrary coefficients A_n satisfies Eq. (1.43) outside the body, the boundary conditions on S_ε (by definition, U^0 and its normal derivative are continuous on S_ε), and the radiation condition. The coefficients A_n can be found from the same equation (1.9), which is necessary for the wave equation inside the body to be satisfied, and have the same form (1.10). Inside the body, we need not isolate the term U^0; that is, we can use the series (1.13), where the B_n are given in (1.14) or, say, (1.15).

The formalism remains unchanged if, apart from the dielectric body, there are some metal surfaces S involved in diffraction (for example, if the dielectric is placed in an open resonator) with boundary conditions of the form $U|_S = 0$ or $\partial U/\partial N|_S = 0$ and with the same conditions on the edges. In this case the same conditions on S and on the edges must be imposed on the eigenfunctions u_n and on the field U^0.

3. Let us show how the technique works in a simple example. Consider a two-dimensional open resonator formed by a circular dielectric cylinder with large permittivity ε. Such resonators are known to admit high-Q "locked" modes. In our notation, this implies that there exist eigenoscillations, that is, solutions of system (1.4) satisfying the radiation condition, for which the denominator in (1.16b) is small. We assume that $\varepsilon \gg 1$ and ka is of the order of 1. Then, to study the resonance, we must seek eigenvalues ε_n with large modulus ($|\varepsilon_n| \gg 1$).

We restrict ourselves to studying the homogeneous problem and the factor (1.16b), that is, the resonator characteristics that are independent of the form of the excitation (see also Section 4.8). The homogeneous problem for the cylinder in vacuum can be reduced to a transcendental equation by a standard procedure. Namely, let us represent the eigenoscillation field, that is, the eigenfunction, in the form

$$u_n^+ = A_n J_m\left(k\sqrt{\varepsilon_n} r\right) \cos m\varphi \tag{1.46a}$$

inside the cylinder and in the form

$$u_n^- = B_n H_m^{(2)}(kr) \cos m\varphi \tag{1.46b}$$

outside the cylinder. Here n stands for the pair consisting of the azimuth index m and the radial index q; the latter specifies the number of the root of the transcendental equation (1.47). On the boundary $r = a$ of the dielectric, the function must be continuous together with the normal derivative; this condition results in the equation

$$\frac{ka\sqrt{\varepsilon_n} J_m'\left(k\sqrt{\varepsilon_n}\right)}{J_m\left(ka\sqrt{\varepsilon_n}\right)} = \frac{ka H_m^{(2)\prime}(ka)}{H_n^{(2)}(ka)}. \tag{1.47}$$

Just the same equation would be given by the k-method, apart from the fact that k would be the unknown (the eigenfrequency) and ε_n would be replaced by ε. It is easier to find ε_n rather than k from Eq. (1.47), since in this case the right-hand side of (1.47) is a constant. We shall consider eigenfunctions corresponding to low values of the azimuth number m, that is, oscillations for which the standing wave consists of two waves of nearly radial direction.

One could also find the so-called "whispering gallery" oscillations in which a wave runs along the cylindrical surface on the interor side and is reflected from the boundary at a grazing angle. In the outside, these oscillations decay very rapidly (nearly exponentially). Such modes also exist for small $|\varepsilon|$ but large ka. The analysis of these modes (which is also based on (1.47)) is somewhat more cumbersome, since it involves the use of the Debye asymptotics of cylindrical functions of large index.

We start from determining ε_n. For large $|ka\sqrt{\varepsilon_n}|$, it follows from (1.47) that $ka\sqrt{\varepsilon_n}$ is close to a root of the function J'_m. We denote the qth root by μ_{mq} ($J'_m(\mu_{mq}) = 0$) and set

$$ka\sqrt{\varepsilon_n} = \mu_{mq} + x \quad (\mu_{mq} \gg 1). \tag{1.48}$$

It follows from (1.47) that $|x| \ll 1$. We restrict ourselves to the computation of x in the first approximation (with respect to $1/\mu_{mq}$), which is sufficient for the rough description of the resonance conditions and for finding the Q-factor. By substituting (1.48) into (1.47) and by neglecting the terms of the order of $(m/\mu_{mq})^2$, we find that $x = -c/\mu_{mq}$, where

$$c = \frac{kaH_m^{(2)\prime}(ka)}{H_m^{(2)}(ka)} \tag{1.49}$$

is the right-hand side of (1.47). Thus, the desired expression for the eigenvalues with large modulus reads

$$\varepsilon_n = \frac{\mu_{mq}^2}{k^2a^2} - \frac{2c}{k^2a^2} + O\left(\frac{1}{\mu_{mq}}\right). \tag{1.50}$$

Formula (1.16b), in conjunction with (1.50), permits one to find the *dependence of the amplitude factor on k for given ε* as well as the *dependence of the amplitude factor on ε for given k*. Both curves can be referred to as *resonance curves*. They are characterized by the values of the Q-factor as well as by the resonance values of frequency and permittivity, respectively.

Let us start from the usual resonance curve ($\varepsilon = \text{const}$). The resonance frequency $\tilde{k}_n$, for which the amplitude factor (1.16b) is maximal, approximately satisfies Eq. (1.40) and is equal to the real part of the eigenfrequency k_n introduced in the k-method, $\tilde{k}_n = \operatorname{Re} k_n$. Finding $\tilde{k}_n$ from (1.47) is more complicated than finding ε_n. Directly from (1.50), we obtain (this approximation is coarser than (1.50) itself)

$$\tilde{k}_n = \frac{\mu_{mq}}{a}\operatorname{Re}\frac{1}{\sqrt{\varepsilon}}. \tag{1.51}$$

The maximum value of the factor (1.41) is approximately equal to

$$\left(-\frac{2\operatorname{Im} c}{k^2a^2} - \operatorname{Im}\varepsilon\right)^{-1}. \tag{1.52}$$

Both terms in the denominator of (1.52) are positive. The resonance amplitude is not infinite, since there are both radiation losses (the first term in the denominator) and dielectric losses (for $\mathrm{Im}\,\varepsilon \neq 0$; they are given by the second term). The half-width δk of the resonance curve can be found from the equation

$$\left|\mathrm{Re}\,\varepsilon_n(\tilde{k}_n + \delta k) - \mathrm{Re}\,\varepsilon_n(\tilde{k}_n)\right| = \mathrm{Im}\,\varepsilon_n(\tilde{k}_n) - \mathrm{Im}\,\varepsilon. \tag{1.53}$$

For the *frequency Q-factor* Q_k, defined by the usual formula

$$Q_k = \frac{\tilde{k}_n}{2\delta k}, \tag{1.54}$$

we obtain

$$Q_k = \frac{\tilde{k}_n \left|\dfrac{d}{dk}\,\mathrm{Re}\,\varepsilon_n(k)\right|}{2[\mathrm{Im}\,\varepsilon_n(\tilde{k}_n) - \mathrm{Im}\,\varepsilon]}. \tag{1.55}$$

For the ε_n given by (1.50), the last formula gives, in the same approximation as was used to write out (1.51),

$$Q_k = \frac{\pi}{4}[J_m^2(\tilde{k}_n a) + N_n^2(\tilde{k}_n a)]\mu_{mq}^2. \tag{1.56}$$

For simplicity, in the last formula we assume that $\mathrm{Im}\,\varepsilon = 0$ and write out $\mathrm{Im}\,c$ in the explicit form. The Q-factor Q_k for the oscillations in question is of the order of the permittivity of the body.

Let us now study the position and the width of the *resonance curve at constant frequency*. According to (1.50), the resonance occurs at

$$\mathrm{Re}\,\varepsilon = \frac{\mu_{mq}^2}{k^2a^2} - \frac{2\,\mathrm{Re}\,c}{k^2a^2}. \tag{1.57}$$

The maximum value of the factor (1.16b) is again approximately equal to (1.52). By analogy with (1.53), we define the half-width $\delta\,\mathrm{Re}\,\varepsilon$ of the resonance curve to be the value of $\mathrm{Re}\,\varepsilon$ for which the real part of the denominator is equal to its imaginary part:

$$|\delta\,\mathrm{Re}\,\varepsilon| = \mathrm{Im}\,\varepsilon_n - \mathrm{Im}\,\varepsilon. \tag{1.58}$$

We define the *permittivity Q-factor* Q_ε by the formula

$$Q_\varepsilon = \frac{\mathrm{Re}\,\varepsilon}{2\delta\,\mathrm{Re}\,\varepsilon}. \tag{1.59}$$

The substitution of (1.50) yields the first approximation of the form

$$Q_\varepsilon = \frac{1}{2}Q_k. \tag{1.60}$$

The Q-factors Q_ε and Q_k are different, since they describe the half-widths of resonance curves in two different experiments, namely, at constant frequency (Q_ε) and at constant permittivity (Q_k). By comparing (1.55) with (1.59), in the leading term we obtain

$$\frac{Q_k}{Q_\varepsilon} = \left| \left. \frac{d \ln \operatorname{Re} \varepsilon_n(k)}{d \ln k} \right|_{k=\tilde{k}_n} \right| . \tag{1.61}$$

According to (1.50), $\varepsilon_n \sim k^{-2}$, and the permittivity Q-factor proves to be half as large as the frequency Q-factor.

1.3 Inhomogeneous Dielectric The First Polarization

In Sections 1.3 and 1.4 we generalize the technique developed in Sectons 1.1 and 1.2 for problems of diffraction on a dielectric body with constant permittivity ε to the case in which $\varepsilon = \varepsilon(\mathbf{r})$ is a function of the coordinates. In this case, the permittivity ε_n of the auxiliary body also proves to be a function of the coordinates and hence no longer plays the role of the eigenvalue of the homogeneous problem. Instead, some numerical parameter occurring in $\varepsilon_n(\mathbf{r})$ is treated as the eigenvalue. The form of the function $\varepsilon_n(\mathbf{r})$ depends on that of $\varepsilon(\mathbf{r})$, just as for $\varepsilon =$ const the shape of the body in the auxiliary homogeneous problem reproduces the shape of the actual body in the main diffraction problem. The results of Sections 1.1 and 1.2 are special cases of the results of this section.

In the first subsection we give the simplest solution for $\varepsilon = \varepsilon(\mathbf{r})$ by applying the technique of differential equations, just as in Sections 1.1 and 1.2. In the second subsection, the same results will be obtained once more with the help of Green's function and integral equations. A more general solution of the problem will be given in Subsections 3–5. The derivation of these solutions by the methods of theory of functions of a complex variable will be presented in Subsection 6.

In the present section we do not discuss the results in detail, since the physical interpretation of the general case $\varepsilon = \varepsilon(\mathbf{r})$ is virtually the same as that for $\varepsilon =$ const.

1. We seek a field $U(\mathbf{r})$ satisfying the equation

$$\Delta U + k^2 \varepsilon U = f, \quad \varepsilon = \varepsilon(\mathbf{r}). \tag{1.62}$$

Let us assume that ε is continuous function of the coordinates and is equal to 1 outside some bounded domain. This assumption permits us not to introduce surfaces on which the continuity conditions (1.44) must be satisfied. However, this assumption does not actually contain any restrictions. A discontinuous function $\varepsilon(\mathbf{r})$ can be viewed as a limit of continuous functons. Then conditions (1.44) arise as the limit form of Eq. (1.62) under the additional requirement that U and $\partial U / \partial N$ remain finite. Hence, the solutions for discontinuous $\varepsilon(\mathbf{r})$, that is, the solutions satisfying (1.44), are limits of solutions obtained in the present section. Since the solutions for continuous $\varepsilon(\mathbf{r})$ do not contain the gradient of $\varepsilon(\mathbf{r})$, we see that all the formulas that will be obtained below obviously remain valid for discontinuous ε, when both (1.62) and (1.44) must be satisfied. The same general trick, which often dramatically reduces the required amount of computations, will be used in Sections 1.4 and 1.6. As to electromagnetic theory, it is just the case of discontinuous ε that is of practical interest there.

Strictly speaking, in electromagnetic theory Eq. (1.62) holds only for two-dimensional problems. If ε is independent of z (in particlar, the interfaces, that is, the boundaries where ε is discontinuous, are cylindrical surfaces parallel to the z-axis) and if the sources f are also independent of z, then the corresponding diffraction problems are known to have two classes of solutions for which $\partial/\partial z \equiv 0$. In the first class (the E-polarization), the nozero components are E_z, H_x,and H_y, whereas in the second class (the H-polarization) these are H_z, E_x, and E_y. For $U = E_z$, the two-dimensional version of Eq. (1.62) holds in problems of the first class. The equations and the conditions on the boundary of the dielectric introduced in Sections 1.1 and 1.2 are also valid for $U = E_z$ in the two-dimensional case for the E-polarization. The wave equation (1.121), which will be considered in the next section, also describes a two-dimensional problem, but for the H-polarization ($U = H_z$). Three-dimensional electromagnetic problems result in field equations more complicated than (1.62) or (1.121) and boundary conditions more complicated than (1.44); in this case it is more convenient to deal with first-order equations, that is, directly with the Maxwell equations. The corresponding technique will be developed in Section 1.6.

However, the scalar technique is much simpler and much more obvious, and it may well be used to study the main features of the method. The scalar technique is no simpler for two-dimensional problems than it is for three-dimensional ones. That is why we consider (1.62) and (1.121) in the three-dimensional case.

Furthermore, these equations also describe the three-dimensional problem on the propagation of acoustic waves in an inhomogeneous medium. If c, the speed of sound, is a function of the coordinates and the density ρ of the medium is constant, then the wave equation for the pressure U has the form (1.62), where

$$k^2\varepsilon(\mathbf{r}) = \frac{\omega^2}{c^2}. \tag{1.63}$$

If both ρ and c vary in such a way that $\rho c^2 =$ const, then the equation for the pressure has the form (1.121), where ε must be replaced by ρ. In the general case, in acoustics it is also more convenient to deal with first-order equations.

In Section 1.5, Eq. (1.62) is treated as a stationary Schrödinger equation.

We shall simultaneously consider the problems on a closed resonator without losses (where either condition (1.1d) or the condition that the normal derivative is zero must be satisfied on some closed surface), a closed resonator with losses (where (1.34) must hold), an open resonator (where U satisfies the radiation condition (0.35)), and a dielectric body in vacuum (where the radiation condition is the only requirement, apart from (1.62), imposed on U).

Let us introduce the field U^0 created by the same sources in the absence of the dielectric,

$$\Delta U^0 + k^2 U^0 = f, \tag{1.64}$$

and the fields u_n satisfying the equation

$$\Delta u_n + k^2 \varepsilon_n(\mathbf{r}) u_n = 0. \tag{1.65}$$

Both U^0 and u_n must satisfy the cited boundary conditions (on the surface S and at infnity). In the following, we define the functions $\varepsilon_n(\mathbf{r})$ in a way such that the main features of the method, namely, explicit expressions for and the resonance character of the coefficients of the expansion

$$U = U^0 + \sum_n A_n u_n, \tag{1.66}$$

will be preserved. The series for U satisfies the boundary conditions termwise, and the A_n will be found from the requirement that the function (1.66) must satisfy Eq. (1.62).

The orthogonality conditions for the eigenfunctions can be found from (1.65). Let us multiply Eq. (1.65) by u_m and the similar equation for u_m by u_n. We subtract one equation from the other and integrate over the entire space. Since the boundary conditions for all u_n are the same, so that the integrals like

$$\int \left(u_n \frac{\partial u_m}{\partial N} - u_m \frac{\partial u_n}{\partial N} \right) dS \tag{1.67}$$

taken over S or over the sphere at infinity are zero, we obtain

$$\int (\varepsilon_n - \varepsilon_m) u_n u_m \, dV = 0. \tag{1.68}$$

Let us substitute the series (1.66) into (1.62). Simple calculations using Eqs. (1.64) and (1.65) give the following series expansion of U^0 in the functions u_n:

$$\sum_n A_n (\varepsilon - \varepsilon_n) u_n = (1 - \varepsilon) U^0. \tag{1.69}$$

In complete analogy with (1.9), this expansion makes sense only where $\varepsilon \neq 1$. The functions ε_n must be equal to 1 wherever $\varepsilon = 1$, that is, outside the body.

Let us multiply (1.69) by u_m and integrate over the entire volume (actually, over the volume occupied by the body). Then we obtain the following system of linear algebraic equations for the coefficients A_n:

$$\sum_n A_n \int (\varepsilon - \varepsilon_n) u_n u_m \, dV = \int (1 - \varepsilon) U^0 u_m \, dV. \tag{1.70}$$

Let us point out just now that in Subsection 3 we shall obtain another system for the A_n from the same equation (1.69); this will result in another system of functions ε_n and hence different functions u_n and coefficients A_n.

Let us now require the coefficient matrix on the left-hand side of (1.70) *to be diagonal* (this requirements provides explicit expresions for the coefficients A_n). According to (1.68), this requirement will be satisfied if the functions ε_n and ε_m are related to ε so that the differences $\varepsilon_n - \varepsilon_m$ and $\varepsilon - \varepsilon_n$ are proportional,

$$\varepsilon - \varepsilon_n = \lambda_{nm}(\varepsilon_n - \varepsilon_m), \tag{1.71}$$

where the proportionality coefficient λ_{nm} must be independent of $\mathbf{r}$. The general solution of this functional equation is given by

$$\varepsilon_n = \gamma + \sigma_n(\varepsilon - \gamma); \tag{1.72}$$

it expresses $\varepsilon_n(\mathbf{r})$ via $\varepsilon(\mathbf{r})$, an arbitrary function $\gamma(\mathbf{r})$, and a number σ_n, which plays the role of an eigenvalue. According to (1.72), all the terms on the left-hand side in (1.69) contain the factor $\varepsilon - \gamma$, whereas the right-hand side is zero for $\varepsilon = 1$. It follows that γ must be equal to 1 outside the body,

$$\gamma = 1 \quad \text{for} \quad \varepsilon = 1. \tag{1.73}$$

In other respects, $\gamma(\mathbf{r})$ is arbitrary, except that it must not coincide with $\varepsilon(\mathbf{r})$. To any choice of γ there corresponds a system of eigenfunctions u_n and eigenvalues σ_n. In this subsection we assume that $\gamma \equiv 1$; possible generalizatons $((\gamma \not\equiv 1))$ are dealt with in Subsection 4. Thus, in the auxiliary problems generating the system of eigenfunctions u_n and eigenvalues σ_n, the permittivity is

$$\varepsilon_n(\mathbf{r}) = 1 + \sigma_n[\varepsilon(\mathbf{r}) - 1]. \tag{1.74}$$

The problem considered in the preceding two sections (where $\varepsilon(\mathbf{r}) = $ const) can be obtained as a special case of this general setting, and in this special case the eigenvalues ε_n are related to the eigenvalues σ_n by the obvious formula

$$\varepsilon_n = 1 + \sigma_n(\varepsilon - 1). \tag{1.75}$$

One can also readily generalize fromula (1.39) for the sign of the imaginary part of the eigenvalue in the presence of nonzero dielectric losses. Let us repeat the computations that were used in the derivation of (1.37), but with the integration carried out over the entire volume. If there are losses in the walls or radiation losses, then on the right-hand side we obtain a positive quantity multiplied by i, whereas on the left-hand side we have the same integral as in (1.37), the only difference being that the difference $\varepsilon_n - \varepsilon_n^*$ is now included in the integrand. If, to simplify the notation, we assume that ε is real and $\varepsilon > 1$, then this equation will imply that $\operatorname{Im}\sigma_n > 0$, which is just the desired generalization of (1.39).

From (1.70) we obtain the explicit expression

$$A_n = \frac{1}{\sigma_n - 1} \frac{\displaystyle\int (\varepsilon - 1) U^0 u_n \, dV}{\displaystyle\int (\varepsilon - 1) u_n^2 \, dV} \tag{1.76}$$

for the coefficients in the expansion of the diffracted field.

Resonances occur at frequencies such that

$$\sigma_n(k) - 1 = 0 \tag{1.77}$$

for some n. This equation is a generalization of (1.17), and one can redaily obtain an expression like (1.18) for $d\sigma_n/dk^2$.

Just as in (1.10b), we can eliminate the field U^0 from the expression for A_n and express the A_n directly via the excitation sources f. The formula

$$\int (\varepsilon_n - 1) U^0 u_n \, dV = -\frac{1}{k^2} \int f u_n \, dV, \tag{1.78}$$

which generalizes (1.11), can readily be obtained from Eq. (1.64) for U^0 and Eq. (1.65) for u_n. The computations for the case in which $\varepsilon(\mathbf{r})$ is continuous are simpler than those in the presence of discontinuities, where one must integrate over V^+ and V^- separately and take account of the boundary conditions on S_ε. The substitution of (1.74) into the right-hand side of (1.78) gives just the integral occurring in the numerator in (1.76). Thus, the desired analog of (1.10b) has the form

$$A_n = \frac{1}{\sigma_n(1-\sigma_n)} \frac{\int u_n f \, dV}{k^2 \int (\varepsilon - 1) u_n^2 \, dV}. \tag{1.79}$$

Finally, inside the body, for $\varepsilon \neq 1$, we can also expand U^0 in a series in the u_n according to (1.69):

$$U^0 = \sum_n A_n (\sigma_n - 1) u_n. \tag{1.80}$$

Thus the total field can also be expanded in a series in the u_n:

$$U = \sum_n B_n u_n, \tag{1.81a}$$

$$B_n = \sigma_n A_n. \tag{1.81b}$$

Formulas (1.80) and (1.81) are an obvious generalization of (1.9) and (1.14).

2. In this subsection we obtain the same results as in Subsection 1 by using a different technique. This will allow us to approach the main formulas (1.74) and (1.76) from a slightly different viewpoint.

According to (1.62) and (1.64), the diffracted field $U - U^0$ satisfies the equation

$$\Delta(U - U^0) + k^2(U - U^0) = -k^2(\varepsilon - 1) U \tag{1.82}$$

and the same boundary condition as U and U^0. The operator on the left-hand side of this equation is the Helmholtz operator in the absence of the dielectric. Let us introduce the Green function $G(\mathbf{r}, \mathbf{r}_1)$ of this operator, that is, the solution of the equation

$$\Delta G + k^2 G = \delta(\mathbf{r} - \mathbf{r}_1) \tag{1.83}$$

with the same boundary conditions. This solution corresponds to the resonator without the dielectric excited by a δ-like source. If there are no bodies except for the dielectric in the original problem, then G is the Green function of vacuum satisfying the radiation conditions. For two- and three-dimensional problems, in this case we have, respectively,

$$G = \frac{i}{4} H_0^{(2)}(k|\mathbf{r} - \mathbf{r}_1|), \quad G = \frac{-1}{4\pi} \frac{e^{-ik|\mathbf{r}-\mathbf{r}_1|}}{|\mathbf{r} - \mathbf{r}_1|}. \tag{1.84}$$

Let us multiply Eq. (1.83) by $U(\mathbf{r}) - U^0(\mathbf{r})$ and Eq. (1.82) by $G(\mathbf{r}, \mathbf{r}_1)$. By integrating with respect to $\mathbf{r}$, we obtain

$$U(\mathbf{r}_1) - U^0(\mathbf{r}_1) = -k^2 \int [\varepsilon(\mathbf{r}) - 1] G(\mathbf{r}, \mathbf{r}_1) U(\mathbf{r})\, dV. \tag{1.85}$$

This well-known integral equation for $U(\mathbf{r})$ is used, for example, in the justification of Born's approximate method, which (as long as only one iteration is considered) amounts to the replacement of the unknown function $U(\mathbf{r})$ in the integrand on the right-hand side by $U^0(\mathbf{r})$.

Formula (1.85) provides a sourcewise representation of the field $U - U^0$ via the kernel $-k^2(\varepsilon - 1)G$. Let us introduce the *eigenfunctions*

$$u_n = -\sigma_n k^2 \int (\varepsilon - 1) G u_n\, dV \tag{1.86}$$

of this kernel. The following three properties of these eigenfunctions readily follow from (1.86).

First, the function $G(\mathbf{r}, \mathbf{r}_1)$ is symmetric with respect to its arguments and hence satisfies the boundary conditions regardless of whether it is viewed as a functon of $\mathbf{r}$ or $\mathbf{r}_1$. Consequently, u_n satisfies the same conditions.

Second, the functions u_n are orthogonal with the weight $\varepsilon - 1$:

$$\int (\varepsilon - 1) u_n u_m\, dV = 0, \quad \sigma_n \neq \sigma_m. \tag{1.87}$$

Finally, we obtain the third property of the eigenfunctions u_n given by Eq. (1.86) by applying the operator $\Delta + k^2$ with respect to the variables $\mathbf{r}_1$ to Eq. (1.86). Since G treated as a function of $\mathbf{r}_1$ satisfies the same equation (1.86), we obtain

$$(\Delta + k^2) u_n(\mathbf{r}_1) = -\sigma_n k^2 \int [\varepsilon(\mathbf{r}) - 1] \delta(\mathbf{r} - \mathbf{r}_1) u_n(\mathbf{r})\, dV. \tag{1.88}$$

Consequently, u_n satisfies the wave equation

$$\Delta u_n + k^2 [1 + \sigma_n(\varepsilon - 1)] u_n = 0 \tag{1.89}$$

and the correct boundary condition, that is, is a solution of the problem on the free oscillations of a body with permittivity $\varepsilon_n(\mathbf{r})$ given by (1.74).

To find the coefficients A_n in (1.66), we substitute this expansion into (1.85), thus obtaining

$$\sum_n A_n\Big(1 - \frac{1}{\sigma_n}\Big)u_n = -k^2 \int (\varepsilon - 1)U^0 G\,dV. \tag{1.90}$$

Let us multiply this equation by $(\varepsilon - 1)u_m$, integrate over the entire volume, and use the orthogonality condition (1.87). In the resulting integral on the right-hand side, we reverse the order of integration and use (1.86) once more. Then for the A_n we obtain the expression (1.76). Formula (1.79) for A_n can be obtained if in (1.90) we replace U^0 by its expression

$$U^0 = \int Gf\,dV \tag{1.91}$$

via the Green function (which can readily be obtained from (1.64) and (1.83)) and again reverse the order of integration.

Just as one might expect, the technique based on integral equations results in the same auxiliary problems in which the permittivity is given by (1.74) and the same formulas for the expansion coefficients as the technique based on differential equations. Here we presented the method based on (1.85) to illustrate one of the main tricks used in the generalized method of eigenoscillations for the construction of the system of eigenfunctions. Namely, the trick is to reduce a diffraction problem to an integral equation (say, like (1.85)) and then introduce the eigenfunctions of the corresponding integral operator (like (1.86)). This method will also be used in Chapter 2.

3. The expression (1.72) for the permittivity $\varepsilon_n(\mathbf{r})$ is not the only one that results in a closed-form expression for the coefficients A_n. There are some other choices of the functions $\varepsilon_n(\mathbf{r})$ for which one can also reduce (1.69) to a diagonal system of linear equations for the A_n (but with a matrix different from (1.70)).

Let us multiply Eq. (1.69) by $\varepsilon_m u_m/\varepsilon$ and integrate over the entire volume. Then the system of equations for the A_n acquires the form

$$\sum_n A_n \int \varepsilon_n\varepsilon_m\Big(\frac{1}{\varepsilon_n} - \frac{1}{\varepsilon}\Big)u_n u_m\,dV = \int \varepsilon_m\Big(1 - \frac{1}{\varepsilon}\Big)u_m U^0\,dV. \tag{1.92}$$

According to (1.68), the matrix of this system will be diagonal if the factor in the integrand on the right-hand side is proportional to $\varepsilon_n - \varepsilon_m$, that is, if

$$\frac{1}{\varepsilon_n} - \frac{1}{\varepsilon} = \lambda_{nm}\Big(\frac{1}{\varepsilon_m} - \frac{1}{\varepsilon_n}\Big). \tag{1.93}$$

Here the numbers λ_{nm} are different from the proportonality coefficients in (1.71).

One of the solutions of the functional equation (1.93) is given by

$$\frac{1}{\varepsilon_n} = 1 + \tilde{\sigma}_n\Big(\frac{1}{\varepsilon} - 1\Big). \tag{1.94}$$

Needless to say, here both unities can also be replaced by an arbitrary function $\gamma(r)$ satisfying (1.73), but we shall not write out the corresponding formulas (cf. (1.131)).

Once we adopt (1.94), for the A_n we obtain the explicit expression

$$A_n = \frac{1}{\tilde{\sigma}_n - 1} \frac{\int \left(1 - \frac{1}{\varepsilon}\right) \varepsilon_n u_n U^0 \, dV}{\left(1 - \frac{1}{\varepsilon}\right) \varepsilon_n u_n^2 \, dV}, \tag{1.95}$$

which also has the resonant denominator $\tilde{\sigma}_n(k) - 1$.

In some respect, the choice of the permittivity $\varepsilon_n(\mathbf{r})$ of the auxiliary body in the form (1.94) is less natural than (1.74) or (1.72) for Eq. (1.62). The orthogonality conditions (1.68) for the functions (1.74) have the form (1.87), where neither ε_n nor ε_m occurs. A similar representation fails to exist for (1.94), which apparently means that, unlike (1.72), (1.94) cannot be obtained from a Fredholm integral equation. In the next section we shall see that the choice of $\varepsilon_n(\mathbf{r})$ in the form (1.94) is in this sense natural for another form of the wave equation.

We have presented the results (1.94) and (1.95) to illustrate various possibilities that can be used in constructing systems of eigenfunctions. For some choices of the function $\varepsilon(\mathbf{r})$, it is easier to solve Eq. (1.65) with $\varepsilon_n(\mathbf{r})$ given by (1.94) rather than (1.74).

4. Let us now find out what additional possibilities for choosing $\varepsilon_n(\mathbf{r})$ arise if one does not set $\gamma(\mathbf{r}) = 1$ in (1.72) (as was done in (1.74)). For $\gamma \not\equiv 1$, the orthogonality condition (1.87) is replaced by

$$\int (\varepsilon - \gamma) u_n u_m \, dV = 0, \quad n \neq m; \tag{1.96}$$

the expression for the coefficients A_n can still be obtained from (1.70) and has the form

$$A_n = \frac{1}{\sigma_n - 1} \frac{\int (\varepsilon - 1) u_n U^0 \, dV}{\int (\varepsilon - \gamma) u_n^2 \, dV}, \tag{1.97}$$

which generalizes (1.76). The expansion of U^0 inside the body, according to (1.69), (1.72), and (1.96), becomes

$$U^0 = \sum_n A_n (\sigma_n - 1) \frac{\varepsilon - \gamma}{\varepsilon - 1} u_n. \tag{1.98}$$

We arrive at an interesting version of the method by choosing $\gamma(\mathbf{r})$ as follows. Inside the body, we set

$$\gamma(\mathbf{r}) = c\varepsilon(\mathbf{r}), \tag{1.99}$$

where the specific value of the number c ($c \neq 1$) is unessential to our purposes; outside the body, Eq. (1.73) must hold. Then, according to (1.72), $\varepsilon_n(\mathbf{r})$ in the auxiliary problem (1.65) is also proportional to $\varepsilon(\mathbf{r})$ (at the points where $\varepsilon(\mathbf{r}) \neq 1$). The proportionality coefficient is equal to $c + \sigma_n(1-c)$ and can be regarded as an eigenvalue instead of σ_n. Let us denote this coefficient by $\overline{k_n^2}/k^2$:

$$\varepsilon_n = \frac{\overline{k_n^2}}{k^2}\varepsilon, \quad \varepsilon \neq 1. \tag{1.100}$$

This is done to emphasize that the version (1.99) of the ε-method is close to the k-method in some respects. Indeed, the equation for the eigenfunctions u_n under (1.100) coincides (only inside the body, that is, where $\varepsilon(\mathbf{r}) \neq 1$) with the equation for the k-method:

$$\Delta u_n + \overline{k_n^2}\varepsilon(\mathbf{r})u_n = 0. \tag{1.101a}$$

The difference between the two methods is as follows: in the k-method, the same equation remains valid outside the body (for $\varepsilon = 1$), whereas in the ε-method, the functions u_n for $\varepsilon = 1$ satisfy the equation

$$\Delta u_n + k^2 u_n = 0, \tag{1.101b}$$

which does not contain the eigenvalue $\overline{k_n^2}$.

This version (1.100) of the ε-method can be interpreted as a straightforward generalization of the method based on formulas (1.4a) and (1.4b) to the case of diffraction on a nonhomogeneous dielectric body. However, the version (1.74) is apparently simpler, since in the version (1.100) $\varepsilon_n(\mathbf{r})$ is described by two distinct formulas for $\varepsilon \neq 1$ and $\varepsilon = 1$, which is not convenient if $\varepsilon(\mathbf{r})$ attains the unit value by varying continuously.

The orthogonality conditions for (1.100) become

$$\int_{V+} \varepsilon u_n u_m \, dV = 0, \quad n \neq m, \tag{1.102}$$

where the integral is taken over the domain in which $\varepsilon \neq 1$. The expression (1.97) for the coefficients A_n acquires the following form after eliminating σ_n:

$$A_n = \frac{k^2}{\overline{k_n^2} - k^2} \frac{\int (\varepsilon - 1) u_n U^0 \, dV}{\int_{V+} \varepsilon u_n^2 \, dV}. \tag{1.103}$$

This is very similar to (1.10a). Resonances occur when the denominator $\overline{k_n^2} - k^2$ is small or even vanishes.

5. Just as in Subsection 2, the technique with $\gamma \neq 1$ can be constructed by the method of integral equations. To this end, we must define the Green functon by the formula

$$\Delta G + k^2\gamma G = \delta(\mathbf{r} - \mathbf{r}_1), \tag{1.104}$$

instead of (1.83), with the corresponding boundary conditions. Then Eq. (1.85) acquires the form

$$U = \int fG\,dV - k^2 \int (\varepsilon - \gamma) UG\,dV. \tag{1.105}$$

If we define σ_n and u_n to be the eigenvalues and the corresponding eigenfunctons of the integral equation

$$u_n = -\sigma_n k^2 \int (\varepsilon - \gamma) G u_n\,dV, \tag{1.106}$$

then for u_n we obtain the differential equaton (1.65) with the same function ε_n (1.72); that is, Eq. (1.106) results in the same functions u_n and σ_n as Eqs. (1.65) and (1.72). However, the distinguished term

$$U^0 = \int fG\,dV \tag{1.107}$$

outside the series in (1.66) no longer satisfies Eq. (1.64). Instead, it satisfies the equation

$$\Delta U^0 + k^2 \gamma U^0 = f, \tag{1.108}$$

which follows from (1.104). Thus, U^0 *is the field that arises in the diffraction of the field of given sources on a body whose permittivity is* $\gamma(r)$. The coefficients A_n become

$$A_n = \frac{1}{\sigma_n - 1} \frac{\displaystyle\int (\varepsilon - \gamma) u_n U^0\,dV}{\displaystyle\int (\varepsilon - \gamma) u_n^2\,dV}. \tag{1.109}$$

It turns out that in this case we cane remove condition (1.73). By analogy with (1.76), the last expression can be reduced to a form similar to (1.79) and containing the sources f themselves rather than U^0:

$$A_n = \frac{1}{\sigma_n(1 - \sigma_n)} \frac{\displaystyle\int u_n f\,dV}{\displaystyle\int (\varepsilon - \gamma) u_n^2\,dV}. \tag{1.110}$$

The apparent contradiction between these results and the results of the preceding subsection ((1.108) is different from (1.64) and (1.109), from (1.97)) is due to the fact that the U^0 in (1.97) is not the same function as the U^0 in (1.109). The difference between these two functions is a field whose sources lie in the domain $\gamma \neq 1$ and which can hence be expanded in a series with respect to the u_n everywhere.

In the choice of the functions u_n and $\varepsilon_n(\mathbf{r})$ given by formula (1.72) for $\gamma \neq 1$, *it is more natural to define the distinguished term* U^0 *according to* (1.108), *that is, as the solution of a diffraction problem for a body whose permittivity is also equal to* γ, than according to (1.64), that is, for a different ε. With the natural definition, we can reduce the A_n to the form (1.110) and enjoy the existence of the simple integral equation (1.105).

6.[1] The series (1.66) and (1.81) can also be obtained by the methods of the *theory of functions of one complex variable.* This approach will allow us to establish a connection between the possibility of representing the solution of the diffraction problem (1.62) in the form of the expansions (1.66) and (1.81) and the behavior of this solution at large absolute values of the permittivity of the body.

Instead of (1.62), let us consider the more general equation

$$\Delta U + k^2[1 + \sigma(\varepsilon - 1)]U = f, \quad \varepsilon = \varepsilon(\mathbf{r}), \tag{1.111}$$

with the same boundary conditions. The solution of this equation depends on σ and will be denoted by $U(\sigma)$. For $\sigma = 1$ it passes into the solution U of the diffraction problem (1.62); that is, $U(1) = U$. For $\sigma = 0$, the function $U(\sigma)$ passes into the field of the same sources in the absence of the dielectric, $U(0) = U^0$.

We are interested in the dependence of $U(\sigma)$ on σ in the entire plane of the complex variable σ. In our present argument, the coordinates of the observation point are viewed as given parameters. The function $U(\sigma)$ is meromorphic in σ, that is, the only possible singularities of $U(\sigma)$ in any finite part of the complex plane are poles. The poles σ_n of the function $U(\sigma)$ coincide with the eigenvalues of problem (1.65) (provided that the $\varepsilon_n(r)$ are defined by (1.74)).

Suppose that we know the poles σ_n of $U(\sigma)$ and the corresponding residues

$$c_n = \lim_{\sigma \to \sigma_n} (\sigma - \sigma_n) U(\sigma). \tag{1.112}$$

In some cases, this knowledge proves to be sufficient for the reconstruction of $U(\sigma)$. Suppose that there exists a sequence of expanding contours on which $|U(\sigma)| \to 0$ at least as rapidly as $1/|\sigma|$ (in what follows, when describing the behavior of $|U(\sigma)|$ as $\sigma \to \infty$, we avoid mentioning these contours explicitly). Then we have the expansion

$$U(\sigma) = \sum_n \frac{c_n}{\sigma - \sigma_n}. \tag{1.113}$$

If $|U(\sigma)|$ is bounded as $|\sigma| \to \infty$, then

$$U(\sigma) = U(0) + \sum_n \frac{\sigma}{\sigma_n} \frac{c_n}{\sigma - \sigma_n}. \tag{1.114}$$

In the literature, these expansions are known as the *Mittag-Leffler expansions of a meromorphic function into simple fractions.*

Let us find the residues c_n. To this end, we represent the function f occurring in (1.111) as the sum of two terms,

$$f = \frac{(\varepsilon - 1)u_n \int f u_n \, dV}{\int (\varepsilon - 1) u_n^2 \, dV} + \tilde{f}, \tag{1.115}$$

[1]This subsection contains results due to A. D. Shatrov (personal communication).

where u_n is the solution of the homogeneous equation that is obtained from (1.111) by setting $\sigma = \sigma_n$ and $f = 0$. We can readily see that

$$\int \tilde{f} u_n \, dV = 0. \tag{1.116}$$

Accordingly, the solution of (1.111) also is represented as the sum of two terms. By virtue of condition (1.116) ($\tilde{f}$ is orthogonal to the solution of the homogeneous equation), the term corresponding to $\tilde{f}$ remains finite as $\sigma \to \sigma_n$ and hence does not contribute to c_n. The term corresponding to the first summand on the right-hand side in (1.115) is equal to

$$\frac{1}{\sigma - \sigma_n} \frac{u_n \displaystyle\int f u_n \, dV}{k^2 \displaystyle\int (\varepsilon - 1) u_n^2 \, dV}, \tag{1.117}$$

whence for c_n we obtain the following expresseion via the eigenfunction:

$$c_n = \frac{u_n \displaystyle\int f u_n \, dV}{k^2 \displaystyle\int (\varepsilon - 1) u_n^2 \, dV}. \tag{1.118}$$

The series (1.113) for $\sigma = 1$ passes into (1.81), and (1.114) passes into (1.66). Thus, the expansion (1.66) is valid whenever $|U(\sigma)|$ is bounded, and the expansion (1.81) holds if $|U(\sigma)| = O(1/|\sigma|)$.

If the source is shielded from the observation point by some substance with permittivity not equal to 1, then $|U(\sigma)| = O(1/|\sigma|)$as $|\sigma| \to \infty$ and the expansion (1.113) (and hence (1.81)) is valid. For the lower half-plane, this is obvious because the transition to the lower half-plane means that the dielectric becomes a metal. The analysis of sample problems admitting closed-form solutions shows that in this case $|U(\sigma)|$ decays in the upper half-plane as well.

If both the source and the observation point lie outside the dielectric body (and are not shielded by this body), then $U(\sigma)$ tends to a nonzero constant. Consequently, (1.113) does not hold; instead, we have (1.114) and hence (1.66). However, we can write out another expansion of $U(1)$ with regard to the fact that if $U(\sigma)$ tends to $U(\infty)$ sufficiently rapidly, then the expansion (1.113) holds for $U(\sigma) - U(\infty)$ and

$$U(1) = U(\infty) + \sum_n \frac{c_n}{1 - \sigma_n}. \tag{1.119}$$

This series differs from (1.66) in that instead of the field of the sources in vacuum we single out the field of the sources in the presence of a metallic body that replaces the dielectric body on which the diffraction occurs. The series (1.119) converges slower than (1.66).

One must separately consider the case in which the dielectric body has an internal cavity bouded by a surface S, and moreover, one of the eigenfrequencies of this cavity corresponding to the boundary condition $U|_S = 0$ coincides with k. In this case, if both the source and the observation point lie inside the cavity, then $|U(\sigma)| \to \infty$ as $|\sigma| \to \infty$. The rate of growth of $|U(\sigma)|$ as $|\sigma| \to \infty$ depends on the behavior of ε near S. The maximum rate of growth, namely, $|\sigma|^{1/2}$, occurs if the permittivity undergoes a jump on S. For a body with $\varepsilon = \text{const}$, this law of growth can be understood as follows. By replacing the number σ in (1.111) by a complex number with $\operatorname{Im}\sigma < 0$, we replace the dielectric by a metal with conductivity proportional to $|\sigma|$. It is well known that the Q-factor of a closed resonator is proportional to the square root of the conductivity of the walls. In this case, $U(1)$ is given by the representation

$$U(1) = U(0) + \left.\frac{dU(\sigma)}{d\sigma}\right|_{\sigma=0} + \sum_n \frac{1}{\sigma_n^2}\frac{c_n}{1-\sigma_n}, \tag{1.120}$$

which is more complicated than (1.113) or (1.114) and in which the distinguished field has no simple physical interpretation.

1.4 Inhomogeneous Dielectric The Second Polarization

In this section we apply the ε-method to the solution of the differential equation

$$\nabla\left(\frac{1}{\varepsilon}\nabla U\right) + k^2 U = f, \quad \varepsilon = \varepsilon(\mathbf{r}), \tag{1.121}$$

with the radiation condition at infinity and with some conditions (of the first, second, or third kind) on the surface S.

In the first three subsectios of this section, we assume that $\varepsilon(\mathbf{r})$ is a continuous function that is equal to 1 outside some bounded domain. In the fourth subsection we consider a problem in which $\varepsilon(\mathbf{r})$ is a discontinuous function, so that there exists some surface S_ε on which the boundary conditions

$$(U^+ - U^-)|_{S_\varepsilon} = 0, \tag{1.122a}$$

$$\left.\left(\frac{1}{\varepsilon}\frac{\partial U^+}{\partial N} - \frac{\partial U^-}{\partial N}\right)\right|_{S_\varepsilon} = 0 \tag{1.122b}$$

must be satisfied and in the exterior of which $\varepsilon = 1$. We already know that these boundary coditions are the limit form of Eq. (1.121) on the surface of discontinuity of $\varepsilon(\mathbf{r})$. The solutions obtained from (1.121) for a continuous $\varepsilon(\mathbf{r})$ satisfy conditions (1.122) on S_ε in the limit as $|\nabla\varepsilon| \to \infty$ on S_ε. Hence it suffices to consider problems with continuous $\varepsilon(\mathbf{r})$ and then pass to the limit as $|\nabla\varepsilon| \to \infty$ in the final formulas if bodies with interfaces are to be considered. This approach simplifies the computations against those which would result if the boundary S_ε with conditions (1.122) were introduced from the very beginning, because one need not separately consider the integrals over V^+ and V^-.

1. Let us repeat the constructions of Section 1.3 for Eq. (1.121) with appropriate modifications. We introduce the eigenfunctions $u_n(\mathbf{r})$ of the equation

$$\nabla\left(\frac{1}{\varepsilon_n}\nabla u_n\right) + k^2 u_n = 0, \quad \varepsilon_n = \varepsilon_n(\mathbf{r}) \tag{1.123}$$

with the corresponding boundary conditions on the metal surface and at infinity. The form of the functions $\varepsilon_n(\mathbf{r})$ will be found later.

Let us find orthogonality conditions similar to (1.68) for the functions u_n. To this end, we multiply Eq. (1.123) by u_m, subtract the same expressions with the indices permuted, and integrate over the volume. Let us use the identity

$$u_m \nabla\left(\frac{1}{\varepsilon_n}\nabla u_n\right) = -\frac{1}{\varepsilon_n}\nabla u_n \nabla u_m + \nabla\left(\frac{1}{\varepsilon_n} u_m \nabla u_n\right). \tag{1.124}$$

The surface integrals arising when we integrate the last summand have the form (1.67) and are zero. The integrals over the sphere at infinity vanish since both u_n and u_m satisfy the radiation conditions, and $\varepsilon_n(\mathbf{r})$ (as will be seen later) satisfies the condition

$$\varepsilon_n(\mathbf{r}) = 1 \quad \text{for} \quad \varepsilon(\mathbf{r}) = 1. \tag{1.125}$$

Thus, the desired relationship (i.e., the *orthogonality condition*) between two eigenfunctions u_n and u_m corresponding to two auxiliary permittivities ε_n and ε_m has the form

$$\int \left(\frac{1}{\varepsilon_n} - \frac{1}{\varepsilon_m}\right) \nabla u_n \nabla u_m \, dV = 0. \tag{1.126}$$

Let us specify U^0 by Eq. (1.64) and represent the solution of Eq. (1.121) by the series (1.66). The series (1.66) satisfies the boundary conditions as well as the wave equation outside the body term by term. The coefficients A_n can be found from the requirement that Eq. (1.121) must be satisfied inside the body, that is, in the domain where $\varepsilon \neq 1$. By substituting (1.66) into (1.121), we obtain the following formula, similar to (1.69):

$$\sum_n A_n \nabla\left[\left(\frac{1}{\varepsilon_n} - \frac{1}{\varepsilon_m}\right)\nabla u_n\right] = \nabla\left[\left(1 - \frac{1}{\varepsilon}\right)\nabla U^0\right]. \tag{1.127}$$

This formula, in particular, implies (1.125).

Then we can obtain a system of linear equations for the coefficients A_n by two different methods, which lead to different expressions for $\varepsilon_n(\mathbf{r})$ and A_n. The first method, given in Subsection 2, corresponds to Subsection 1.3.1, and the second, given in Subsection 3, to Subsection 1.3.3.

2. Let us multiply Eq. (1.127) by u_m and integrate over the volume with regard to (1.124). The surface integrals occurring in this computation are zero in view of (1.125). We obtain the system

$$\sum_n A_n \int \left(\frac{1}{\varepsilon} - \frac{1}{\varepsilon_n}\right) \nabla u_n \nabla u_m \, dV = \int \left(1 - \frac{1}{\varepsilon}\right) \nabla U^0 \nabla u_n \, dV. \tag{1.128}$$

The matrix of this system will be diagonal if the integrand in the coefficients is proportional to the integrand in the orthogonality conditions, that is, if Eq. (1.93) is satisfied. Then we can express $\varepsilon_n(\mathbf{r})$ via $\varepsilon(\mathbf{r})$ and introduce the eigenvalue σ_n by the formula

$$\frac{1}{\varepsilon_n(\mathbf{r})} = 1 + \sigma_n \left(\frac{1}{\varepsilon(\mathbf{r})} - 1 \right) \tag{1.129}$$

(cf. (1.94)). Furthermore, just as in (1.87), we can rewrite the orthogonality conditions in the form

$$\int \left(1 - \frac{1}{\varepsilon} \right) \nabla u_n \nabla u_m \, dV = 0, \quad n \neq m, \tag{1.130}$$

which does not contain ε_n and ε_m. We could even introduce an arbitrary function $\gamma(\mathbf{r})$ satisfying Eq. (1.73) and write out a more general solution of the functional equation (1.93):

$$\frac{1}{\varepsilon_n} = \frac{1}{\gamma} + \sigma_n \left(\frac{1}{\varepsilon} - \frac{1}{\gamma} \right). \tag{1.131}$$

In particular, if we take γ to be proportional to ε (for $\varepsilon \neq 1$), that is, if we adopt (1.99), then ε_n will be proportional to ε (see (1.100)), so that the wave equation for u_n inside the body acquires the form

$$\nabla \left(\frac{1}{\varepsilon} \nabla u_n \right) + \overline{k_n^2} u_n = 0, \tag{1.132}$$

similar to (1.101a). Here the eigenvalue is denoted by $\overline{k_n^2}$, in the same way as in the end of Subsection 1.3.4. Just as in the preceding section, Eq. (1.132) does not differ from the equation for the eigenfunctions in the k-method but is satisfied only inside the body, that is, in the domain where $\varepsilon \neq 1$. Outside the body, u_n satisfies Eq. (1.101b). In the k-method, u_n would be described by Eq. (1.132), which contains $\overline{k_n^2}$, inside as well as outside the body.

We shall not write out the orthogonality conditions, the expressions for A_n, etc. for the more general solution (1.131) and restrict ourselves to the analysis of the simplest form (1.129) of the function $\varepsilon_n(\mathbf{r})$.

The imaginary part of the permittivity $\varepsilon_n(\mathbf{r})$ is nonnegative in the mean over the body, just as it is in the problems considered in the previous section. Indeed, by multiplying Eq. (1.123) by u_n^* and by using (1.124) once more, we readily obtain

$$\frac{1}{\varepsilon_n} |\nabla u_n|^2 = k^2 |u_n|^2 + \nabla \left(\frac{1}{\varepsilon_n} u_n^* \nabla u_n \right). \tag{1.133}$$

Let us subtract this expression from its complex comjugate and integrate over the volume. Then on the right-hand side we obtain the imaginary part of the integral $\int u_n (\partial u_n^* / \partial N) \, dS$ taken over the impedance surface and the sphere at infinity. This quantity is always nonnegative (and is zero for closed resonator without losses), and so

$$\int \frac{\operatorname{Im} e_n}{|\varepsilon_n|^2} |\nabla u_n|^2 \, dV \geq 0. \tag{1.134}$$

The physical meaning of this property of permittivity of auxiliary bodies was considered in Section 1.2.

If $\varepsilon_n(\mathbf{r})$ is chosen according to (1.129), then the amplitudes A_n are given by

$$A_n = \frac{1}{\sigma_n - 1} \frac{\int \left(1 - \frac{1}{\varepsilon}\right) \nabla U^0 \nabla u_n \, dV}{\int \left(1 - \frac{1}{\varepsilon}\right) (\nabla u_n)^2 \, dV}. \tag{1.135}$$

The resonance denominator has the same form $\sigma_n(k) - 1$ as in (1.76).

Just as before, we can eliminate the field U^0 from (1.135) by expressing U^0 via the sources f that generate this field. The analog of (1.78) which must be used here can be obtained by integrating the formula

$$\nabla \left[u_n \nabla U^0 - \frac{1}{\varepsilon_n} U^0 \nabla u_n\right] - \left(1 - \frac{1}{\varepsilon_n}\right) \nabla u_n \nabla U^0 = u_n f, \tag{1.136}$$

which can readily be derived from (1.121) and (1.124), and has the form

$$\int \left(1 - \frac{1}{\varepsilon_n}\right) \nabla u_n \nabla U^0 \, dV = -\int f u_n \, dV. \tag{1.137}$$

Now we can use Eq. (1.129) to obtain the desired formula, similar to (1.79) (that is, containing f instead of U^0):

$$A_n = \frac{1}{\sigma_n(1 - \sigma_n)} \frac{\int u_n f \, dV}{\int \left(1 - \frac{1}{\varepsilon}\right) (\nabla u_n)^2 \, dV}. \tag{1.138}$$

Inside the body (that is, for $\varepsilon \neq 1$), we can expand U^0 in a series with respect to the u_n, that is, represent the solution U in the form (1.13) and thus eliminate U^0 from all the formulas completely. The coefficients of the series expansion of U^0 with respect to u_n can readily be obtained from the orthogonality condition (1.130). These coefficients differ from A_n (1.135) by the absence of the first factor, so that for this polarization formulas (1.80) and (1.81) remain valid.

The formulas of this subsection can also be obtained if we replace the differential equation (1.121) by the corresponding integral equation and repeat the argument of Subsection 1.3.2. Equation (1.121) can be rewritten in the form

$$\Delta(U - U^0) + k^2(U - U^0) = \nabla \left[\left(\frac{1}{\varepsilon} - 1\right) \nabla U\right]. \tag{1.139}$$

On introducing the Green function in the absence of the dielectric via (1.83), we obtain the integro-differential equation

$$U(\mathbf{r}_1) - U^0(\mathbf{r}_1) = -\int G(\mathbf{r}, \mathbf{r}_1 \nabla \left[\left(\frac{1}{\varepsilon} - 1\right) \nabla U\right] dV. \tag{1.140}$$

We can introduce the functions u_n as the *eigenfunctions of the homogeneous equation* corresponding to (1.140):

$$u_n = -\sigma_n \int G\nabla \left[\left(\frac{1}{\varepsilon} - 1\right)\nabla u_n\right] dV. \tag{1.141}$$

One can readily verify that these functions satisfy the same equation (1.123) with the permittivity (1.129), that is, coincide with the u_n introduced above. In this case the orthogonality conditions (1.130) do not contain ε_n and ε_m. The technique of differential equations in this problem is simpler than that of Eq. (1.140), since Eq. (1.140) is much more complicated than (1.85).

3. Let us now use (1.127) to derive a system of equations for A_n other than (1.128). This will result in a different set of functions $\varepsilon_n(\mathbf{r})$, for which the matrix of the system will also be diagonal.

Let the functions $\tilde{u}_m$ be determined by the equation

$$\nabla\tilde{u}_m = \frac{\varepsilon}{\varepsilon_m}\nabla u_m. \tag{1.142}$$

Let us multiply Eq. (1.127) by $\tilde{u}_m$ and integrate over the entire volume. After the transformation (1.124), the function $\tilde{u}_m$ will occur in the integrand only in the combinaion $\nabla\tilde{u}_m$, so that the system for A_n, according to Eq. (1.142), will contain only u_n and the functions $\tilde{u}_n$ will not be involved in the subsequent manipulations. This system has the form

$$\sum_n A_n \int \left(\frac{1}{\varepsilon} - \frac{1}{\varepsilon_n}\right)\frac{\varepsilon}{\varepsilon_m}\nabla u_n \nabla u_m \, dV = \int \left(1 - \frac{1}{\varepsilon}\right)\frac{\varepsilon}{\varepsilon_m}\nabla U^0 \nabla u_m \, dV. \tag{1.143}$$

It will be diagonal if the functions $\varepsilon_n(\mathbf{r})$, according to (1.126), satisfy the functional equation

$$\frac{\varepsilon_n - \varepsilon}{\varepsilon_n - \varepsilon_m} = \lambda_{nm}\left(\frac{1}{\varepsilon_n} - \frac{1}{\varepsilon_m}\right). \tag{1.144}$$

This equation coincides with (1.71), and the solution has the form

$$\varepsilon_n = 1 + \tilde{\sigma}(\varepsilon - 1), \tag{1.145}$$

where $\tilde{\sigma}_n$ is the eigenvalue of the problem (introduced in another way than σ_n in (1.129) was introduced). Here we again do not consider an arbitrary function $\gamma(\mathbf{r})$. The choice of permittivity in the auxiliary problem in the form (1.145) is *less natural than* (1.129) *for* Eq. (1.121). For the case (1.145), the orthogonality condition contains ε_n and ε_m; moreover, there is no elementary way to construct an integral or an integro-dfferential operator with eigenvalues $\tilde{\sigma}_n$ and eigenfunctions u_n, not to say that some transformations that do pass for the eigenfunctions corresponding to the case (1.129) now fail. However, the formal construction of the series (1.66) remains possible, and the coefficients

$$A_n = \frac{1}{\tilde{\sigma}_n - 1}\frac{\displaystyle\int \frac{\varepsilon - 1}{\varepsilon_n}\nabla U^0 \nabla u_n \, dV}{\displaystyle\int \frac{\varepsilon - 1}{\varepsilon_n^2}(\nabla u_n)^2 \, dV} \tag{1.146}$$

have the same structure as the coefficients (1.135).

4. Let us now consider the limit case in which $\varepsilon(\mathbf{r})$ has a discontinuity on some boundary S_ε. Then conditions (1.122) must be satisfied on that boundary. To avoid complicated notation, we suppose that the permittivity is constant in the body, $\varepsilon(\mathbf{r}) = \text{const} \neq 1$, and is equal to 1 outside the body. Thus in this subsection we seek the solution of the equations

$$\frac{1}{\varepsilon}\Delta U + k^2 U = f \quad \text{in} \quad V^+, \tag{1.147a}$$

$$\Delta U + k^2 U = f \quad \text{in} \quad V^- \tag{1.147b}$$

with the boundary conditions (1.122) on S_ε and with the usual conditions on the metal surface and at infinity. This problem differs from the corresponding problem in Sections 1.1 and 1.2 by the boundary condition for the normal derivative on S_ε; that is, the difference resides in formulas (1.1c) and (1.122b).

Equations (1.123) for u_n inside (V^+) and outside (V^-) the body coincide with (1.4a) and (1.4b), respectively. Moreover, ε_n (1.129) is also constant in V^+ and is equal to 1 in V^-, that is, undergoes a jump on S_ε. The boundary conditions for u_n on S_ε are the limit form of Eq. (1.123), that is, have the form

$$(u_n^+ - u_n^-)|_{S_\varepsilon} = 0, \quad \left(\frac{1}{\varepsilon_n}\frac{\partial u_n^+}{\partial N} - \frac{\partial u_n-}{\partial N}\right)\bigg|_{S_\varepsilon} = 0. \tag{1.148}$$

In the diffraction problem considered in this subsection, permittivity occurs not only in Eq. (1.147a) but also in the boundary condition (1.122). According to (1.148), *the eigenvalue ε_n of the homogeneous problem also occurs in Eq.* (1.4a) *as well as in the boundary condition.* Hence *the homogeneous problem for u_n, just as before, describes free oscillations of a body with permittivity ε_n*, that is, *has an independent physical meaning.*

The eigenvalue ε_n of the homogeneous problem (1.4a,b), (1.148) is related to the eigenvalue σ_n of the more general problem (1.121) by Eq. (1.129), where now ε_n and ε are numbers, and

$$\sigma_n = \frac{\varepsilon}{\varepsilon_n}\frac{\varepsilon_n - 1}{\varepsilon - 1}. \tag{1.149}$$

As usual, we seek the solution of the diffraction problem (1.147), (1.122) in the form (1.66), where U^0 (the field of the same sources in the absence of the dielectric body) must satisfy Eq. (1.64) and must be continuous *together with the normal derivative* on S_ε (thus, U^0 must not satisfy condition (1.122b)).

In the basic fomulas of this section, the passage to $|\nabla\varepsilon| \to \infty$ is trivial in that it does not require the computation of any indeterminacies. Indeed, the formulas do not contain $\nabla\varepsilon$, whereas ∇u_n and ∇u_m occur only in integrands and are finite everywhere, including the surface S_ε itself, according to (1.148).

The orthogonality conditions (1.126) have the form

$$\int_{V+} \nabla u_n \nabla u_m \, dV = 0, \quad n \neq m. \tag{1.150}$$

According to (1.149), the coefficients A_n given by (1.135) are equal to

$$A_n = \frac{\varepsilon_n(\varepsilon - 1)}{\varepsilon_n - \varepsilon} \frac{\int_{V+} \nabla U^0 \nabla u_n \, dV}{\int_{V+} (\nabla u_n)^2 \, dV}. \tag{1.151}$$

Needless to say, this formula can also be obtained from (1.146) for another choice of $\varepsilon_n(\mathbf{r})$ (see (1.145)). If $\varepsilon(\mathbf{r})$ = const, then in both cases we have $\varepsilon_n(\mathbf{r})$ = const, so that the eigenfunctions u_n for (1.129) and (1.145) coincide. If, according to (1.145), we substitute

$$\tilde{\sigma}_n = \frac{\varepsilon_n - 1}{\varepsilon - 1} \tag{1.152}$$

into (1.145), then (1.135) passes into (1.151).

If we express the numerator in (1.151) via the integral of the exciting sources according to (1.137), then we obtain the following analog of Eq. (1.138), which does not contain U^0:

$$A_n = \frac{\varepsilon_n^2(\varepsilon - 1)}{(\varepsilon_n - 1)(\varepsilon - \varepsilon_n)} \frac{\int f u_n \, dV}{\int_{V+} (\nabla u_n)^2 \, dV}. \tag{1.153}$$

It has the same meaning as Eq. (1.10b)

By substituting (1.149) into (1.81), we can also obtain an expression via ε_n of the coefficients of the expansion in V^+ of the total field with respect to the u_n.

Needless to say, the basic formulas of this subsection, namely, Eqs. (1.150), (1.151), and (1.153), can be obtained directly from Eqs. (1.147) and the boundary conditions (1.122) for the unknown field and from Eqs. (1.4a,b) and the boundary conditions (1.148) for u_n. However, the computations required there are more cumberome.

1.5 Quantum-Mechanical Problem on Elastic Scattering on a Quasistationary Level

Here we deal with a problem which is a quantum-mechanical analog of the problem on the diffraction on a body with variable permittivity, considered in Section 1.3. Mathematically, the analogy between the two problems is not destroyed by the fact that the permittivity $\varepsilon(\mathbf{r})$, which is equivalent to the potential field $U(\mathbf{r})$,

$$\varepsilon = 1 - \frac{U}{k^2}, \tag{1.154}$$

depends on the frequency. Indeed, the frequency is a constant parameter, which is the same in all equations.

Quantum-mechanical scattering theory is somewhat simpler than diffraction theory, since in scattering theory the incident field is always a plane wave and the scattered field is only sought remote from the body. In the most interesting case of resonance scattering by a quasistationary level, the quantum-mechanical problem corresponds to an electromagnetic problem in which the near-surface layer of the body in question is characterized by large negative values of $\varepsilon(\mathbf{r})$. In this case, one term in the sum (1.66), which represents the external field, is dominating, while the other terms are small. Accordingly, the desired scattering pattern is given by a simple formula.

The discrete technique given below differs from the Sturm method, traditionally used in scattering theory, in the character of the auxiliary problem (the eigenfunctions satisfy the correct conditions at infinity, and the eigenvalue is not equal to the coupling constant) as well as in that only part of the solution is expanded in a series. It is these specific features that allow one, in the case of scattering by a quasistationary level, to obtain explicit expresions for the fields and scattering patterns and to develop an efficient numerical technique for an arbitrary form of the barrier.

The scattering pattern can also be determined with the help of another version of the generalized eigenoscillation method, namely, the s-method (see Section 2.5). This will be done in Section 4.2. The homogeneous problem of the s-method essentially coincides with the well-known quantum-mechanical problem of finding the scattering matrix. For a spherically symmetric potential, the latter problem can be solved numerically either by integrating the differential equation directly or by using the integral equation of the s-method. In Section 4.2 we shall compare numerical results produced by the two methods to illustrate the accuracy provided by the corresponding approximate formulas.

The present section and Section 4.2 can be read independently of the rest of the book. There are only few references to other sections, and some of the basic formulas are repeated. The time-dependent factor is assumed to be $\exp(-i\omega t)$, *contrary to what is used in the rest of the book.*

1. The nonrelativistic steady-state problem on the elastic scattering of a beam of particles with energy k^2 by a potential $U(\mathbf{r})$ can be stated as follows: Find a solution Ψ of the three-dimensional Schrödinger equation

$$\Delta\Psi + (k^2 - U)\Psi = 0 \tag{1.155}$$

such that Ψ is everywhere finite and satisfies the asymptotic condition

$$\Psi|_{\mathbf{r}\to\infty} \simeq e^{ikz} + \Phi(\varphi,\theta)\frac{e^{ik\mathbf{r}}}{k\mathbf{r}}, \tag{1.156}$$

where $\Phi(\varphi,\theta)$ is the desired scattering pattern.

As is customary in quantum mechanics, we restrict ourselves to determining the function $\Phi(\varphi,\theta)$, that is, study the situation in which both the source and the observation point are at infinity. The same technique can be applied to find the Green function of the Schrödinger equation.

Throughout the following we assume that $U(\mathbf{r}) \equiv 0$ outside some sphere of finite radius, so that the equivalent body occupies a finite volume. Apparently, the technique can be generalized to cover the case of potentials that tend to zero as $r \to \infty$.

2. Let us introduce a system of functions ψ_n determined by the equation

$$\Delta\psi_n + [k^2 - U_n(\mathbf{r})]\psi_n = 0, \tag{1.157}$$

the condition that ψ_n is finite at $r = 0$, and the condition

$$\psi_n|_{\mathbf{r}\to\infty} \simeq \Phi_n(\varphi, \theta)\frac{e^{ik\mathbf{r}}}{k\mathbf{r}} \tag{1.158}$$

(where the functions Φ_n are not given and must be found simultaneously with ψ_n). Here $U_n(\mathbf{r})$ is the auxiliary potential given by

$$U_n(\mathbf{r}) = \sigma_n U(\mathbf{r}) + (1 - \sigma_n)\overline{U}(\mathbf{r}). \tag{1.159}$$

In this problem $\sigma_n(k) = \sigma_n'(k) + i\sigma_n''(k)$ plays the role of an eigenvalue.

The real function $\overline{U}(\mathbf{r})$ can in general be chosen arbitrarily save that there must be no resonances in the problem with this potential. To ensure that the terms corresponding to this function in the subsequent expansions take account of the nonresonance background as completely as possible, it is convenient to set

$$\overline{U}(\mathbf{r}) = \begin{cases} U_{\max}, & r \leq r_m(\varphi, \theta), \\ U(r), & r > r_m(\varphi, \theta), \end{cases} \tag{1.160}$$

in each direction, where $r_m(\varphi, \theta)$ is the radial coordinate of the point at which the potential U is maximal on the corresponding ray and $U_{\max}$ is the value of the maximum. Let $\overline{\Psi}$ (and, accordingly, $\overline{\Phi}$) be the solution of problem (1.155), (1.156) with U replaced by $\overline{U}$.

The eigenfunctions ψ_n are orthogonal in the sense that

$$\int (U - \overline{U})\psi_n\psi_m \, dV = 0, \quad n \neq m. \tag{1.161}$$

The functions ψ_n and $\overline{\Psi}$ are related by the equation

$$\int (U - \overline{U})\psi_n\overline{\Psi} \, dV = -\frac{\sigma_n''}{\sigma_n}\Phi_n(\pi) \int (U - \overline{U})|\psi_n|^2 \, dV \tag{1.162}$$

which can readily be obtained from the corresponding equations with regard to the fact that all conditions at infinity contain the same number k. Here we adopt the normalization

$$\int_0^{2\pi} \int_0^{\pi} |\Phi_n|^2 \sin\theta \, d\theta \, d\varphi = 4\pi, \tag{1.163}$$

and $\Phi_n(\pi)$ stands for the value of the eigenpattern in the direction of the incident wave.

We seek the solution of problem (1.155), (1.156) in the form of the series

$$\Psi = \overline{\Psi} + \sum_n A_n\psi_n. \tag{1.164}$$

The diargams Φ, $\overline{\Phi}$, and Φ_n are related by the obvious formula

$$\Phi = \overline{\Phi} + \sum_n A_n \Phi_n. \tag{1.165}$$

The function Ψ represented in the form (1.164) satisfies condition (1.156) termwise for arbitrary values of the coefficients A_n. These coefficients must be determined from the requirement that Ψ satisfies Eq. (1.155). We substitute (1.164) into (1.155) and use the orthogonality relation (1.161) together with identity (1.162) to obtain

$$A_n = \frac{\sigma_n''}{1-\sigma_n} \Phi_n(\pi) \frac{\displaystyle\int (U-\overline{U})|\psi_n|^2 \, dV}{\displaystyle\sigma_n \int (U-\overline{U})\psi_n^2 \, dV}. \tag{1.166}$$

Formulas (1.156)–(1.166) give the desired solution of the problem. If $\overline{U}(\mathbf{r})$ is sufficiently close to $U(\mathbf{r})$ on the outer part of the barrier, then

$$|\sigma_n''| \ll |1 - \sigma_n'| \tag{1.167}$$

remote from the resonance (the quasistationary level), so that all the coefficients A_n are small and the solution is very well described by the first terms in (1.164) and (1.165). Under the resonance conditions, near some value $k = k_*$, one of the eigenvalues (say, σ_m) does not satisfy (1.167). The corresponding coefficient A_m is not small, the series is reduced to a single term, and the dependence of the field on k is mainly determined by the function

$$\lambda_m(k) = \frac{1-\sigma_m'}{\sigma_m''}. \tag{1.168}$$

The root k_* of the equation $\lambda_m(k) = 0$ is the center of the resonance curve, that is, the quasistationary level (real). The resonance width $2\delta k$ (the band of values of k in which the scattered field strongly depends on the energy) is determined by the rate of change of $\lambda_m(k)$ at the quasistationary level; namely,

$$\delta k = \left\{ \left. \frac{d\lambda_m}{dk} \right|_{k=k_*} \right\}^{-1}. \tag{1.169}$$

3. The efficiency of the above technique depends on how one finds σ_m and, eventually, $\lambda_m(k)$. The main computational advantage of the method is in the fact that σ_m can be determined directly by a simple variational procedure. Since the ψ_n do not grow at infinity, we can write out the complex-valued functional

$$R(\psi) = \frac{\displaystyle\int [(k^2 - U)\psi^2 - (\nabla\psi)^2] \, dV}{\displaystyle\int (U-\overline{U})\psi^2 \, dV}, \tag{1.170}$$

which is stationary at the eigenfunctions of problem (1.157), (1.158) and assumes the values σ_n at the stationary points. The integral in the numerator is understood here as the limit of the correponding expression as $\operatorname{Im} k \to +0$. The admissible functions for (1.170) must be everywhere continuous and have the same asymptotic dependence (1.158) on r as the eigenfunctions.

It may be convenient not to require the admissible functions to be continuous on the surface S of discontinuity of the potential (if there is such a surface, which is a certain idealization for a quantum-mechanical problem). To preserve the stationary properties of the functional in this case, it suffices to add the integral

$$\int_S (\psi^- - \psi^+)\left(\frac{\partial \psi^-}{\partial N} + \frac{\partial \psi^+}{\partial N}\right) dS \tag{1.171}$$

to the numerator of (1.170), where the superscripts "+" and "−" indicate the limit values of functions on different sides of the discontinuity surface. The normal is assumed to point from "−" to "+."

4. In the preceding, the differential statement of the problem for ψ_n was used. However, one can readily show that the functions ψ_n are also solutions of a homogeneous integral equation whose kernel is proportional to the Green function corresponding to the potential $\overline{U}(\mathbf{r})$. More precisely, if $G(\mathbf{r}_1, \mathbf{r}_2)$ satisfies the equation

$$\Delta G + (k^2 - \overline{U})G = \delta(\mathbf{r}_1 - \mathbf{r}_2) \tag{1.172}$$

and condition (1.158), then for our choice of $\overline{U}_n(\mathbf{r})$ the equation

$$\psi_n = -\sigma_n \int (U - \overline{U}) G \psi_n \, dV \tag{1.173}$$

holds. This equation, which coincides with the Lippmann–Schwinger equation in the *distorted wave metod*, is not selfadjoint for real k. The system of functions ψ_n generated by this kernel is by no means complete, since their asymptotics contains only outgoing waves. For example, the functions e^{ikz} and $\overline{\Psi}$ cannot be expanded in ψ_n. However, the difference $\Psi - \overline{\Psi}$, whose asymptotics also contains only outgoing waves, can be represented via this kernel and hence can be expanded in a series with respect to the ψ_n (see (1.164)).

1.6 Dielectric Body. The Maxwell Equations

The technique developed in the preceding sections can be transferred to the Maxwell equations almost automatically. Just as in the scalar case, the solution is sought in the form of series expansions with respect to solutions of certain homogeneous problems; the coefficients of these series exhibit resonance behavior. The properties of the eigenfunctions are derived directly from the equations defining these functions. In the general case of an inhomogeneous dielectric, to obtain explicit expressions for the coefficients, we must require the permittivity in the homogeneous problems to be related in a certain way to the actual permittivity, and so on.

However, the vector problem has some specific features. If we apply the usual k-method to closed resonators, then the series obtained by this method must be supplemented by gradient terms, whose source is the divergence of currents. The series obtained in other versions of the generalized method do not require including such terms, because these terms are already contained in the distinguished term (the field of the same sources in the absence of the body or in the presence of another body). Needless to say, the same is true of the series describing the field in an open resonator.

Throughout the preceding, we used the following formula to relate two solutions of wave equations by transforming volume integrals to surface ones:

$$\operatorname{div}(\alpha \mathbf{A}) = \alpha \operatorname{div} \mathbf{A} + \mathbf{A} \operatorname{grad} \alpha. \tag{1.174}$$

Now we shall use the analog of this formula given by

$$\operatorname{div}[\mathbf{AB}] = \mathbf{B} \operatorname{rot} \mathbf{A} - \mathbf{A} \operatorname{rot} \mathbf{B}, \tag{1.175}$$

which permits one to relate two solutions of the Maxwell equations. An example of such a relationship is the Lorentz lemma; many of the transformations below are analogs of calculations used in the derivation of this lemma.

In Subsections 1 and 2 we consider diffraction on a body with constant ε; thus, the technique developed in Sections 1.1 and 1.2 will be transferred to the Maxwell equations. We omit the physical analysis, which was given when we considered the scalar problem. In Subsections 3 and 4 we consider the inhomogeneous dielectric. The results of Subsections 3 and 4 imply all the formulas in Subsections 1 and 2. We could follow the scheme of Section 1.4 by first studying the case of an inhomogeneous dielectric and then passing to the limit in the final formulas of Subsections 3 and 4; this would result in less cumbersome calculations. However, the approach adopted in our exposition gives a logically simpler solution of the problem stated in Subsection 1, which is mostly of interest in practical applications.

1. We seek the solution of the diffraction problem for the field generated by given currents $\mathbf{j}^{(e)}$ and $\mathbf{j}^{(m)}$ on a body, occupying a given domain V^+, with permittivity ε and permeability 1. The desired fields $\{\mathbf{E}, \mathbf{H}\}$ satisfy the equations

$$\operatorname{rot} \mathbf{H} - ik\varepsilon \mathbf{E} = \frac{4\pi}{c} \mathbf{j}^{(e)}, \tag{1.176a}$$

$$\operatorname{rot} \mathbf{E} + ik \mathbf{H} = -\frac{4\pi}{c} \mathbf{j}^{(m)} \tag{1.176b}$$

in V^+, the equations

$$\operatorname{rot} \mathbf{H} - ik \mathbf{E} = \frac{4\pi}{c} \mathbf{j}^{(e)}, \tag{1.177a}$$

$$\operatorname{rot} \mathbf{E} + ik \mathbf{H} = -\frac{4\pi}{c} \mathbf{j}^{(m)} \tag{1.177b}$$

in V^- (that is, outside the body), and the continuity conditions

$$E_t^+ - E_t^- = 0, \quad H_t^+ - H_t^- = 0 \tag{1.178}$$

on the boundary S_ε of the body, where t is an arbitrary direction tangent to S_ε. If there are also some other bodies in the field, for example, metallic or impedance surfaces S, then the fields $\{\mathbf{E},\mathbf{H}\}$ must satisfy the condition $E_t = 0$ or the Leontovich condition on these surfaces. If the dielectric is not contained in a closed resonator, then $\{\mathbf{E},\mathbf{H}\}$ must additionally satisfy the radiation conditions

$$\begin{aligned} E_\theta &= H_\varphi = \Phi_1(\theta,\varphi)\frac{e^{-ikR}}{kR}\left[1+O\Big(\frac{1}{kR}\Big)\right], \\ H_\theta &= -E_\varphi = \Phi_2(\theta,\varphi)\frac{e^{-ikR}}{kR}\left[1+O\Big(\frac{1}{kR}\Big)\right]; \end{aligned} \tag{1.179}$$

note that if $\mathbf{j}^{(e)}$ and $\mathbf{j}^{(m)}$ are located at infinity, then the radiation condition must be imposed on the fields $\{\mathbf{E}-\mathbf{E}^0, \mathbf{H}-\mathbf{H}^0\}$, where the fields $\{\mathbf{E}^0,\mathbf{H}^0\}$ will be defined below in (1.180). These conditions are valid for all the fields introduced in what follows and will not be stated explicitly each time.

Let us introduce the field $\{\mathbf{E}^0,\mathbf{H}^0\}$ generated by the same sources in the absence of the dielectric body, that is, satisfying the equations

$$\operatorname{rot}\mathbf{H}^0 - ik\mathbf{E}^0 = \frac{4\pi}{c}\mathbf{j}^{(e)}, \tag{1.180a}$$

$$\operatorname{rot}\mathbf{E}^0 + ik\mathbf{H}^0 = -\frac{4\pi}{c}\mathbf{j}^{(m)} \tag{1.180b}$$

in the entire space $V^+ + V^-$. Needless to say, $\{\mathbf{E}^0,\mathbf{H}^0\}$ also satisfy conditions (1.178) on S_ε.

We define the eigenfunctions $\{\mathbf{e}_n,\mathbf{h}_n\}$ as the fields satisfying the homogeneous equations

$$\operatorname{rot}\mathbf{h}_n - ik\varepsilon_n\mathbf{e}_n = 0, \tag{1.181a}$$

$$\operatorname{rot}\mathbf{e}_n + ik\mathbf{h}_n = 0 \tag{1.181b}$$

on V^+, the homogeneous equations

$$\operatorname{rot}\mathbf{h}_n - ik\mathbf{e}_n = 0, \tag{1.182a}$$

$$\operatorname{rot}\mathbf{e}_n + ik\mathbf{h}_n = 0 \tag{1.182b}$$

on V^-, the boundary conditions

$$e^+_{nt} - e^-_{nt} = 0, \quad h^+_{nt} - h^-_{nt} = 0 \tag{1.183}$$

on S_ε, and conditions (1.179). The field $\{\mathbf{e}_n,\mathbf{h}_n\}$ describes the eigenoscillations at a given frequency of a body occupying the volume V^+ and having the permittivity ε_n. The numbers ε_n are the eigenvalues of problem (1.181)–(1.183).

The eigenfunctions corresponding to distinct eigenvalues ε_n are orthogonal in the sense that

$$\int_{V^+} \mathbf{e}_n\mathbf{e}_m\, dV = 0, \quad n \neq m. \tag{1.184}$$

To prove this, let us add the first equation in (1.181) multiplied by $\mathbf{e}_m$ to the second equation multiplied by $\mathbf{h}_m$. From the resulting equation, we subtract the same equation with the indices n and m transposed. With regard to (1.175), we obtain

$$ik(\varepsilon_n - \varepsilon_m)\mathbf{e}_n\mathbf{e}_m = \operatorname{div}\{[\mathbf{e}_n\mathbf{h}_m] - [\mathbf{e}_m\mathbf{h}_n]\}. \tag{1.185}$$

Let us integrate this equation over V^+. Then on the left-hand side we obtain the integral (1.184), and on the right-hand side we obtain the surface integral

$$\int \{[\mathbf{e}_n\mathbf{h}_m]_N - [\mathbf{e}_m\mathbf{h}_n]_N\}\, dS \tag{1.186}$$

over S_ε, in which the fields are taken on the inner side of S_ε. Let us perform the same procedure for Eqs. (1.182). Since they do not contain ε_n, it follows that in V^- the fields $\{\mathbf{e}_n, \mathbf{h}_n\}$ and $\{\mathbf{e}_m, \mathbf{h}_m\}$ satisfy a relation that is simpler than (1.185); namely, the right-hand side of (1.185) is zero in V^-. The integration of this relation over V^- yields two surface integrals (1.186), one over S_ε and the other over the sphere at infinity (and also over S if any other bodies are present). The integrals over the sphere at infinity vanish by virtue of the radiation conditions, and the integrals over S are also zero. Consequently, the integral (1.186) over the outer side of S_ε is also zero. This integral contains only the field components tangent to S_ε. However, according to (1.183), these components are continuous on S_ε, and hence the same integral over the inner side of S_ε is also zero. This implies the orthogonality conditions (1.184).

The imaginary part of ε_n is nonnegative, that is, condition (1.39) holds. To verify this, let us multiply Eqs. (1.181) and the conjugate equations corresponding to the same n by $\{\mathbf{e}_n^*, -\mathbf{h}_n^*\}$ and $\{\mathbf{e}_n, -\mathbf{h}_n\}$, respectively, and sum the resultant equations. Then we obtain the following identity, similar to (1.185):

$$(\varepsilon_n - \varepsilon_n^*)|\mathbf{e}_n|^2 = \frac{1}{ik}\operatorname{div}\{[\mathbf{h}_n\mathbf{e}_n^*] + [\mathbf{h}_n^*\mathbf{e}_n]\}. \tag{1.187}$$

Next, let us integrate this equation over V^+ and the similar equation (with zero left-hand side) in V^- over V^- and use (1.183). Then we obtain

$$\operatorname{Im}\varepsilon_n \cdot k \int_{V+} |\mathbf{e}_n|^2\, dV = \operatorname{Re}\int [\mathbf{e}_n^*\mathbf{h}_n]_N\, dS. \tag{1.188}$$

The integral of the right-hand side over the sphere at infinity or an impedance surface is positive. It is equal to radiation losses or the losses on the impedance surface. It follows that relation (1.39) remains valid in the vector case as well. The eigenvalues ε_n are real only for a body placed in a closed resonator without losses in the walls.

2. We seek the solution of problem (1.176), (1.179) in the form of the series

$$\mathbf{E} = \mathbf{E}^0 + \sum_n A_n\mathbf{e}_n, \tag{1.189a}$$

$$\mathbf{H} = \mathbf{H}^0 + \sum_n A_n\mathbf{h}_n. \tag{1.189b}$$

In contrast with the series used (for closed resonators) in the ordinary k-method, *both series in* (1.189) *have the same coefficients.* Indeed, the substitution of (1.189a) into (1.176b) with regard to (1.180) and (1.181) implies (1.189b).

For any values of the coefficients, these series satisfy Eq. (1.176b) in V^+, both Eqs. (1.177) in V^-, and the boundary conditions (1.178) and (1.179). The coefficients A_n are found from the requirement that the series (1.189) must satisfy Eq. (1.176a) in V^+. By substituting (1.189) into (1.176a), we obtain

$$\sum_n A_n(\varepsilon_n - \varepsilon)\mathbf{e}_n = (\varepsilon - 1)\mathbf{E}^0 \quad \text{in} \quad V^+. \tag{1.190}$$

With regard to (1.184), this implies the desired explicit expression for the coefficients A_n:

$$A_n = \frac{\varepsilon - 1}{\varepsilon_n - \varepsilon} \frac{\displaystyle\int_{V+} \mathbf{E}^0 \mathbf{e}_n \, dV}{\displaystyle\int_{V+} (e_n)^2 \, dV}. \tag{1.191}$$

This formula is similar to (1.10a).

Just as in Section 1.1, we can transform the numerator in (1.191) so that A_n will be expressed via the currents $\mathbf{j}^{(e)}$ and $\mathbf{j}^{(m)}$ rather than the field $\mathbf{E}^0$ generated by these currents. To this end, we must multiply Eqs. (1.180) by $\{\mathbf{e}_n, \mathbf{h}_n\}$, multiply Eqs. (1.181) by $\{\mathbf{E}^0, \mathbf{H}^0\}$, subtract the second pair of equations from the first pair, and integrate the resultant equation

$$ik(\varepsilon_n - 1)\mathbf{E}^0\mathbf{e}_n = \frac{4\pi}{c}\mathbf{j}^{(e)}\mathbf{e}_n - \frac{4\pi}{c}\mathbf{j}^{(m)}\mathbf{h}_n + \operatorname{div}\{[\mathbf{e}_n\mathbf{H}^0] - [\mathbf{E}^0\mathbf{h}_n]\} \tag{1.192}$$

over V^+. Then we must perform the same procedure with Eqs. (1.180) and (1.182). As a result, we obtain an equation similar to (1.192) but with zero left-hand side; we must integrate this equation over V^-. By repeating the argument following (1.186), we see that the surface integrals drop out, and the integral in the numerator in (1.191) is given by

$$\int_{V+} \mathbf{E}^0\mathbf{e}_n \, dV = \frac{1}{ik(\varepsilon_n - 1)} \frac{4\pi}{c} \int (\mathbf{j}^{(e)}\mathbf{e}_n - \mathbf{j}^{(m)}\mathbf{h}_n) \, dV. \tag{1.193}$$

Here the integral on the right-hand side is taken over the entire space.

Finally, let us reduce the series (1.189) to a form that does not contain the distinguished terms $\{\mathbf{E}^0, \mathbf{H}^0\}$. In general (that is, for an arbitrary arrangement of the sources), this transformation is impossible in V^-, but in V^+ the field $\mathbf{E}^0$ can be expanded with respect to $\mathbf{e}_n$; this expansion is given by (1.190). By substituting this expansion into (1.189a), we obtain, by analogy with (1.13) and (1.14),

$$\mathbf{E} = \sum_n B_n \mathbf{e}_n, \tag{1.194a}$$

$$B_n = \frac{1 - \varepsilon_n}{1 - \varepsilon} A_n. \tag{1.194b}$$

In this way, we can obtain a formula for V^+ in which $\{\mathbf{E}^0, \mathbf{H}^0\}$ does not occur at all. According to (1.189a), (1.191), (1.193), and (1.194), this expansion for $\mathbf{E}$ reads

$$\mathbf{E} = \frac{1}{ik}\frac{4\pi}{c}\sum_n \frac{1}{\varepsilon_n - \varepsilon} \frac{\int (\mathbf{j}^{(e)}\mathbf{e}_n - \mathbf{j}^{(m)}\mathbf{h}_n)\, dV}{\int_{V+} (\mathbf{e}_n)^2\, dV} \mathbf{e}_n. \tag{1.195}$$

Moreover, in V^+ it is possible to express $\mathbf{H}$ by a series that does not contain the distinguished term $\mathbf{H}^0$. To this end, one must apply the operator rot to (1.190) and use (1.180b) and (1.181b). The substitution of the resultant series for $\mathbf{H}^0$ into (1.189b) yields

$$\mathbf{H} = -\frac{1}{ik}\frac{4\pi}{c}\mathbf{j}^{(m)}\frac{1}{\varepsilon - 1} + \sum_n B_n \mathbf{h}_n. \tag{1.196}$$

3. Now suppose that *the permittivity of the body is a continuous function of the coördinates and is equal to* 1 *outside a finite domain.* We seek the solution of the Maxwell equations

$$\operatorname{rot}\mathbf{H} - ik\varepsilon(\mathbf{r})\mathbf{E} = \frac{4\pi}{c}\mathbf{j}^{(e)}, \tag{1.197a}$$

$$\operatorname{rot}\mathbf{E} + ik\mathbf{H} = -\frac{4\pi}{c}\mathbf{j}^{(m)}. \tag{1.197b}$$

The field $\{\mathbf{E}^0, \mathbf{H}^0\}$ is still determined by Eqs. (1.180). Just as in scalar problems, the requirement that $\varepsilon(\mathbf{r})$ must be continuous can be removed in the final expressions, since the boundary conditions (1.178) are just the limit form of Eqs. (1.197) on the surface of discontinuity of $\varepsilon(\mathbf{r})$.

We define the eigenfunctions $\{\mathbf{e}_n, \mathbf{h}_n\}$ in which the diffracted field $\{\mathbf{E} - \mathbf{E}^0, \mathbf{H} - \mathbf{H}^0\}$ will be expanded as the fields of eigenoscillations of auxiliary bodies with permittivities $\varepsilon_n(\mathbf{r})$. In other words, we subject $\{\mathbf{e}_n, \mathbf{h}_n\}$ to the equations

$$\operatorname{rot}\mathbf{h}_n - ik\varepsilon_n(\mathbf{r})\mathbf{e}_n = 0, \tag{1.198a}$$

$$\operatorname{rot}\mathbf{e}_n + ik\mathbf{h}_n = 0. \tag{1.197b}$$

The functions $\varepsilon_n(\mathbf{r})$ will be determined in what follows by the same method as in Sections 1.3 and 1.4.

The orthogonality conditions for two eigenfunctions readily follow from (1.185) after the integration over the entire space. The integral of the right-hand side is zero, so that

$$\int (\varepsilon_n - \varepsilon_m)\mathbf{e}_n\mathbf{e}_m\, dV = 0. \tag{1.199}$$

The functions $\varepsilon_n(\mathbf{r})$ are complex-valued, and their imaginary parts are positive in the average. More precisely, it follows from (1.187) that

$$\int \operatorname{Im}\varepsilon_n |\mathbf{e}_n|^2\, dV \geq 0. \tag{1.200}$$

We seek the solution of the diffraction problem in the form of the series (1.189). These series satisfy Eq. (1.197b) termwise (for an arbitrary choice of the coefficients A_n). By analogy with (1.190), Eq. (1.197a) yields

$$\sum_n A_n(\varepsilon_n - \varepsilon)\mathbf{e}_n = (\varepsilon - 1)\mathbf{E}^0. \tag{1.201}$$

The only difference between this formula and (1.190) is that in (1.201) ε and ε_n are functions of $\mathbf{r}$. Formulas (1.199) and (1.201) form a basis of the subsequent transformations.

4. In this subsection we obtain *two different systems of equations for two systems of coefficients* A_n by multiplying Eq. (1.201) by two different systems of functions followed by integrating over the entire volume. From the requirement that the matrices of these systems must be diagonal, we find two possible forms of the functions $\varepsilon_n(\mathbf{r})$. The first of these two methods corresponds to the main solution of Section 1.3, and the second, to the main solution of Section 1.4.

First, let us multiply Eq. (1.201) by $\mathbf{e}_m$ and integrate over the entire volume. Then for the A_n we obtain the system

$$\sum_n A_n \int (\varepsilon_n - \varepsilon)\mathbf{e}_n\mathbf{e}_m \, dV = \int (\varepsilon - 1)\mathbf{E}^0\mathbf{e}_m \, dV. \tag{1.202}$$

For the matrix of this system to be diagonal, it is necessary that the functions $\varepsilon_n(\mathbf{r})$, according to (1.199) satisfy the functional equation (1.71). We restrict ourselves to the simplest solutions (1.74) of this equations. Possible generalizations, which result from the more general solution (1.72), (1.73) with an arbitrary function $\gamma(\mathbf{r})$, can be found according to the method of Section 1.3. Just as in the scalar case, considered in Section 1.5, such generalizations are especially efficient if $\{\mathbf{E}^0, \mathbf{H}^0\}$ is chosen to be the field resulting from the diffraction by a body with permittivity $\gamma(\mathbf{r})$.

If the $\varepsilon_n(\mathbf{r})$ are given by formulas (1.74) with the eigenvalues σ_n, then for the A_n we obtain the explicit expression

$$A_n = \frac{1}{\sigma_n - 1} \frac{\int (\varepsilon - 1)\mathbf{E}^0\mathbf{e}_n \, dV}{\int (\varepsilon - 1)(\mathbf{e}_n)^2 \, dV}, \tag{1.203}$$

similar to (1.76). For the problem considered in the beginning of this section, Eq. (1.203) passes into (1.191).

One can obtain a formula that does not contain $\mathbf{E}^0$ for A_n from (1.203) by integrating (1.192) over the entire volume. As is always the case for problems with continuous $\varepsilon(\mathbf{r})$, the divergent terms drop out readily, and one need not consider the integrals over V^+ and V^- separately. By substituting the result of the integration into (1.203) and by using (1.74), we obtain the formula

$$A_n = \frac{1}{(\sigma_n - 1)\sigma_n} \frac{4\pi}{ikc} \frac{\int (\mathbf{j}^{(e)}\mathbf{e}_n - \mathbf{j}^{(m)}\mathbf{h}_n) \, dV}{\int (\varepsilon - 1)(\mathbf{e}_n)^2 \, dV}, \tag{1.204}$$

which is similar to (1.138). Finally, inside the body, where $\varepsilon \neq 1$, we can replace $\mathbf{E}^0$ by the expansion 1.201 in (1.189a). Then for $\mathbf{E}$ we obtain the series

$$\mathbf{E} = \sum_n A_n \sigma_n \mathbf{e}_n, \tag{1.205}$$

which is similar to (1.81) and generalizes (1.194). By applying the operation rot to (1.205), we obtain a series of the type (1.196) for $\mathbf{H}$.

Let us now use the second method for obtaining a system of equations for A_n. Here we must multiply (1.201) by $\varepsilon_m \mathbf{e}_m$. To make it obvious that the two methods are equipollent, we first introduce the *induction* $\mathbf{d}_n$ by the formula

$$\mathbf{d}_n = \varepsilon_n \mathbf{e}_n. \tag{1.206}$$

Then the main formulas (1.199) and (1.201) acquire the form

$$\int \left(\frac{1}{\varepsilon_n} - \frac{1}{\varepsilon_m} \right) \mathbf{d}_n \mathbf{d}_m \, dV = 0, \tag{1.207}$$

$$\sum_n A_n \left(\frac{1}{\varepsilon_n} - \frac{1}{\varepsilon} \right) \mathbf{d}_n = \left(\frac{1}{\varepsilon} - 1 \right) \mathbf{E}^0. \tag{1.208}$$

Let us multiply Eq. (1.208) by $\mathbf{d}_m$ and integrate; then we obtain a system for A_n similar to (1.128). This system admits a closed-form solution provided that the $\varepsilon_n(\mathbf{r})$ satisfy the functional equation (1.93). We again restrict ourselves to the solution (1.129) for $\varepsilon_n(\mathbf{r})$ and do not introduce the function $\gamma(\mathbf{r})$.

For this choice of $\varepsilon_n(\mathbf{r})$, we have

$$A_n = \frac{1}{\sigma_n - 1} \frac{\int \left(\frac{1}{\varepsilon} - 1 \right) \mathbf{E}^0 \mathbf{d}_n \, dV}{\int \left(\frac{1}{\varepsilon} - 1 \right) (\mathbf{d}_n)^2 \, dV}. \tag{1.209}$$

This formula is similar to (1.135). For the special case in which $\varepsilon(\mathbf{r})$ is constant in V^+, this formula implies Eq. (1.191).

To obtain and expression free of $\mathbf{E}^0$ for A_n, we must substitute the expression (1.129) for $\varepsilon_n(\mathbf{r})$ into (1.192) and integrate over the entire volume. Then on the left-hand side we obtain the same integral as in (1.209), the divergent term drops out as usual, and for A_n we obtain the expression

$$A_n = \frac{1}{(\sigma_n - 1)\sigma_n} \frac{4\pi}{ikc} \frac{\int \left(\mathbf{j}^{(e)} \mathbf{e}_n - \mathbf{j}^{(m)} \mathbf{h}_n \right) dV}{\int \left(\frac{1}{\varepsilon} - 1 \right) (\mathbf{d}_n)^2 \, dV}, \tag{1.210}$$

which is similar to (1.138) and has the same structure as (1.204).

By expanding $\mathbf{E}^0$ according to (1.208) (at the points where $\varepsilon \neq 1$), we obtain the expansion

$$\mathbf{E} = \frac{1}{\varepsilon} \sum_n A_n \sigma_n \mathbf{d}_n \tag{1.211}$$

of the total field, which is similar to (1.205).

5. Now let us briefly consider the possible ways of applying the method to the diffraction problem for a body with *permeability* μ *different from* 1.

If $\varepsilon \equiv 1$ in the entire space, then we can apply the technique described above for the case $\mu \equiv 1$, $\varepsilon \neq 1$ with the following obvious substitutions: ε_n must be replaced by μ_n, formulas (1.74) and (1.94) must be replaced by the formulas

$$\mu_n = 1 + \sigma_n(\mu - 1), \quad \frac{1}{\mu_n} = 1 + \widetilde{\sigma}_n \left(\frac{1}{\mu} - 1 \right), \tag{1.212}$$

$\mathbf{e}_n$ in formulas like (1.184) and (1.188) must be replaced by $\mathbf{h}_n$, and $\mathbf{h}_n$ must be replaced by $-\mathbf{e}_n$. This technique will logically be referred to as the *μ-method.*

Now suppose that both $\varepsilon \not\equiv 1$ and $\mu \not\equiv 1$ in the body. We first restrict ourselves to the simplest solution, obtained by applying the ε-method and the μ-method consecutively. First, we must repeat all the considerations of the present section, including μ in Eqs. (1.180) determining $\{\mathbf{E}^0, \mathbf{H}^0\}$ and in Eqs. (1.181) and (1.182) or, respectively, Eq. (1.198) determining $\{\mathbf{e}_n, \mathbf{h}_n\}$. In other words, we define $\{\mathbf{E}^0, \mathbf{H}^0\}$ as *the fields of the same sources, not in vacuum, but in the presence of the same body with* $\varepsilon = 1$ *and with the actual value of* μ. The fields $\{\mathbf{e}_n, \mathbf{h}_n\}$ are the fields of eigenoscillations of the bodies with parameters ε_n and μ. Then all formulas providing the series expansion of $\{\mathbf{E}, \mathbf{H}\}$, as well as the expressions for the coefficients of this series, remain valid.

The field $\{\mathbf{E}^0, \mathbf{H}^0\}$ itself in this case is a solution of some diffraction problem. A similar situation occurs, say, in the diffraction problem for a dielectric body in the presence of a metallic surface S (see Section 1.2). In that problem, $\{\mathbf{E}^0, \mathbf{H}^0\}$ was the field due to the diffraction on the surface S alone, without the body. *The solution of the original diffraction problem for two bodies* (ε *and* S) *was represented as the sum of two parts*, namely, the solution $\{\mathbf{E}^0, \mathbf{H}^0\}$ of the nonhomogeneous problem for one body $((S))$ and the sum of solutions $\{\mathbf{e}_n, \mathbf{h}_n\}$ of homogeneous problems for the two bodies. The problem considered here can be referred to as the problem on the (ε, μ)-body; here the situation is the same except that instead of S we have the same body with permeability μ.

In turn, the field $\{\mathbf{E}^0, \mathbf{H}^0\}$ can be found by the μ-method. As a result, the total field consists of the field of the same sources in vacuum plus two series. In the first series, the eigenfunctions describe eigenoscillations of bodies with the parameters $1, \mu_n$. In the second series, the eigenfunctions pertain to the body with the parameters ε_n, μ. It is only the second series that contains a resonance term, whose amplitude has $(\varepsilon_n - \varepsilon)$ in the denominator. If the first series also contains a large term (an "alient" resonance), then this term can be eliminated by the method indicated in Section 1.1.

To describe the consecutive application of several methods, the following notation proves useful. By (ε, μ) we denote the field occurring in the diffraction on the body with these parameters, and by (ε_n, μ) we denote the field of eigenoscillations. Then the field of given sources in vacuum is $(1, 1)$, and the symbolic description of the whole procedure reads

$$(\varepsilon, \mu) = (1, 1) + \sum_n (1, \mu_n) + \sum_n (\varepsilon_n, \mu). \tag{1.213}$$

We can apply the same two methods to the (ε, μ)-problem in the opposite order. Then the solution will have the form

$$(\varepsilon, \mu) = (1, 1) + \sum_n (\varepsilon_n, 1) + \sum_n (\varepsilon, \mu_n). \tag{1.214}$$

The resonance term is again contained in the last sum. It will be different from the resonance term in the last sum in (1.213). Specifically, in (1.214) the terms of the second sum describe the eigenoscillations of an auxiliary body (the "comparison body") with the actual ε and with $\mu = \mu_n$ containing the eigenvalue. In (1.213), that comparison body has the actual value of μ, and the eigenvalue occurs in ε_n.

We could state the homogeneous problems differently, by introducing the eigenvalue into μ_n in one of the problems (where ε is equal to the actual function $\varepsilon(\mathbf{r})$) and by treating the other homogeneous problem in the opposite way. In this symmetric form, the solution becomes

$$(\varepsilon, \mu) = (1, 1) + \sum_n (\varepsilon_n, \mu) + \sum_n (\varepsilon, \mu_n). \tag{1.215}$$

The resonance terms occur in both sums simultaneously. This method (for another problem with two parameters) will be used below in Section 2.4.

We see that different expansions result in different resonance terms; this phenomenon will be encountered several times later on in this book. The resonance curves, understood as the dependence of the factors like

$$\frac{1}{\varepsilon_n(k) - \varepsilon}, \quad \frac{1}{\mu_n(k) - \mu} \tag{1.216}$$

on the frequency (for simplicity, we assume that $\varepsilon, \mu = \text{const}$), are different in different methods. The behavior of various resonance curves corresponding to the same problem in different methods will be discussed in more detail in Section 4.1; so far, we only note that both functions in (1.216) describe the vicinity of the resonance in high-Q systems equally well. In each case we must apply the method that is most suitable for the calculations, primarily for the calculation of the eigenvalues. For example, if the eigenvalues can be found directly from a transcendental equation, then the choice of one of the methods (1.213), (1.214), or (1.215) is determined by how ε and μ occur in the transcendental equation. If it is easier to find ε than μ from this equation, then the method (1.213) must be used, and so forth.

* * *

In this chapter, the generalized eigenoscillation method was applied to problems of diffraction on dielectric bodies, including bodies whose permittivity depends on the coordinates. The scheme for constructing the solution is the same in all the cases. First we introduce equations for the eigenfunctions and establish the orthogonality conditions for these functions. For bodies of constant permittivity ε, the eigenvalues are the permittivities ε_n of bodies of the same shape (comparison bodies) in which undamped oscillations with a given frequency of the sources is possible. For bodies with variable $\varepsilon(\mathbf{r})$, the comparison bodies also have variable $\varepsilon_n(\mathbf{r})$. The form of these functions is found from the condition that the amplitudes in the expansion of the diffracted field with respect to the eigenfunctions should admit explicit expressions. Then various formulas for these amplitudes are given, including a formula that contains the exciting currents instead of the incident field. Outside the body, the diffracted field is expanded in series in the eigenfunctions. Inside the body, one can also expand the total field. The technique is also applied to quantum-mechanical scattering problems.

Chapter 2

Spectral Parameter in Boundary Conditions

The common feature of all versions of the generalized eigenoscillation method presented in this chapter is that the spectral parameter is introduced into the boundary conditions of the homogeneous problems rather than into the equations (which was the case with the k- and ε-methods). To this end, one considers an auxiliary body having the same shape as the body in the original problem and replaces the actual boundary conditions of the diffraction problem by some *auxiliary conditions* that contain a parameter playing the role of the spectral parameter. For example, in the w-method (Section 2.1) one prescribes a boundary condition of the impedance type on the surface of the body, and the eigenvalues w_n of the corresponding homogeneous problem are just the impedance values of the auxiliary body for which there exist nontrivial solution at a given frequency. In all methods described in the present chapter, each eigenfunction must satisfy the same equation as the diffracted field (that is, the homogeneous equation with the actual frequency) and the same conditions at infinity (except for the method given in Section 2.5). Thus, the representation of the desired field in the form of the expansion (1.66) satisfies the equation in the diffraction problem and the radiation conditions (if they are imposed) term by term for an arbitrary choice of the coefficients A_n. These coefficients are determined from the remaining condition saying that the desired field must satisfy the actual boundary conditions. At that point, one uses the orthogonality relations that hold in this situation.

In all methods given in the present chapter, the eigenfunctions are orthogonal *on the surface where the auxiliary boundary conditions are prescribed.* Consequently, the dimension of the series representing the actual field is less by one than in the ε- and k-methods. (For two-dimensional problems, the series are one-dimensional, etc.) This can be explained as follows. In the ε- and k-methods, the conditions of the original problem are violated in the homogeneous problem in some domain (where the eigenfunctions satisfy an equation different from that for the diffracted field). The orthogonality holds with respect to this very domain, and the dimension of the series coincides with that of the diffraction problem. In the versions of the method given in the present chapter, the conditions of the original problem are violated only on the surface; the orthogonality conditions hold on this very surface, and the dimension of the series is less by one. (Needless to say, by a rearrangement of terms one can convert any series to a one-dimensional series, but in specific problems the notion of the dimension of a series does not lead to a misunderstanding.)

If there are no losses in the diffraction problem except possibly for those on the surface of the body, then the homogeneous problem (which is independent of the actual boundary conditions) is, as a rule, selfadjoint, and the eigenvalues are real. In the general case, the homogeneous problem is nonselfadjoint and the eigenvalues are complex. The sign of the imaginary parts of these eigenvalues corresponds to energy release from the surface of the auxiliary body. This energy is spent to maintain undamped oscillations at the actual frequency in the absence of the actual sources. The auxiliary boundary conditions in this case describe some *active film* (i.e., a film with negative losses) that has the shape of the boundary of the body and radiates proportionally to the square of the field magnitude at the film.

The expansion coefficients A_n in these methods, as well as in the methods described in Chapter 2, are of resonance nature. The resonance occurs when one of the eigenvalues (viewed, say, as a function of the frequency) is close to the spectral parameter value in the diffraction problem. For example, for the case of diffraction on a body with the impedance condition

$$\left(U + w \frac{\partial U}{\partial N} \right) \Big|_S = 0,$$

this happens when the eigenimpedance w_n is close to the actual value of w. In this case, the corresponding amplitude of the expansion is large, and even one term of the series describes the diffracted field fairly well.

Just as in Chapter 1, we start from scalar problems and then briefly describe in Section 2.6 how the technique can be generalized to the Maxwell equations.

2.1 Spectral Parameter in Impedance Boundary Conditions (the *w*-Method)

The method considered in this section applies to diffraction problems for bodies and surfaces (which need not necessarily be closed) with impedance boundary conditions. In particular, these conditions include the vanishing of either the field or its normal derivative on the boundary of the body. We assume the boundaries of the bodies to be finite. The method applies equally well to closed and open systems.

We start from the application of the w-method to diffraction problems for bodies with closed boundaries.

1. Suppose that we seek a function $U(\mathbf{r})$ that satisfies the equation

$$\Delta U + k^2 U = f \tag{2.1}$$

in the domain V and the impedance condition

$$U + w\frac{\partial U}{\partial N} = 0 \tag{2.2}$$

on the boundary S (here N is the outward normal on the boundary). If V is infinite, then the diffracted field must additionally satisfy the radiation condition (0.35).

The impedance w is a complex number in general. Its imaginary part characterizes the absorption of energy by the walls ($\operatorname{Im} w \leq 0$). If these losses are absent, then w is real. This is the case, say, if the surface of the body is fast-periodically corrugated. The first- and the second-kind boundary conditions $U|_S = 0$ and $\left.\frac{\partial U}{\partial N}\right|_S = 0$ are the limit cases of (2.2) with $w = 0$ and $w = \infty$, respectively.

We seek the solution of the problem in the form of the series (1.66):

$$U = U^0 + \sum_n A_n u_n. \tag{2.3}$$

Here, just as in the preceding, U^0 is the field of the sources f in vacuum,

$$\Delta U^0 + k^2 U^0 = f, \tag{2.4}$$

and the functions u_n in which the diffracted field is expanded are the eigenfunctions of the homogeneous problem. These functions must satisfy the equation

$$\Delta u_n + k^2 u_n = 0 \tag{2.5}$$

for the diffracted field in V, the impedance boundary conditions

$$\left.\left(u_n + w_n\frac{\partial u_n}{\partial N}\right)\right|_S = 0 \tag{2.6}$$

with a spectral parameter on S, and the radiation conditions if the system is open.

Condition (2.6) can be imposed on a part of the boundary. Then on the remaining part of the boundary the functions U^0 and u_n must satisfy the actual boundary conditions. If the diffraction occurs not only on S, then the diffraction problem (2.1), (2.2) and problems (2.4)–(2.6) for the fields U^0 and u_n must be equipped with the same additional conditions on the other surfaces or bodies. When introducing U^0, we can assume that there are some other bodies outside V. For example, if we solve the excitation problem for a closed resonator, then U^0 can be a solution of the excitation problem for a larger resonator rather than satisfy the radiation condition. The formal technique also remains the same if U^0 is created not only by the sources f occurring in the diffraction problem but also by some other sources located outside V. The ambiguity in the choice of U^0 may prove useful, say, if one knows the solution of the diffraction problem for a body closely similar to the one that must be studied.

The quantities w_n have the dimension of length. They will be referred to as the *eigenimpedances*. They play the role of the eigenvalues of the homogeneous problem.

The physical meaning of this problem is as follows: it describes undamped eigenoscillations of the body at the frequency k of the diffraction problem in the absence of the actual sources. These oscillations are possible only for certain values w_n of the spectral parameter. We have already mentioned that the homogeneous problem is selfadjoint (and the eigenimpedances are real) even if there are losses on the walls S in the original problem. However, if the losses in the walls are accompanied by some other losses (say, radiation losses), then the homogeneous problem will be nonselfadjoint and the eigenimpedances will be complex numbers with

$$\operatorname{Im} w_n > 0. \tag{2.7}$$

This condition can readily be obtained by applying Green's formula to the domain V for the functions u_n and u_n^* and means that there is energy release (proportional to the square of the field) on the surface, which maintains undamped oscillations. In this interpretation, the boundary condition (2.6) describes some opaque active energy-releasing film on the surface of the body.

The eigenfunctions thus introduced and their normal derivatives are orthogonal on the surface S. To verify this, it suffices to apply Green's second formula in V to u_n and u_m. The corresponding integral over S will always be zero,

$$\int_S \left(u_n \frac{\partial u_m}{\partial N} - u_m \frac{\partial u_n}{\partial N} \right) dS = 0. \tag{2.8}$$

For open systems, the proof also uses the radiation condition. Now we can eliminate either the functions themselves or their derivatives from (2.8) with the help of (2.6); thus we find that in the absence of degeneration ($w_n \neq w_m$) one has

$$\int_S u_n u_m \, dS = 0, \quad n \neq m, \tag{2.9}$$

$$\int_S \frac{\partial u_n}{\partial N} \frac{\partial u_m}{\partial N} \, dS = 0, \quad n \neq m. \tag{2.10}$$

These orthogonality conditions permit one to apply the usual procedure of finding the A_n from the remaining requirement that the desired field must satisfy the boundary condition (2.2):

$$\left[U^0 + \sum_n A_n u_n + w \left(\frac{\partial U^0}{\partial N} + \sum_n A_n \frac{\partial u_n}{\partial N} \right) \right] \bigg|_S = 0. \tag{2.11}$$

We multiply (2.11) by u_m, integrate over S, and use (2.9), thus obtaining

$$A_n = \frac{1}{w_n - w} \frac{\int_S \left(U^0 + w \frac{\partial U^0}{\partial N} \right) \frac{\partial u_n}{\partial N} \, dS}{\int_S \left(\frac{\partial u_n}{\partial N} \right)^2 dS}. \tag{2.12}$$

For the special case in which the desired field must satisfy the condition $U = 0|_S$, it suffices to set $w = 0$ in (2.12), and then we find

$$A_n = \frac{1}{w_n} \frac{\int_S U^0 \frac{\partial u_n}{\partial N}\, dS}{\int_S \left(\frac{\partial u_n}{\partial N}\right)^2 dS}. \tag{2.13}$$

On the other hand, if the normal derivative of the total field must vanish on S (this corresponds to the passage to the limit as $w \to \infty$), then it is expedient to introduce a different denominator into (2.12), and we obtain

$$A_n = w_n \frac{\int_S u_n \frac{\partial U^0}{\partial N}\, dS}{\int_S u_n^2\, dS}. \tag{2.14}$$

According to (2.5), the eigenimpedances w_n are functions of k. The resonance occurs if one of the eigenimpedances is close to the actual impedance of the diffraction problem, that is, $w_n(k)$ is close to w. In particular, for the conditions

$$U|_S = 0 \quad \text{or} \quad \left.\frac{\partial U}{\partial N}\right|_S = 0 \tag{2.15}$$

this happens when w_n is small or large, respectively.

For closed resonators without losses ($\operatorname{Im} w = 0$), the roots of the equation

$$w_n(k) - w = 0 \tag{2.16}$$

are real. Accordingly, there are infinite resonances at the frequencies equal to these roots. These resonance frequencies coincide with the eigenfrequencies in the k-method. For resonators with absorbing walls ($\operatorname{Im} w < 0$), the resonances are finite, and the roots of (2.16) are complex and coincide with the complex eigenfrequencies in the k-method. For open systems, the resonances are always finite (the imaginary parts of w_n and w have opposite signs), and the roots of (2.16) are also complex.

2. In this subsection we consider the generalization of the w-method to *bodies with variable impedance.* If the impedance is a function of the point s on the surface S of the body ($w = w(s)$), that is, if Eq. (2.2) is replaced by the boundary condition

$$\left.\left(U + w(s)\frac{\partial U}{\partial N}\right)\right|_S = 0, \tag{2.17}$$

then the impedance w_n in the auxiliary problem must also be a function of s, just as in the ε-method with $\varepsilon = \varepsilon(\mathbf{r})$ the permittivity ε_n of the auxiliary body is a function of $\mathbf{r}$. Accordingly, we set

$$\left.\left(u_n + w_n(s)\frac{\partial u_n}{\partial N}\right)\right|_S = 0, \tag{2.18}$$

where $w_n(s)$ is a function of s that depends on $w(s)$ and contains a numerical parameter, which will be used as the spectral parameter. The form of the function $w_n(s)$ needed to state the homogeneous problem is determined by the same requirement that was used in Section 1.3 to find the functions $\varepsilon_n(\mathbf{r})$ (1.94) or (1.72): the system of linear algebraic equations for the A_n must be diagonal. This system can be obtained by substituting (2.3) into the boundary condition (2.17), multiplying by u_m, and integration over S and has the form (cf. (1.92))

$$\sum_n A_n \int_S \left(\frac{1}{w} - \frac{1}{w_n}\right) u_n u_m \, dS = -\int_S \left(\frac{1}{w}U^0 + \frac{\partial U^0}{\partial N}\right) u_m \, dS. \tag{2.19}$$

Let us compare this system with the orthogonality condition

$$\int_S \left(\frac{1}{w_n} - \frac{1}{w_m}\right) u_n u_m \, dS = 0 \tag{2.20}$$

that can be obtained from (2.8) by the substitution (2.18) (this condition generalizes (2.9)). Obviously, if $w(s)$ and $w_n(s)$ satisfy the functional equation (1.93) (with $\varepsilon, \varepsilon_n$ replaced by w, w_n), then system (2.19) becomes diagonal. The solution of this equation for $w_n(s)$ has the same form (1.94) as for $\varepsilon_n(\mathbf{r})$. It contains the eigenvalue and an arbitrary function.

Following Section 1.3, we can use another orthogonality condition, which generalizes (2.10) rather than (2.9), and write out a different system of equations for the A_n. We can diagonalize this system by relating $w_n(s)$ to $w(s)$ by a formula similar to (1.72). Needless to say, the eigenvalues and eigenfunctions (and hence the coefficients A_n) will be different in this case. We shall not write out the corresponding formulas; this will be done in Section 2.4 in a more general case. Here we only mention a *special case* in which the arbitrary function occurring in $w_n(s)$ is chosen in a specific way. By analogy with (1.100), we can define $w_n(s)$ by the formula

$$w_n(s) = \lambda_n w(s), \tag{2.21}$$

where λ_n is the eigenvalue. For this choice of $w_n(s)$, the eigenfunctions u_n will be orthogonal with weight $1/w(s)$ according to (2.20), and the coefficients A_n will have the resonance denominator $\lambda_n - 1$.

For infinitely thin screens (that is, bodies with nonclosed boundaries), the field in the diffraction problem, as well as the eigenfunctions of the homogeneous problems, must satisfy the energy finiteness conditions near the edges, which are known to be equivalent to the condition that the integrals

$$\int_V |u|^2 \, dV, \quad \int_V |\nabla u|^2 \, dV \tag{2.22}$$

must be finite. All the preceding considerations of this section remain valid, but if we understand the integrals over S as integrals over one side of the surface, then the formulas will look differently. For example, for the first boundary condition in (2.15),

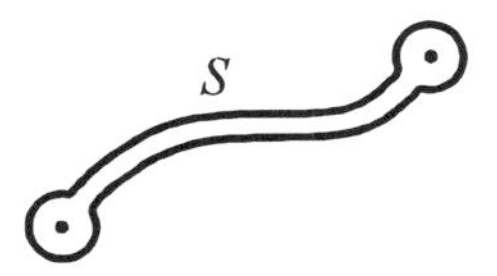

Figure 2.1

in formula (2.13) we must replace $\partial u_n/\partial N$ by $\partial u_n^-/\partial N - \partial u_n^+/\partial N$ and $(\partial u_n/\partial N)^2$ by $(\partial u_n^-/\partial N)^2 + (\partial u_n^+/\partial N)^2$, where the $\partial u^\pm/\partial N$ are the values of the derivatives on the opposite sides of S and N is directed from "$-$" to "$+$". This obvious generalization can be obtained either by completing S to a closed surface and requiring that u_n and $\partial u_n/\partial N$ must be continuous on the completion (and satisfy (2.6) on both sides of S) or by the passage to the limit from a body without edges (see Fig. 2.1) to a nonclosed surface.

3. One possible algorithm for finding the eigenelements of the auxiliary homogeneous problems of the w-method is based on integral equations (with simple kernels) for the eigenfunctions. The integration in these equations is carried out over the surface S, that is, over the domain where the auxiliary boundary condition is imposed; hence, the dimension of these equations is by one less than the dimension of the corresponding homogeneous problem. For bodies with closed boundaries, these equations are especially easy to obtain. For example, let us derive these equations for the exterior problem (2.5), (2.6). To this end, we apply Green's second formula in the domain V to the eigenfunction u_n and the Green function G (1.84) of the point source in vacuum. Since both u_n and G satisfy the radiation conditions, we see that the integral over the sphere at infinity in the resulting formula vanishes. In the integral over S, according to (2.6), we replace $\partial u_n/\partial N$ by $-u_n/w_n$, thus obtaining an expression for the eigenfunction u_n at an arbitrary point of the domain V via the values of the eigenfunction on S. To obtain the desired equation, one must place the observation point r on the surface S (we must have in mind that in this case the double layer potential gives the additional term $u_n/2$). This procedure results in the following *integral equation of the second kind*, in which *the eigenvalue* w_n (*more precisely, the reciprocal value* $1/w_n$) *is a factor in one of the two terms constituting the kernel*:

$$\frac{1}{2}u_n(r) = \int_S u_n(r')\left(\frac{\partial G}{\partial N_{r'}}(r,r') + \frac{1}{w_n}G(r,r')\right)\,dS_{r'}. \tag{2.23}$$

Here N is the exterior normal to V.

The corresponding equation for the interior problem is different in that G is a real-valued function; namely, $G = -\dfrac{\cos k|r-r'|}{4\pi|r-r'|}$ for the spatial problem and $G = \dfrac{1}{4}N_0(k|r-r'|)$ for the two-dimensional problem. As was indicated above, the eigenvalues w_n are real in this case.

For nonclosed screens S, it is also possible to obtain integral equations of the second kind on S for the eigenfunctions. We do not present these equations here, since they have complicated kernels and are difficult to use for finding the eigenelements.

4.[1] Just as in the ε-method, the main results of the w-method can be obtained from the theory of functions of complex variable. For example, consider the problem studied in Subsection 1. We momentarily introduce the notation $v = U - U^0$ for the diffracted field. The field v satisfies the homogeneous wave equation and the boundary condition

$$\frac{\partial v}{\partial N} + \frac{1}{w} v = g, \tag{2.24}$$

on S, where $g = -\partial U^0/\partial N - U^0/w$. Here the function g plays the role of an external force, that is, the function f of the original problem. Consider the solution $v(\alpha)$ of the homogeneous wave equation (2.5) with the boundary condition

$$\frac{\partial v(\alpha)}{\partial N} + \alpha v(\alpha) = g \tag{2.25}$$

on S. The function $v(\alpha)$ passes into v for $\alpha = \dfrac{1}{w}$. It has poles at the points $\alpha_n = \dfrac{1}{w_n}$, where the w_n are the eigenvalues of problem (2.5), (2.6).

Let us find the residue c_n of the function $v(\alpha)$ at the pole α_n. To this end, we represent g in the form

$$g = u_n \frac{\int_S g u_n \, dS}{\int_S u_n^2 \, dS} + \tilde{g}, \tag{2.26}$$

so that $\tilde{g}$ satisfies $\int_S \tilde{g} u_n \, dS = 0$. Accordingly, $v(\alpha)$ is represented as the sum of two terms; the term containing $\tilde{g}$ remains finite as $\alpha \to \tilde{\alpha}_n$ and does not contribute to c_n. The second term is equal to

$$\frac{1}{\alpha - \alpha_n} \frac{u_n \int_S g u_n \, dS}{\int_S u_n^2 \, dS}. \tag{2.27}$$

Thus for c_n we have the expression

$$c_n = \frac{u_n \int_S g u_n \, dS}{\int_S u_n^2 \, dS} = -\frac{\alpha_n u_n \int_S g \frac{\partial u_n}{\partial N} \, dS}{\int_S \left(\frac{\partial u_n}{\partial N} \right)^2 dS}. \tag{2.28}$$

Suppose that in the *complex plane* of the variable α there is an *expanding sequence of contours* on which $v(\alpha)$ decays at least as rapidly as $1/|\alpha|$. Then

$$v(\alpha) = \sum_n \frac{c_n}{\alpha - \alpha_n}. \tag{2.29}$$

[1]This subsection was written by A. D. Shatrov and contains his results.

By substituting c_n into this formula, we obtain

$$v(\alpha) = \sum_n \frac{\alpha_n u_n}{\alpha_n - \alpha} \cdot \frac{\displaystyle\int_S g \frac{\partial u_n}{\partial N}\, dS}{\displaystyle\int_S \left(\frac{\partial u_n}{\partial N}\right)^2 dS}. \tag{2.30}$$

By setting $\alpha = 1/w$ in (2.30), we obtain the expansion (2.3) with the coefficients (2.12).

As $|\alpha| \to \infty$, the boundary condition (2.25) acquires the limit form $v|_S = 0$. Hence the assumption that $v(\alpha) = O(1/|\alpha|)$ as $|\alpha| \to \infty$ (which is necessary for the validity of (2.29)) always holds except for the extraordinary case of a closed resonator in which one of the eigenfrequencies corresponding to the boundary condition $u|_S = 0$ coincides with k (that is, one of the eigenvalues w_n of problem (2.5), (2.6) is zero). In this case, $v(\alpha)$ does not tend to zero as $|\alpha| \to \infty$, and before applying the expansion (2.29) we must subtract $v_1 = \lim_{|\alpha| \to \infty} v(\alpha)$ from $v(\alpha)$. Let U_0 be the solution (which is known to exist in this case) of the homogeneous wave equation with the boundary condition $v|_S = 0$. Then

$$v_1 = U_0 \frac{\displaystyle\int_S g \frac{\partial U_0}{\partial N}\, dS}{\displaystyle\int_S \left(\frac{\partial U_0}{\partial N}\right)^2 dS}. \tag{2.31}$$

Formally, this expression coincides with the general term of the series (2.30) for $\alpha_m = \infty$. The difference $v(\alpha) - v_1$ tends to zero appropriately as $|\alpha| \to \infty$, and the expansion (2.29) can be applied to this difference. As a result, for $v(\alpha)$ we obtain the same series (2.30) where the terms corresponding to the poles α_n are supplemented by the term (2.31).

2.2 Spectral Parameter in Transmission Conditions (the ρ-Method). Metallic and Semitransparent Surfaces

In this section we consider two complementary versions of the generalized method that permit one to construct solutions of diffraction problems for closed and nonclosed metallic surfaces. In Section 2.3 these methods will be applied to diffraction problems for dielectric bodies. The novelty in these methods as compared with the w-method is that in the auxiliary homogeneous problem we impose boundary conditions that have the meaning of *transmission conditions*. For bodies with closed boundaries, this implies that a relationship between the exterior and the interior domain is established, and for bodies with nonclosed boundaries (infinitely thin screens), this implies a relationship between the fields on the opposite sides of the screen. These conditions can be interpreted as saying that the boundary is a semitransparent screen, whereas the impedance boundary conditions of the w-method imply complete insulation (shielding) of the domain in question from the remaining volume, that is, describe an opaque film following the shape of the body. Thus, the auxiliary homogeneous problem of the ρ-method is posed in the entire space (for the case of closed boundaries, for the exterior and the interior domain simultaneously). Accordingly, the eigenelements of the auxiliary problem permit one to construct solutions for both the interior and the exterior diffraction problem, and the eigenvalues treated as functions of the frequency contain information about the resonances in both problems.

The advantage of the first version of the ρ-method for metallic bodies is in the fact that here we have a particularly simple integral equation of the second kind for the eigenfunctions, in which the integration extends over the surface of the body.

1. Let us first present an application of the ρ-method to diffraction problems for closed metallic surfaces. We do not distinguish between exterior and interior problems, since the formal technique for constructing the solutions is the same in both cases. Let us state the diffraction problem. It is required to find a function $U(\mathbf{r})$ that satisfies the equation

$$\Delta U + k^2 U = f, \tag{2.32}$$

where f represents the given external sources, outside some closed surface S, the radiation condition at infinity, and one of the following boundary conditions on S: either

$$U|_S = 0, \tag{2.33a}$$

or

$$\left.\frac{\partial U}{\partial N}\right|_S = 0. \tag{2.33b}$$

The domain outside S may be infinite (diffraction on a body of finite volume, in particular, the excitation of an open resonator), and then we deal with the exterior problem. This domain can also be finite (the excitation of a closed resonator), and then we deal with an interior problem. The radiation condition in the latter case must be dropped.

We seek the solution of the problem in the form

$$U = U^0 + \sum_n A_n u_n, \tag{2.34}$$

where U^0 is the incident field and the u_n are the eigenfunctions of an auxiliary homogeneous problem, which is now posed for the entire space, that is, for the exterior domain (V^-) and the interior domain (V^+) simultaneously. This problem is obtained as follows. We replace the metallic surface S by a semitransparent film of the same shape and impose a certain relationship on S between the boundary values of the exterior and interior fields. Associated with the two versions of the ρ-method are two types of boundary conditions in the homogeneous problem, which are designed to solve the diffraction problem with the boundary conditions (2.33a) and (2.33b), respectively. Specifically, for the diffraction problem with the condition $U|_S = 0$, the boundary conditions on S in the corresponding homogeneous problem read

$$(u_n^+ - u_n^-)|_S = 0, \quad \left(\frac{\partial u_n^+}{\partial N} - \frac{\partial u_n^-}{\partial N} - \frac{u_n}{\rho_n} \right) \bigg|_S = 0. \tag{2.35a}$$

If in the diffraction problem the field must satisfy condition (2.33b), then in the auxiliary problem we use the boundary conditions

$$\left(\frac{\partial \tilde{u}_n^+}{\partial N} - \frac{\partial \tilde{u}_n^-}{\partial N} \right) \bigg|_S = 0, \quad \left(\tilde{u}_n^+ - \tilde{u}_n^- + \frac{1}{\tilde{\rho}_n} \frac{\partial u_n}{\partial N} \right) \bigg|_S = 0. \tag{2.35b}$$

The signs "+" and "−" indicate the two sides of the surface S; the direction of the normal N is from "−" to "+." The method in which the auxiliary problem is posed with the help of conditions (2.35a) will be referred to as the first version of the ρ-method, while conditions (2.35b) pertain to the second version. The eigenfunctions of both versions (needless to say, they are different) must also satisfy the homogeneous wave equation

$$\Delta u_n + k^2 u_n = 0 \tag{2.36}$$

and the radiation conditions. The proportionality coefficients ρ_n and $\tilde{\rho}_n$ in the boundary conditions (2.35a) and (2.35b) play the role of the eigenvalues of the homogeneous problems.

One can readily show that the auxiliary problems describe systems in which on the film surface S there is energy release compensating for the radiation losses. In other words, the semitransparent film is an active radiating element, and it is due to this factor that for certain values of ρ and $\tilde{\rho}$ undamped oscillations at a given frequency k (that is, no trivial solutions of the homogeneous problems) are possible. The eigenvalues must have positive imaginary part. To verify this, it suffices to write out the energy balance for the homogeneous problems. One can do that by applying Green's formula to the functions u_n and u_n^* in the domains V^+ and V^-, subtracting the results from each other, and using (2.35a) in one of the cases and (2.35b) in the other. Then for the first version we obtain

$$\frac{\operatorname{Im} \rho_n}{|\rho_n|^2} \int_S |u_n|^2 \, dS = \operatorname{Im} \int_{S_\infty} u_n \frac{\partial u_n^*}{\partial N} \, dS, \tag{2.37a}$$

whereas for the second version we have

$$\frac{\operatorname{Im} \tilde{\rho}_n}{|\tilde{\rho}_n|^2} \int_S \left| \frac{\partial \tilde{u}_n}{\partial N} \right|^2 dS = \operatorname{Im} \int_{S_\infty} \tilde{u}_n \frac{\partial \tilde{u}_n^*}{\partial N} \, dS. \tag{2.37b}$$

Since the integrals over the sphere at infinity S_∞ occurring on the right-hand sides in (2.37a) and (2.37b) are positive (they represent the radiation losses), it follows that

$$\operatorname{Im} \rho_n > 0, \quad \operatorname{Im} \tilde{\rho}_n > 0. \tag{2.38}$$

The eigenfunctions of the first version, as well as the normal derivatives of the eigenfunctions of the second version, are orthogonal with respect to integration over the closed surface S. One can readily prove these properties by applying the second Green formula (0.20) to two functions with distinct indices n and m in the interior domain V^+ and the exterior domain V^-, by subtracting the results from each other, and by using the fact that the eigenfunctions satisfy the radiation conditions. (In this case, the integral over the sphere at infinity drops out.) Then, according to the boundary conditions (2.35a) of the first version, we have

$$\int_S u_n u_m \, dS = 0, \quad n \neq m, \tag{2.39a}$$

and the use of (2.35b) gives

$$\int_S \frac{\partial \tilde{u}_n}{\partial N} \frac{\partial \tilde{u}_m}{\partial N} \, dS = 0, \quad n \neq m. \tag{2.39b}$$

The orthogonality conditions permit one to obtain explicit one-term expressions for the coefficients A_n in the representation (2.34) of the desired field. Indeed, for the diffraction problem with the boundary condition (2.33a) one must require that

$$\left(U^0 + \sum_n A_n u_n\right)|_S = 0. \tag{2.40a}$$

We multiply this equation by u_m, then integrate over S with regard to (2.39a), and introduce the notation $v_n = \left(\frac{\partial u_n^+}{\partial N} - \frac{\partial u_n^-}{\partial N}\right)\Big|_S$, thus obtaining

$$A_n = -\frac{\int_S U^0 v_n \, dS}{\rho_n \int_S v_n^2 \, dS}. \tag{2.41a}$$

For the diffraction problem with the zero normal derivative on S, we must have

$$\left(\frac{\partial U^0}{\partial N} + \sum_n A_n \frac{\partial \tilde{u}_n}{\partial N}\right)\Big|_S = 0. \tag{2.40b}$$

Using the orthogonality conditions (2.39b), we find that

$$A_n = \frac{1}{\tilde{\rho}_n} \frac{\int_S \frac{\partial U^0}{\partial N} \tilde{v}_n \, dS}{\int_S \tilde{v}_n^2 \, dS}, \tag{2.41b}$$

where $\tilde{v}_n = (\tilde{u}_n^+ - \tilde{u}_n^-)|_S$.

By transforming the integrals occurring in the numerators in (2.41a) and (2.41b), we can express the A_n by formulas containing the sources f themselves rather than the

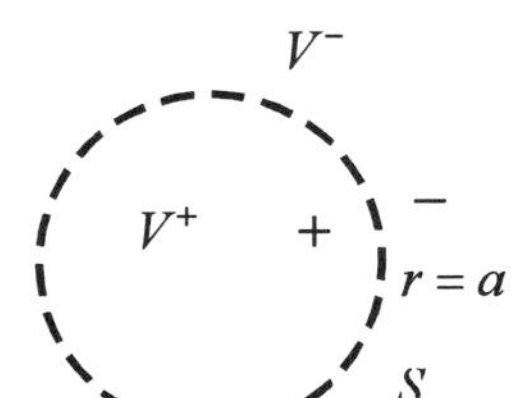

Figure 2.2

incident field. To this end, we apply Green's formula (0.20) to the functions U^0 and u_n in the domains V^+ and V^-, subtract the results from each other, and use the radiation conditions, the boundary conditions (2.35a) or (2.35b), and the fact that, by definition, U^0 and $\dfrac{\partial U^0}{\partial N}$ are continuous on S. Then for the numerators in (2.41a) and (2.41b) we obtain, respectively,

$$\int_S U^0 v_n \, dS = \int_V u_n f \, dV, \tag{2.41c}$$

$$\int_S \frac{\partial U^0}{\partial N} \widetilde{v}_n \, dS = -\int_V \widetilde{u}_n f \, dV. \tag{2.41d}$$

When applying the ρ- method to diffraction problems for metallic bodies with closed boundaries, one must have in mind that the homogeneous problems are posed for the entire space, and their eigenelements can be used for both interior and exterior problems. Hence the eigenvalues may vanish exactly (which implies an infinite resonance) at the frequencies that are resonant for the interior domain. Needless to say, these infinite resonances are physically meaningless for the solution of the exterior problem, and they are also absent in the expression (2.34) for the field, since the numerator of the corresponding Fourier coefficient is zero at these frequencies. However, this shows that the roots of the equation $\rho_n(k) = 0$ or $\widetilde{\rho}_n(k) = 0$ may have nothing to do with the resonance properties of an open resonator and that the frequency dependence of the A_n can also be determined by the form of the Fourier coefficients. In problems for infinitely thin nonclosed screens, that is, bodies without interior cavities, to which the ρ-method also applies, this is not the case, and the small absolute value of ρ_n or $\widetilde{\rho}_n$ invariably means that we are close to an actual resonance.

Let us illustrate the properties of the functions u_n and $\widetilde{u}_n$ and the numbers ρ_n and $\widetilde{\rho}_n$ for the example of a circular problem (a cylinder of radius a with one of the conditions $U|_S = 0$ or $\left.\dfrac{\partial U}{\partial N}\right|_S = 0$; see Fig. 2.2). In this case, the solution (2.34) is just the expansion provided by the usual separation of variables. For this case, we have the following expression for the eigenfunctions of the first version inside and outside the cylinder:

$$u_n = H_n^{(2)}(ka) J_n(kr) \cos n\varphi, \quad r \leq a, \tag{2.42}$$

$$u_n = H_n^{(2)}(kr) J_n(ka) \cos n\varphi, \quad r \geq a. \tag{2.43}$$

The eigenvalues are given by

$$\rho_n = \frac{i\pi a}{2} H_n^{(2)}(ka) J_n(ka). \tag{2.44a}$$

The eigenfunctions and the eigenvalues of the second version for the same case are given by

$$\widetilde{u}_n = H_n^{(2)\prime}(ka) J_n(kr) \cos n\varphi, \quad r \leq a, \tag{2.45}$$

$$\widetilde{u}_n = H_n^{(2)}(kr) J_n'(ka) \cos n\varphi, \quad r \geq a, \tag{2.46}$$

$$\widetilde{\rho}_n = \frac{i\pi a k^2}{2} H_n^{(2)\prime}(ka) J_n'(ka). \tag{2.44b}$$

For simplicity, here we have restricted ourselves to solutions that are even functions of φ. As n increases, the factor $\frac{1}{\rho_n}$ grows at the same rate as n, and the decay of the coefficients A_n, which is necessary for the convergence of the series (2.34), is provided by the fast decay of the Fourier coefficients, that is, the numerator in (2.41a). For example, for the exterior problem with the plane incident wave $U^0 = \exp(-ikr\cos\varphi)$, we have $A_n \sim \frac{1}{H_n^{(2)}(ka)}$, and for $n \gg ka$ the coefficients A_n decay exponentially as $n \to \infty$.

The eigenvalues ρ_n, which simultaneously describe the interior and the exterior problem, vanish at the resonance frequencies for the interior of the cylinder; in our examples, these frequencies are given by the roots of the equations $J_n(ka) = 0$ and $J_n'(ka) = 0$. In the expression (2.34) for the field, these resonances, which are superfluous for the exterior problem, are absent. According to (2.41), (2.43), and (2.46), the denominator in the expression for A_n vanishes at these frequencies, and the corresponding eigenfunction does not occur in the expression for the diffracted field.

2. The application of the first version for metallic bodies can be interpreted as the construction of the solution of an *integral equation of the first kind*. Indeed, the solution of problem (2.32), (2.33a) is known to satisfy the relation

$$U = U^0 - \int_S \frac{\partial U}{\partial N} G\, dS, \tag{2.47}$$

where G is the vacuum Green function (1.84). In particular, on the surface S of the body we have, by (2.33a),

$$\int_S \frac{\partial U}{\partial N} G\, dS = U^0. \tag{2.48}$$

This is the well-known integral equation of the first kind for $\left.\frac{\partial U}{\partial N}\right|_S$, that is, for the current. Le us use this equation to construct the system of functions u_n.

We introduce some system of functions v_n on S by the integral equation of the second kind

$$-\rho_n v_n = \int_S v_n G\, dS. \tag{2.49}$$

(In the following it will be shown that the functions v_n and the eigenvalues ρ_n coincide with those given above.) The kernel G is symmetric but is not Hermitian and hence the orthogonality conditions have the form

$$\int_S v_n v_m \, dS = 0, \qquad n \neq m. \tag{2.50}$$

Let us expand $\partial U/\partial N$ in a series in the v_n:

$$\frac{\partial U}{\partial N} = \sum_n A_n v_n. \tag{2.51}$$

Then from (2.48) with regard to (2.50) we obtain the expression

$$A_n = -\frac{1}{\rho_n} \frac{\int_S U^0 v_n \, dS}{\int_S v_n^2 \, dS} \tag{2.52}$$

for the A_n, which coincides with (2.41a).

Now let us define the functions u_n in the entire domain (rather than only on S) by the formula

$$u_n = -\int_S v_n G \, dS. \tag{2.53}$$

These functions satisfy the wave equation (2.36) and the radiation condition. Needless to say, they do not satisfy condition (2.33a) on S. By (2.53) and (2.49), on S we have

$$u_n = \rho_n v_n. \tag{2.54}$$

Finally, according to (2.47), (2.51), and (2.53), the formal solution of problem (2.32), (2.33a) is given by the series (2.34) with the coefficients (2.52).

It is easy to see that the definitions of the functions u_n and v_n and the numbers ρ_n given in this and in the preceding subsection are completely equivalent. Indeed, according to definition (2.53), the functions u_n satisfy the wave equation and the radiation conditions, and moreover, they have the same boundary properties as the single layer potential, since the kernel G has the same singularity as $|\mathbf{r} - \mathbf{r}'| \to 0$ as the kernels $\ln |\mathbf{r} - \mathbf{r}'|$ and $-\frac{1}{|\mathbf{r} - \mathbf{r}'|}$ for two- or three-dimensional problems, respectively. Consequently, the values of u_n on both sides of S are the same, whereas the normal derivatives have a saltus with magnitude proportional to v_n. By (2.54), we have $v_n = u_n/\rho_n$. Hence, the functions u_n satisfy also the boundary conditions (2.35a) on S.

Needless to say, the integral equation (2.49) for the eigenfunctions of the first version can also be obtained directly from the differential statement of the homogeneous problem. To this end, it suffices to write out the well-known expression for the solution of the wave equation via the jumps of the function and its normal derivative on S and use the boundary conditions (2.35a).

For the eigenfunctions of the second version, this argument also provides an integral equation of the second kind on the surface S, but that equation is more complicated.

Indeed, the boundary values of each function u that satisfies the wave equation (2.36) in the entire space and the radiation condition, together with the boundary values of the normal derivative of u, satisfy the following relations on the closed surface S:

$$u^+ - u^- = 2\int_S \left[(u^+ + u^-)\frac{\partial G}{\partial N} - \left(\frac{\partial u^+}{\partial N} + \frac{\partial u^-}{\partial N}\right) G\right] dS, \tag{2.55}$$

$$u^+ + u^- = 2\int_S \left[(u^+ - u^-)\frac{\partial G}{\partial N} - \left(\frac{\partial u^+}{\partial N} - \frac{\partial u^-}{\partial N}\right) G\right] dS. \tag{2.56}$$

To obtain these formulas, one must apply Green's formula (0.20) in the exterior domain (V^-) and the interior domain (V^+) to the function u and the vacuum Green function G, pass to the limit as the point of observation tends to S (here one should use the properties of the single and double layer potentials), and then consider the sum and the difference of the results. Using the boundary conditions (2.35b) in (2.55) and (2.56), we obtain

$$-\frac{1}{\widetilde{\rho}_n}\frac{\partial \widetilde{u}_n}{\partial N_r} = 2\int_S \left[(\widetilde{u}_n^+ + \widetilde{u}_n^-)\frac{\partial G}{\partial N_{r'}} - 2\frac{\partial \widetilde{u}_n}{\partial N_{r'}} G\right] dS_{r'}, \tag{2.57}$$

$$\widetilde{u}_n^+ + \widetilde{u}_n^- = -\frac{2}{\widetilde{\rho}_n}\int_S \frac{\partial \widetilde{u}_n}{\partial N_{r'}}\frac{\partial G}{\partial N_{r'}} dS_{r'}. \tag{2.58}$$

By substituting the expression (2.58) for $\widetilde{u}_n^+ + \widetilde{u}_n^-$ into (2.57) and by reversing the order of integration in the first term, we obtain the above-mentioned integral equation for the normal derivative of the eigenfunction $\widetilde{u}_n$:

$$\frac{1}{4}\frac{\partial \widetilde{u}_n}{\partial N_r}(r) = \int_S \frac{\partial \widetilde{u}_n}{\partial N_{r'}}\left[\int_S \frac{\partial G}{\partial N_{r'}}(r'', r')\frac{\partial G}{\partial N_{r''}}(r, r'')\, dS_{r''} + \widetilde{\rho}_n G\right] dS_{r'}. \tag{2.59}$$

The points r, r', and r'' belong to the surface S.

3. The homogeneous problems of the ρ-method can readily be used for constructing the solutions of diffraction problems in which the boundary conditions (2.33a), (2.33b) on the common boundary of the domains V^+ and V^- are replaced by the *transmission conditions* of the form

$$(U^+ - U^-)|_S = 0, \tag{2.60a}$$

$$\left.\left(\frac{\partial U^+}{\partial N} - \frac{\partial U^-}{\partial N} - \frac{U}{\rho}\right)\right|_S = 0 \tag{2.60b}$$

or by the conditions

$$\left.\left(\frac{\partial U^+}{\partial N} - \frac{\partial U^-}{\partial N}\right)\right|_S = 0, \tag{2.61a}$$

$$\left.\left(U^+ - U^- + \frac{1}{\rho}\frac{\partial U}{\partial N}\right)\right|_S = 0, \tag{2.61b}$$

where ρ and $\widetilde{\rho}$ are given constants. Conditions of this kind occur in diffraction problems where the boundary S is a semitransparent film rather than a continuous metallic surface. If the surface S is a continuously bent fast-periodic lattice of metal bands, then the corresponding diffraction problem can also be reduced to a problem with such boundary conditions. In this case, the numbers ρ and $\widetilde{\rho}$ are parameters of the lattice; that is, they characterize the relative dimensions of the conductors. If the conductors in the lattice are ideal, then these parameters are real:

$$\rho = -\frac{p}{2\pi} \ln \sin \frac{\pi q}{2}, \tag{2.62a}$$

$$\widetilde{\rho} = \frac{\pi}{2\rho} \frac{1}{\ln \cos \dfrac{\pi q}{2}}. \tag{2.62b}$$

Here p is the period and q the filling ratio of the lattice.

One should use the first version of the method for the boundary conditions (2.60) and the second version for the boundary conditions (2.61). The series (2.34) term by term satisfies condition (2.60a) in the former case and condition (2.61a) in the latter case. Recall that in the representation (2.34) U^0 is the field of the sources in vacuum and hence is continuous together with the normal derivative on S.

Conditions (2.60b) and (2.61b), with regard to the orthogonality conditions (2.39a) and (2.39b), respectively, give

$$A_n = \frac{\rho_n}{\rho - \rho_n} \frac{\int_S U^0 u_n \, dS}{\int_S u_n^2 \, dS}, \tag{2.63a}$$

$$A_n = \frac{\widetilde{\rho}_n}{\widetilde{\rho} - \widetilde{\rho}_n} \frac{\int_S \dfrac{\partial U^0}{\partial N} \dfrac{\partial \widetilde{u}}{\partial N} \, dS}{\int_S \left(\dfrac{\partial \widetilde{u}_n}{\partial N} \right)^2 dS}. \tag{2.63b}$$

The denominators in these formula never vanish, since ρ and $\widetilde{\rho}$ are real (or, in the presence of losses in the conductors, have negative imaginary parts), whereas the eigenvalues have been proved to have positive imaginary parts. The absence of infinite resonances is natural, since in this case the diffraction problem pertains to an open resonator. The maximum values of the amplitudes A_n are practically inversely proportional to the imaginary parts of the eigenvalues and are attained at the frequencies given by the roots of the equations

$$\rho - \operatorname{Re} \rho_n(k) = 0 \tag{2.64a}$$

or

$$\widetilde{\rho} - \operatorname{Re} \widetilde{\rho}_n(k) = 0. \tag{2.64b}$$

These frequencies are the resonance frequencies.

It is easy to generalize the above results to the case in which ρ and $\widetilde{\rho}$ in condition (2.60) and (2.61) are functions of the coordinates on S (that is, $\rho = \rho(s)$ and $\widetilde{\rho} = \widetilde{\rho}(s)$). This is the case in the diffraction problem if the diffracting surface S is of variable transparency (say, is a band lattice with variable filling ratio). In this case, just as in the w-method with $w = w(s)$, the quantities ρ_n and $\widetilde{\rho}_n$ in the conditions of the homogeneous problems (2.35) are functions of s, and the role of the spectral parameter is played by some parameter occurring in these functions. The possible form of the functions $\rho_n(s)$ and $\widetilde{\rho}(s)$ is determined by formulas of the type of (1.72) and (1.94). The general problem will be considered in more detail in Section 2.4.

The application of this method to *diffraction problems on nonclosed surfaces* does not encounter any essential difficulties. In this case, all the above formulas remain valid, with the exception of the integral equation (2.59) in the second version. The substitute of this equation is more complicated and we do not present it here. Formulas like (2.41) and (2.63) undergo modifications similar to those for the corresponding formulas in Section 2.1; these formulas can be obtained, say, by passing to the limit in the case of a very thin body (see Fig. 2.1). Finally, recall that in the ρ-method for nonclosed surfaces, the smallness of the eigenvalues always corresponds to an actual resonance.

Statements of the problems for nonclosed bodies must also include the finiteness-of-energy requirement near sharp edges.

2.3 Spectral Parameter in Transmission Conditions (the ρ-Method). Dielectric Bodies

1. Let us apply the ρ-method to the solution of the diffraction problem for a body with parameters ε, μ independent of the coordinates. The problem is as follows: Find a field satisfying the equations

$$\Delta U^+ + k^2 \varepsilon\mu U^+ = f^+ \quad \text{in} \quad V^+, \tag{2.65}$$

$$\Delta U^- + k^2 U^- = f^- \quad \text{in} \quad V^-, \tag{2.66}$$

(here V^+ is the interior and (V^-) the exterior of the body, see Fig. 2.3), the radiation condition at infinity, and the transmission conditions

$$(U^+ - U^-)|_S = 0, \tag{2.67a}$$

$$\left(\eta \frac{\partial U^+}{\partial N} - \frac{\partial U^-}{\partial N}\right)\bigg|_S = 0 \tag{2.67b}$$

on the common boundary S of the domains V^+ and V^-. In the two-dimensional electromagnetic problem, $\eta = 1/\varepsilon$ for the H-polarization ($U = H_z$) and $\eta = 1/\mu$ for the E-polarization ($U = E_z$). In acoustics, η is the reciprocal of the density of the body. The methods of this section could readily be developed for three-dimensional electromagnetic problem by analogy with what will be done in Section 2.6 for other versions of the ρ-method.

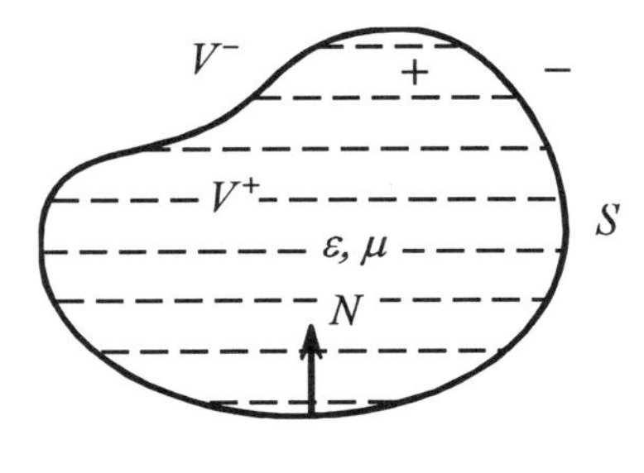

Figure 2.3

We represent the solution of the above diffraction problem everywhere by the *sum of two series*

$$U^{\pm} = U^0_{\pm} + \sum_n A_n u_n^{\pm} + \sum_n B_n \widetilde{u}_n^{\pm}. \tag{2.68}$$

Here U^0_- is the field of the sources f^- in vacuum, and U^0_+ is the field of the sources f^+ in the free space with parameters ε and μ. The functions U^0_+ and U^0_- "do not feel" the boundary S and satisfy Eqs. (2.65) and (2.66) in the entire space, that is,

$$\Delta U^0_+ + k^2\varepsilon\mu U^0_+ = f^+, \tag{2.69}$$

$$\Delta U^0_- + k^2 U^0_- = f^-. \tag{2.70}$$

Thus, U^0_+ and U^0_- are distinct functions defined everywhere in $V^+ + V^-$. It obviously follows from the above definition of U^0 that $U^0_+ = 0$ for $f^+ = 0$ and $U^0_- = 0$ for $f^- = 0$.

The functions u_n and $\widetilde{u}_n$ with respect to which the field $U - U^0$ is expanded are the eigenfunctions of two auxiliary homogeneous problems that differ from each other by the transmission conditions on S. The functions u_n satisfy the generalized conditions

$$(u_n^+ - u_n^-)|_S = 0, \tag{2.71a}$$

$$\left(\eta \frac{\partial u_n^+}{\partial N} - \frac{\partial u_n^-}{\partial N} - \frac{u_n}{\rho_n}\right)\bigg|_S = 0 \tag{2.71b}$$

of the first version, and the functions $\widetilde{u}_n$ satisfy the generalized conditions

$$\left(\eta \frac{\partial \widetilde{u}_n^+}{\partial N} - \frac{\partial \widetilde{u}_n^-}{\partial N}\right)\bigg|_S = 0, \tag{2.72a}$$

$$\left(\widetilde{u}_n^+ - \widetilde{u}_n^- + \frac{1}{\widetilde{\rho}_n}\frac{\partial \widetilde{u}_n^+}{\partial N}\right)\bigg|_S = 0 \tag{2.72b}$$

of the second version. Apart from these conditions, u_n and $\widetilde{u}_n$ must satisfy the homogeneous equations

$$\Delta u_n^+ + k^2\varepsilon\mu u_n^+ = 0 \quad \text{in} \quad V^+, \tag{2.73a}$$

$$\Delta u_n^- + k^2 u_n^- = 0 \quad \text{in} \quad V^- \tag{2.73b}$$

and the radiation conditions at infinity.

The numbers ρ_n in problem (2.71), (2.73) and $\tilde{\rho}_n$ in problem (2.72), (2.73) are the eigenvalues. We postpone the proof of formulas (2.38) until the end of the section.

The functions u_n and $\tilde{u}_n$ differ from those introduced in Section 2.2 in that they satisfy the wave equation with ε and μ in V^+ and that these parameters also occur in the boundary conditions.

By applying the second Green formula (0.20) in V^+ and V^- with regard to the boundary conditions (2.71) and (2.72) and the radiation conditions, we can readily see that the eigenfunctions u_n, as well as the normal derivatives of the functions $\tilde{u}_n$, are orthogonal in the sense that

$$\int_S u_n u_m \, dS = 0, \quad n \neq m, \tag{2.74}$$

$$\int_S \frac{\partial \tilde{u}_n}{\partial N} \frac{\partial \tilde{u}_m}{\partial N} \, dS = 0, \quad n \neq m. \tag{2.75}$$

These conditions permit one to obtain explicit expressions for the coefficients A_n and B_n in the expansion (2.68). Indeed, let us subject the desired field expressed by the series (2.68) to the boundary conditions (2.67) of the diffraction problem. (The equations in V^+ and V^-, as well as the radiation conditions, are satisfied by the series (2.68) term by term.) Since the functions u_n satisfy condition (2.67a) and the functions $\tilde{u}_n$ satisfy condition (2.67b), it follows that the coefficients A_n and the functions u_n do not occur in the equation resulting from the substitution of the expansion (2.68) into (2.67a), and likewise, the functions $\tilde{u}_n$ and the coefficients B_n do not occur in the equation resulting from the substitution of the same expansion into (2.67b). Consequently, we obtain two independent systems of equations for the desired amplitudes, namely,

$$\left(U_+^0 - U_-^0 + \sum_n B_n (\tilde{u}_n^+ - \tilde{u}_n^-) \right) \Big|_S = 0, \tag{2.76}$$

$$\left(\eta \frac{\partial U_+^0}{\partial N} - \frac{\partial U_-^0}{\partial N} + \sum_n A_n \left(\eta \frac{\partial u_n^+}{\partial N} - \frac{\partial u_n^-}{\partial N} \right) \right) \Big|_S = 0. \tag{2.77}$$

Let us use the boundary conditions for the homogeneous problems, then multiply (2.76) by $\frac{\partial \tilde{u}_m^+}{\partial N}$ and (2.77) by u_m, and finally integrate the result over S. In view of the orthogonality conditions (2.74) and (2.75), we obtain

$$A_n = \rho_n \frac{\displaystyle\int_S \left(\frac{\partial U_-^0}{\partial N} - \frac{\partial U_+^0}{\partial N} \eta \right) u_n \, dS}{\displaystyle\int_S u_n^2 \, dS}, \tag{2.78}$$

$$B_n = \tilde{\rho}_n \frac{\displaystyle\int_S (U_+^0 - U_-^0) \frac{\partial \tilde{u}_n^+}{\partial N} \, dS}{\displaystyle\int_S \left(\frac{\partial \tilde{u}_n^+}{\partial N} \right)^2 dS}. \tag{2.79}$$

Thus, the resonance occurs when the moduli of the eigenvalues ρ_n and $\tilde{\rho}_n$ treated, say, as functions of the frequency, become large. As usual, the conditions of the homogeneous problems in this case become close to the conditions of the actual problem. The resonance frequencies of both problems are the same, since the boundary conditions of the actual problem are obtained from (2.71) and (2.72) by passing to the limit as $|\rho_n| \to \infty$ and $|\tilde{\rho}_n| \to \infty$. (Note that the homogeneous problems for u_n and $\tilde{u}_n$ coincide in the limit.) Recall that in the problems of the preceding section, the resonance occurs as $\rho_n \to 0$ and $\tilde{\rho}_n \to 0$.

The form of Eqs. (2.78) and (2.79) remains the same if the desired field must also satisfy some other conditions, say, if on some surface we must have $U = 0$ (a resonator consisting of a dielectric body and, for the case of the E-polarization, some metallic screens). Needless to say, the same conditions must be satisfied by the fields $U^0_\pm$ and the functions u_n and $\tilde{u}_n$.

2. One can seek the solution of the diffraction problem (2.65)–(2.67) for a dielectric body in the form of a series in the eigenfunctions of *only one homogeneous problem*, namely, either in the form

$$U^\pm = U^0_\pm + \sum_n A_n u_n^\pm \tag{2.80}$$

or in the form

$$U^\pm = U^0_\pm + \sum_n B_n \tilde{u}_n^\pm, \tag{2.81}$$

where u_n and $\tilde{u}_n$ are the same functions as in Subsection 1. However, determining $U^0_\pm$ is now a much more complicated procedure. For example, if the sources f are located outside the body (in V^-), then in this domain we can, by analogy with the preceding subsection, choose U^0_- to be the field of these sources in the absence of the body (the incident field). Then U^0_+ must satisfy the equation

$$\Delta U^0_+ + k^2 \varepsilon\mu U^0_+ = 0 \tag{2.82}$$

inside the body (in V^+). However, if we use the representation (2.80) of the desired field, then U^0_+ must coincide with the known field U^0_- on the boundary S,

$$(U^0_+ - U^0_-)|_S = 0. \tag{2.83}$$

(The normal derivatives $\partial U^0_+/\partial N$ and $\partial U^0_-/\partial N$ in this case, of course, need not coincide on S.) Thus, U^0_+ for (2.80) must be found from the Dirichlet problem. If we use the expansion (2.81), then the normal derivative of U^0_+ on S proves to be given,

$$\left(\frac{\partial U^0_+}{\partial N} - \frac{1}{\eta} \frac{\partial U^0_-}{\partial N} \right) \Bigg|_S = 0. \tag{2.84}$$

In this case $(U^0_+ - U^0_-)|_S \neq 0$, and U^0_+ for (2.81) must be found from the Neumann problem. There will be nothing new if all the sources are located in V^+. The definitions of U_+^0 and U_-^0 in this case are merely interchanged, that is, U_+^0 is a known function and U_-^0 must be found from the Dirichlet or Neumann problem (now in the exterior domain). If the sources are simultaneously present in both domains, then, generally speaking, *the solution of two Dirichlet or Neumann problems is needed.* Furthermore, in each of the domains U^0 is the sum of two terms, namely, the field of the sources in the same domain in the absence of the boundary and the solution of the Dirichlet or Neumann problem corresponding to the sources outside the domain.

For this definition of U^0, the series (2.80) term by term satisfies the actual condition (2.67a), and the series (2.81) term by term satisfies (2.67b). Hence, by imposing condition (2.67b) on (2.80) and condition (2.67a) on (2.81) and by using the corresponding orthogonality condition (2.74) or (2.75), we obtain the same expressions (2.78) and (2.79) for A_n and B_n. However, these coefficients are the Fourier coefficients of a series different from (2.68), and the function U^0 in (2.78) and (2.79) has a different meaning.

Note that in some cases there is no need to solve the above-mentioned Dirichlet or Neumann problems. For example, if we deal with an external excitation and only the field outside the body is of interest to us, then we need not seek U_+^0, which can be eliminated from the expressions for the amplitudes with the help of the Green formula (0.20). The corresponding formulas read

$$A_n = \rho_n \frac{\displaystyle\int_S \left(u_n \frac{\partial U_-^0}{\partial N} - \eta U_-^0 \frac{\partial u_n^+}{\partial N} \right) dS}{\displaystyle\int_S u_n^2 \, dS}, \tag{2.85}$$

$$B_n = \tilde{\rho}_n \eta \frac{\displaystyle\int_S \left(\tilde{u}_n \frac{\partial U_-^0}{\partial N} - U_-^0 \frac{\partial \tilde{u}_n^-}{\partial N} \right) dS}{\displaystyle\int_S \left(\frac{\partial \tilde{u}_n^-}{\partial N} \right)^2 dS}. \tag{2.86}$$

Here U_-^0 and $\partial U_-^0/\partial N$ are the field created on S by the sources in the absence of the body and the normal derivative of this field, respectively.

3. One can show that the eigenfunctions of the homogeneous problem (2.71), (2.73) satisfy a system of homogeneous integral equations on S. For example, for the case $\eta = 1$ ($\mu = 1$ and the E-polarization is considered) this system has the form

$$u_n(r) = \int_S \left[u_n(r') \left(\frac{\partial F}{\partial N_{r'}} - \frac{1}{\rho_n} G^- \right) - v_n(r') F \right] dS_{r'}, \tag{2.87}$$

$$v_n(r) = \int_S \left[u_n(r') \left(\frac{\partial^2 F}{\partial N_r \partial N_{r'}} - \frac{1}{\rho_n} \frac{\partial G^-}{\partial N_r} \right) - v_n(r') \frac{\partial F}{\partial N_r} \right] dS_{r'} + \frac{1}{2\rho_n} u_n(r), \tag{2.88}$$

where $v_n = \partial u_n^+/\partial N$, $F = G^+ - G^-$, and G^+ and G^- are the Green functions of the point source in free space with the wave numbers $k\sqrt{\varepsilon}$ and k, respectively. The possibility of writing out these equations is due to the existence of integral equations (also on the surface of the body) for the original diffraction problem.

Note that for arbitrary ε the eigenfunctions u_n introduced in this section provide also the solution of the exterior and the interior diffraction problems for a metallic body with the boundary condition (2.33a) according to (2.34) and (2.41a). However, the simplest integral equation (2.49) related to diffraction problem for metallic bodies can be obtained from (2.87) at $\varepsilon = 1$ ($F \equiv 0$). The eigenfunctions $\widetilde{u}_n$ permit one to solve the diffraction problem for metallic bodies with conditions (2.33b). Needless to say, for $\varepsilon = \mu = 1$ the eigenelements of the homogeneous problems introduced in this section coincide with those introduced in Section 2.2 for diffraction by metallic bodies.

4. By way of example, let us illustrate the formulas of this section for the diffraction problem for a circular dielectric cylinder of radius a with permittivity ε ($\mu \equiv 1$) under the condition that the field, as well as its normal derivative, is continuous on the boundary ($\eta = 1$). The eigenfunctions and eigenvalues of the homogeneous problems (2.71), (2.73) and (2.72), (2.73) are given, respectively, by

$$u_n = H_n^{(2)}(ka)J_n\left(k\sqrt{\varepsilon}r\right)\cos n\varphi, \quad r \leq a, \tag{2.89a}$$

$$u_n = J_n\left(k\sqrt{\varepsilon}a\right)H_n^{(2)}(kr)\cos n\varphi, \quad r \geq a, \tag{2.89b}$$

$$\widetilde{u}_n = H_n^{(2)\prime}(ka)J_n\left(k\sqrt{\varepsilon}r\right)\cos n\varphi, \quad r < a, \tag{2.90a}$$

$$\widetilde{u}_n = \sqrt{\varepsilon}J_n'\left(k\sqrt{\varepsilon}a\right)H_n^{(2)}(kr)\cos n\varphi, \quad r > a, \tag{2.90b}$$

$$\rho_n = \frac{J_n(k\sqrt{\varepsilon}a)H_n^{(2)}(ka)}{k(J_n(k\sqrt{\varepsilon}a)H_n^{(2)\prime}(ka) - \sqrt{\varepsilon}J_n'(k\sqrt{\varepsilon}a)H_n^{(2)}(ka))}, \tag{2.91}$$

$$\widetilde{\rho}_n = \frac{k\sqrt{\varepsilon}J_n'(k\sqrt{\varepsilon}a)H_n^{(2)\prime}(ka)}{J_n(k\sqrt{\varepsilon}a)H_n^{(2)\prime}(ka) - \sqrt{\varepsilon}J_n'(k\sqrt{\varepsilon}a)H_n^{(2)}(ka)}. \tag{2.92}$$

If the incident wave is plane ($U_-^0 = e^{-ikr\cos\varphi}$), then we can readily obtain the following formulas for the coefficients A_n and B_n in the representations (2.80) and (2.81):

$$A_n = \frac{(\sqrt{\varepsilon}J_n(ka)J_n'(k\sqrt{\varepsilon}a) - J_n'(ka)J_n(k\sqrt{\varepsilon}a))2(-i)^n}{C_n(ka,\varepsilon)(1+\delta_{0n})J_n(k\sqrt{\varepsilon}a)}, \tag{2.93}$$

$$B_n = \frac{(\sqrt{\varepsilon}J_n(ka)J_n'(k\sqrt{\varepsilon}a) - J_n'(ka)J_n(k\sqrt{\varepsilon}a))2(-i)^n}{C_n(ka,\varepsilon)(1+\delta_{0n})J_n'(k\sqrt{\varepsilon}a)}, \tag{2.94}$$

where

$$C_n(ka,\varepsilon) = J_n(k\sqrt{\varepsilon}a)H_n^{(2)\prime}(ka) - \sqrt{\varepsilon}J_n'(k\sqrt{e}a)H_n^{(2)}(ka).$$

Furthermore, the solutions of the Dirichlet problem for (2.80) and the Neumann problem for (2.81) are given, respectively, by the formulas

$$U_+^0 = \frac{J_0(ka)}{J_0(k\sqrt{\varepsilon}a)}J_0(k\sqrt{\varepsilon}r) + 2\sum_{n=1}(-i)^n\frac{J_n(ka)}{J_n(k\sqrt{\varepsilon}a)}J_n(k\sqrt{\varepsilon}r)\cos n\varphi, \tag{2.95}$$

$$U_+^0 = \frac{J_0'(ka)}{J_0'(k\sqrt{\varepsilon}a)}J_0(k\sqrt{\varepsilon}r) + +2\sum_{n=1}(-i)^n\frac{J_n'(ka)}{J_n'(k\sqrt{\varepsilon}a)}J_n(k\sqrt{\varepsilon}r)\cos n\varphi. \tag{2.96}$$

We have shown that both series (2.80) and (2.81) can be used in solving a diffraction problem for a dielectric body, and moreover, none of them is preferable. In some respect, these representations supplement each other. Indeed, for a frequency at which the Dirichlet problem becomes unsolvable (in this simple example, this happens for k satisfying the equation $J_n(k\sqrt{\varepsilon}a) = 0$), the application of the expansion (2.80) becomes inconvenient, since one has to treat indeterminacies of the form $\infty - \infty$ in the series for the interior field. In this case, it is almost invariably advisable to use the representation (2.81). Conversely, if the Neumann problem is unsolvable (in our example, this happens at the frequencies satisfying the equation $J_n'(k\sqrt{\varepsilon}a) = 0$), then one has to use the expansion (2.80) in the functions u_n.

5. It was mentioned in Subsection 1 that $\operatorname{Im}\rho_n > 0$ and $\operatorname{Im}\widetilde{\rho}_n > 0$. Let us prove the first of these inequalities; the proof of the second inequality is similar. Let us apply the first Green formula (0.17) in the domain V^+ to the functions u_n and u_n^* and then interchange these functions. Then we obtain

$$\int_S u_n^+ \frac{\partial u_n^{+*}}{\partial N}\, dS = -\int_{V^+} u_n \Delta u_n^*\, dV - \int_{V^+} |\nabla u_n|^2\, dV, \tag{2.97}$$

$$\int_S u_n^{+*} \frac{\partial u_n^{+}}{\partial N}\, dS = -\int_{V^+} u_n^* \Delta u_n\, dV - \int_{V^+} |\nabla u_n|^2\, dV. \tag{2.98}$$

If we now divide Eq. (2.97) by ε^*, Eq. (2.98) by ε, and subtract the results from each other, then the first terms on the right-hand side disappear (according to the equations for u_n and u_n^*), and we have (for $\mu \equiv 1$)

$$\int_S \left(u_n^+ \frac{1}{\varepsilon^*} \frac{\partial u_n^{+*}}{\partial N} - u_n^{+*} \frac{1}{\varepsilon} \frac{\partial u_n^+}{\partial N} \right) dS = \left(\frac{1}{\varepsilon} - \frac{1}{\varepsilon^*} \right) \int_{V^+} |\nabla u_n|^2\, dV. \tag{2.99}$$

An application of the second Green formula (0.20) in the domain V^- to the same functions yields

$$\int_S \left(u_n^- \frac{\partial u_n^{-*}}{\partial N} - u_n^{-*} \frac{\partial u_n^-}{\partial N} \right) dS = -\int_{S_\infty} \left(u_n^- \frac{\partial u_n^{-*}}{\partial N} - u_n^{-*} \frac{\partial u_n^-}{\partial N} \right) dS. \tag{2.100}$$

Let us subtract (2.100) from (2.99) with regard to the boundary conditions (2.71). We find that

$$\frac{\operatorname{Im}\rho_n}{|\rho_n|^2} \int_S |u_n|^2\, dS = -\frac{\operatorname{Im}\varepsilon}{|\varepsilon|^2} \int_{V^+} |\nabla u_n|^2\, dV + \operatorname{Im} \int_{S_\infty} u_n^- \frac{\partial u_n^{-*}}{\partial N}\, dS. \tag{2.101}$$

The terms on the right-hand side in (2.101) represent the sum of dielectric losses in the body and radiation losses, which are positive. It follows that $\operatorname{Im}\rho_n > 0$.

2.4 Spectral Parameter in Transmission Conditions of the General Form

1. The method presented in this section provides the solution of the diffraction problem for surfaces with *boundary condition of a more general form*, which occur in electromagnetic theory and acoustics. In the limit cases, these conditions pass into the boundary conditions of the methods considered earlier in this chapter; in this sense, the technique considered here can be viewed as a generalization of these methods. Let us state the diffraction problem. It is required to find a field satisfying the wave equation

$$\Delta U + k^2 U = f \tag{2.102}$$

everywhere outside a surface S, the radiation conditions at infinity, and the boundary conditions

$$\left[U^+ - U^- - \alpha\left(\frac{\partial U^+}{\partial N} + \frac{\partial U^-}{\partial N}\right)\right]\Big|_S = 0, \tag{2.103a}$$

$$\left[U^+ + U^- - \beta\left(\frac{\partial U^+}{\partial N} - \frac{\partial U^-}{\partial N}\right)\right]\Big|_S = 0 \tag{2.103b}$$

on the surface S (which we assume to be closed).

Boundary conditions of this kind describe, for example, diffraction on fast-periodic metallic lattices (the period is small compared with the wavelength) made of solid conductors. Then α and β are parameters characterizing the relative dimensions of the conductors and the form of their cross-sections. These parameters are real if the conductors are ideal.

The passage to the limit forms of the boundary conditions (2.103a) and (2.103b) is obvious. The boundary conditions for a continuous ideal screen ($U|_S = 0$ or $\left.\frac{\partial U}{\partial N}\right|_S = 0$) correspond to $\alpha = \beta = 0$ or $\frac{1}{\alpha} = \frac{1}{\beta} = 0$, respectively. The impedance conditions are obtained for $\alpha = \beta = w$. (In this case, the boundary conditions (2.103) split into conditions for V^+ and V^-, and these domains prove to be uncoupled.) The conditions (2.60) for a semitransparent surface are obtained for $\alpha = 0$ and $\beta = 2\rho$. Conditions (2.61) follow for $\frac{1}{\beta} = 0$ and $\alpha = -\frac{1}{2\widetilde{\rho}}$. Finally, the continuity conditions for the field and its derivative are obtained for $\alpha = \frac{1}{\beta} = 0$.

The solution of the above diffraction problem will be sought in the form of the *sum of two series,*

$$U = U^0 + \sum_n A_n u_n + \sum_n B_n \widetilde{u}_n. \tag{2.104}$$

Here U^0 is the field of the actual sources f in vacuum, and u_n and $\widetilde{u}_n$ are the eigenfunctions of two auxiliary homogeneous problems, which differ from each other only in the statement of the boundary conditions on S. For the functions u_n, we preserve the condition of the actual problem (2.103a), and the coefficient of the difference of the derivatives in (2.103b) will be treated as the spectral parameter. Thus, we have

$$\left[u_n^+ - u_n^- - \alpha\left(\frac{\partial u_n^+}{\partial N} + \frac{\partial u_n^-}{\partial N}\right)\right]\Big|_S = 0, \tag{2.105a}$$

$$\left[u_n^+ + u_n^- - \beta_n\left(\frac{\partial u_n^+}{\partial N} - \frac{\partial u_n^-}{\partial N}\right)\right]\Big|_S = 0. \tag{2.105b}$$

Conversely, for the functions $\widetilde{u}_n$ we preserve the actual condition (2.103b) and introduce the spectral parameter via (2.103a):

$$\left[\widetilde{u}_n^+ - \widetilde{u}_n^- - \alpha_n \left(\frac{\partial \widetilde{u}_n^+}{\partial N} + \frac{\partial \widetilde{u}_n^-}{\partial N}\right)\right]\Bigg|_S = 0, \tag{2.106a}$$

$$\left[\widetilde{u}_n^+ + \widetilde{u}_n^- - \beta \left(\frac{\partial \widetilde{u}_n^+}{\partial N} - \frac{\partial \widetilde{u}_n^-}{\partial N}\right)\right]\Bigg|_S = 0. \tag{2.106b}$$

Furthermore, the eigenfunctions of both problems must satisfy the homogeneous equations

$$\Delta u_n + k^2 u_n = 0, \tag{2.107}$$

$$\Delta \widetilde{u}_n + k^2 \widetilde{u}_n = 0 \tag{2.108}$$

in the entire space and the radiation conditions at infinity.

By applying the Green formula (0.20) to eigenfunctions with distinct indices n and m in the interior and the exterior domains, by subtracting the results from each other, and by eliminating, say, the normal derivatives according to the boundary conditions, we obtain the orthogonality conditions for the first homogeneous problem (2.105), (2.107) in the form

$$\int_S \psi_n \psi_m \, dS = 0, \quad n \neq m, \tag{2.109}$$

and for the second problem (2.106), (2.108) in the form

$$\int_S \varphi_n \varphi_m \, dS = 0, \quad n \neq m, \tag{2.110}$$

where

$$\psi_n = u_n^+ + u_n^-, \quad \varphi_n = \widetilde{u}_n^+ - \widetilde{u}_n^-.$$

Thus, the orthogonality conditions on S are satisfied by the sums of the limit values of u_n on both sides of S and the jumps of $\widetilde{u}_n$ on S.

It is easy to see that all homogeneous problems of the surface methods considered earlier are special cases of the homogeneous problems of the present method. Indeed, the statement of the first version of the ρ-method follows in (2.105) we set $\alpha = 0$ (then $\rho_n = \beta_n/2$). The second version of the ρ-method follows if in (2.106) we set $1/\beta = 0$ (then $\widetilde{\rho}_n = -\dfrac{1}{2\alpha_n}$). Finally, to obtain the w-method, it suffices, say, to declare that the eigenvalue w_n in (2.105) is assumed to be the coefficients of the sum and the difference of the derivatives.

Let us find the coefficients A_n and B_n. To this end, we subject the expansion (2.104) to the boundary conditions (2.103). Since the eigenfunctions u_n, by definition, satisfy (2.103a) and the $\widetilde{u}_n$ satisfy (2.103b), we see that the equation for the coefficients resulting from (2.103a) will not contain A_n, and the equation for the coefficients resulting from (2.103b) will not contain B_n. Using (2.106a) and (2.105b) once more, we obtain

$$\sum_n B_n \left(1 - \frac{\alpha}{\alpha_n}\right) \varphi_n = 2\alpha \frac{\partial U^0}{\partial N}\bigg|_S, \tag{2.111}$$

$$\sum_n A_n \left(1 - \frac{\beta}{\beta_n}\right) \psi_n = -2U^0\big|_S. \tag{2.112}$$

Let us multiply (2.111) by φ_m and (2.112) by ψ_m and integrate over S with regard to the orthogonality conditions (2.109) and (2.110). Then we find that

$$A_n = \frac{2\beta_n}{\beta - \beta_n} \frac{\int_S U^0 \psi_n \, dS}{\int_S \psi_n^2 \, dS}, \tag{2.113}$$

$$B_n = \frac{2\alpha\alpha_n}{\alpha_n - \alpha} \frac{\int_S \frac{\partial U^0}{\partial N} \varphi_n \, dS}{\int_S \varphi_n^2 \, dS}. \tag{2.114}$$

Everything that was said about the resonance conditions in Sections 2.1 and 2.2 for the w- and ρ-methods fully pertains also to the general case. Namely, a resonance occurs when the eigenvalues, treated, say, as functions of the frequency, become close to the corresponding parameter values in the diffraction problem, that is, when $\beta_n(k)$ is close to β and $\alpha_n(k)$ is close to α. Since there are radiation losses in the system, it follows that the resonances are always finite. The eigenvalues have positive imaginary parts (the proof can be carried along the same lines as in Section 2.2), and hence the equations

$$\beta_n(k) - \beta = 0, \tag{2.115}$$

$$\alpha_n(k) - \alpha = 0 \tag{2.116}$$

do not have real roots (we have $\operatorname{Im}\alpha$, $\operatorname{Im}\beta \leq 0$). The positiveness of the imaginary parts of the eigenvalues means that there is energy release on S compensating for the radiation losses; otherwise, the undamped oscillations would be impossible.

2. We can seek the solution of the diffraction problem (2.102), (2.103) in the form of the expansion in the eigenfunctions of only one homogeneous problem, that is, either in the form

$$U = U^0 + \sum_n A_n u_n, \tag{2.117}$$

or in the form

$$U = U^0 + \sum_n B_n \tilde{u}_n. \tag{2.118}$$

However, in this case U^0 is not a very simple field (that is, the field of the sources f in the free space) and must satisfy complicated boundary conditions on S. Indeed, by substituting the expansion (2.117) into (2.103a) and the expansion (2.118) into (2.103b), we find that U^0 must satisfy the condition

$$\left[U_+^0 - U_-^0 - \alpha\left(\frac{\partial U_+^0}{\partial N} + \frac{\partial U_-^0}{\partial N}\right)\right]\Bigg|_S = 0 \tag{2.119}$$

on S for the first representation and the condition

$$\left[U_+^0 + U_-^0 - \beta\left(\frac{\partial U_+^0}{\partial N} - \frac{\partial U_-^0}{\partial N}\right)\right]\Bigg|_S = 0 \tag{2.120}$$

for the second representation. We can ensure the validity of these conditions by choosing U^0 as follows. If the sources f are located only in V^-, then in this domain the field U_-^0 can be assumed to be the field of these sources in vacuum, that is, a known function. Then the definition of U_+^0 will obviously be the third boundary value problem for the homogeneous equation

$$\Delta U_+^0 + k^2 U_+^0 = 0 \tag{2.121}$$

with the nonhomogeneous boundary condition (for U^0 in (2.117))

$$\left(U_+^0 - \alpha\frac{\partial U_+^0}{\partial N}\right)\Bigg|_S = \left(U_-^0 + \alpha\frac{\partial U_-^0}{\partial N}\right)\Bigg|_S \tag{2.122}$$

or (for U^0 in (2.118))

$$\left(U_+^0 - \beta\frac{\partial U_+^0}{\partial N}\right)\Bigg|_S = -\left(U_-^0 + \beta\frac{\partial U_-^0}{\partial N}\right)\Bigg|_S. \tag{2.123}$$

The right-hand sides of (2.122) and (2.123) are given functions.

If the sources are located only in V^+, then we have the third boundary value problem in the exterior domain. Furthermore, U_+^0 will be the known field of these sources in free space. If the sources are present in both domains, then U^0 in each of the domains is the sum of two terms, one of which is the field in vacuum generated by the sources located in the corresponding domain, and the other is the solution of the third boundary value problem corresponding to the sources located outside this domain.

We can also define U^0 differently. If the sources are located, say, in V^-, then we can set $U_+^0 \equiv 0$. Then, according to (2.122) or (2.123), U_-^0 is the solution of the diffraction problem for the equation

$$\Delta U_-^0 + k^2 U_-^0 = f \tag{2.124}$$

with the impedance boundary condition

$$\left(U_-^0 + \alpha\frac{\partial U_-^0}{\partial N}\right)\Bigg|_S = 0 \tag{2.125}$$

on S, or, respectively, with the condition

$$\left(U_-^0 + \beta \frac{\partial U_-^0}{\partial N}\right)\bigg|_S = 0. \tag{2.126}$$

On finding U^0, we can determine the coefficients A_n and B_n by substituting (2.117) into (2.103b) and (2.118) into (2.103a) with regard to the corresponding orthogonality conditions. Then for the coefficients A_n and B_n we obtain the same formulas (2.113) and (2.114) with the new functions U^0 and $\dfrac{\partial U^0}{\partial N}$.

3. It is not difficult to generalize the results obtained here to the case in which in the actual boundary conditions (2.103a) and (2.103b) the quantities α and β are functions of the coordinates on the boundary S, that is, $\alpha = \alpha(S)$ and $\beta = \beta(S)$. Physically, this means that the surface on which the diffraction occurs is of variable transparency. For example, this can be a lattice with variable step or variable filling ratio.

In this case, the quantities α_n and β_n in the boundary conditions must also be functions of the point s on the surface S, and the spectral parameter is some parameter occurring in these functions. If we set

$$\frac{1}{\beta_n} = \gamma(s) + \sigma_n\left(\frac{1}{\beta(s)} - \gamma(s)\right), \tag{2.127}$$

$$\frac{1}{\alpha_n} = \gamma(s) + \tilde{\sigma}_n\left(\frac{1}{\alpha(s)} - \gamma(s)\right), \tag{2.128}$$

where $\gamma(s)$ is, generally speaking, an arbitrary function (it can be taken proportional to $\dfrac{1}{\beta(s)}$ (respectively, $\dfrac{1}{\alpha(s)}$)), which is not equal to $\dfrac{1}{\beta(s)}$ (respectively, $\dfrac{1}{\alpha(s)}$), and σ_n and $\tilde{\sigma}_n$ are the eigenvalues of the corresponding problems, then the orthogonality conditions for the eigenfunctions will contain a weight,

$$\int_S \left(\frac{1}{\beta} - \gamma\right)\psi_n\psi_m\,dS = 0, \quad n \neq m, \tag{2.129}$$

$$\int_S \left(\frac{1}{\alpha} - \gamma\right)\varphi_n\varphi_m\,dS = 0, \quad n \neq m, \tag{2.130}$$

and the coefficients A_n and B_n in the expansions (2.104), (2.117), and (2.118) will be given by the expressions

$$A_n = \frac{2}{\sigma_n - 1}\frac{\displaystyle\int_S \frac{1}{\beta}U^0\psi_n\,dS}{\displaystyle\int_S (1/\beta - \gamma)\psi_n^2\,dS}, \tag{2.131}$$

$$B_n = \frac{2}{1 - \tilde{\sigma}_n}\frac{\displaystyle\int_S \frac{\partial U^0}{\partial N}\varphi_n\,dS}{\displaystyle\int_S (1/\alpha - \gamma)\varphi_n^2\,dS}. \tag{2.132}$$

These formulas show that the resonance phenomena display themselves when the eigenvalues are close to 1. It can be proved that the sign of the imaginary parts of σ_n and $\widetilde{\sigma}_n$ corresponds to energy release.

4. Finally, let us write out *integral equations* for the eigenfunctions introduced in this section. This can readily be done as follows. Let us use formulas (2.55) and (2.56), relating the sums and the differences of the fields and their normal derivatives on the closed surface S. By substituting the boundary conditions (2.105) and (2.106) into these formulas, we obtain the following system of two homogeneous integral equations of the second kind on the surface S:

$$\varphi = 2\int_S \left(\psi \frac{\partial G}{\partial N} - \frac{1}{\alpha}\varphi G\right) dS, \tag{2.133}$$

$$\psi = 2\int_S \left(\varphi \frac{\partial G}{\partial N} - \frac{1}{\beta}\psi G\right) dS. \tag{2.134}$$

Here, just as in (2.110), ψ and φ are the sums and the differences of the limit values of the eigenfunctions on the opposite sides of S, and G is the vacuum Green function. By setting either $\alpha = \alpha_n$ or $\beta = \beta_n$ in these equations, we obtain integral equations either for the homogeneous problem (2.106), (2.108) or for problem (2.105), (2.107). The same system, in particular, implies the integral equations of the w- and ρ-methods for bodies with closed boundaries.

In conclusion, let us note that all the results of the present section remain valid for diffraction problems for bodies with nonclosed boundaries, except for the integral equations (2.133) and (2.134) for the eigenfunctions, which are more complicated in this case. Furthermore, the statement of diffraction problems for such surface must necessarily include the finiteness-of-energy condition near sharp edges.

2.5 Spectral Parameter in Conditions at Infinity (the s-Method)

The version of the method given in this section applies to *diffraction problems in open systems.* The auxiliary homogeneous problem proves to be real and can be reduced to a *real integral equation* provided that only radiation losses occur in the diffraction problem. This is due to the following rule, which has already been discussed for closed problems. If losses of only one type are present in the system, then one can always make the corresponding auxiliary problem real by introducing the spectral parameter just for the domain where the losses occur. More precisely, the spectral parameter must be introduced via the diffraction problem parameter responsible for these losses. In the s-method, the spectral parameter of the homogeneous problem (which corresponds to the diffraction problem with only radiation losses) will be introduced into the conditions at infinity imposed on the eigenfunction. The physical meaning of these conditions is as follows. There is an eigenwave convergent from infinity and an eigenwave diffracted by the body. The angular dependences of the convergent and divergent waves, determined by the shape and other properties of the body, must coincide (up to complex conjugation). The diffracted-to-incident wave amplitude ratio plays the role of the spectral parameter. The energy balance condition implies that in each pair of counterwaves the powers of the waves must be equal. Thus, in contrast with the other versions of the method considered in the present chapter, the spectral parameter of the homogeneous problem is introduced via asymptotic conditions on the sphere of infinite radius containing the body rather than via auxiliary boundary conditions.

1. The s-method applies to diffraction problems for bodies with various properties. In this subsection, we describe the specific features of the s-method using an example of a scalar problem in which the diffraction of the incident field U^0 occurs on a semi-transparent surface S (not necessarily closed) characterized by a parameter ρ. Consider the wave equation

$$\Delta U + k^2 U = f \tag{2.135}$$

with the boundary conditions

$$(U^+ - U^-)|_S = 0, \tag{2.136a}$$

$$\left.\left(\frac{\partial U^+}{\partial N} - \frac{\partial U^-}{\partial N} - \frac{U}{\rho}\right)\right|_S = 0 \tag{2.136b}$$

on the surface S and the radiation conditions at infinity. The parameter ρ characterizes the transparency of S. In the limit case $\rho = 0$, conditions (2.136) describe an ideal metallic surface with the boundary condition $U|_S = 0$.

The functions in which we expand the diffracted field in this section are not the eigenfunctions of a simple homogeneous problem, but they can be expressed via such eigenfunctions. We start from stating this homogeneous problem.

Consider the nontrivial solutions u_n of the homogeneous wave equation

$$\Delta u_n + k^2 u_n = 0 \tag{2.137}$$

with the conditions (the same as for the diffraction problem)

$$(u_n^+ - u_n^-)|_S = 0, \tag{2.138a}$$

$$\left.\left(\frac{\partial u_n^+}{\partial N} - \frac{\partial u_n^-}{\partial N} - \frac{u_n}{\rho}\right)\right|_S = 0 \tag{2.138b}$$

and with the following condition at infinity (in the three-dimensional case):

$$u_n \simeq \left(\Phi_n^* \frac{e^{ikr}}{kr} + s_n \Phi_n \frac{e{-}ikr}{kr}\right) \frac{1}{1+s_n}. \tag{2.139a}$$

Here the symbol $\simeq$ has the same meaning as in the statement of the radiation condition (0.35) in the form $u \simeq \Phi \dfrac{e^{-ikr}}{kr}$. To complete the statement of the problem, we impose the condition

$$\Phi_n^*(\varphi, \theta) = \Phi_n(\pi + \varphi, \pi - \theta) \tag{2.139b}$$

on the functions Φ_n. For a two-dimensional problem, condition (2.139) must be represented in the form

$$u_n \simeq \left(\Phi_n^* \frac{e^{ikr}}{\sqrt{kr}} + s_n \Phi_n \frac{e^{-ikr}}{\sqrt{kr}}\right) \frac{1}{1+s_n}, \tag{2.140a}$$

and furthermore,

$$\Phi_n^*(\varphi) = i\Phi_n(\varphi + \pi). \tag{2.140b}$$

In (2.139) and (2.140), the angular dependences of the two waves do not coincide, but they are complex conjugate. The numbers s_n are the eigenvalues of the problem. According to (2.139) and (2.140), each of the corresponding functions u_n is the sum of a convergent wave and a wave diffracted by the body with the angular dependence described by the function Φ_n, which will be referred to as the *scattering eigenpattern* of the body. The eigenvalues s_n, given by the divergent-to-convergent wave amplitude ratios, can be viewed as the *reflection eigencoefficients.* In the absence of losses in the body (that is, for real ρ), all the eigenvalues are equal to 1 in modulus, which is a consequence of the fact that the homogeneous problem, as usual, has an independent physical meaning and the energy balance must hold in this problem (the power carried by the incident wave is equal to the power carried by the diffracted wave). The collection of the numbers s_n forms the *scattering matrix.*

Let us reduce the homogeneous problem (2.137)–(2.140) to an integral equation on the boundary S of the body. We shall see that the equation is real, which is rather convenient when one seeks the eigenelements or studies their properties.

We seek u_n as the sum of two single layer potentials

$$u_n = \int_S v_n (G^* + s_n G)\, dS. \tag{2.141}$$

Here G in the vacuum Green function (1.84), and v_n is an unknown function on S. The representation (2.141) ensures the validity of Eq. (2.137), the boundary condition (2.138a) (with regard to the continuity of the single layer potential across the integration surface), and the asymptotic behavior (2.139) or (2.140) as $r \to \infty$. It remains to satisfy condition (2.138b). In view of the properties of derivatives of the single layer potential, the normal derivative undergoes a jump equal to

$$\frac{\partial u_n^+}{\partial N} - \frac{\partial u_n^-}{\partial N} = -(1+s_n) v_n. \tag{2.142}$$

Hence, according to (2.138b), we have

$$v_n = -\frac{u_n}{\rho(1+s_n)}. \tag{2.143}$$

By substituting this into (2.141), we obtain an expression for the eigenfunction u_n everywhere via its values on S. By letting the observation point tend to S, we obtain the desired homogeneous integral equation of the second kind

$$-\rho(1+s_n) u_n = \int_S u_n (G^* + s_n G)\, dS. \tag{2.144}$$

Needless to say, this equation can also be obtained by a more standard method. Namely, let us apply the Green formulas (0.20) in the entire space to the functions u_n and $G^* + s_n G$. The integrals over the sphere at infinity will be zero by virtue of (2.139b) and (2.140b).

Let us consider the properties of Eq. (2.144). It is easily seen that in the absence of losses on S ($\operatorname{Im}\rho = 0$) the eigenvalues s_n are equal to 1 in modulus. (This also follows from the differential statement of the problem). To verify this, it suffices to multiply Eq. (2.144) by u_n^*, integrate over S, and express s_n from the resultant equation:

$$s_n = -\frac{\rho \int_S |u_n|^2 \, dS + \int_S \int_S u_n(p) u_n^*(q) G^*(q,p) \, dS_p \, dS_q}{\rho \int_S |u_n|^2 \, dS + \int_S \int_S u_n(p) u_n^*(q) G(q,p) \, dS_p \, dS_q}. \tag{2.145}$$

With regard to the symmetry of the kernel G, it is obvious that the numerator and the denominator in (2.145) are complex conjugate, so that $|s_n| = 1$. We represent s_n in the form $s_n = e^{2i\delta_n}$ ($\operatorname{Im}\delta_n = 0$). One can readily verify that the eigenfunctions of Eq. (2.144), that is, the values of u_n on S, can also be assumed to be real (and we do so in the following). To this end, we write out the complex conjugate of (2.144) and take account of the fact that $s_n^* = 1/s_n$. Then we see that u_n^* also satisfies Eq. (2.144), which implies that u_n is a real function up to a constant factor. The above properties permit us to conclude that Eq. (2.144) is real. Indeed, let us multiply both parts of Eq. (2.144) by $e^{-i\delta_n}$. Then we obtain the equation

$$-\rho u_n \cos\delta_n = \int_S u_n \operatorname{Re}(e^{i\delta_n} G) \, dS, \tag{2.146}$$

which has the form

$$4\pi\rho u_n = \int_S u_n \frac{\cos kR + \lambda_n \sin kR}{R} \, dS \tag{2.147a}$$

in the three-dimensional problem and the form

$$-4\rho u_n = \int_S u_n (N_0(kR) - \lambda_n J_0(kR)) \, dS \tag{2.137b}$$

in the two-dimensional problem. Here $\lambda_n = \tan\delta_n$ plays the role of an eigenvalue, and R is the distance between the observation point and the integration point. According to (2.147), the continuation of u_n to the entire space is also real.

It is only in the special case of a plane incident wave (see subsection 3) that the total field can be expanded in the u_n. *In the general case, a different system of functions must be used.* We use a system of functions each of which satisfies the radiation condition. Namely, we take functions proportional to those terms in u_n (2.141) and (2.144) which satisfy the radiation condition. We denote these functions by u_n^s. They are defined everywhere via the values of u_n on the boundary S of the body as follows:

$$u_n^s = \int_S u_n G \, dS. \tag{2.148}$$

According to this definition and (2.144), at each point in space we have the relation

$$-\rho(1+s_n)u_n = u_n^{s^*} + s_n u_n^s. \tag{2.149}$$

Using this relation on S or (which is the same) Eq. (2.144) and taking account of the symmetry of the Green function G, we can show that the functions u_n and u_n^s satisfy the following orthogonality conditions on S:

$$\int_S (\rho u_n + u_n^s) u_m \, dS = 0, \quad n \neq m. \tag{2.150a}$$

These orthogonality conditions can be rewritten with regard to (2.149) as

$$\int_S u_m \operatorname{Im} u_n^s \, dS = 0, \quad n \neq m. \tag{2.150b}$$

Let us now proceed to the solution of the diffraction problem (2.135), (2.136). We represent the desired field U everywhere by the series

$$U = U^0 + \sum_n A_n u_n^s, \tag{2.151}$$

where U^0 is the field of the sources f in vacuum,

$$\Delta U^0 + k^2 U^0 = f. \tag{2.152}$$

We shall see in Chapter 4 that in studying high-Q open resonators it is expedient to use the possibility (inherent in all versions of the generalized method) of defining U^0 as the field occurring in the diffraction by some auxiliary body. Here we shall write out the expressions for the A_n for the simplest definition of U^0.

By the definition of U^0 and u_n^s, the series (2.151) satisfies Eq. (2.135), the boundary condition (2.136a), and the radiation conditions. Let us satisfy the remaining condition (2.136b) with regard to the fact that

$$\left(\frac{\partial u_n^{s+}}{\partial N} - \frac{\partial u_n^{s-}}{\partial N} \right) \Bigg|_S = -u_n|_S \tag{2.153}$$

by virtue of (2.148). Then we obtain

$$\left(\sum_n A_n (\rho u_n + u_n^s) \right) \Bigg|_S = -U^0|_S. \tag{2.154}$$

Using the orthogonality conditions (2.150a), we find the desired coefficients of the series in the form

$$A_n = -\frac{\displaystyle\int_S U^0 u_n \, dS}{\displaystyle\int_S (\rho u_n + u_n^s) u_n \, dS}. \tag{2.155a}$$

With regard to (2.149), we can rewrite this formula in the form

$$A_n = -\frac{(1+s_n)}{2i}\frac{\int_S U^0 u_n\,dS}{\int_S u_m \operatorname{Im} u_n^s\,dS}. \tag{2.155b}$$

Note that the expression (2.155) contains the functions u_n^s as well as u_n.

We see from (2.155) that for each given n the resonance occurs at frequencies for which $\operatorname{Im} u_n^s$ is small. In this case, by virtue of (2.149), the functions u_n^s and $-\rho u_n$ are close to each other, which means (see (2.153)) that u_n^s approximately satisfies the actual boundary condition (2.136b).

Thus the solution of the diffraction problem (2.135), (2.136) is given by formulas (2.151), (2.152), (2.147), (2.148), and (2.155).

2. Let us now consider an application of the s-method to the problem of *diffraction by a dielectric body.* Consider the wave equation

$$\Delta U + k^2 \varepsilon U = f, \tag{2.156}$$

where ε is a continuous function of the coordinates different from 1 in a finite domain, with the radiation conditions at infinity. Recall that the two-dimensional version of this problem corresponds to the E-polarization ($U = E_z$), whereas the three-dimensional version does not have any electromagnetic analog. We define the eigenfunctions of the corresponding homogeneous problem as nontrivial solutions of the equation

$$\Delta u_n + k^2 \varepsilon u_n = 0 \tag{2.157}$$

with the asymptotic condition (2.139) or (2.140) $r \to \infty$. (This is just the condition used to introduce the eigenvalues s_n.) Let us reduce this homogeneous problem to an integral equation. We seek u_n as the sum of two volume potentials in which the integration extends over the domain occupied by the body ($\varepsilon \neq 1$), namely,

$$u_n = \int v_n (G^* + s_n G)\,dV, \tag{2.158}$$

where G is the vacuum Green function. This representation ensures the validity of the asymptotic condition (2.139), (2.140) for an arbitrary real (up to a constant factor) function v_n. We determine this function from the condition that u_n must satisfy Eq. (2.157). By applying the operator $(\Delta + k^2\varepsilon)$ to (2.158), we obtain

$$v_n = -\frac{k^2(\varepsilon - 1)}{1 + s_n} u_n. \tag{2.159}$$

On substituting this value of v_n into (2.158), we find that

$$(1 + s_n) u_n = k^2 \int (1 - \varepsilon) u_n (G^* + s_n G)\,dV. \tag{2.160}$$

Equation (2.160) is the *desired integral equation*[2] for u_n if the observation point is in the dielectric body. At the same time, this equation is a formula for calculating u_n outside the body via the values of u_n in the region where $\varepsilon \neq 1$.

For $\operatorname{Im} \varepsilon = 0$, the integral equation (2.160) has the same properties as Eq. (2.144) with $\operatorname{Im} \rho = 0$. Namely, the eigenvalues are equal to 1 in modulus, and the eigenfunctions can be assumed to be real. Hence Eq. (2.160) is a real equation and can be rewritten in the form

$$4\pi u_n = k^2 \int (\varepsilon - 1) u_n \frac{\cos kR + \lambda_n \sin kR}{R} \, dV \tag{2.161a}$$

for the three-dimensional problem and in the form

$$-4u_n = k^2 \int (\varepsilon - 1) u_n (N_0(kR) - \lambda_n J_0(kR)) \, dV \tag{2.161b}$$

for the two-dimensional problem; here R and λ_n have the same meaning as in (2.147).

Next, just as in the preceding subsection, we introduce everywhere defined functions u_n^s by setting

$$u_n^s = k^2 \int (1 - \varepsilon) u_n G \, dV, \tag{2.162}$$

so that, according to (2.160),

$$(1 + s_n) u_n = u_n^{s^*} + s_n u_n^s. \tag{2.163}$$

Let us note some properties of the functions u_n^s. It is obvious from the definition that they all satisfy the radiation conditions. One can readily verify by applying the operator $(\Delta + k^2 \varepsilon)$ to (2.162) that the u_n^s satisfy the differential identity

$$\Delta u_n^s + k^2 \varepsilon u_n^s = k^2 (1 - \varepsilon)(u_n - u_n^s). \tag{2.164}$$

Finally, by using the symmetry of G in (2.160), we can show that u_n and u_n^s satisfy the orthogonality conditions

$$\int (\varepsilon - 1)(u_n - u_n^s) u_m \, dV = 0, \quad n \neq m. \tag{2.165}$$

If we now represent the solution of the diffraction problem by the series (2.151), where u_n^s is defined in (2.162), then the radiation conditions will be satisfied term by term, and the substitution into Eq. (2.156) with regard to (2.164) yields

$$\sum_n A_n (\varepsilon - 1)(u_n - u_n^s) = (\varepsilon - 1) U^0. \tag{2.166}$$

[2]The integral equation (2.160) has the eigenvalue $s_n = -1$, for which the corresponding eigenfunctions do not occur in the differential statement. If all the sources in the diffraction problem are contained in V^-, then these functions are not used in the expansion (see Chapter 5).

In (2.166) we use the orthogonality (2.165), thus obtaining the following expression for the coefficients A_n:

$$A_n = \frac{\int (\varepsilon - 1) U^0 u_n \, dV}{\int (\varepsilon - 1)(u_n - u_n^s) u_n \, dV}. \tag{2.167a}$$

Using (2.163), we can rewrite this formula in the form

$$A_n = -\frac{(1 + s_n)}{2i} \frac{\int (\varepsilon - 1) U^0 u_n \, dV}{(\varepsilon - 1) u_m \operatorname{Im} u_n^s \, dV}. \tag{2.167b}$$

Formulas (2.151), (2.152), (2.161), and (2.167) provide the solution of the diffraction problem for a dielectric body.

It follows from the expressions (2.167) that the resonance occurs when the functions u_n^s and u_n are close to each other ($\operatorname{Im} u_n^s$ is small). According to (2.164), this means that u_n^s approximately satisfies the equation for the diffracted field. Thus, as is always the case, the resonance occurs when the conditions for one of the functions in which the expansion is carried out become close to the actual conditions of the diffraction problem.

Finally, note that although Eqs. (2.160) and (2.161) are integral equations with integrals over the volume of the dielectric body, the sums in the representation of the desired field according to the s-method are two-dimensional for three-dimensional problems and one-dimensional for two-dimensional diffraction problems. This is due to the fact that the spectral parameter is introduced on a surface (in this case, on the surface of the sphere at infinity).

3. As long as the diffraction of a plane wave is considered, a different solution method can be used. Namely, instead of expanding the total field in the u_n^s, one can expand directly in the eigenfunctions u_n without separating the incident field, that is, use the representation

$$U = \sum_n A_n u_n. \tag{2.168}$$

This method applies to diffraction by a dielectric body as well as to diffraction by a surface S. Since the formal construction of the solution is essentially the same for both problem, we consider them simultaneously and understood u_n and u_n^s either as the functions determined by formulas (2.160) and (2.162) (the problem with a dielectric body) or as the functions (2.144) and (2.148) (the problem with a semitransparent surface). The representation (2.168) differs from (2.151) in two respects. First, the functions u_n satisfy the conditions of the diffraction problem, which are violated by the u_n^s. (In the problem on a dielectric body, the u_n satisfy the equation of the diffraction problem, and in the problem on a transparent surface, they satisfy the actual boundary conditions. The corresponding functions u_n^s do not satisfy these conditions.) Second, the functions u_n contain the incoming wave, and the u_n^s do not contain the incoming wave, whereas in the diffraction of a plane wave the incoming field is present everywhere. Hence, in (2.168) we *must not single out* U^0, whereas in (2.151) *this is obligatory*. Thus, in the series (2.151), which satisfies the radiation conditions for any A_n, these coefficients are found from the requirement that the equation (or the boundary conditions) of the diffraction problem must be satisfied. On the contrary, in the series (2.168), where these requirement are satisfied automatically, the coefficients A_n must be found from the radiation condition, that is, from the requirement that the total field at large distances should be representable in the form

$$U \simeq e^{-ikz} + \frac{e^{-ikr}}{kr}\Phi(\varphi,\theta), \tag{2.169}$$

where $\Phi(\varphi,\theta)$ is the desired scattering pattern.

We seek the coefficients A_n from the relations between the opposite waves on a sphere of sufficiently large radius on which the representation (2.139) is valid. To find these coefficients, we need the orthogonality of the eigenpatterns Φ_n. Let us show that they satisfy the Hermitian orthogonality condition

$$\int_\Omega \Phi_n\Phi_m^*\,d\Omega = 0, \quad n \neq m. \tag{2.170}$$

For the case of diffraction by a dielectric body, we prove this by applying the Green formula (0.20) in the entire space to the functions u_m and u_n^s. Since u_n^s satisfies Eq. (2.164), we obtain

$$\int_{S_\infty}\left(u_m\frac{\partial u_n^s}{\partial N} - u_n^s\frac{\partial u_m}{\partial N}\right)dS = k^2\int(1-\varepsilon)(u_n - u_n^s)u_m\,dV. \tag{2.171}$$

According to (2.162), (2.163), and the asymptotic condition (2.139a) or (2.140a), the left-hand side is proportional to $\int_\Omega \Phi_n\Phi_m^*\,d\Omega$, whereas the right-hand side is zero for $n \neq m$ by (2.165). This implies (2.170).

For the case of diffraction by a semitransparent surface S, we prove (2.170) by applying the Green formula (0.20) for u_m and u_n^s separately in V^+ and V^-, considering the difference of the results, and using the boundary conditions (2.138) and (2.153) together with the continuity of u_n^s on S. Then we obtain

$$\int_{S_\infty}\left(u_m\frac{\partial u_n^s}{\partial N} - u_n^s\frac{\partial u_m}{\partial N}\right)dS = -\frac{1}{\rho}\int_S u_m(\rho u_n + u_n^s)\,dS, \tag{2.172}$$

whence, according to (2.139a) (or (2.140a)), (2.148), and (2.150), relation (2.170) follows readily.

Let us now proceed to the calculation of the coefficients A_n so as to find the desired pattern Φ. We illustrate the calculations for a two-dimensional problem and give only the definitive formulas for the three-dimensional case in the end of this section.

Since the representations (2.168) and (2.169) of the total field must coincide for $kr \gg 1$, we have, according to (2.140a),

$$\frac{e^{ikr}}{\sqrt{kr}}\sum_n\frac{A_n Phi_n^*}{1+s_n} + \frac{e^{-ikr}}{\sqrt{kr}}\sum_n\frac{s_nA_n\Phi_n}{1+s_n} = e^{-ikr\cos\varphi} + \frac{e^{-ikr}}{\sqrt{kr}}\Phi. \tag{2.173}$$

Let us multiply this equation by Φ_m or Φ_m^*, integrate over the angular variable, and use the orthogonality condition (2.170). Then we obtain

$$\frac{e^{ikr}}{\sqrt{kr}}\frac{A_m}{1+s_m} + \frac{e^{-ikr}}{\sqrt{kr}}\sum_n \frac{s_n A_n}{1+s_n}\int_0^{2\pi} \Phi_n \Phi_m \, d\varphi$$
$$= \int_0^{2\pi} \Phi_m e^{-ikr\cos\varphi}\, d\varphi + \frac{e^{-ikr}}{\sqrt{kr}}\int_0^{2\pi} \Phi\Phi_m \, d\varphi, \tag{2.174a}$$
$$\frac{e^{ikr}}{\sqrt{kr}}\sum_n \frac{A_n}{1+s_n}\int_0^{2\pi} \Phi_n^* \Phi_m^* \, d\varphi + \frac{e^{-ikr}}{\sqrt{kr}}\frac{s_m A_m}{1+s_m}$$
$$= \int_0^{2\pi} \Phi_m^* e^{-ikr\cos\varphi}\, d\varphi + \frac{e^{-ikr}}{\sqrt{kr}}\int_0^{2\pi} \Phi\Phi_m^* \, d\varphi. \tag{2.174b}$$

For the eigenpatterns we admit the normalization $\int_0^{2\pi} |\Phi_n|^2 \, d\varphi = 1$. The first terms on the right-hand sides in (2.174a) and (2.174b) can be computed by the stationary phase method. They are equal to

$$\int_0^{2\pi} \Phi_m e^{-ikr\cos\varphi}\, d\varphi$$
$$= \sqrt{\frac{2\pi}{kr}}\left(e^{i(kr-\pi/4)}\Phi_m(\pi) + e^{-i(kr-\pi/4)}\Phi_m(0)\right) + O\left(\frac{1}{kr}\right), \tag{2.175a}$$
$$\int_0^{2\pi} \Phi_m^* e^{-ikr\cos\varphi}\, d\varphi =$$
$$= \sqrt{\frac{2\pi}{kr}}\left(e^{i(kr-\pi/4)}\Phi_m^*(\pi) + e^{-i(kr-\pi/4)}\Phi_m^*(0)\right) + O\left(\frac{1}{kr}\right). \tag{2.175b}$$

Now by matching the coefficients of $\exp(ikr)$ in (2.174a), we obtain

$$A_n = \sqrt{2\pi}e^{-i\pi/4}(1+s_n)\Phi_n(\pi). \tag{2.176}$$

On the other hand, if we match the coefficients of $\exp(-ikr)$ in (2.174b) and take account of (2.140b) and the expression for A_n, then we find that the Fourier coefficient of the series expansion of the pattern Φ in the Φ_n is equal to A_n:

$$\int_0^{2\pi} \Phi\Phi_n^* \, d\varphi = A_n. \tag{2.177}$$

Thus

$$\Phi = \sqrt{2\pi}e^{-i\pi/4}\sum_n (1+s_n)\Phi_n(\pi)\Phi_n(\varphi). \tag{2.178}$$

The last formulas solve the problem considered in this subsection.

2.6 Metallic and Semitransparent Surfaces The Maxwell Equations

The vector generalization of the methods developed in this chapter is not related to any essential difficulties. Just as in scalar problems, the solution is sought in the form of series in the eigenfunctions of some homogeneous problems, whose statements differ from those for the scalar case only in the form of the auxiliary boundary conditions on the surface S. The spectral parameter is still introduced into these auxiliary conditions, and the coefficients in the expansion of the unknown field are determined from the requirement that the actual boundary conditions must hold. These coefficients are of resonance nature. The sums in the representation of the solution are two- rather than three-dimensional (the latter case occurs in the method of eigenfrequencies). This is due to the fact that the eigenfunctions are orthogonal on a surface. Furthermore, the sums do not contain *gradient terms*, since the incident field is always written out separately. The series for **E** and **H** have *the same coefficients.*

In this section we consider diffraction problems for ideal metallic surfaces as well as for surfaces with impedance properties. Moreover, we present possible generalizations to the case of semitransparent surfaces. The surfaces S on which diffraction occurs are assumed to be closed (this restriction can be removed) and not to stretch to infinity.

1. In the diffraction problem for a metallic body, one seeks the solution of the nonhomogeneous Maxwell equations

$$\operatorname{rot}\mathbf{H} - ik\mathbf{E} = \frac{4\pi}{c}\mathbf{j}^{(e)}, \tag{2.179}$$

$$\operatorname{rot}\mathbf{E} + ik\mathbf{H} = -\frac{4\pi}{c}\mathbf{j}^{(m)}, \tag{2.180}$$

under the following boundary conditions on the surface S of the body:

$$E_t|_S = 0, \tag{2.181a}$$

$$E_\tau|_S = 0, \tag{2.181b}$$

where t and τ are mutually orthogonal unit vectors tangent to S and forming a right-handed trihedral with the outward normal to V. In the exterior problem, the fields $(\mathbf{E}, \mathbf{H})$ must additionally satisfy the radiation condition (1.179).

To construct the solution of this diffraction problem, various versions of the generalized method can be used. We start from the *impedance method.* For example, let us construct the solution of the exterior problem. (For the interior problem, the solution is completely similar.) The corresponding homogeneous problem, which determines the system of eigenelements for the expansion of the desired solution, can be stated as follows. We introduce the eigenvalues w_n of the homogeneous problem with the help of the auxiliary impedance boundary conditions

$$(e_{nt} - w_n h_{n\tau})|_S = 0, \tag{2.182a}$$

$$(e_{n\tau} + w_n h_{nt})|_S = 0, \tag{2.182b}$$

which must be satisfied on S. The eigenfunctions must satisfy the homogeneous Maxwell equations

$$\operatorname{rot}\mathbf{h}_n - ik\mathbf{e}_n = 0, \tag{2.183}$$

$$\operatorname{rot}\mathbf{e}_n + ik\mathbf{h}_n = 0 \tag{2.184}$$

in the domain and the radiation conditions at infinity.

The functions $\{\mathbf{e}_n, \mathbf{h}_n\}$ thus defined are orthogonal in the following sense:

$$\int_S (e_{nt}e_{mt} + e_{n\tau}e_{m\tau})\, dS = 0, \quad n \neq m, \tag{2.185a}$$

$$\int_S (h_{nt}h_{mt} + h_{n\tau}h_{m\tau})\, dS = 0, \quad n \neq m. \tag{2.185b}$$

One can readily obtain these conditions by applying the Lorentz lemma

$$\int_S \{[\mathbf{e}_n\mathbf{h}_m]_N - [\mathbf{e}_m\mathbf{h}_n]_N\}\, dS = 0 \tag{2.186}$$

to the exterior domain and by using the boundary conditions (2.182) on S and the conditions at infinity.

We seek the solution of the diffraction problem (2.179)–(2.181) in the form of series in the eigenfunctions of the homogeneous problem (2.182)–(2.184), with the incident field $\{\mathbf{E}^0, \mathbf{H}^0\}$ treated as a separate term. We define the incident field as the field created in free space by the sources $\mathbf{j}^{(e)}$ and $\mathbf{j}^{(m)}$ and do not touch possible generalizations. If we represent the desired electric field in the form

$$\mathbf{E} = \mathbf{E}^0 + \sum_n A_n\mathbf{e}_n, \tag{2.187}$$

then, according to (2.179), (2.180), (2.183), and (2.184), the magnetic field will be represented by the series with the same coefficients,

$$\mathbf{H} = \mathbf{H}^0 + \sum_n A_n\mathbf{h}_n. \tag{2.188}$$

The representations (2.187) and (2.188) satisfy equations (2.179) and (2.180) of the diffraction problem and the radiation conditions term by term for arbitrary values of the coefficients. The coefficients A_n must be found from the requirement that the desired fields should satisfy the boundary conditions (2.181):

$$\left(E_t^0 + \sum_n A_n e_{nt}\right)\Big|_S = 0, \tag{2.189a}$$

$$\left(E_\tau^0 + \sum_n A_n e_{n\tau}\right)\Big|_S = 0, \tag{2.189b}$$

whence, by using (2.182) and the orthogonality conditions (2.185b), we obtain

$$A_n = \frac{\displaystyle\int_S (E_\tau^0 h_{nt} - E_t^0 h_{n\tau})\, dS}{\displaystyle w_n \int_S (h_{nt}^2 + h_{n\tau}^2)\, dS}. \tag{2.190}$$

The eigenfunctions of problem (2.182)–(2.184) also permit one to construct the solution of the diffraction problem for an impedance surface. In this case, conditions (2.181) on S must be replaced by the conditions

$$(E_t - wH_\tau)\,|_S = 0, \tag{2.191a}$$

$$(E_\tau + wH_t)\,|_S = 0. \tag{2.191b}$$

One can readily show that in this case the coefficients A_n in the representations (2.188) and (2.187) are given by

$$A_n = \frac{\int_S [E_t^0 h_{n\tau} - E_\tau^0 h_{nt} - w(H_t^0 h_{nt} + H_\tau^0 h_{n\tau})]\, dS}{(w - w_n)\int_S (h_{nt}^2 + h_{n\tau}^2)\, dS}. \tag{2.192}$$

2. For nonclosed surfaces, the solutions of the above diffraction problems can be constructed differently, by expanding the diffracted field in the eigenfunctions of *two* (rather than one) *auxiliary problems* which differ from each other in the statement of the boundary conditions on S. Specifically, in each of these two homogeneous problems, the spectral parameter is introduced into only one auxiliary boundary condition, whereas the other condition remains the same as in the diffraction problem (all the other requirements imposed on the eigenfunctions remain the same).

For the first homogeneous problem, we impose the conditions

$$(e_{nt} - w_n h_{n\tau})|_S = 0, \tag{2.193a}$$

$$(e_{n\tau} + wh_{nt})|_S = 0, \tag{2.193b}$$

and for the second homogeneous problem, we impose the conditions

$$(\widetilde{e}_{nt} - w\widetilde{h}_{n\tau})|_S = 0, \tag{2.194a}$$

$$(\widetilde{e}_{n\tau} + \widetilde{w}_n \widetilde{h}_{nt})|_S = 0. \tag{2.194b}$$

The eigenvalues of the first problem are the numbers w_n, and the eigenvalues of the second problem are the numbers $\widetilde{w}_n$. Furthermore, the orthogonality conditions for the boundary conditions (2.193) have the form

$$\int_S h_{n\tau} h_{m\tau}\, dS = 0, \quad n \neq m, \tag{2.195}$$

and for the boundary conditions (2.194), the orthogonality conditions read

$$\int_S \widetilde{h}_{nt} \widetilde{h}_{mt}\, dS = 0, \quad n \neq m. \tag{2.196}$$

To obtain (2.195) and (2.196), one must apply the Lorentz lemma to the volume in question and use the boundary conditions together with the radiation conditions.

We represent the desired fields **E** and **H** as the sums of two series

$$\mathbf{E} = \mathbf{E}^0 + \sum_n A_n \mathbf{e}_n + \sum_n B_n \tilde{\mathbf{e}}_n, \tag{2.197}$$

$$\mathbf{H} = \mathbf{H}^0 + \sum_n A_n \mathbf{h}_n + \sum_n B_n \tilde{\mathbf{h}}_n. \tag{2.198}$$

Here again $\{\mathbf{E}^0, \mathbf{H}^0\}$ is the incident field. The coefficients in the series for **E** and **H** are the same. These coefficients can be found from the requirement that the expansions (2.197) and (2.198) must satisfy the boundary conditions (2.191) of the diffraction problem. Obviously, in accordance with the boundary conditions of the auxiliary problems, each of the boundary conditions (2.191) contains the eigenfunctions of only one homogeneous problem. Hence the system of equations for the coefficients splits into two independent systems, and we obtain

$$E_t^0 + \sum_n A_n e_{nt} = w\left(H_\tau^0 + \sum_n A_n h_{n\tau}\right)\Big|_S, \tag{2.199}$$

$$E_\tau^0 + \sum_n B_n \tilde{e}_{n\tau} = -w\left(H_t^0 + \sum_n B_n \tilde{h}_{nt}\right)\Big|_S. \tag{2.200}$$

These formulas, together with the boundary conditions (2.193) and (2.194) and the orthogonality conditions (2.195) and (2.196) permit one to find the coefficients of the expansions:

$$A_n = \frac{1}{w - w_n} \frac{\int_S (E_t^0 - wH_\tau^0) h_{n\tau}\, dS}{\int_S h_{n\tau}^2\, dS}, \tag{2.201}$$

$$B_n = \frac{1}{\tilde{w}_n - w} \frac{\int_S (E_\tau^0 + wH_t^0) \tilde{h}_{nt}\, dS}{\int_S \tilde{h}_{nt}^2\, dS}. \tag{2.202}$$

Obviously, for $w = 0$ the series (2.197) and (2.198) with coefficients (2.201) and (2.202) give the solution of the diffraction problem for a metallic body.

3. Diffraction problems for metallic bodies and semitransparent screens, as well as problems involving dielectric bodies, can be treated by the *method of surface electric current*, which is a generalization to the vector case of the first version of the ρ-method. This method can be described as follows. The unknown field $\{\mathbf{E}, \mathbf{H}\}$ is represented in the form of series in the eigenfunctions of the homogeneous problem in which the transmission boundary conditions

$$(e_{nt}^+ - e_{nt}^-)|_S = 0, \tag{2.203a}$$

$$(e_{n\tau}^+ - e_{n\tau}^-)|_S = 0, \tag{2.203b}$$

$$(\rho_n^{(e)}(h_{nt}^+ - h_{nt}^-) - e_{n\tau})|_S = 0, \tag{2.204a}$$

$$(\rho_n^{(e)}(h_{n\tau}^+ - h_{n\tau}^-) + e_{nt})|_S = 0 \tag{2.204b}$$

must be satisfied on the surface S of the body. The eigenfunctions $\{\mathbf{e}_n, \mathbf{h}_n\}$ are defined in the entire space and must satisfy the homogeneous Maxwell equations (for the problem on a dielectric with the corresponding values of ε and μ) and the radiation conditions. The numbers $\rho_n^{(e)}$ play the role of eigenvalues of the homogeneous problem. In the usual way, one can show that $\operatorname{Re}\rho_n^{(e)} > 0$. (This follows from the energy balance for the homogeneous problem.) The eigenfunctions prove to be orthogonal in the sense of (2.185a). For the problem on diffraction by an infinitely thin screen S with the boundary conditions

$$
\begin{aligned}
(E_t^+ - E_t^-)|_S &= 0, && (2.205a)\\
(E_\tau^+ - E_\tau^-)|_S &= 0, && (2.205b)\\
[\rho^{(e)}(H_t^+ - H_t^-) - E_\tau]|_S &= 0, && (2.206a)\\
[\rho^{(e)}(H_\tau^+ - H_\tau^-) + E_t]|_S &= 0, && (2.206b)
\end{aligned}
$$

the coefficients of the series (2.187) and (2.188) are given by

$$
A_n = \frac{\rho_n^{(e)}}{\rho^{(e)} - \rho_n^{(e)}} \frac{\int_S (E_\tau^0 e_{n\tau} + E_t^0 e_{nt})\, dS}{\int_S (e_{nt}^2 + e_{n\tau}^2)\, dS}. \qquad (2.207)
$$

The value of A_n for a metallic body (not necessarily infinitely thin) can be obtained from (2.207) by setting $\rho^{(e)} = 0$.

For diffraction by a dielectric body, one can also obtain the expressions for the coefficients A_n easily, but the fields $\{\mathbf{E}^0, \mathbf{H}^0\}$ are determined in this case by a slightly more complicated procedure (the same is true in the scalar case).

The vector analog of the second version of the ρ-method is called the *method of surface magnetic current.* It differs from the method of surface electric current only in the form of the boundary conditions of the homogeneous problem: conditions (2.203) and (2.204) on S are replaced by the conditions

$$
\begin{aligned}
(h_{nt}^+ - h_{nt}^-)|_S &= 0, && (2.208a)\\
(h_{n\tau}^+ - h_{n\tau}^-)|_S &= 0, && (2.208b)\\
[\rho_n^{(m)}(e_{nt}^+ - e_{nt}^-) + h_{n\tau}]|_S &= 0, && (2.209a)\\
[\rho_n^{(m)}(e_{n\tau}^+ - e_{n\tau}^-) - h_{nt}]|_S &= 0. && (2.209b)
\end{aligned}
$$

Here the eigenvalues $\rho_n^{(m)}$ again satisfy $\operatorname{Re}\rho_n^{(m)} > 0$. The orthogonality condition has the form (2.185b).

This method allows one to construct solutions of diffraction problems for dielectric bodies and semitransparent screens. Finally, note that, just as in Subsection 2, one can construct solutions of various diffraction problems by expanding the desired field in a series in the eigenfunctions of two (rather than one) homogeneous problems, in each of which the spectral parameter is introduced into only one boundary condition (which is not the same for both problems) and the other conditions remain the same as in the actual problem.

One can generalize the method given in Subsections 2 and 3 to construct solutions of diffraction problems for a nonclosed surface with the more general conditions

$$\begin{aligned} E_t^+ - E_t^- &= \alpha(H_\tau^+ + H_\tau^-), \quad E_t^+ + E_t^- = \beta(H_\tau^+ - H_\tau^-), \\ E_\tau^+ - E_\tau^- &= \widetilde{\alpha}(H_t^+ + H_t^-), \quad E_\tau^+ + E_\tau^- = \widetilde{\beta}(H_t^+ - H_t^-). \end{aligned}$$

To this end, one must consider four homogeneous problems, in each of which one of the coefficients $\alpha, \ldots$ is replaced by a spectral parameter. The solution is sought in the form of the sum of four series with the same coefficients for **E** and **H**, etc. We do not present the corresponding formulas here.

In this section the generalized method of eigenoscillations was applied to diffraction problems for bodies (dielectric or metallic) with interfaces and for screens of general type (impedance, semitransparent, etc.). The problems for the eigenfunctions contain the spectral parameter in the boundary conditions, and these eigenfunctions are orthogonal on the boundary. The conditions at infinity are also treated as boundary conditions. In problems for high-Q resonators, the nonresonant background can be summed efficiently by choosing U^0 to be the solution of the auxiliary diffraction problem for a simpler body.

Chapter 3

Variational Technique

To solve homogeneous problems arising in the generalized method of eigenoscillations, we need to work out a universal technique suitable for all versions of the generalized method of eigenoscillations used for studying resonant systems of arbitrary geometry. Such a technique may be developed on the variational background. To this end, we must state homogeneous problems in variational terms so that this statement will be completely equivalent to that in differential terms. More precisely, to each homogeneous problem we must assign a functional such that all stationary points of this functional (and only these points) are eigenfunctions of the original problem. Typically, most homogeneous problems arising in the generalized method of eigenoscillations contain the spectral parameter in the boundary conditions. Hence it is necessary to develop a special technique for constructing functionals such that these conditions do not constrain the class of admissible functions for these functionals[1] (that is, these conditions are natural for these functionals). If this is the case, then it is possible to enlarge this class so that all boundary conditions of the problem become natural, that is, solve the problem of constructing *complete functionals* [1].

In this chapter we describe the corresponding technique and use it to construct functionals for all versions of the generalized method of eigenoscillations. For simplicity, in all special cases, we deal with only one of the two implications comprising the equivalence; namely, we prove that the variational statement of the problem follows from the differential statement. The problem of complete equivalence will be studied in Section 3.5 for a multiparameter problem that contains a majority of versions of the generalized method of eigenoscillations as special cases. For the other problems (the s-method and vector problems), the proof of the complete equivalence is similar to that given in Section 3.5; therefore, we do not repeat this proof.

The variational technique is mainly used for calculating the eigenvalues of homo-

[1] Since the existing terminology varies, we agree that the widest set of functions on which a functional preserves its stationary properties will be called the *class of admissible functions*. Sometimes we refer to some functions of this class as *trial functions* to emphasize the fact that these functions are close to an eigenfunction and that the straightforward substitution of these functions into the functional allows one to calculate the desired eigenvalue with sufficient accuracy. The functions used as a basis in the class of admissible functions will be called the *coordinate functions*.

geneous problems. By using the property of a functional to be stationary, one can substitute a function close to its eigenfunction (with an error of the order of δ) and calculate the eigenvalue with accuracy $O(\delta^2)$. If the eigenfunction is not known (even approximately), then we must use direct numerical (Ritz type) methods for finding the stationary points of the functional. The proof of convergence of these methods for some classes of functionals that are not positive-definite (the more so for complex-valued functionals) is a rather difficult mathematical problem. In Section 3.7 we consider some theoretical aspects of this problem. In Section 4.1 we present some model examples that illustrate the convergence of the method.

3.1 Basic Principles of Construction of Stationary Functionals for Generalized Eigenvalue Problems

In this section we discuss general methods for constructing functionals that are stationary on the eigenfunctions of homogeneous boundary value problems. The main attention is paid to modified functionals for which given boundary conditions are natural.[2]. The construction of functionals with such properties is vital for developing a variational technique for homogeneous problems with a spectral parameter in the boundary conditions. Indeed, if the parameter value (an eigenvalue) in not known in advance, then we cannot subject the trial functions to boundary conditions containing this parameter. Even if a boundary condition does not contain a spectral parameter, it is also expedient to make this condition natural; this permits one to use a single system of coordinate functions in numerical calculations independently of the form of the domain in question. For coordinate functions, it may even be difficult to satisfy the continuity conditions for a function and its normal derivative on the interface between two media; for instance, such a situation occurs in solving exterior problems posed in the entire space if each coordinate function must be bounded both inside and outside some domain and satisfy the radiation condition.

To modify functionals means to equip them with some integrals over the boundary of the domain in question. The possibility of this operation was already indicated in [3]: "by adding the corresponding... boundary integrals to a given integral, we obtain a possibility to change the natural boundary conditions of the problem significantly..." This possibility was realized in some special problems of electromagnetic theory in [35].

Historically, the first method to construct functionals with the desired properties for homogeneous problems in the generalized method of eigenoscillations was the method of indeterminate coefficients. In what follows, we propose and justify a more sophisticated method, which involves a purposeful search for the form of the additional surface integrals and is less laborious than the method of indeterminate coefficients. The the-

[2]Following [3], we say that boundary conditions are natural if the admissible functions need not satisfy these conditions for the functional to be stationary.

oretical backgrounds of the approach proposed are treated in Subsection 1 in a rather general form. Then we discuss some specific features arising in applications of this method to homogeneous problems of the generalized method of eigenoscillations, in particular, to exterior problems with a radiation condition.

1. First, we suggest a constructive proof of the possibility to obtain stationary functionals for which all boundary conditions (or some of these conditions) are natural. Here we restrict our consideration to scalar problems, since the generalization to the vector case is obvious.

Let us consider a linear boundary value problem comprised by an equation

$$A[u] = 0 \tag{3.1}$$

specified in a domain V and some boundary conditions

$$l_i(u)|_S = 0, \qquad i = 1, 2, \ldots, m. \tag{3.2}$$

Consider the inner product [3]

$$(u_1, u_2) = \int_V u_1 u_2^* \, dV. \tag{3.3}$$

Then we can introduce the adjoint problem

$$A^*[\overline{u}] = 0, \tag{3.4}$$

$$l_i^*(\overline{u})|_S = 0, \qquad i = 1, 2, \ldots, m. \tag{3.5}$$

The fact that these two problems are adjoints of each other just means that the difference

$$I_S(u, \overline{u}) = (A[u], \overline{u}) - (u, A^*[\overline{u}]) \tag{3.6}$$

is zero for any functions u and $\overline{u}$ satisfying the boundary conditions (3.2) and (3.5), respectively. In the sequel, we need the following representation of I_S for the case in which u and $\overline{u}$ do not satisfy these conditions:

$$I_S(u, \overline{u}) = \int_S \sum_{i=1}^{m} [\hat{l}_i(u) p_i^*(\overline{u}) + \hat{l}_i^{\,*}(\overline{u}) p_i(u)] \, dS. \tag{3.7}$$

Here $\hat{l}_i$, p_i, $\hat{l}_i^{\,*}$, and p_i^* are some linear differential expressions such that $\hat{l}_i$ and $\hat{l}_i^{\,*}$ either coincide with l_i and l_i^*, respectively, or can be derived from these expressions. More precisely, we have $\hat{l}_i(u)|_S = 0$ or $\hat{l}_i^{\,*}(\overline{u})|_S = 0$ provided that all conditions (3.2) or (3.5) are satisfied. In each special case, these expressions can readily be found by integrating the first term in (3.6) by parts and by carrying out some elementary

[3] The asterisk on a function denotes complex conjugation, while the asterisk on an operator denotes the adjoint operator with respect to the inner product (3.3).

transformations. As a rule, these transformations are not unique and there exist various representations (3.7).

Usually (e.g., see [35]), the functional

$$L_V(u, \overline{u}) = (A[u], \overline{u}) \tag{3.8}$$

is assigned to the homogeneous problem (3.1), (3.2); this is a stationary functional defined on the solution of this problem in the class of admissible functions satisfying conditions (3.2). The fact that the functional is stationary means that if u and $\overline{u}$ are solutions of problems (3.1), (3.2) and (3.4)), (3.5), respectively, then the first variation of the functional $L_V(u, \overline{u})$, that is, the principal linear part

$$\delta L_V = (A[u], \delta\overline{u}) + (A[\delta u], \overline{u}) \tag{3.9}$$

of its increment on the functions $u+\delta u$ and $\overline{u}+\delta\overline{u}$ vanishes if δu satisfies conditions (3.2). It follows from Eq. (3.1) that the first term in (3.9) vanishes for arbitrary $\delta\overline{u}$. By definition (3.6), the second term is equal to $I_S(\delta u, \overline{u})$, and we have (see (3.7))

$$\delta L_V = \int_S \sum_{i=1}^{m} \hat{l}_i(\delta u) p_i^*(\delta\overline{u})\, dS. \tag{3.10}$$

It follows from the above-mentioned properties of $\hat{l}_i$ that it suffices to have $l_i(\delta u)|_S = 0$ for the integrand in (3.10) to vanish.

Our aim is to add a surface integral to the functional (3.8) so that δu need not satisfy conditions (3.2) for the functional to be stationary. Obviously, the first variation of this surface integral must be equal to $-\delta L_V$. The integral

$$L_S(u, \overline{u}) = -\int_S \sum_{i=1}^{m} \hat{l}_i(u) p_i^*(\overline{u})\, dS \tag{3.11}$$

has this property. Indeed, taking into account (3.10), we obtain

$$\delta L_S = -\delta L_V - \int_S \sum_{i=1}^{m} \hat{l}_i(u) p_i^*(\delta\overline{u})\, dS, \tag{3.12}$$

where the second term is zero (for any $\delta\overline{u}$), since $\hat{l}_i(u)|_S = 0$.

Thus we have constructed the functional

$$L(u, \overline{u}) = L_V(u, \overline{u}) + L_S(u, \overline{u}), \tag{3.13}$$

which is stationary on the solutions of the homogeneous problem. All boundary conditions are natural for this functional. If admissible functions are subjected to some of the boundary conditions, then the additional term (3.11) can be simplified by using these conditions, so that fewer expressions of the form $\hat{l}_i(u)$ will occur in it.

The additional term L_S, which ensures that the boundary conditions are natural, is not unique. For instance, it can contain terms of the form

$$\int_S \hat{l}_i(u) \hat{l}_j^*(\overline{u})\, dS$$

(with arbitrary coefficients) whose first variation is zero. The nonuniqueness follows from the fact that the representation (3.7) itself is not unique. This allows one to assign a family of functionals with the same properties of being stationary to each homogeneous problem. In what follows, we shall use this ambiguity in the construction of the variational technique for homogeneous problems of the generalized method of eigenoscillations.

2. In the argument in the preceding section, it is important that one can vary the solution of the adjoint problem (that is, choose the increment $\delta\overline{u}$) arbitrarily; that is, one need not impose any conditions on $\delta\overline{u}$ for the functionals (3.8) and (3.13) to be stationary. Since the relation between u and $\overline{u}$ (that is, between the solutions of the original and adjoint problems) can usually be written out explicitly, it is convenient to choose the increments of these functions with regard to this relation rather than independently. In what follows, we assume that only δu is arbitrary and express $\delta\overline{u}$ via δu.

The homogeneous problems arising in the generalized method of eigenoscillations and studied in this chapter possess the property

$$\overline{u} = u^*; \tag{3.14}$$

that is, the solution of the adjoint problem is the complex conjugate of u. It follows from the definition of the inner product (3.3) that the functional L_V depends only on u. With regard to the special form of the operator A,

$$Au \equiv \triangle u + k^2 \varepsilon u, \tag{3.15}$$

we can rewrite this functional as

$$L_V(u) = \int_V (\triangle u + k^2 \varepsilon u) u \, dV. \tag{3.16}$$

In practice (when dealing with interior problems), it is convenient to transform this functional by using Green's formula to the symmetric form

$$L_V(u) = \int_V [k^2 \varepsilon u^2 - (\nabla u)^2] \, dV + \int_S u \frac{\partial u}{\partial N} \, dS, \tag{3.17}$$

where N is the outward normal to V. Some specific difficulties arising in the application of this transformation to exterior homogeneous problems of the generalized method of eigenoscillations will be studied in Subsection 3.

Here we indicate another obvious property of functionals of the form (3.13). According to formulas (3.8) and (3.11), these functionals vanish at their stationary points (that is, on the eigenfunctions of the problem in question). If the spectral parameter λ in the equation or in the boundary conditions of the problem occurs linearly in $L(u, \overline{u})$, then we can formally equate $L(u, \overline{u})$ with zero, solve the resulting relation for λ, and thus obtain a stationary functional for the eigenvalues in the form of a ratio of two bilinear (quadratic) functionals. This procedure is justified by the fact that the class of admissible functions for the original functional (3.13) is not constrained by any conditions containing λ. By applying the method of Lagrange multipliers (with λ being the indeterminate multiplier) to the functional thus obtained, we return to (3.13).

3. In the exterior problems in all versions of the generalized method of eigenoscillations (except for the s-method), the eigenfunctions must satisfy the radiation condition. This condition is also necessary for the existence of the integral in (3.17), and hence we cannot make this condition natural. Since we assume that the frequency k is given and the bodies and the boundaries in question all lie in a finite part of space, it is relatively easily to impose the radiation condition on the trial functions. However, in this case one cannot use a functional like (3.17) directly. Instead, one must consider the limit

$$L_V(u) = \lim_{R\to\infty} \left\{ \int_{V_R} [k^2 \varepsilon u^2 - (\nabla u)^2], dV + \int_{S_R} u \frac{\partial u}{\partial N}\, dS \right\}, \tag{3.18}$$

where V_R is the finite part of V bounded by the sphere of radius R and S_R is the surface of this sphere.

One can simplify the calculation of this limit as follows. The volume integral contains only one term that does not vanish as $R \to \infty$. This term is proportional to e^{-2ikR} and is compensated for by the integral over S_R, which ensures the existence of the limit. But this term disappears if k is a complex number with $k'' = \operatorname{Im} k < 0$. Thus, instead of introducing the second integral into (3.18), we can assume that k is complex and then pass to the limit as $k'' \to -0$.

Note that we need not actually perform any calculations with complex k; it suffices to ignore all integrals over the sphere at infinity arising in the integration by parts.

In the following, we shall not write the integral in the form (3.18). We formally extend the volume V^- to infinity and assume that the integral is calculated by the above method.

3.2 Spectral Parameter in the Equation

In this section, by using the ε-method (Chapter 1), we construct functionals that are stationary on the eigenfunctions of homogeneous problems. For the first polarization, these problems contain a spectral parameter only in the equation. Therefore, in general one can impose the boundary conditions on trial functions. Nevertheless, we shall try to make the boundary conditions as natural as possible so as to have more freedom in choosing the coordinate functions in numerical calculations. Needless to say, we can always simplify the final form of the functional if we find some boundary conditions that can be satisfied by these functions.

We shall consider the case of the first polarization in detail and thus illustrate the general scheme for constructing stationary functionals. For other statements of homogeneous problems of the ε-method, we shall only write out the final results in the form of a table.

1. Let us consider the homogeneous problem of the ε-method which appears in the study of a closed resonator with metallic walls and with a dielectric body inside the resonator. It is required to find an ε for which there exists a nontrivial solution of the equation

$$\begin{aligned} (\triangle u + k^2 \varepsilon u) &= 0 \quad \text{in } V^+, && (3.19a)\\ (\triangle u + k^2 u) &= 0 \quad \text{in } V^- && (3.19b) \end{aligned}$$

such that both u and its normal derivative are continuous on the boundary S_ε of the dielectric, that is,

$$\begin{aligned} \left(u^+ - u^-\right)\Big|_{S_\varepsilon} &= 0, && (3.20a)\\ \left(\frac{\partial u^+}{\partial N} - \frac{\partial u^-}{\partial N}\right)\Bigg|_{S_\varepsilon} &= 0, && (3.20b) \end{aligned}$$

and the condition

$$u|_S = 0 \tag{3.21}$$

is satisfied on the boundary S of the resonator. In this problem the permittivity ε is a spectral parameter, and the frequency k is assumed to be given.[4]

This is a selfadjoint problem. According to Section 3.1, the functional that is stationary on the solution of this problem (we require that the boundary conditions must be satisfied) has the form

$$\begin{aligned} L_V(u) &= \int_{V^+} (\triangle u + k^2 \varepsilon u) u \, dV + \int_{V^-} (\triangle u + k^2 u) u \, dV \\ &= k^2 \varepsilon \int_{V^+} u^2 dV + k^2 \int_{V^-} u^2 dV - \int_V (\nabla u)^2 dV \qquad (3.22)\\ &\qquad (V = V^+ + V^-). \end{aligned}$$

The first variation of this functional on a function $u + \delta u$, where u satisfies all conditions of the problem, has the form

$$\delta L_V(u) = \int_{S_\varepsilon} (\delta u^+ - \delta u^-) \frac{\partial u^+}{\partial N} \, dS - \int_S \delta u \frac{\partial u}{\partial N} \, dS. \tag{3.23}$$

The normal in this formula is the outward normal to the domain V^-. It follows from (3.20b) that $\partial u^+ / \partial N$ in the first integral can be replaced by $\partial u^- / \partial N$ or by any linear combination of these derivatives in which the sum of coefficients is equal to one.

It follows from (3.23) that for δL_V to vanish, δu must satisfy conditions (3.20a) and (3.21); condition (3.20b) turns out to be natural. By using the form (3.23) of the

[4]In this chapter we consider only homogeneous problems and thus omit the subscript n on the solutions. This is justified by the fact that, in some cases, the statements of problems and the functionals corresponding to these problems are the same for different versions of the generalized method of eigenoscillations and do not depend on the way of introducing the spectral parameter.

first variation, one can readily write out the terms complementary to L_V and ensure that all other conditions also become natural. The final form of the functional is

$$L(u) = L_V(u) - 2\int_{S_\varepsilon}(u^+ - u^-)\frac{\partial u^+}{\partial N}\,dS + 2\int_S u\frac{\partial u}{\partial N}\,dS. \tag{3.24}$$

We have already pointed out that, by equating the right-hand side of (3.24) with zero and by solving this relation for ε, we obtain the following functional in the form of the ratio of two quadratic expressions:

$$\mathcal{E}(u) = \left\{2\int_{S_\varepsilon}(u^+ - u^-)\frac{\partial u^+}{\partial N}\,dS - 2\int_S u\frac{\partial u}{\partial N}\,dS\right\}\Big/\left\{k^2\int_{V+} u^2 dV\right\}. \tag{3.25}$$

The stationary values of this functional coincide with the eigenvalues ε_n of the corresponding homogeneous problem.

The functional (3.24) can also be used for solving the homogeneous problem arising in the method of eigenfrequencies. In this case, ε is a given number and k is the desired eigenvalue. Accordingly, the Ritz method applied to (3.24) in both cases yields the same transcendental equation $D(k, \varepsilon) = 0$, where ε is the desired variable in one case and k is the desired variable in the other case. If the actual value of permittivity of the body is complex, then we obtain a complex equation in the method of eigenfrequencies and a real equation in the ε-method. All versions of the generalized method of eigenoscillations are characterized by the fact that the functionals obtained for interior problems can be used for calculating the eigenfrequencies of the corresponding regions. In what follows, we do not pay special attention to this fact.

2. If the resonator walls are not ideally metallic, then condition (3.21) for the eigenfunctions is replaced by the impedance boundary condition

$$\left(u + w\frac{\partial u}{\partial N}\right)\bigg|_S = 0, \tag{3.26}$$

where the impedance w may be complex and need not be constant on S. Obviously, the functional (3.24) cannot be used for solving problems with such boundary conditions, and the last term in this functional must be changed. First, note that as we pass from (3.15) to (3.16), the integral over S does not disappear; it is retained in the form

$$-\int_S w\left(\frac{\partial u}{\partial N}\right)^2 dS. \tag{3.27}$$

By adding the first variation of this term to the second term in (3.23), we obtain

$$-2\int_S\left[\delta u + w\frac{\partial(\delta u)}{\partial N}\right]\frac{\partial u}{\partial N}\,dS. \tag{3.28}$$

This expression can be compensated for by the variation of some integral over S, which can readily be constructed. By combining this integral with (3.27), we obtain

$$L_S = \int_S\left[2u\frac{\partial u}{\partial N} + w\left(\frac{\partial u}{\partial N}\right)^2\right] dS. \tag{3.29}$$

It is this integral that must replace the last term in (3.24) for problems with condition (3.26). By construction, this condition is natural for the functional thus obtained. It follows from (3.29) that we can pass to the limit as $w \to 0$ in this functional, whence we again arrive at the functional (3.24).

Obviously, we can add the surface integral of the squared left-hand side of (3.26) with an arbitrary weight to the integral (3.29). In particular, in this manner we can construct a functional that depends linearly on $1/w$ rather than w. We shall write out this functional in the next section. Here we only note that we can pass to the limit as $w \to \infty$ in this functional; this procedure transforms (3.26) into the condition $\partial u/\partial N = 0$. This boundary condition describes a metallic wall in the case of the second polarization, and therefore, this functional is less convenient for problems of the first polarization.

In exterior homogeneous problems of the ε-method, which appear in the study of diffraction on a dielectric body in vacuum, one must replace condition (3.21) or (3.26) for the eigenfunctions by the radiation condition. The functionals for such problems can be obtained from the functionals constructed above by omitting the last term and by extending the integral over V^- to infinity. The admissible functions must satisfy the radiation conditions, and the integral over V^- must be understood in the sense described in the end of the preceding section.

In contrast with interior problems, an integral of this type makes sense for exterior problems only in the ε-method and cannot be used in the method of eigenfrequencies. This is due to the fact that the eigenfunctions of the eigenfrequency method increase at infinity and fail to have expressions of the form (3.18). Essentially, this is one of the most important advantages of all versions of the generalized method of eigenoscillations, where the frequency in exterior homogeneous problems remains real and thus there is a possibility of constructing the same variational technique for interior and exterior problems.

3. The functionals for problems with variable $\varepsilon(\mathbf{r})$, as well as for problems describing the case of the second polarization, can be constructed according to the same scheme. Some new considerations occur only in the case of the second polarization with a piecewise continuous ε. In this problem, it is necessary to define a suitable inner product on which the original functional $L_V(u)$ depends. The corresponding result can be obtained if we first study the case of a continuous $\varepsilon(\mathbf{r})$ and then pass to the limit. However, it is much easier to write out a linear combination of the expressions (3.3) for the domains V^+ and V^- and then find the coefficients of these expressions from the condition that the problem is selfadjoint. We shall not perform all these computations but only write out the final results for various statements of problems of the ε-method in the form of a table (see Table 3.1). All boundary conditions are natural for the functionals in this table.

Table 3.1

Homogeneous Problem	Functional
$\Delta u + k^2\varepsilon u = 0$ in V^+, $\Delta u + k^2 u = 0$ in V^-, $u^+ = u^-$ on S_ε, $\dfrac{\partial u^+}{\partial N} = \dfrac{\partial u^-}{\partial N}$ on S_ε	$L(u) = k^2\varepsilon \int_{V+} u^2 dV + k^2 \int_{V-} u^2 dV - \int_{V^+ + V^-} (\nabla u)^2 dV - 2\int_{S_\varepsilon} (u^+ - u^-) \dfrac{\partial u^+}{\partial N} dS$
$\Delta u + k^2\{1 + \lambda[\varepsilon(\mathbf{r}) - 1]\}u = 0$ in V	$L(u) = \lambda k^2 \int_V (\varepsilon - 1) u^2 dV + \int_V [k^2u^2 - (\nabla u)^2] dV$
$\Delta u + k^2\varepsilon u = 0$ in V^+ , $\Delta u + k^2 u = 0$ in V^- , $u^+ = u^-$ on S_ε, $\dfrac{1}{\varepsilon}\dfrac{\partial u^+}{\partial N} = \dfrac{\partial u^-}{\partial N}$ on S_ε	$L(u) = \varepsilon \left\{ \int_{V-} [k^2u^2 - (\nabla u)^2] dV + k^2 \int_{V+} u^2 dV \right\} - \int_{V+} (\nabla u)^2 dV - 2\int_{S_\varepsilon} (u^+ - u^-) \dfrac{\partial u^+}{\partial N} dS$
$\nabla \left\{ \left[1 + \lambda \left(\dfrac{1}{\varepsilon(\mathbf{r})} - 1 \right) \right] \nabla u \right\} + k^2 u = 0$ in V	$L(u) = \lambda \int_V u \left[\left(\dfrac{1}{\varepsilon} - 1 \right) \nabla u \right] dV + \int_V [k^2u^2 - (\nabla u)^2] dV$

3.3 Spectral Parameter in Boundary Conditions

In this section we construct stationary functionals for homogeneous problems posed in Chapter 2. All these problems contain the spectral parameter in the boundary conditions. It is important for the construction of stationary functionals that these conditions must be natural; It is only in this case that we can choose trial functions to be independent of the desired eigenvalues.

For each homogeneous problem, we can construct a family of functionals that are stationary on the eigenfunctions of this problem; any two of these functionals differ from each other by the surface integral of a quadratic form of the left-hand sides of the boundary conditions with arbitrary weight coefficients. We are chiefly interested in the functionals from this family which contain the spectral parameter only in the first or minus first power. By equating this functional with zero and by solving the resulting equation for the spectral parameter, we obtain a unique stationary expression for the eigenvalues in the form of the ratio of two quadratic functionals.

1. Let us consider the homogeneous problem of the w-method (Section 2.1)

$$\Delta u + k^2 u = 0 \quad \text{in } V, \tag{3.30}$$

$$\left(u + w\frac{\partial u}{\partial N}\right)\bigg|_{S_w} = 0 \tag{3.31}$$

with spectral parameter w. If V is an open domain, then u must additionally satisfy the radiation condition.

In the preceding section we already studied problems with condition (3.31), where w was a given constant (or function). The additional term that makes this condition natural has the form (3.28). By setting $\varepsilon = 1$ in accordance with (3.30), we rewrite this functional in the complete form with this term as

$$L_1(u) = \int_V [k^2u^2 - (\nabla u)^2]\, dV + 2\int_{S_w} u\frac{\partial u}{\partial N}\, dS + w\int_{S_w}\left(\frac{\partial u}{\partial N}\right)^2 dS. \tag{3.32}$$

This functional satisfies all conditions imposed on functionals of this type. Namely, it is stationary on the eigenfunctions of the homogeneous problem (3.30), (3.31), and the boundary condition (3.31) is natural for it. The functional (3.32) can also be used in exterior problems of the w-method; in this case, admissible functions must satisfy the radiation condition and the integral over the infinite volume must be understood in the sense explained in the end of Section 3.1.

The functional $L_1(u)$ is zero at its stationary points; this allows us to write out the following functional for the eigenvalues:

$$W_1(u) = \frac{\displaystyle\int_V [(\nabla u)^2 - k^2u^2]\, dV - 2\int_{S_w} u\frac{\partial u}{\partial N}\, dS}{\displaystyle\int_{S_w}\left(\frac{\partial u}{\partial N}\right)^2 dS}. \tag{3.33}$$

One can also write out a functional different from (3.32) for the homogeneous problem (3.30), (3.31). To this end, it suffices to subtract the integral

$$\frac{1}{w}\int_{S_w}\left(u + w\frac{\partial u}{\partial N}\right)^2 dS, \tag{3.34}$$

whose first variation is obviously zero, from $L_1(u)$. As a result, we obtain the following functional, already known in the literature [15]:

$$L_2(u) = \int_V [k^2u^2 - (\nabla u)^2]\, dV - \frac{1}{w}\int_{S_w} u^2\, dS. \tag{3.35}$$

Here the spectral parameter occurs in the minus first power, which allows us, just as before, to equate the right-hand side of the functional with zero, solve the resulting expression obtained for w, and thus obtain another functional for the eigenvalues.

It is obvious how one must choose the coefficient of the integral in (3.34). Speciically, this method must eliminate the w-dependent term from (3.32). We could obtain the functional (3.35) by straightforward calculations if in the constructions of the preceding section, instead of (3.27), we used the integral

$$-\int_{S_w} u^2/w\, dS,$$

which coincides with (3.27) on the eigenfunction. This procedure is obviously equivalent to the subtraction of (3.34) from the final result.

The functionals (3.32) and (3.35) can also be used for solving the interior problem (3.30), (3.31) treated as a homogeneous problem of finding the eigenfrequencies k_n. In this case, as was already pointed out in the preceding section, we can pass to the limit as $w \to 0$ in the functional (3.32) and, accordingly, as $w \to \infty$ in the functional (3.35).

Consequently, in the w-method it is more convenient to use the first functional for calculating small eigenvalues (say, in the cases of the first polarization in diffraction problems with a metallic boundary S_w) and the second functional for calculating large eigenvalues w_n (in the case of the second polarization for metals). Otherwise, the denominators in the functionals for w_n will be small, which spoils the relative accuracy of the calculations. Similar difficulties arise when we use these functionals for calculating stationary points in the Ritz method.

For homogeneous problems of the w-method, which arise in problems of diffraction on bodies with variable surface impedance, we have to modify the functionals (3.32) and (3.35) slightly. Namely, we must include the coefficient of the last term in these functionals in the integrand and replace w by the representation of the variable eigenimpedance via the actual surface impedance. One of such functionals is given in Table 3.2.

2. Let us consider the homogeneous problem of the ρ-method

$$\Delta u + k^2 u = 0 \quad \text{in } V, \tag{3.36}$$

$$\left(u^+ - u^-\right)\Big|_{S_\rho} = 0, \tag{3.37a}$$

$$\left(\frac{\partial u^+}{\partial N} - \frac{\partial u^-}{\partial N} - \frac{1}{\rho}u\right)\Big|_{S_\rho} = 0. \tag{3.37b}$$

For simplicity, we consider only the exterior problem. In the case of a closed resonator, by using the above terms, we can take account of the condition on the exterior boundary S.

Let us rewrite the functional L_V with regard to the boundary conditions (3.37) in the form

$$L_V(u) = \int_V [k^2u^2 - (\nabla u)^2]\, dV - \frac{1}{\rho}\int_{S_\rho} (u^+)^2 dS \tag{3.38}$$

Table 3.2

Homogeneous Problem	Functionals
$\triangle u + k^2 u = 0$ in V, $u + w\frac{\partial u}{\partial N} = 0$ on S_w	$L_1(u) = \int_V [k^2u^2 - (\nabla u)^2]dV + 2\int_{S_w} u\frac{\partial u}{\partial N}dS + w\int_{S_w}\left(\frac{\partial u}{\partial N}\right)^2 dS$
	$L_2(u) = \int_V [k^2u^2 - (\nabla u)^2]dV - \frac{1}{w}\int_{S_w} u^2 dS$
$\triangle u + k^2 u = 0$ in V, $u + \lambda w(s)\frac{\partial u}{\partial N} = 0$ on S_w	$L(u) = \int_V [k^2u^2 - (\nabla u)^2]dV + 2\int_{S_w} u\frac{\partial u}{\partial N}dS + \lambda\int_{S_w} w\left(\frac{\partial u}{\partial N}\right)^2 dS$
$\triangle u + k^2 u = 0$ in $V^+ + V^-$, $u^+ = u^-$ on S_ρ, $\frac{\partial u^+}{\partial N} - \frac{\partial u^-}{\partial N} - \frac{1}{\rho}u = 0$ on S_ρ	$L_1(u) = \int_V [k^2u^2 - (\nabla u)^2]dV - 2\int_{S_\rho}(u^+ - u^-)\frac{\partial u}{\partial N}^- dS - \frac{1}{\rho}\int_{S_\rho}(u^+)^2 dS$
	$L_2(u) = \int_V [k^2u^2 - (\nabla u)^2]dV - 2\int_{S_\rho}\left(u^+\frac{\partial u^+}{\partial N} - u^-\frac{\partial u^-}{\partial N}\right)dS + \rho\int_{S_\rho}\left(\frac{\partial u^+}{\partial N} - \frac{\partial u^-}{\partial N}\right)^2 dS$
$\triangle u + k^2 u = 0$ in $V^+ + V^-$, $\frac{\partial u^+}{\partial N} = \frac{\partial u^-}{\partial N}$ on S_ρ, $u^+ - u^- + \frac{1}{\rho}\frac{\partial u}{\partial N} = 0$ on S_ρ	$L_1(u) = \int_V [k^2u^2 - (\nabla u)^2]dV - 2\int_{S_\rho}(u^+ - u^-)\frac{\partial u^+}{\partial N}dS - \frac{1}{\rho}\int_{S_\rho}\left(\frac{\partial u^+}{\partial N}\right)^2 dS$
	$L_2(u) = \int_V [k^2u^2 - (\nabla u)^2]dV + \rho\int_{S_\rho}(u^+ - u^-)^2 dS$
$\triangle u + k^2 u = 0$ in $V^+ + V^-$, $u^+ - u^- = \alpha\left(\frac{\partial u^+}{\partial N} + \frac{\partial u^-}{\partial N}\right)$ on S, $u^+ - u^- = \beta\left(\frac{\partial u^+}{\partial N} - \frac{\partial u^-}{\partial N}\right)$ on S	$L_1(u) = \int_V [k^2u^2 - (\nabla u)^2]dV - 2\int_S\left(u^+\frac{\partial u^+}{\partial N} - u^-\frac{\partial u^-}{\partial N}\right)dS + \frac{\alpha}{2}\int_S\left(\frac{\partial u^+}{\partial N} + \frac{\partial u^-}{\partial N}\right)^2 dS + \frac{\beta}{2}\int_S\left(\frac{\partial u^+}{\partial N} - \frac{\partial u^-}{\partial N}\right)^2 dS$
	$L_2(u) = \int_V [k^2u^2 - (\nabla u)^2]dV - \frac{1}{2\alpha}\int_S(u^+ - u^-)^2 dS - \frac{1}{2\beta}\int_S(u^+ + u^-)^2 dS$

and calculate the first variation of this functional on the eigenfunction as follows:

$$\begin{aligned}\delta L_V &= 2\int_{S_\rho}\left[\delta u^+\left(\frac{\partial u^+}{\partial N}-\frac{1}{\rho}u^+\right)-\delta u^-\frac{\partial u^-}{\partial N}\right]dS \\ &= 2\int_{S_\rho}(\delta u^+-\delta u^-)\frac{\partial u^-}{\partial N}dS.\end{aligned} \tag{3.39}$$

This expression is zero provided that condition (3.37a) is satisfied by the trial functions; the second boundary condition turns is natural with regard to the special choice of the form of the last term in (3.38). Condition (3.37a) is natural (this follows from (3.39)) by virtue of the same additional term as in the ε-method (see (3.23)). Finally, the functional for the homogeneous problem of the ρ-method can be rewritten as

$$L_1(u) = \int_V [k^2u^2-(\nabla u)^2]\,dV - 2\int_{S_\rho}(u^+-u^-)\frac{\partial u^-}{\partial N}\,dS - \frac{1}{\rho}\int_{S_\rho}(u^+)^2 dS. \tag{3.40}$$

The admissible functions for this functional can be discontinuous on S_ρ; at least, the normal derivatives of these functions can be discontinuous, since the desired eigenfunctions possess the same property.

The form of the last term in (3.40) shows that this functional is of little use in seeking small eigenvalues. To find a functional convenient in this case, it suffices to add the term

$$\rho\int_{S_\rho}\left(\frac{\partial u^+}{\partial N}-\frac{\partial u^-}{\partial N}-\frac{1}{\rho}u^+\right)^2 dS \tag{3.41}$$

to (3.40). As a result, we obtain

$$\begin{aligned}L_2(u) &= \int_V [k^2u^2-(\nabla u)^2]\,dV - 2\int_{S_\rho}\left(u^+\frac{\partial u^+}{\partial N}-u^-\frac{\partial u^-}{\partial N}\right)dS \\ &\quad + \rho\int_{S_\rho}\left(\frac{\partial u^+}{\partial N}-\frac{\partial u^-}{\partial N}\right)^2 dS.\end{aligned} \tag{3.42}$$

By analogy with the w-method, for problems with variable transparency $\rho(S)$ of the surface, in the functionals obtained, one must replace ρ by the eigentransparency expressed via the actual transparency and move the function thus obtained to the integrand.

If we study systems with nonclosed semitransparent boundaries, then it is convenient to close this boundary by imposing the continuity conditions on u and $\partial u/\partial N$ on a complement S_1 of the boundary. After these auxiliary conditions have been made natural with the help of an additional integral over S_1 (see the preceding section), we can choose trial functions in V^+ and V^- independently.

It is obvious how to construct functionals for other versions of homogeneous problems of the ρ-method. The final results, as well as the functionals for homogeneous problems with transmission conditions of the general form, are given in Table 3.2.

3.4 Spectral Parameter in Asymptotic Conditions at Infinity

In this section, following the above scheme, we construct functionals for homogeneous problems arising in the s-method (Section 2.5). In these problems, the spectral parameter is contained in the asymptotic condition at infinity, which relates the radiation patterns of the convergent and divergent waves. Therefore, it is impossible to impose this condition on admissible functions. On the other hand, it is also impossible to do this in a natural way (as was the case with the radiation condition), since if this asymptotics becomes weaker, then the integrals over an infinite region will diverge. A linear substitution allows us to transfer the spectral parameter s to an auxiliary condition that relates the values of the eigenpatterns in opposite angular directions. As soon as the above-mentioned asymptotic condition does not contain the spectral parameter s, we can impose this condition on the admissible functions.

The functionals obtained by this procedure are real. It is most convenient to represent them in a form containing a real parameter λ uniquely determined by the complex spectral parameter s.

1. Let us start from the two-dimensional problem of the s-method which arises in solving the diffraction problem for a dielectric body with variable permittivity (Subsection 2.5.2). In this problem, the eigenfunctions must satisfy the equation

$$\triangle u + k^2 \varepsilon u = 0 \tag{3.43}$$

with a given frequency k and with a continuous real $\varepsilon(\mathbf{r})$ equal to 1 outside a finite domain. Let us replace the pattern $\Phi(\varphi)$, considered earlier, by a new function $\Psi(\varphi)$ according to the formula

$$\Phi(\varphi) = \frac{1+s}{s}\,\Psi(\varphi). \tag{3.44}$$

Then the asymptotic conditions (2.140) become

$$u|_{r\to\infty} \simeq \frac{1}{\sqrt{kr}}\left[\Psi^*(\varphi)\mathrm{e}^{ikr} + \Psi(\varphi)\mathrm{e}^{-ikr}\right], \tag{3.45}$$

$$\Psi(\pi+\varphi) + is\Psi^*(\varphi) = 0. \tag{3.46}$$

In this representation, the first condition does not contain the spectral parameter and can be imposed on the admissible functions, whereas the function Ψ remains arbitrary. Now we need to take into account the pattern of the eigenfunction, and it is not convenient to use the representation of the functional in the limit form (3.17). We shall rewrite the functional in the form (3.16) as

$$L_V(u) = \int_V (\triangle u + k^2 \varepsilon u) u \, dV. \tag{3.47}$$

One can readily verify that if a function u satisfies condition (3.45), then the divergent terms in (3.47) cancel, so that this integral exists.

Let us reduce the first variation δL_V by Green's formula to the following integral over the infinitely remote circle:

$$\delta L_V = \int_{s_\infty} \left[\frac{\partial(\delta u)}{\partial r} u - \delta u \frac{\partial u}{\partial r}\right] ds. \tag{3.48}$$

It follows from condition (3.45) for u and δu that

$$\delta L_V = 2i \int_0^{2\pi} [\delta\Psi^*(\varphi)\Psi(\varphi) - \delta\Psi(\varphi)\Psi^*(\varphi)]\, d\varphi, \tag{3.49}$$

where $\delta\Psi$ denotes the variation of Ψ corresponding to δu. Finally, by using condition (3.46) for the eigenpattern Ψ, we obtain

$$\delta L_V = 2s^* \int_0^{2\pi} [\delta\Psi(\pi + \varphi) + is\, \delta\Psi^*(\varphi)]\Psi(\varphi)\, d\varphi. \tag{3.50}$$

Obviously, for δL_V to vanish for arbitrary Ψ, $\delta\Psi$ must also satisfy (3.46). To make this condition natural, we supplement the functional L_V with an integral with respect to the angle. The form of this integral readily follows from (3.50). As a result, we obtain

$$L_1(u) = L_V(u) - 2\int_0^{2\pi} [s^*\Psi(\pi + \varphi) + i\Psi^*(\varphi)]\Psi(\varphi) d\varphi. \tag{3.51}$$

Now the admissible functions must satisfy only the asymptotic condition (3.45), that is, be the real parts of arbitrary functions that satisfy the radiation condition.

The functional (3.51) itself is complex-valued, and its imaginary part is equal to

$$-i\int_0^{2\pi} |\Psi(\pi + \varphi) + is\Psi^*(\varphi)|^2\, d\varphi. \tag{3.52}$$

Here the integrand is the squared absolute value of the left-hand side of condition (3.46). Therefore, one can subtract this integral from the functional without violating its stationary properties. We can further modify this functional so that it will contain the real number $\lambda = i(1 - s)/(1 + s)$ instead of the complex number s (see Section 2.5). This can be achieved by using $i + \lambda$ instead of i as the coefficient of the integral (3.52) subtracted from (3.51). Finally, the functional acquires the form

$$\begin{aligned} L(u) &= \int_V (\Delta u + k^2\varepsilon u)u\, dV \\ &\quad + 2\,\mathrm{Re} \int_0^{2\pi} [\lambda\Psi^*(\varphi) + (i\lambda - 1)\Psi(\pi + \varphi)]\Psi(\varphi)\, d\varphi. \end{aligned} \tag{3.53}$$

Since the spectral parameter λ occurs in (3.53) linearly, we obtain the following functional for the eigenvalues λ_n by equating $L(u)$ with zero and by solving the resulting relation for λ:

$$\Lambda(u) = \frac{2\,\mathrm{Re} \int_0^{2\pi} \Psi(\varphi)\Psi(\pi + \varphi)\, d\varphi - \int_V (\Delta u + k^2\varepsilon u)u\, dV}{2\int_0^{2\pi} \left\{|\Psi(\varphi)|^2 - \mathrm{Im}[\Psi(\varphi)\Psi(\pi + \varphi)]\right\} d\varphi}. \tag{3.54}$$

When studying systems with semitransparent boundaries by the s-method, one can make the boundary conditions of the form (3.37) natural in the standard way, just as in the preceding section.

2. In the three-dimensional case, all arguments remain the same. Hence we present only the assumptions and the final result. Once the function Ψ is introduced by formula (3.44), the asymptotic conditions at infinity acquire the form

$$u|_{r\to\infty} \simeq \frac{1}{kr}\left[\Psi^*(\varphi,\vartheta)\mathrm{e}^{ikr} + \Psi(\varphi,\vartheta)\mathrm{e}^{-ikr}\right], \tag{3.55}$$

$$\Psi(\pi+\varphi,\pi-\vartheta) - s\Psi^*(\varphi,\vartheta) = 0. \tag{3.56}$$

The functional that is stationary on the eigenfunctions has the following form provided that conditions (3.55) are satisfied:

$$\begin{aligned} L(u) \;=\; & \int_V (\Delta u + k^2\varepsilon u) u dV \\ & + \frac{2}{k}\,\mathrm{Re}\int_0^{2\pi}\int_0^{\pi}[\lambda\Psi^*(\varphi,\vartheta) + (\lambda+i)\Psi(\pi+\varphi,\pi-\vartheta)]\Psi(\varphi,\vartheta) \\ & \times \sin\vartheta\, d\vartheta d\varphi. \end{aligned} \tag{3.57}$$

Just as in the two-dimensional case, this readily implies a functional of the form (3.54) for the eigenvalues.

In conclusion, we point out that it is obviously possible to represent the asymptotics of admissible functions in the form

$$u|_{r\to\infty} \simeq \frac{1}{kr}\left[C_1(\varphi,\vartheta)\cos kr + C_2(\varphi,\vartheta)\sin kr\right], \tag{3.58}$$

where C_1 and C_2 are arbitrary real functions related to the pattern Ψ by the formula

$$\Psi = \frac{1}{2}(C_1 + iC_2). \tag{3.59}$$

3.5 Multiparameter Problems. Restricted Classes of Admissible Functions

Homogeneous problems arising in the generalized method of eigenoscillations contain at least two parameters, namely, the frequency k and a parameter chosen as the spectral parameter. In some cases already considered, there were even three parameters, for instance, k, ε, and w in the problem considered in in Subsection 3.2.1. This example shows that an increase in the number of parameters does not make the construction of the variational technique more complicated. Furthermore, the functionals already known for the multiparameter problem turn out to be universal in the sense that their stationary properties are independent of which of the parameters plays the role of the spectral parameter. Such functionals can be used for solving homogeneous problems arising in several versions of the generalized method of eigenoscillations.

In this section we consider a scalar homogeneous problem containing several parameters, each of which can equally be chosen to be the spectral parameter. For this problem, we write out a functional that yields the basic functionals constructed in this chapter (except for the functionals of the s-method) as special cases. We shall prove the direct and converse theorems stating that this universal functional is stationary on the eigenfunctions of the homogeneous problem. This functional can be constructed by the same standard technique that has already been demonstrated in this chapter many times, and therefore, we omit the corresponding calculations.

The general statement of the problem treated in this section does not cover the s-method, since the asymptotic conditions at infinity used in this statement are alternative to the radiation conditions used in the majority of versions of the generalized method of eigenoscillations. The proof of the converse theorem for functionals of the s-method is similar to that presented in this section and is therefore omitted.

In Subsections 3 and 4 we consider several pecularities that arise in variation of the functional in a restricted class of admissible functions. Namely, we consider the case in which these functions satisfy the equation in the entire domain or in some part of this domain. In this case, the functional may have stationary points that do not coincide with any eigenfunctions of the homogeneous problem in question. If admissible functions satisfy the equation only in a part of the domain, then we can eliminate such stationary points by modifying the functional with the use of transformations that have already been discussed. In the class of functions that satisfy the equation in the entire region, such transformations are of no use; all functionals of the class considered have additional stationary points. In this case, it is expedient to use the variational technique based on the fact that the mean-square discrepancy in the boundary conditions is stationary (the least squares method).

1. Suppose that the following equation is satisfied in some domain V, not necessarily closed:

$$\triangle u + k^2[1 + \lambda(\varepsilon - 1)]u = 0, \tag{3.60}$$

where ε is, in general, a piecewise continuous function of the coordinates. The statement of the problem for a constant or piecewise constant ε follows from (3.60) as a special case. Next, we assume that there are several surfaces that are boundaries of V or lie inside V and that the following boundary conditions are satisfied on these surfaces:

$$u + w\frac{\partial u}{\partial N} = 0 \tag{3.61}$$

on a closed surface S_w (here we consider the outward normal to V);

$$u^+ - u^- \;=\; 0, \tag{3.62a}$$

$$\frac{\partial u^+}{\partial N} - \frac{\partial u^-}{\partial N} - \frac{1}{\rho}u \;=\; 0 \tag{3.62b}$$

on a surface S_ρ;

$$\frac{\partial u^+}{\partial N} - \frac{\partial u^-}{\partial N} \;=\; 0, \tag{3.63a}$$

$$u^+ - u^- + \frac{1}{\tilde{\rho}}\frac{\partial u}{\partial N} = 0 \qquad (3.63b)$$

on a surface $S_{\tilde{\rho}}$. The surfaces S_ρ and $S_{\tilde{\rho}}$ may be nonclosed. In exterior problems, the function u must additionally satisfy the radiation condition.

There may be several surfaces of each type, and the parameter in the boundary conditions may take different values on each of these surfaces. For instance, the class of surfaces S_ρ includes surfaces on which ε is discontinuous but u and $\partial u/\partial N$ must be continuous ($\rho = \infty$). In our further considerations, it does not matter which parameter in the problem is spectral. The only exception is the frequency k.

Let us assign the functional

$$\begin{aligned} L(u) = & \int_V \{k^2[1+\lambda(\varepsilon-1)]u^2 - (\nabla u)^2\}\, dV + \int_{S_w}\left[2u\frac{\partial u}{\partial N} + w\left(\frac{\partial u}{\partial N}\right)^2\right] dS \\ & - \int_{S_\rho}\left[\frac{1}{4\rho}(u^+ + u^-)^2 + (u^+ - u^-)\left(\frac{\partial u^+}{\partial N} + \frac{\partial u^-}{\partial N}\right)\right] dS \\ & + \tilde{\rho}\int_{S_{\tilde{\rho}}}(u^+ - u^-)^2 dS \end{aligned} \qquad (3.64)$$

to the above homogeneous problem. The admissible functions of this functional must be bounded everywhere, the functions u and $\partial u/\partial N$ can have discontinuities of any type on the boundaries S_ρ and $S_{\tilde{\rho}}$, and the radiation conditions must be satisfied at infinity if the domain V is open.

Let us show that for a function u to be an eigenfunction of the homogeneous problem (3.60)–(3.63), it is necessary and sufficient that the functional be stationary on u in the class of admissible functions described above.

First, we prove the necessity. Suppose that u satisfies all conditions of the homogeneous problem. Let us verify that in this case the functional (3.64) is stationary on this function. To this end, we consider a variation δu of the function u and write out the corresponding first variation of the functional:

$$\begin{aligned} \delta L = & \; 2\int_V \{\Delta u + k^2[1+\lambda(\varepsilon-1)]u\}\delta u\, dV + 2\int_{S_w}\frac{\partial(\delta u)}{\partial N}\left(u + w\frac{\partial u}{\partial N}\right) dS \\ & - \int_{S_\rho}\left\{(\delta u^+ + \delta u^-)\left[\frac{1}{2\rho}(u^+ + u^-) - \left(\frac{\partial u^+}{\partial N} - \frac{\partial u^-}{\partial N}\right)\right]\right. \\ & \qquad \left. + \left[\frac{\partial(\delta u^+)}{\partial N} + \frac{\partial(\delta u^-)}{\partial N}\right](u^+ - u^-)\right\} dS \\ & - 2\int_{S_{\tilde{\rho}}}\left\{\delta u^-\left[\tilde{\rho}(u^+ - u^-) + \frac{\partial u^-}{\partial N}\right] - \delta u^+\left[\tilde{\rho}(u^+ - u^-) + \frac{\partial u^+}{\partial N}\right]\right\} dS. \end{aligned} \qquad (3.65)$$

Each of the integrals in this expression vanishes separately in view of the fact that u satisfies the equation and the corresponding boundary conditions. Hence the functional (3.64) is stationary on the eigenfunctions of problem (3.60)–(3.63).

To prove the sufficiency, we assume that the first variation of the functional (3.65) vanishes for some values of the parameters k, λ, w, ρ, and $\tilde{\rho}$ and for an arbitrary δu from the class of admissible functions. We shall show that in this case u is a solution of problem (3.60)–(3.63) with the same parameter values.

Since δu is arbitrary, we first choose it so that all surface integrals in (3.65) disappear. To this end, it suffices to require that δu vanishes on both sides of the surface $S_{\tilde{\rho}}$, its normal derivative is zero on S_w, and the values of δu, as well as $\dfrac{\partial(\delta u)}{\partial N}$, on the opposite sides of S_ρ differ only in sign. Then only the integral over the domain V remains in (3.65). Although δu satisfies the above boundary conditions, it is still possible to choose this function arbitrarily outside the boundaries. Therefore, it readily follows from the condition $\delta L_V = 0$ that the expression in braces in the integral over V is identically zero, that is, u satisfies Eq. (3.60).

Now we shall successively weaken the conditions imposed on δu in the following way: first, we allow $\dfrac{\partial(\delta u)}{\partial N}$ to be nonzero on S_w, then we allow $\dfrac{\partial(\delta u^+)}{\partial N} + \dfrac{\partial(\delta u^-)}{\partial N}$ and $\delta u^+ + \delta u^-$ to be nonzero on S_ρ and δu^- to be nonzero on $S_{\tilde{\rho}}$, and finally, we allow δu^+ to be nonzero on $S_{\tilde{\rho}}$. Then it follows from the condition $\delta L = 0$ and the fact that a nonzero function can be chosen arbitrarily on the corresponding surface that u satisfies all boundary conditions of the homogeneous problem.

It follows from the preceding that the only distinction for infinite domains is that all functions in question must satisfy the radiation condition and the integral over V is understood in the sense discussed in the end of Section 3.1.

2. The functional (3.64) possesses the same properties as similar functionals studied in the preceding sections. In particular, by equating (3.64) with zero and by solving this equation for one of the parameters, we obtain a functional for the corresponding eigenvalues in the form of the ratio of two quadratic functionals. Just as previously, this procedure is justified by the fact that $L(u)$ vanishes on these eigenfunctions and their increments are not subjected to any conditions containing a parameter that can be regarded as a spectral parameter. By applying the method of Lagrange multipliers to the functional thus obtained, we return to the functional (3.64), whose property of being stationary has already been proved.

So far it has been assumed that admissible functions need not satisfy any boundary conditions. In specific problems one often succeeds in subjecting admissible functions to some conditions that do not contain any spectral parameter. In this case, the functional (3.64) is simplified while preserving the above-established property of being stationary. The question as to whether the admissible functions *a priori* satisfy the equation must be studied separately. This will be done in the next subsection.

Obviously, the functional (3.64) does not exhaust all possibilities arising in the construction of the variational technique for the homogeneous problem (3.60)–(3.63). The functional (3.64) can be supplemented with integrals of quadratic forms of the left-hand sides of the boundary conditions, taken over the corresponding surfaces, with any weight coefficients. In this case, the stationary points of the functional are preserved; only admissible passages to the limit in the boundary conditions can be changed. The functional (3.64) in the form written here admits the following basic passages to the limit: from (3.61) to the condition $u = 0$ (as $w \to 0$), from (3.63) to $\partial u/\partial N = 0$ (as $\tilde{\rho} \to 0$), and from (3.62) to the condition that u and $\partial u/\partial N$ are continuous (as $\rho \to \infty$).

3. In practice, it is often convenient to restrict the consideration to admissible functions that satisfy the equation of the homogeneous problem in some part V_1 of the domain V. This is possible if, say, V_1 is a domain filled with a homogeneous ($\varepsilon = \text{const}$) dielectric with permittivity ε given in the homogeneous problem, in particular, with $\varepsilon \equiv 1$. In this case, we can readily write out the general solution of the Helmholtz equation in V_1 at a given frequency. This general solution can be used as the definition of the class of admissible functions. However, in this case, the assumptions used to prove the second part (sufficiency) of the statement of Subsection 1 can be violated. Namely, there are some frequencies k at which the increment δu in (3.65) cannot take arbitrary values at the boundary. In the restricted class of admissible functions, the condition that the functional is stationary need not be sufficient for the corresponding function u to be an eigenfunction of the problem in question. In other words, the functional acquires additional stationary points that are not related to the desired eigenvalues and eigenfunctions.

It follows from well-known theorems on the existence of solutions to boundary value problems that at the eigenfrequencies there appear some restrictions on the behavior of a function at the boundary and, in this case, some orthogonality conditions must necessarily be satisfied for given boundary values of this function (in our case, δu) and the corresponding eigenfunction of the problem of the eigenfrequency method.

We shall illustrate this fact by a special example of the homogeneous problem of the ε-method. In Section 3.2, the functional (3.24), which is stationary on the eigenfunctions of this problem, was considered for the case in which the domain considered is closed and the eigenfunction must vanish on the exterior boundary S. The conditions on both boundaries (S and S_ε) are natural for this functional. Nevertheless, for simplicity, we subject the admissible functions to the boundary condition $u|_S = 0$.

If we consider only the functions that satisfy Eq. (3.19b) in V^- and apply Green's formula to (3.24) in V^-, then the integral over this domain disappears and the functional acquires the form

$$L(u) = \int_{V+} [k^2 \varepsilon u^2 - (\nabla u)^2]\, dV + \int_{S_\varepsilon} \left[u^- \frac{\partial u^-}{\partial N} + 2 \frac{\partial u^+}{\partial N} (u^+ - u^-) \right] dS. \tag{3.66}$$

Taking into account the fact that δu also satisfies Eq. (3.19b) and the above boundary condition on S, we obtain

$$\int_{S_\varepsilon} \frac{\partial(\delta u^-)}{\partial N} u^-\, dS = \int_{S_\varepsilon} \delta u^- \frac{\partial u^-}{\partial N}\, dS. \tag{3.67}$$

Thus we can write the first variation of L in the form

$$\delta L = 2 \int_{V+} (\Delta u + k^2 \varepsilon u) \delta u\, dV - 2 \int_{S_\varepsilon} \left[\frac{\partial(\delta u^+)}{\partial N} (u^+ - u^-) \right. \tag{3.68}$$

$$\left. - \; \delta u^- \left(\frac{\partial u^+}{\partial N} - \frac{\partial u^-}{\partial N} \right) \right] dS. \tag{3.69}$$

Just as previously, δL vanishes on the eigenfunctions of the problem. However, now we cannot claim that each function u for which $\delta L = 0$ is a solution of the homogeneous problem. As to the domain V^+, the choice of the function δu is rather arbitrary in this domain, so that we can separately equate the integral over V^+ and the first term in the integral over S_ε in (3.69) with zero and thus verify that u satisfies the equation of the problem in V^+ and the continuity condition on S_ε (the equation in V^- and the condition on S are satisfied by virtue of the choice of admissible functions). It remains to satisfy the condition that $\partial u/\partial N$ is continuous on S_ε. To this end, in view of the form of the last term in the integral over $S_\varepsilon$in (3.69), δu^- must be arbitrary on S_ε. However, if k is an eigenfrequency and v is the corresponding eigenfunction of the homogeneous problem of the eigenfrequency method in the domain V^- with the boundary condition $v|_{S+S_\varepsilon} = 0$, then the freedom in the choice of the function δu^- in (3.69) is constrained by the condition

$$\int_{S_\varepsilon} \delta u^- \frac{\partial v}{\partial N}\, dS = 0. \tag{3.70}$$

Hence (3.69) vanishes at such frequencies not only if the derivatives are continuous on S_ε but also if

$$\left(\frac{\partial u^+}{\partial N} - \frac{\partial u^-}{\partial N}\right)\Bigg|_{S_\varepsilon} = \frac{\partial v}{\partial N}\Bigg|_{S_\varepsilon}. \tag{3.71}$$

Such functions are additional stationary points of the functions considered. These points can correspond to arbitrary values of ε. In particular, for any ε the first variation vanishes on the function

$$u = \begin{cases} 0 & \text{in } V^+, \\ v & \text{in } V^-. \end{cases} \tag{3.72}$$

One can readily see that additional stationary points disappear if the functional is supplemented with the term

$$B \int_{S_\varepsilon} \left(\frac{\partial u^+}{\partial N} - \frac{\partial u^-}{\partial N}\right)^2 dS \tag{3.73}$$

with an arbitrary coefficient B. Indeed, in this case the last term in (3.69) becomes

$$2 \int_{S_\varepsilon} \left\{\delta u^- + B\left[\frac{\partial(\delta u^+)}{\partial N} - \frac{\partial(\delta u^-)}{\partial N}\right]\right\} \left(\frac{\partial u^+}{\partial N} - \frac{\partial u^-}{\partial N}\right) dS, \tag{3.74}$$

and now the expression in square brackets can be made arbitrary, since the values of $\partial(\delta u^+)/\partial N$ can be chosen arbitrarily. Hence the first variation vanishes only if $\partial u/\partial N$ is continuous on S_ε.

It should be noted that additional stationary point can appear not only at some discrete values of frequency (as in the above example). For instance, if a spectral parameter λ occurs in one of the two transmission conditions $l_1(\lambda, u^+, u^-)|_{S_\lambda} = 0$ (just as in the ρ-method), then the term in the first variation ensuring the validity of the second transmission condition $l_2(u^+, u^-)|_{S_\lambda} = 0$ can be of the form

$$\int_{S_\lambda} \hat{l}(\lambda, \delta u^-) l_2(u^+, u^-)\, dS \tag{3.75}$$

with some $\hat{l}(\lambda, \delta u^-)$. Obviously, $\hat{l}(\lambda, \delta u^-)|_{S_\lambda}$ cannot be made arbitrary for k and λ such that there exists a nontrivial solution of the problem in V^- with the boundary condition $\hat{l}(\lambda, u^-)|_{S_\lambda} = 0$. However, this problem for any given k can generate a rich spectrum of eigenvalues $\lambda = \hat{\lambda}_n$, which are in no way related to the eigenvalues of the original problem. To these $\hat{\lambda}_n$ there correspond additional stationary points of the functional. It is important that in this case one can also modify the functional in a trivial way, by making the first term in (3.75) contain the value of u on the other side of the boundary S_λ, where there are no restrictions on the admissible functions.

4. A special situation arises if the admissible functions satisfy the equation in the entire domain. Then, in the case of closed domains, all functionals constructed previously have additional stationary points and it is impossible to eliminate them by any additional surface integrals. For instance, in the example of the preceding subsection, it is impossible to choose $\delta u^-|_{S_\varepsilon}$ in the variation (3.69) arbitrarily for some values of frequencies and to choose $\partial(\delta u^+)/\partial N|_{S_\varepsilon}$ arbitrarily for other values of frequencies. Moreover, there are some restrictions on $\partial(\delta u^+)/\partial N$ for any k and the corresponding values of $\hat{\varepsilon}_n$, which are eigenvalues of the problem of the ε-method stated only for the domain V^+ with the boundary condition $\partial u^+/\partial N|_{S_\varepsilon} = 0$. No matter how we try to change δu-dependent factors in δL with the help of additional surface integrals, one can always pose a problem of the ε-method (with the corresponding boundary conditions on S_ε) so that these restrictions remain in the problem.

In other words, for any k, for all modifications of the functional $L(u)$, there exists a set of values $\hat{\varepsilon}_n$ (distinct from the desired eigenvalues ε_n of the problem in question) and of the corresponding functions on which the functional is stationary. For admissible functions satisfying the equation in the entire domain V, it is necessary to construct special functionals. In classical selfadjoint problems of diffraction theory (the eigenfrequency method), such functionals are well known in the literature, namely, they are integrals of the sum of squared left-hand sides of the boundary conditions (possibly, with weight coefficients). This approach is usually called the *least squares method* [49]; some versions of this method are well known under the names of the partial region method [36], the Treftz method [32], etc.

3.6 Vector Problems. Bodies of Revolution

In the study of three-dimensional electromagnetic systems with the help of the generalized method of eigenoscillations we encounter homogeneous vector problems stated in Sections 1.6 and 2.6. For these problems, one can readily construct a variational technique by using the method developed and illustrated in detail in the preceding sections

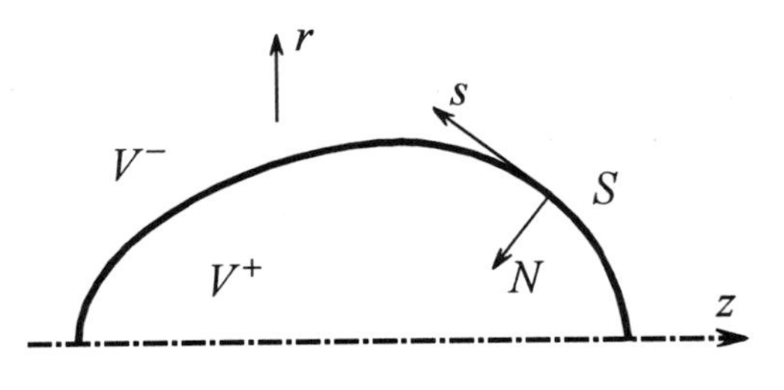

Figure 3.1

of this chapter. Here we do not discuss the whole construction but only present the final results in Table 3.3. Note that a similar variational technique for the ε-method was used in [29].

In this section, we study the practically important case in which the body in question is axisymmetric, that is, is a body of revolution. Generally speaking, one can study diffraction on bodies of revolution by using the general variational technique for vector problems (this technique has already been discussed). However, just as in the method of eigenfrequencies, we can state homogeneous problems of the generalized method of eigenoscillations for bodies of revolution in terms of two scalar functions (potentials) related, say to the azimuth components of the field. If we substitute the expressions for all three components of the vector **E** (or **H**) in terms of these potentials into the functionals of Table 3.3, then we do not obtain the desired simplification, since in this case a decrease in the number of unknown functions leads to an increase in the order of the derivatives in the integrand. In the literature, stationary functionals are known for the interior homogeneous problem of the method of eigenoscillations in the two-potential statement [12]. In what follows, we shall construct similar functionals for several homogeneous problems of the generalized method of eigenoscillations.

1. We start from the homogeneous problem of the w-method, which comprises homogeneous equations of the form (3.63b) and the boundary conditions (3.69). This problem can be stated both for a closed domain (V^+ in Fig. 3.1) and for an open domain (V^-). To study both cases simultaneously, we use double symbols like "$\pm$."

We consider the potential functions $u(r,z)$ and $v(r,z)$ related to the φ-components of the electromagnetic field by the formulas

$$E_\varphi(r,z,\varphi) = \frac{1}{r}\,u(r,z)e^{im\varphi}, \qquad H_\varphi(r,z,\varphi) = \frac{1}{r}\,v(r,z)e^{im\varphi}. \tag{3.76}$$

It readily follows from the Maxwell equations that in the domain $V^\pm$ (that is, in the cross-section of the body) the functions u and v must satisfy the system of equations

$$\frac{\partial}{\partial r}\left(\frac{r}{\alpha^2}\frac{\partial u}{\partial r}\right) + \frac{r}{\alpha^2}\frac{\partial^2 u}{\partial z^2} + \frac{u}{r} - \frac{2mkr}{\alpha^4}\frac{\partial v}{\partial z} \;=\; 0, \tag{3.77a}$$

$$\frac{\partial}{\partial r}\left(\frac{r}{\alpha^2}\frac{\partial v}{\partial r}\right) + \frac{r}{\alpha^2}\frac{\partial^2 v}{\partial z^2} + \frac{v}{r} + \frac{2mkr}{\alpha^4}\frac{\partial u}{\partial z} = 0, \tag{3.77b}$$

Table 3.3

Homogeneous Problem	Functionals
$\nabla \times (\nabla \times \mathbf{E}) - k^2\varepsilon\mathbf{E} = 0$ in V^+, $\nabla \times (\nabla \times \mathbf{E}) - k^2\mathbf{E} = 0$ in V^-, $(\mathbf{E}^+ - \mathbf{E}^-) \times \mathbf{N} = 0$ on S_ε, $(\nabla \times \mathbf{E}^+ - \nabla \times \mathbf{E}^-) \times \mathbf{N} = 0$ on S_ε	$L(\mathbf{E}) = \int_{V^+ + V^-} (\nabla \times \mathbf{E})^2 dV - k^2\varepsilon \int_{V^+} (\mathbf{E})^2 dV - k^2 \int_{V^-} (\mathbf{E})^2 dV - 2\int_{S_\varepsilon} \left[(\nabla \times \mathbf{E}^+) \times (\mathbf{E}^+ - \mathbf{E}^-)\right]_N dS$
$\nabla \times (\nabla \times \mathbf{E}) - k^2\mathbf{E} = 0$ in V, $\left[\left(\mathbf{N} + \frac{w}{k}\nabla\right) \times \mathbf{E}\right] \times \mathbf{N} = 0$ on S_w	$L_1(\mathbf{E}) = \int_V \left[(\nabla \times \mathbf{E})^2 - k^2\mathbf{E}^2\right] dV + \frac{k}{w}\int_{S_w} (\mathbf{E} \times \mathbf{N})^2 dS$
	$L_2(\mathbf{E}) = \int_V \left[(\nabla \times \mathbf{E})^2 - k^2\mathbf{E}^2\right] dV + 2\int_{S_w} [(\nabla \times \mathbf{E}) \times \mathbf{E}]_N \, dS - \frac{w}{k}\int_{S_w} [(\nabla \times \mathbf{E}) \times \mathbf{N}]^2 \, dS$
$\nabla \times (\nabla \times \mathbf{E}) - k^2\mathbf{E} = 0$ in $V^+ + V^-$, $(\mathbf{E}^+ - \mathbf{E}^-) \times \mathbf{N} = 0$ on S_ρ, $\left[\frac{\rho}{k}(\nabla \times \mathbf{E}^+ - \nabla \times \mathbf{E}^-) + \mathbf{E} \times \mathbf{N}\right] \times \mathbf{N} = 0$ on S_ρ	$L_1(\mathbf{E}) = \int_{V^+ + V^-} \left[(\nabla \times \mathbf{E})^2 - k^2\mathbf{E}^2\right] dV - 2\int_{S_\rho} \left[(\nabla \times \mathbf{E}^-) \times (\mathbf{E}^+ - \mathbf{E}^-)\right]_N dS + \frac{k}{\rho}\int_{S_\rho} (\mathbf{E}^+ \times \mathbf{N})^2 dS$
	$L_2(\mathbf{E}) = \int_{V^+ + V^-} \left[(\nabla \times \mathbf{E})^2 - k^2\mathbf{E}^2\right] dV + 2\int_{S_\rho} \left[(\nabla \times \mathbf{E}^-) \times \mathbf{E}^- - (\nabla \times \mathbf{E}^+) \times \mathbf{E}^+\right]_N dS - \frac{\rho}{k}\int_{S_\rho} \left[(\nabla \times \mathbf{E}^+ - \nabla \times \mathbf{E}^-) \times \mathbf{N}\right]^2 dS$

where $\alpha^2 = (kr)^2 - m^2$. We choose orthogonal unit vectors $\mathbf{t}$ and τ in the boundary conditions (3.69) so that they coincide with the directions of φ and $\mp s$, respectively. Then we can rewrite the boundary conditions for u and v on S in the form

$$\alpha^2 u \mp krwp = 0, \qquad \alpha^2 wv \pm krq = 0 \tag{3.78}$$

where

$$p = r\frac{\partial u}{\partial N} + \frac{m}{k}\frac{\partial v}{\partial s}, \qquad q = r\frac{\partial v}{\partial N} - \frac{m}{k}\frac{\partial u}{\partial s}. \tag{3.79}$$

It follows from (3.76) that u and v must vanish on the z-axis:

$$u(0,z) = v(0,z) = 0. \tag{3.80}$$

The following radiation conditions must be satisfied for the exterior problem:

$$\left.\begin{Bmatrix} u \\ v \end{Bmatrix}\right|_{r^2+z^2\to\infty} \simeq \begin{Bmatrix} \Phi_1(\varphi) \\ \Phi_2(\varphi) \end{Bmatrix} e^{-ik\sqrt{r^2+z^2}}. \tag{3.81}$$

The homogeneous problem (3.77)–(3.81) is selfadjoint with respect to the inner product (for exterior problems, up to complex conjugation)

$$\left(\begin{Bmatrix} u_1 \\ v_1 \end{Bmatrix}, \begin{Bmatrix} u_2 \\ v_2 \end{Bmatrix}\right) = \int_{V\pm} (u_1u_2 + v_1v_2)\, dV. \tag{3.82}$$

Therefore, the functional $L_V(u,v)$ can be constructed in the standard way. Specifically, by multiplying the left-hand side of (3.77a) by u and that of (3.77)b by v, by adding the results, and by integrating over $V^{\pm}$, we obtain

$$\begin{aligned} L_V(u,v) \;=\; \int_{V\pm} \Bigg\{ & \left[\frac{\partial}{\partial r}\left(\frac{r}{\alpha^2}\frac{\partial u}{\partial r}\right) + \frac{r}{\alpha^2}\frac{\partial^2 u}{\partial z^2} + \frac{u}{r}\right] u \\ & + \left[\frac{\partial}{\partial r}\left(\frac{r}{\alpha^2}\frac{\partial v}{\partial r}\right) + \frac{r}{\alpha^2}\frac{\partial^2 v}{\partial z^2} + \frac{v}{r}\right] v \\ & + \frac{2mkr}{\alpha^4}\left(\frac{\partial u}{\partial z}v - \frac{\partial v}{\partial z}u\right)\Bigg\}\, dV. \end{aligned} \tag{3.83}$$

We claim that this functional is stationary on the eigenfunctions of the homogeneous problem (3.77)–(3.81) in the class of functions satisfying all required boundary conditions. To verify this fact, we write out the first variation of this functional on the functions $u + \delta u$ and $v + \delta v$. Subjecting u and v to all conditions of the problem and δu and δv to condition (3.80) on the z-axis and the radiation condition (3.81) (for the exterior problem), we obtain

$$\delta L_V = -\int_S \left[(krw\delta p \mp \alpha^2 \delta u)\frac{p}{\alpha^4} + (\alpha^2 w \delta v \pm kr\delta q)\frac{v}{kr\alpha^2}\right] dS, \tag{3.84}$$

where δp and δq are the increments of the functions p and q due to the increments δu and δv. The expressions in parentheses in the integrand in (3.84) vanish for δu and δv satisfying conditions (3.78), since these expressions are proportional to the left-hand sides of these conditions.

By representing δL_V in the form (3.84), we can readily write out an additional term that ensures that the boundary conditions (3.78) are natural. Finally, the functional has the form

$$L(u,v) = L_V(u,v) + \int_S \left[w\left(\frac{1}{kr} v^2 + \frac{kr}{\alpha^4} p^2 \right) \mp \frac{1}{\alpha^2}(up - vq) \right] dS. \qquad (3.85)$$

For this functional to be stationary on the solution of problem (3.77)–(3.81), the admissible functions need only to satisfy conditions (3.80) and the radiation condition (3.81) (for the exterior problems). These conditions ensure the existence of the integral over $V^{\pm}$ in (3.83) and therefore cannot be made natural.

In addition to (3.85), it is possible to construct other functionals with the same properties; these functionals differ from (3.85) by an integral over S of squares or cross-products of the left-hand sides of the boundary conditions with some weight factors. Furthermore, the first term in (3.85) can be rewritten in the symmetric form

$$\begin{aligned} L_V(u,v) \;=\; & \int_{V^{\pm}} \left\{ \frac{1}{r}(u^2 + v^2) - \frac{r}{\alpha^2}[(\nabla u)^2 + (\nabla v)^2] \right. \\ & \left. + \frac{2mkr}{\alpha^4}\left(v\frac{\partial u}{\partial z} - u\frac{\partial v}{\partial z} \right) \right\} dV \\ & \mp \int_S \frac{r}{\alpha^2}\left(u\frac{\partial u}{\partial N} + v\frac{\partial v}{\partial N} \right) dS. \end{aligned} \qquad (3.86)$$

For exterior problems, the integral over V^- in (3.86) has the same meaning as that in the scalar case considered above (namely, as the limit as $k'' \to -0$).

2. In the special case of symmetric oscillations ($m = 0$), there are two types of oscillations, E_{on} and H_{on}, for which $H_\varphi = 0$ or $E_\varphi = 0$, respectively. This follows from the fact that for $m = 0$ the homogeneous problem (3.77)–(3.81) splits into two independent scalar problems for the functions u and v. The first of these problem (for E_{on}-oscillations) has the form

$$\frac{\partial}{\partial r}\left(\frac{1}{r}\frac{\partial u}{\partial r} \right) + \frac{1}{r}\frac{\partial^2 u}{\partial z^2} + \frac{k^2}{r}u \;=\; 0 \quad \text{in } V^{\pm}, \qquad (3.87)$$

$$\left. \left(ku \mp w\frac{\partial u}{\partial N} \right) \right|_S \;=\; 0, \qquad (3.88)$$

$$u(0,z) \;=\; 0. \qquad (3.89)$$

The only distinction of the homogeneous problem for H_{on}-oscillations from (3.87)–(3.89) is that u is replaced by v and w by $1/w$. Therefore, in general, it suffices to solve one of these problems. The solution of the other problem can be obtained from this equivalence.

The functional (3.85) for $m = 0$ is the sum of two terms, one of which depends only on u and the other only on v. Therefore, for problem (3.87)–(3.89) we need to write only the first term. After obvious transformations, the corresponding functional acquires the form

$$\begin{aligned} L_1(u) &= \int_{V\pm} \frac{1}{r}[k^2u^2 - (\nabla u)^2]\, dV \mp 2\int_S \frac{1}{r} u \frac{\partial u}{\partial N} dS \\ &\quad + \frac{w}{k}\int_S \frac{1}{r}\left(\frac{\partial u}{\partial N}\right)^2 dS. \end{aligned} \tag{3.90}$$

The other functional, which depends only on v, is

$$L_2(v) = \int_{V\pm} \frac{1}{r}[k^2v^2 - (\nabla v)^2]\, dV + kw\int_S \frac{1}{r} v^2 dS. \tag{3.91}$$

Taking into account the above-mentioned equivalence of homogeneous problems for E_{0n}- and H_{0n}-oscillations, we see that this functional differs from (3.90) by the integral of the square of the left-hand side of condition (3.88) with weight coefficient $1/(krw)$.

To analyze resonators with finite impedance of walls, one can use any of these two functionals for calculating oscillations of both types. However, if the actual impedance is small (or zero in the case of an ideal metal), that is, one is interested in small (in absolute value) eigenvalues w_n, then it is expedient to use the functional (3.90) only for E_{0n}-oscillations and the functional (3.91) for H_{0n}-oscillations.

3. Now let us consider the homogeneous problem of the ρ-method stated in Subsection 2.6.3 for studying resonators with semitransparent walls. By introducing the potentials u and v according to formula (3.76), we reduce this problem for axially symmetric bodies to system (3.77) (satisfied both in V^+ and in V^-) with the conditions

$$\begin{cases} u^+ - u^- = 0, \\ \dfrac{\partial v^+}{\partial N} - \dfrac{\partial v-}{\partial N} = 0, \\ \alpha^2 u - kr\rho(p^+ - p^-) = 0, \\ \alpha^2\rho(v^+ - v^-) + krq = 0 \end{cases} \tag{3.92}$$

on the boundary S of these domains, condition (3.80) on the z-axis, and condition (3.81) at infinity.

For L_V we must take the same functional (3.83) with the only distinction that the integral in this functional must be extended to $V = V^+ + V^-$. The first variation of this functional on the solution of the problem has the form

$$\begin{aligned} \delta L_V &= \int_S \frac{1}{\alpha^2}\left\{\left[\frac{kr\rho}{\alpha^2}(\delta p^- - \delta p^+) + \delta u^+\right](p^+ - p^-) + (\delta u^+ - \delta u^-)p^-\right. \\ &\quad \left.\left[\frac{\alpha^2\rho}{kr}(\delta v^+ - \delta v^-) + \delta q^-\right](v^+ - v^-) + (\delta q^+ - \delta q^-)v^+\right\} dS. \end{aligned} \tag{3.93}$$

If the admissible functions satisfy conditions (3.92), then $\delta L_V = 0$, since each of the terms (except for the last one) contains the increment of the left-hand side of some boundary condition as a factor; the last term contains a factor that readily vanishes by virtue of the first two conditions in (3.92).

Finally, the functional for which conditions (3.92) are natural has the form

$$\begin{aligned} L(u,v) &= \int_V \left\{ \frac{1}{r}(u^2+v^2) - \frac{r}{\alpha^2}[(\nabla u)^2 + (\nabla v)^2] + \frac{2mkr}{\alpha^2}\left(v\frac{\partial u}{\partial z} - u\frac{\partial v}{\partial z} \right) \right\} dV \\ &\quad + \int_S \frac{1}{\alpha^2} \Big[2(u^- p^- - u^+ p^+) \\ &\qquad + \frac{m}{k}\left(v^+ \frac{\partial u^+}{\partial s} + u^+ \frac{\partial v^+}{\partial s} - v^- \frac{\partial u^-}{\partial s} - u^- \frac{\partial v^-}{\partial s} \right) \Big] dS \\ &\quad + \rho \int_S \left[\frac{kr}{\alpha^4}(p^+ - p^-)^2 + \frac{1}{kr}(v^+ - v^-)^2 \right] dS. \end{aligned} \tag{3.94}$$

Just as in the w-method, the case $m = 0$ is described by two independent homogeneous problems. The first of these problems (for E_{0n}-oscillations) combines Eq. (3.87) in the domains V^+ and V^-, the boundary conditions

$$u^+ - u^- = 0, \qquad \frac{\partial u^+}{\partial N} - \frac{\partial u^-}{\partial N} - \frac{k}{\rho}u = 0 \tag{3.95}$$

on S, and the corresponding conditions on the z-axis and at infinity. The functional that is stationary on the solutions of this problem is one of the two independent terms in the general functional (3.94), namely,

$$\begin{aligned} L_1(u) &= \int_V \frac{1}{r}[k^2u^2 - (\nabla u)^2]\, dV + 2\int_S \frac{1}{r}\left(u^- \frac{\partial u^-}{\partial N} - u^+ \frac{\partial u^+}{\partial N} \right) dS \\ &\quad + \frac{\rho}{k}\int_S \frac{1}{r^2}\left(\frac{\partial u^+}{\partial N} - \frac{\partial u^-}{\partial N} \right)^2 dS. \end{aligned} \tag{3.96}$$

In the second problem, which describes H_{0n}-oscillations, the boundary conditions have the form

$$\frac{\partial v^+}{\partial N} - \frac{\partial v^-}{\partial N} = 0, \qquad v^+ - v^- + \frac{1}{kr}\frac{\partial v}{\partial N} = 0, \tag{3.97}$$

and the functional of this problem is

$$L_2(v) = \int_V \frac{1}{r}[k^2v^2 - (\nabla v)^2]\, dV + \frac{\rho}{k}\int_S \frac{1}{r}(v^+ - v^-)^2 dS. \tag{3.98}$$

Both functionals (3.96) and (3.98), just as the general functional (3.94), are convenient in seeking small or finite eigenvalues and hence can mostly be used for studying resonators with metallic walls ($\rho = 0$ is the actual value of transparency) or with small transparency. Obviously, these functionals can be modified so that they will contain ρ^{-1} and will be convenient in seeking large eigenvalues.

4. As was pointed out in Section 2.6, after u is replaced by v and v by u, the equations of the homogeneous problem remain unchanged, while the boundary conditions (2.203), (2.204) pass into (2.208), (2.209). Thus, for a homogeneous problem with conditions of the last type, one can readily use the above-constructed functional with regard to the change of variable discussed above.

For homogeneous problem of the ε-method in which the permittivity is the spectral parameter, the equations for the potentials introduced by formulas (3.76) are more complicated than those in other versions of the generalized method of eigenoscillations. Namely, now the spectral parameter ε occurs in the equation nonlinearly via the function $\alpha^2 = k^2\varepsilon r^2 - m^2$. The boundary conditions also contain ε. Therefore, the functionals related to such problems are substantially more complicated. However, these functionals can also be obtained by using the same standard technique.

In conclusion, we note that the homogeneous problem of the ε-method for bodies of revolution can be stated in a simpler way (at least for $m \neq 0$) if the potential functions are introduced via the components H_r and H_z (or E_r and E_z) rather than by formulas (3.76).

3.7 Justification of the Ritz Method for Stationary Functionals in the Generalized Method of Eigenoscillations[5]

In the preceding sections of this chapter, we constructed functionals that are stationary on the eigenfunctions of the homogeneous problems arising in the generalized method of eigenoscillations. These functionals, as a rule, are not positive definite but are often complex-valued. It is a difficult mathematical problem to justify that the Ritz method or any other direct methods can be applied to such functionals. In this section we study some aspects of this problem. Namely, we state conditions under which it is guaranteed that the method converges for many of the functionals constructed above. Beside, we consider examples in which the violation of these conditions may be a source of wrong results.

The following two difficulties arise in the study of this problem: first, the functionals treated in the generalized method of eigenoscillations, as a rule, correspond to operators that are not selfadjoint; second, they contain surface integrals of the normal derivative of the desired functions. Furthermore, the spectral parameter occurs in some of these functionals nonlinearly.

In general, the Ritz method does not converge unless there are some additional assumptions on the system of subspaces $\{E_n\}_{n=1}^{\infty}$ generated by the coordinate functions. However, the convergence usually takes place for the natural situation in which a basis of real functions can be chosen in each of the subspaces E_n. In this case one can solve the problem by using the following observation: all functionals constructed above can

[5]This section is written by V. Dikarev and contains his results [4].

be obtained from some symmetric bilinear functionals. This symmetry compensates for the loss of selfadjointness of the operators and helps one to overcome the first of the above-mentioned difficulties. The second difficulty can be eliminated (unfortunately, not always) with the help of a formal method (see Subsection 5).

In this section, we widely use the properties of Sobolev spaces H_s. All required facts of the theory of these spaces can be found in Chapter 5.

1. In what follows, we consider symmetric bilinear functionals that analytically depend on a spectral parameter (denoted by λ) and are defined on linear function spaces. To emphasize the dependence on the spectral parameter, we include this parameter in the list of arguments of the functional. Recall that a functional $L(u, v; \lambda)$ is said to be *symmetric* if for any u, v in the domain D_L we have $L(u, v; \lambda) = L(v, u; \lambda)$ and the form $L(u, v^*)$ is bilinear.

It is well known that to each bounded bilinear functional L defined on an arbitrary Hilbert space H, one can assign a bounded linear operator A defined on the entire H by means of the relation

$$L(u, v^*) = (Au, v)_H, \qquad u, v \in H.$$

(The symbol $(\cdot,\cdot)_H$ stands for the inner product in H.)

Suppose that $L(u, v; \lambda)$ is a bounded symmetric functional. We seek a stationary function u of the form $L(u, u; \lambda)$ by the Ritz method. An approximation u_n to u is sought in the form

$$u_n = \sum_{k=1}^{n} c_k \varphi_k, \tag{3.99}$$

where $\{\varphi_k\}_{k=1}^n$ is a basis in some subspace $E_n \subset H$. The assumption that $E_n = E_n^*$ is very important in what follows. This means that if $f \in E_n$, then we also have $f^* \in E_n$. It is well known that the Ritz method requires that $L(u_n, u_n; \lambda)$ must be stationary for $u_n \in E_n$. By equating the variation $\delta L(u_n, u_n; \lambda)$ with zero, we obtain the equation

$$\delta L(u_n, u_n; \lambda) = L(u_n, \delta u_n; \lambda) + L(\delta u_n, u_n; \lambda) = 0, \tag{3.100}$$

where δu_n is a function from E_n. We set $\delta u_n = \varphi_k$, $k = 1, \ldots, n$, in this relation and take into account the symmetry of the functional $L(u, v; \lambda)$; then we obtain $L(u_n, \varphi_k; \lambda) = 0$. As a result, we have

$$(A(\lambda)u_n, \varphi_k^*) = 0, \qquad k = 1, \ldots, n, \quad u_n \in E_n. \tag{3.101}$$

Here $\{\varphi_k\}_{k=1}^n$ is a basis in E_n, and $A(\lambda)$ is the operator function generated by the functional $L(u, v; \lambda)$. It follows from (3.101) that the approximation u_n can be obtained by solving the equation

$$A(\lambda)u = 0 \tag{3.102}$$

by the Galerkin method. Thus, in the case of symmetric functionals, the above-mentioned scheme allows us to reduce the Ritz method to the Galerkin method. By using the convergence conditions for the Galerkin method [32, 27], we can justify a criterion for the Ritz method to apply to the functionals $L(u, u; \lambda)$. To state this criterion, we need some additional notation and definitions.

Let us consider a sequence $\{E_n\}_{n=1}^{\infty}$ of finite-dimensional subspaces in H. We denote the spectra of Eqs. (3.101) and (3.102) by Ω_n and Ω, respectively. Recall that the set of values of λ for which a homogeneous equation containing λ as a spectral parameter has nontrivial solutions is called the *spectrum of this homogeneous equation.* The *spectrum of a functional* is the spectrum of the operator-function generated by this functional. In the complex plane of λ, we consider an arbitrary finite closed domain K that contains finitely many points from Ω. The spectra Ω_n are said to *converge* to the spectrum Ω if spectral parts $\Omega_n \cap K$ converge, with regard to their multiplicity, to the spectral parts $\Omega \cap K$.

Now we shall state a criterion for the Ritz method to apply to symmetric functionals $L(u, v; \lambda)$. Assume that the functional in question has the form

$$L(u, v; \lambda) = L_0(u, v; \lambda) + L_1(u, v; \lambda), \tag{3.103}$$

where the functionals L_0 and L_1 satisfy the following conditions.

1. $L_0(u, v; \lambda)$ and $L_1(u, v; \lambda)$ are holomorphic in λ in some domain G of the complex plane.

2. The functional L_1 generates a compact operator $A_1(\lambda)$ for any $\lambda \in G$ (such functionals are said to be *compact*).

3. there exists a $\sigma = \sigma(\lambda) > 0$ such that for any $\lambda \in G$ one has

$$|L_0(u, \bar{u}; \lambda)| \geq \sigma \|u\|_H^2 \tag{3.104}$$

for all $u \in H$, where $\| \cdot \|_H$ is the norm in H.

4. There exists a $\lambda_0 \in G$ such that the form $L(u, v; \lambda_0)$ generates an invertible operator $A(\lambda_0)$.

If a functional satisfies conditions 1–4, then its spectrum is discrete in G (see [10, p. 39]), that is, the accumulation points of the spectrum can only lie on the boundary of G.

A sequence of finite-dimensional subspaces $\{E_n\}_{n=1}^{\infty} \subset H$ is said to be *dense in the limit* in H if, for sufficiently large n, any element $f \in H$ can be approximated as close as possible by appropriate elements $u_n \in E_n$. This definition means that $\lim_{n\to\infty} \|u_n - f\|_H = 0$.

Proposition 3.1 *Suppose that the functional $L(u, v; \lambda)$ satisfies conditions 1–4 and $\{E_n\}_{n=1}^{\infty}$ is any sequence dense in the limit in H and such that $E_n = E_n^*$. Then the following assertions hold:*

(i) *for sufficiently large n, the spectra Ω_n are discrete in G;*

(ii) *the spectra Ω_n converge to the spectrum Ω in G.*

Thus if conditions 1–4 are satisfied, then the Ritz method can be used for calculating the spectrum of the functional $L(u, v; \lambda)$. Note that a result close to Proposition 3.1 is contained in [47].

Now we shall indicate a class of functionals such that if we calculate the eigenfunctions of these functionals by the Ritz method, then we may obtain wrong results. Assume that $L(u, v; \lambda)$ is again a symmetric functional, the form $L(u, u^*; \lambda)$ is real for some $\lambda = \lambda_0$, and the domain D_L satisfies the condition $D_L = D_L^*$. We also assume that there exists a Hilbert space H and infinite-dimensional linear manifolds $F_1, F_2 \subset D_L$ such that D_L is everywhere dense in H, the functional $L(u, v; \lambda_0)$ is bounded in H, and $L(u, u^*; \lambda_0) le 0$ or $L(u, u^*; \lambda_0) \geq 0$ if $u \in F_1$ or $u \in F_2$, respectively.

Proposition 3.2 *Suppose that $L(u, v; \lambda_0)$ satisfies the above conditions. Then there exists a dense-in-the-limit sequence of finite-dimensional subspaces $\{E_n\}_{n=1}^{\infty} \subset D_L$ that satisfies the conditions $E_n \subset E_{n+1}$ and $E_n = E_n^*$, $n = 1, 2, \ldots$, such that the equations*

$$(A(\lambda_0)u_n \,, \varphi_k^*) = 0, \qquad k = 1, \ldots, n, \quad n = 1, 2, \ldots,$$

where $\{\varphi_k\}_{k=1}^{n}$ is a basis in E_n, have nontrivial solutions for all n.

Thus, λ_0 is a limit point of the spectra Ω_n of the approximate equations (3.101). However, it may happen that λ_0 does not belong to the spectrum of the functional $L(u, v; \lambda)$.

In the sequel, to verify condition 3, we shall use the following criterion for the functional L_1 to be compact. Let $H_q(V)$ and $H_s(V)$ be the Sobolev spaces of functions defined on a bounded domain V, and let $q < s$. If there exists a constant C such that

$$|L_1(u, v)| < C\|u\|_q \|v\|_s \tag{3.105}$$

for all $u, v \in H_s(V)$, then $L(u, v)$ is compact in $H_s(V)$ (here $\|\cdot\|_s$ and $\|\cdot\|_q$ are the norms in the spaces $H_s(V)$ and $H_q(V)$, respectively).

2. The above results allow us to justify the applicability of the Ritz method to some of our functionals. As a simple example, let us consider the functional that arises when the ε-method is used in the homogeneous problem with losses in the walls (see Subsection 3.2.2). In our notation, this functional can be represented in the form

$$L(u, v; \varepsilon) = \int_V \nabla u \nabla v \, dV - k^2 \int_{V-} uv \, dV - k^2 \varepsilon \int_{V+} uv \, dV + \int_S \frac{uv}{w} \, dS. \tag{3.106}$$

Here V is a bounded domain in R^3, $V^+ \subset V$, $V^- = V \setminus V^+$, S is the boundary of V, and w is, generally speaking, a complex-valued function. We set $H = H_1(V)$ and

$$\begin{aligned} L_0 &= \int_V (\nabla u \nabla v + uv) \, dV, \\ L_1 &= -\int_V uv \, dV - k^2 \int_{V-} uv \, dV - k^2 \varepsilon \int_{V+} uv \, dV + \int_S \frac{uv}{w} \, dS. \end{aligned} \tag{3.107}$$

Obviously, $L_0(u, v)$ is a bounded form. Moreover, $L_0(u, u^*) = \|u\|_H^2$, that is, we have (3.104). Let us verify that the form L_1 is compact. For example, let us verify that the last term in (3.107) is compact. Using the properties of Sobolev spaces, we obtain

$$\begin{aligned} \left| \int \frac{uv}{w} dS \right| &\leq \min_{s \in S} |w^{-1}(s)| \|u\|_{H_0(S)} \|v\|_{H_0(S)} \\ &\leq C_1 \|u\|_{H_\epsilon(S)} \|v\|_{H_\epsilon(S)} \leq C_2 \|u\|_{H_q(V)} \|v\|_{H_q(V)}, \end{aligned} \tag{3.108}$$

where C_1 and C_2 are constants and $1/2 < q < 1$. The desired compactness follows from the estimate (3.108), the inequality $q < 1$, and an earlier remark (inequality (3.105)).

Condition 4 can readily be verified.

Thus we find that the spectra Ω_n of the Ritz functionals converge to the spectrum of the functional (3.106).

Just in the same way, we can verify that the Ritz method can be used for the functionals in the second and fourth rows in Table 3.1, as well as for the functionals in the second and fourth rows of Table 3.2 in the class of continuous functions and for the functionals in the seventh and ninth rows in Table 3.2.

3. Let us study the functional in the third row Table 3.1, which is related to the ε-method in the problem on the second polarization. If we restrict ourselves to the class of continuous functions and slightly change the form of the functional, then we can rewrite it in the form

$$L(u, v; \varepsilon) = \varepsilon^{-1} \int_{V+} \nabla u \nabla v \, dV + \int_{V-} \nabla u \nabla v \, dV - k^2 \int_V uv \, dV. \tag{3.109}$$

Here the domains V^+ and V^- are the same as in the preceding example, $H = H_1(V)$, and k is real.

Let us verify that the spectrum of the functional (3.109) is real. Assume that u_0 is the eigenfunction corresponding to the eigenvalue ε_0. Then $L(u_0, v; \varepsilon_0) = 0$ for all $v \in H$. If we set $v = u_0^*$ here, then it follows from (3.109) that

$$\varepsilon_0^{-1} = \left\{ k^2 \int_V |u_0|^2 dV - \int_{V-} |\nabla u_0|^2 dV \right\} \left[\int_{V+} |\nabla u_0|^2 dV \right]^{-1}. \tag{3.110}$$

Should we have

$$\int_{V+} |\nabla u_0|^2 dV = 0,$$

we would obtain $u_0 \equiv \text{const}$ in V^+ and hence $u_0 \equiv 0$ in V, just as it follows from the boundary conditions and the uniqueness theorem for the Cauchy problem in the case of elliptic equations.

Therefore,

$$\int_{V+} |\nabla u_0|^2 dV \neq 0$$

and ε_0 is real.

Now let us study the convergence of the Ritz method for the functional (3.109). We set

$$\begin{aligned} L_0 &= \varepsilon^{-1} \int_{V+} (\nabla u \nabla v + uv) \, dV + \int_{V-} (\nabla u \nabla v + uv) \, dV, \\ L_1 &= -k^2 \int_V uv \, dV - \varepsilon^{-1} \int_{V+} uv \, dV - \int_{V-} uv \, dV. \end{aligned} \tag{3.111}$$

Just as in the preceding example, we verify that the form L_1 is compact and the functional L_0 is bounded. Let us study $L_0(u, u^*; \varepsilon)$. We set

$$\int_{V+} (|\nabla u|^2 + |u|^2)\, dV = A, \qquad \int_{V-} (|\nabla u|^2 + |u|^2)\, dV = B. \tag{3.112}$$

Then

$$L_0(u, u^*; \varepsilon) = A\varepsilon^{-1} + B.$$

If ε is not a negative real number, then we have $\varepsilon = |\varepsilon| e^{i\varphi}$, where $|\varphi| < \pi$. In this case, for $\alpha = e^{i\varphi/2}$ we have

$$\begin{aligned} |L_0(u, \bar{u}; \varepsilon)| &= |\alpha L_0(u, \bar{u}; \varepsilon)| \geq \operatorname{Re}[\alpha L_0(u, \bar{u}; \varepsilon)] \\ &= A \operatorname{Re} |\varepsilon|^{-1} e^{-i\varphi/2} + B \operatorname{Re} \alpha \geq \min(\operatorname{Re} \alpha, \operatorname{Re} \alpha |\varepsilon|^{-1}) \|u\|_H. \end{aligned} \tag{3.113}$$

With the help of a representation similar to (3.110), one can verify that the spectra of the Ritz equations of the functional (3.109) are real. It follows from Proposition 3.1 that the positive parts of these spectra converge to the positive part of the spectrum of the functional (3.109).

It turns out that this statement does not hold for the negative part of the spectrum of the functional (3.109). More exactly, we shall show that any nonpositive point ε_0 is the limit point for points in the spectra of Ritz equations under an appropriate choice of a complete real system of coordinate functions.

Let us use Proposition 3.2. One can readily verify that there exists a sequence of infinitely differentiable functions $\{\varphi_l(x)\}_{l=1}^{\infty}$ with disjoint supports in V^- such that

$$\int_{V-} |\nabla \varphi_l|^2\, dV - k^2 \int_{V-} |\varphi_l|^2\, dV > 0.$$

The functions $\varphi_l(x)$ are orthogonal in the of metric $H_1(V)$. We take F_2 to be the linear span of the system $\{\varphi_l(x)\}_{l=1}^{\infty}$. Next, we set F_1 equal to the set of all infinitely differentiable functions that vanish outside V^+.

Thus all conditions of Proposition 3.2 are satisfied. All positive points of the spectrum of the functional (3.109) can be calculated as the limit points for the spectra of the Ritz equations, while any $\varepsilon \leq 0$ (not necessarily a point of the spectrum of the functional (3.109)) can be a limit poit of the spectra of the Ritz equations along some expanding complete system of subspaces.

4. In the preceding sections of this chapter, we constructed functionals for which all boundary conditions of the boundary value problem (or at least some of these conditions) are natural. This fact is crucial if the boundary conditions contain the spectral parameter. It is very important to us that the condition $D_L = D_L^*$ (the selfadjointness of the domain of the functional) is always satisfied for such functionals; this is a necessary condition for the applicability of Propositions 3.1 and 3.2. Let us consider an example which shows that for some functionals thus obtained any point of the real axis can be the limit point of the spectra of the Ritz equations. The boundary conditions that do not contain partial derivatives of the desired function must be treated with extreme care.

Let

$$L(u, v; \lambda) = \int_V (\nabla u \nabla v - \lambda uv)\, dV - \int_{S_1} \left(u\frac{\partial v}{\partial n} + v\frac{\partial u}{\partial n} \right) dS, \tag{3.114}$$

where V is a bounded volume with piecewise smooth boundary S and S_1 is an open part of S. For $w = 0$, this functional corresponds to the functional in the first row in Table 3.2. The Dirichlet condition $u|_{S_1} = 0$ is natural for the functional (3.114). Let us show that any point on the real axis of the complex plane of λ can be the limit point for the Ritz equations of this functional along an appropriate dense-in-the-limit sequence of subspaces $\{E_n\}$ $(E_n^* = E_n)$. Again we shall use Proposition 3.2. Let $H = H_1(V)$.

We construct the manifold F_2 of infinitely differentiable functions that vanish on S just as in the case of the functional (3.109). To construct F_1, we choose a smooth part $S_2 \subset S_1$ of the surface. It is known that a sufficiently small neighborhood of $S_2 \subset V$ (that is, a part of V adjacent to S_2) can be parametrized by the coordinates $\sigma_1, \sigma_2, \sigma_3$, where σ_3 is the distance from a point x to S_2 along the normal, $0 \le \sigma_3 \le a$, and σ_1, σ_2 are coordinates on S_2.

On S_2 we consider a sequence of smooth nonnegative compactly supported functions $\{\varphi_k(\sigma_1, \sigma_2)\}_{k=1}^{\infty}$ with disjoint supports and nonnegative functions $f_k(\sigma_3)$ that are smooth and vanish for $\sigma_3 \ge a$. We set $\Psi_k = \varphi_k(\sigma_1, \sigma_2) f_k(\sigma_3)$. Let us verify that the functions $f_k(\sigma_3)$ can be chosen so that the inequalities

$$\int_V (|\nabla \Psi_k|^2 - \lambda |\Psi_k|^2)\, dV < 2 \int_{S_2} \Psi_k \frac{\partial \Psi_k}{\partial n}\, dS \tag{3.115}$$

will be satisfied for all k. We have

$$|\nabla \Psi_k|^2 = \sum_{i=1}^{3} \left(\frac{\partial \Psi_k}{\partial x_i} \right)^2 \le C_3 \sum_{i=1}^{3} \left(\frac{\partial \Psi_k}{\partial \sigma_i} \right)^2 = C_3 [f_k'^2 \varphi_k^2 + f_k^2 |\nabla_2 \varphi_k|^2]. \tag{3.116}$$

Therefore,

$$\begin{aligned} &\int_V (|\nabla \Psi_k|^2 - \lambda |\Psi_k|^2)\, dV \\ &\quad < C_3 \int_V \{f_k'^2 \varphi_k^2 + f_k^2 [|\lambda| \varphi_k^2 + |\nabla_2 \varphi_k|^2]\} \frac{D(x,y,z)}{D(\sigma_1, \sigma_2, \sigma_3)}\, d\sigma_1 d\sigma_2 d\sigma_3 \\ &\quad < C_4 \int_0^a \{f_k'^2 + f_k^2\}\, d\sigma_3, \end{aligned} \tag{3.117}$$

where C_4 depends only on C_3, φ_k, and $|\lambda|$. The following lemma allows us to choose f_k such that inequality (3.115) holds.

Lemma 3.1 *For each $C > 0$, there exists an infinitely differentiable real function $f(t)$, $0 \le t < \infty)$, such that $f(t) = 0$ for $t \ge a$ and*

$$\min\{f'^2(0)\ ; -f(0)f'(0)\} \ge C \int_0^{\infty} [f'^2(t) + f^2(t)]\, d\sigma_3. \tag{3.118}$$

It follows from this lemma that

$$2\int_{S_2} \Psi_k \frac{\partial \Psi_k}{\partial n}\, dS = -2f_k'(0)f_k(0)\int_{S_2} \varphi_k^2\, d\sigma_1 d\sigma_2 > C_5 \int_0^\infty (f_k'^2 + f_k^2)\, d\sigma_3 ,$$

where C_5 is a constant; in the lemma, the constant C is chosen so that

$$2C\int_{S_2} \Psi_k^2\, d\sigma_1 d\sigma_2 > C_5.$$

For F_1, just as previously, we can take the linear span of the functions $\{\Psi_k\}_{k=1}^{\infty}$. Now it follows from Proposition 3.2 that the limit points of the spectra of Ritz equations fill the entire real axis. Since the spectra of Ritz equations are also real, any other limit points cannot exist.

5. In the preceding sections, we used the possibility of modifying functionals by supplementing them with quadratic forms and cross-products of the left-hand sides of the boundary conditions with arbitrary coefficients. Let us consider such a functional and examine how its properties depend on these coefficients. Let us consider the following general functional, which arises in problems of the w-method:

$$\begin{aligned} L(u,v;w) &= \int_V (\nabla u \nabla v - k^2 uv)dV \\ &\quad + \int_S \left[\frac{uv}{w} + D\left(u + w\frac{\partial u}{\partial n}\right)\left(v + w\frac{\partial v}{\partial n}\right)\right] dS. \end{aligned} \tag{3.119}$$

Here $D = D(w)$ is assumed to be an analytic function and w is the spectral parameter. Assume that the elements of the space H are ordered pairs $F = \{f, g\}$, where f and g range over $H_1(V)$ and $H_0(S)$, respectively. We define the inner product of two elements F_1 , $F_2 \in H$ as follows:

$$(F_1\ , F_2) = (f_1\ ,\ f_2)_{H_i(V)} + (g_1\ , g_2)_{H_0(S)}.$$

If $h \in V$ is a smooth function whose normal derivative exists on S, then the set of all pairs of the form $\{h, \partial h/\partial n|_S\}$ is dense in H. We assign a functional $N(v_1, v_2)$, where $v_1 = \{u_1, \tilde{u}_{n_1}\}$ and $v_2 = \{u_2, \tilde{u}_{n_2}\}$, defined on H, to the functional (14) by setting

$$\begin{aligned} N &= \int_V (\nabla u_1 \nabla u_2 - k^2 u_1 u_2)\, dV \\ &\quad + \int_S \left[\frac{u_1 u_2}{w} + D(u_1 + w\tilde{u}_{n_1})(u_2 + w\tilde{u}_{n_2})\right] dS. \end{aligned} \tag{3.120}$$

Let us choose

$$\begin{aligned} L_0 &= \int_V (\nabla u_1 \nabla u_2 + u_1 u_2)\, dV + Dw^2 \int_S \tilde{u}_{n_1}\tilde{u}_{n_2}\, dS, \\ L_1 &= \int_S \left[\frac{u_1 u_2}{w} + D(u_1 u_2 + w u_2 \tilde{u}_{n_1} + w u_1 \tilde{u}_{n_2})\right] dS - \int_V u_1 u_2\, dV. \end{aligned} \tag{3.121}$$

The functional $L_0(v_1, v_2)$ is bounded. Repeating the argument used in the analysis of the form (3.109), we can readily verify that if Dw^2 is not negative, then the form (3.121) with $v_2 = v_1^*$ satisfies Proposition 3.1. On the other hand, if Dw^2 negative, then condition 4 does not hold. The convergence of the Ritz process is not yet studied in this case.

By equating the first variation of the functional (3.120) with zero, we obtain the following conditions on the stationary point $\{u, \tilde{u}_n\}$ of this functional:

$$\begin{aligned} \triangle u + k^2 u &= 0 \qquad \text{in } V, \\ \left.\left(w^{-1} + \frac{\partial u}{\partial n} + D(u + w\tilde{u}_n\right)\right|_S &= 0, \qquad u + w\tilde{u}_n|_S = 0. \end{aligned}$$

The last two relations are equivalent to

$$\left.\frac{\partial u}{\partial n}\right|_S = \tilde{u}_n, \qquad \left.\left(u + w\frac{\partial u}{\partial n}\right)\right|_S = 0. \tag{3.122}$$

Thus the spectra of the functionals (3.119) and (3.120) coincide, and the first term of the stationary pair of the functional (3.120) is a stationary function for the functional (3.119).

Let us consider the solutions $\Psi_k = \{\varphi_k, \partial\varphi_k/\partial n\}$ of the Ritz equations for the functional (3.120) on finite-dimensional subspace of pairs of the form $\{\varphi, \partial\varphi/\partial n\}$, where φ are smooth functions. Obviously, on such subspaces the spectra of the Ritz equations of the functionals (3.119) and (3.120) coincide, and the φ_k satisfy the Ritz equations for the functional (3.119). By applying Proposition 3.1, we see that $\Omega_n \to \Omega$.

We can obtain the functional (3.32) from the functional (3.119) by setting $D = -w^{-1}$. The spectrum of the functional (3.32) is discrete. All nonpositive points of this functional can be calculated as the limit points for the spectra of the Ritz equations. If we intend to calculate the positive part of the spectrum, then, in general, the Ritz method cannot be used, since the limit points of the spectra of Ritz equations fill the entire positive part of the real axis.

In conclusion, we note that while the results of this section show that there may be some "superfluous" limit points of the spectra of Ritz equations, it is not necessary to abandon this method completely. Obviously, this property is depends on the choice of coordinate functions. If the choice is successful, then the spectra do not contain superfluous points. In particular, this fact is supported by the numerical results given in Section 4.1.

In this chapter we constructed functionals that are stationary on the solutions of homogeneous problems arising in various versions of the generalized method of eigenoscillations. These functionals are similar to the well-known functionals of Rayleigh type in the problem of eigenfrequencies of a closed resonator. However, in general, our functionals are not positive definite and are even complex-valued in some problems. We also studied some problems concerning the applicability of the Ritz method to such functionals.

Chapter 4

Applications to Specific Problems

In the preceding chapters we discussed the foundations of a method for solving diffraction problems, which we called the generalized method of eigenoscillations. In the beginning of this chapter (Section 4.1), we consider examples that illustrate different versions of this method. The examples were chosen to be sufficiently simple, so that it is possible to compare the obtained results with the exact solution.

In the subsequent sections we deal with problems that either cannot be solved by other methods at all, or can be solved, but in a more complicated way or less completely. The problems are grouped according to the versions of the generalized method used in the solution. Only in Section 4.2, which is also partly of methodical character, we solve a quantum-mechanical problem by two methods, namely, the s-method and the ε-method. The comparison of numerical results allows us to estimate the error of approximate formulas in which, apart from U^0, only one term of the series is taken into account at resonance. In this problem, as well as in problems on open and closed resonators containing a dielectric body (Section 4.3), we use the variational technique of the ε-method for calculating the eigenvalues.

In Sections 4.4–4.8 we employ both versions of the ρ-method. In Section 4.4 we numerically solve the integral equation (2.49) for two parallel mirrors. Thus we generalize the theory of open resonators of this type to the case in which the mirror sizes are comparable with the distance between the mirrors and with the wavelength. The mirrors may be of nonzero transparency. In Section 4.5 we give the analytic solution of the problem on the shift of real eigenfrequencies of a resonator with semitransparent walls with respect to the eigenfrequencies of a resonator with metallic walls. The differential statement of homogeneous problems of the ρ-method (2.35a) was used. The analytic expression of the function $\rho_n(k)$ near the eigenfrequencies of a closed resonator and the phase velocity of leaky waves in a waveguide with semitransparent walls were calculated. We also used the possibility pointed out, say, in Section 1.5 to take the solution of some other simpler diffraction problem as U^0. This method is also used in most of the subsequent sections.

In Sections 4.6 and 4.7 we consider two-dimensional problems on metallic resonators of arbitrary shape with a small hole and problems on the phase velocities of leaky waves in waveguides with longitudinal slots for E- and H-waves. These problems required some modification of the ρ-method; namely, a weight function that describes the electrostatic singularity on the edges of the slot was introduced into the homogeneous integral equation for the field of eigenoscillations on the slot.

The last problem examined by the ρ-method is considered in Section 4.8. This is the problem on a resonator made of a material with large ε. It turns out that for such an open resonator (in contrast with a resonator with a semitransparent wall) not only the Q-factor but also the shift of the resonance frequency depends on the bodies present near the resonator.

In Sections 4.9 and 4.10 we use the s-method in which solving a problem with radiation is reduced to solving a real integral equation with integrals over the surface of the body. Numerical solutions of this equation are found for several two-dimensional open resonators. We describe the decay of leaky waves in waveguides of several shapes with semitransparent walls or a longitudinal slot. In this method we need not first solve the problem on the resonator of the same shape with ideally conducting walls, find the Green function for the exterior domain (which is very difficult), etc. Hence in these sections the problems on resonators and waveguides with slots are solved without the usual restrictions that the slot is complemented with an infinite flange and that the interior region is of a simple shape.

4.1 Methodical Examples

The technique developed in the first two chapters of this book is illustrated in Subsections 4.1.1–4.1.6 by an example of an elementary one-dimensional problem. This problem can be solved explicitly. Thus, by using various versions of the generalized method of eigenoscillations, we can obtain expansions of the solution in infinite series with respect to different functions. The comparison of these solutions allows us, in particular, to illustrate the relations between the resonance curves that describe the amplitude of the resonant term in different expansions for one and the same problem. In Subsections 4.1.7–4.1.9 we employ the stationary functionals from Chapter 3 for finding the eigenvalues of two homogeneous problems (also one-dimensional). Since these eigenvalues can readily be obtained by straightforward calculations, these examples allow us to estimate the practical rate of convergence of the Ritz method in the case of complex-valued functionals.

Let us consider a one-dimensional (the field in independent of y and z) problem of exciting a closed resonator partially filled with a homogeneous dielectric. Suppose that we need to solve the equations

$$\frac{d^2U}{dx^2} + k^2\varepsilon U = 2k\delta(x), \quad |x| < a, \tag{4.1a}$$

$$\frac{d^2U}{dx^2} + k^2 U = 0, \quad a < |x| < b, \tag{4.1b}$$

with the condition that U and $\frac{dU}{dx}$ are continuous at $|x| = a$ and the condition

$$U(\pm b) = 0. \tag{4.2}$$

This problem is equivalent to the problem of finding oscillations of a loaded string and *has the explicit solution*

$$\begin{aligned} U(x) &= \begin{cases} \dfrac{\sqrt{\varepsilon}\sin k(b-a)\cos k\sqrt{\varepsilon}(a-|x|)}{D} \\ \quad + \dfrac{\cos k(b-a)\sin k\sqrt{\varepsilon}(a-|x|)}{D}, & |x| \le a, \\ \sqrt{\varepsilon}\sin k(b-|x|)/D, & a \le |x| \le b, \end{cases} \\ D &= \varepsilon \sin k\sqrt{\varepsilon}a \sin k(b-a) - \sqrt{\varepsilon}\cos k\sqrt{\varepsilon}a \cos k(b-a). \end{aligned} \tag{4.3}$$

In what follows, we shall derive this solution by various methods. Since this problem is symmetric with respect to $x = 0$, in all homogeneous problems we can restrict our consideration to eigenfunctions even with respect to x.

1. We start from the k-method. The eigenfunctions of the homogeneous problem corresponding to this method are

$$u_n(x) = \cos k_n\sqrt{\varepsilon}x, \qquad |x| \le a, \tag{4.4a}$$

$$u_n(x) = \frac{\cos k_n\sqrt{\varepsilon}a}{\sin k_n(b-a)}\sin k_n(b-|x|), \quad a \le |x| \le b, \tag{4.4b}$$

and the eigenvalues k_n are determined from the transcendental equation

$$\sqrt{\varepsilon}\sin k\sqrt{\varepsilon}a \sin k(b-a) - \cos k\sqrt{\varepsilon}a \cos k(b-a) = 0. \tag{4.5}$$

The solution of this problem can be written in the form

$$U(x) = \sum_n \frac{2k\sin^2 k_n(b-a)u_n(x)}{(k^2-k_n^2)[a\varepsilon\sin^2 k_n(b-a) + (b-a)\cos^2 k_n\sqrt{\varepsilon}a]}. \tag{4.6}$$

Obviously, (4.6) is just another representation of (4.3), namely, this is the decomposition of (4.3), as of a function of k, into partial fractions.

2. If we solve this problem by the ε-method, then the corresponding homogeneous equation coincides with (4.1b) outside the interval $|x| < a$. This equation may have only one solution that satisfies (4.2); therefore, for $a \le |x| \le b$, all eigenfunctions of the ε-method coincide, that is,

$$u_n(x) = \sin k(b-|x|), \quad a \le |x| \le b. \tag{4.7}$$

The problem can be reduced to the homogeneous equation

$$\frac{d^2u_n}{dx^2} + k^2\varepsilon_n u_n = 0, \quad |x| < a, \tag{4.8}$$

for the eigenvalues ε_n with the impedance boundary conditions

$$u_n \pm \frac{1}{k} \tan k(b-a) \frac{du_n}{dx}\bigg|_{x=\pm a} = 0.$$

The eigenfunctions of this problem are

$$u_n(x) = \frac{\sin k(b-a)}{\cos k\sqrt{\varepsilon_n}a} \cos k\sqrt{\varepsilon_n}x. \tag{4.9}$$

The eigenvalues ε_n are determined by the same transcendental equation (4.5), where k is now a given number and ε is a desired variable. It is *always* easier to solve (4.5) for ε than for k. The problem is significantly simpler if ε is complex in the original equation (4.1). In this case Eq. (4.5) becomes complex in the k-method and *remains real* in the ε-method.

Since the source in our problem is in the interior of the dielectric, in the ε-method we need not separate U^0 both for $|x| \le a$ and $|x| > a$. Then the solution has the form

$$U(x) = \sum_n \frac{4\sqrt{\varepsilon_n} \cos \sqrt{\varepsilon_n}a \cdot u_n(x)}{(\varepsilon - \varepsilon_n) \sin k(b-a)(2ka\sqrt{\varepsilon_n} + \sin 2k\sqrt{\varepsilon_n}a)}. \tag{4.10}$$

Just as in (4.6), the series (4.10) is the decomposition of the function $U(x)$ into partial fractions. However, now we consider (4.3) as a function of ε for a given k, so that the decomposition is performed with respect to the roots ε_n of Eq. (4.5).

Along with (4.10), we can write the solution in the following different form if we first separate U^0 as the solution of the nonhomogeneous problem with $\varepsilon = 1$:

$$U^0(x) = \frac{\sin k(|x| - b)}{\cos kb}. \tag{4.11}$$

In this case the complete solution has the form

$$U(x) = U^0(x) + \sum_n \frac{(\varepsilon - 1)4\sqrt{\varepsilon_n} \cos k\sqrt{\varepsilon_n}a \cdot u_n(x)}{(\varepsilon_n - 1)(\varepsilon - \varepsilon_n) \sin k(b-a)(2ka\sqrt{\varepsilon_n} + \sin 2k\sqrt{\varepsilon_n}a}. \tag{4.12}$$

The series in (4.12) is the decomposition of the function $U - U^0$ into partial fractions if $U - U^0$ is considered as a function of the argument $\varepsilon - 1$.

If the problem parameters are such that $\sin k(b-a) = 0$, then, according to the transcendental equation (4.5), we simultaneously have $\cos k\sqrt{\varepsilon_n}a = 0$. In (4.10) and in the sum in (4.12), we have indeterminacy, but this indeterminacy can readily be evaluated, since we have

$$\frac{\cos k\sqrt{\varepsilon_n}a}{\sin k(b-a)} = \sqrt{\varepsilon_n} \frac{\sin k\sqrt{\varepsilon_n}a}{\cos k(b-a)}. \tag{4.13}$$

Moreover, if $\cos kb = 0$, then the entire expression (4.12) becomes indeterminate, since in this case one of the ε_n is equal to 1 and the series becomes infinite. Simultaneously,

Table 4.1

M	1	2	3	4	7	10
(4.6)	0.69	1.05	0.986	1.02	0.977	0.973
(4.10)	0.72	1.06	0.97	1.01	0.996	1.002
(4.12)	0.97	1.002	0.9996	1.0001	1.0000	1.0000

U^0 also becomes infinite. The solution $U(x)$ is determined by the difference of two infinite values, while the representation (4.10) remains valid.

Numerical results for the field obtained by formulas (4.6), (4.10), and (4.12) are shown in Table 4.1. (The computations were carried out by A. V. Golubyatnikov.) The calculations were performed at the point $x = a$ for $ka = 1$, $kb = 2$, and $\varepsilon = 2.5 - i$. This means that a resonator with low values of the Q-factor is considered and there are no resonant terms in the corresponding representations. The table contains the values of $|U_M(a)/U(a)|$, where U_M is the result of calculation by the above-mentioned formulas with M terms retained in the sums. As it agrees with the general theory, formula (4.12) is most efficient, that is, this formula gives a practically exact result even if only one term in the sum is retained (in addition to the field $U^0(x)$ separated above). The comparison of the results obtained by formulas (4.6) and (4.10) shows that the ε-method is preferable (in contrast with the k-method) even if we do not separate U_0 in advance. Furthermore, we gain this advantage not only because the series converge more rapidly but also because in the ε-method we need to seek real roots (rather than complex roots in the k-method) of the transcendental equation (4.5).

3. Now let us solve our problem using the versions of the generalized method described in Chapter 2. As was already noted, the dimension of the series in these methods is less by 1. Thus there are no series at all in our case, and we readily obtain an explicit solution.

Let us start from the w-method. Here the auxiliary problem comprises the homogeneous equation with the actual values of k and ε, the conditions that u_n and $\dfrac{du_n}{dx}$ are continuous on $|x| = a$, and the condition

$$u_n \pm w_n \frac{du_n}{dx} = 0 \quad \text{at} \quad x = \pm b. \tag{4.14}$$

This problem has only one eigenfunction that is even in x:

$$u_1(x) = \begin{cases} \cos k\sqrt{\varepsilon}x, & |x| \le a, \\ \cos k\sqrt{\varepsilon}a \cos k(|x| - a) - \sqrt{\varepsilon} \sin k\sqrt{\varepsilon}a \sin k(|x| - a), & \\ & a \le |x| \le b. \end{cases} \tag{4.15}$$

The corresponding eigenvalue is

$$w_1 = \frac{1}{k} \frac{\sqrt{\varepsilon} \sin k\sqrt{\varepsilon}a \sin k(b - a) - \cos k\sqrt{\varepsilon}a \cos k(b - a)}{\sqrt{\varepsilon} \sin k\sqrt{\varepsilon}a \cos k(b - a) + \cos k\sqrt{\varepsilon}a \sin k(b - a)}. \tag{4.16}$$

To solve the original nonhomogeneous problem, we also need to know an arbitrary solution U^0 of this problem without condition (4.2). It is convenient to take the solution

$$U^0(x) = \begin{cases} \frac{1}{\sqrt{\varepsilon}} \sin k\sqrt{\varepsilon}|x|, & |x| \le a, \\ \frac{1}{\sqrt{\varepsilon}} \sin ka\sqrt{\varepsilon} \cos k(|x|-a) + \cos k\sqrt{\varepsilon}a \sin k(|x|-a), & \\ & a \le |x| \le b. \end{cases} \tag{4.17}$$

If we know U^0, then, according to Section 2.1, we can represent the solution of the original problem in the form

$$U(x) = U^0(x) + \frac{[\sqrt{\varepsilon} \cos k\sqrt{\varepsilon}a \sin k(b-a) + \sin k\sqrt{\varepsilon}a \cos k(b-a)]u_1(x)}{kw_1\sqrt{\varepsilon}[\sqrt{\varepsilon} \sin k\sqrt{\varepsilon}a \cos k(b-a) + \cos k\sqrt{\varepsilon}a \sin k(b-a)]}. \tag{4.18}$$

Replacing U^0, u_1, and w_1 in (4.18) by their explicit expressions (4.17), (4.15), and (4.16), we obtain the same solution (4.3) of the nonhomogeneous problem. In (4.18) all resonant properties are described by the coefficient of $u_1(x)$, more precisely, by the zeros of w_1 treated as a function of k. The numerator of w_1 coincides with the left-hand side of Eq. (4.5); thus, the zeros coincide with the eigenfrequencies k_n, which are the roots of this equation. But *now we need not solve a transcendental equation*; to find an eigenvalue, for a given k it suffices only to *calculate* the *left-hand side of this equation* normalized in the required way. Next, if we know this eigenvalue, then we can readily write out the solution of the problem with *any impedance condition* for $|x| = b$ (instead of (4.2)). To this end, in (4.18) it suffices to replace w_1 by $w_1 - w$, where w is the actual value of the impedance in the boundary condition.

4. Now let us consider the first version of the ρ-method. This method can be applied to our problem in two different ways by introducing an eigenvalue into the boundary condition either at $|x| = b$ (see Section 2.2) or at $|x| = a$ (see Section 2.3). Here we shall follows the method studied in Section 2.3.

In this method, first of all, we must solve the homogeneous equation with condition (4.2) and the following conditions at $|x| = a$:

$$u_n\big|_{|x|=a-0} - u_n\big|_{|x|=a+0} = 0, \tag{4.19a}$$

$$\left.\frac{du_n}{dx}\right|_{|x|=a-0} - \left.\frac{du_n}{dx}\right|_{|x|=a+0} \pm \frac{1}{\rho_n} u_n(\pm a) = 0. \tag{4.19b}$$

Just as in the w-method, this homogeneous problem has only one eigenfunction that is even in x:

$$u_1(x) = \begin{cases} \cos k\sqrt{\varepsilon}x, & |x| \le a, \\ \frac{\cos k\sqrt{\varepsilon}a}{\sin k(b-a)} \sin k(b-|x|), & a \le |x| \le b. \end{cases} \tag{4.20}$$

The corresponding eigenvalue is

$$\rho_1 = \frac{1}{k}\frac{\cos k\sqrt{\varepsilon}a \sin k(b-a)}{\sqrt{\varepsilon}\sin k\sqrt{\varepsilon}a \sin k(b-a) - \cos k\sqrt{\varepsilon}a \cos k(b-a)}. \tag{4.21}$$

As follows from Section 2.3, to find U^0, we first need to solve Eq. (4.1a) for $|x| \leq a$ with any condition at $|x| = a$. The simplest way is to take

$$U^0(x) = \frac{1}{\sqrt{\varepsilon}} \sin k\sqrt{\varepsilon}|x|, \quad |x| \leq a. \tag{4.22a}$$

Next, we need to take the value of this solution at the point $|x| = a$ and then solve the Dirichlet problem for the region $a \leq |x| \leq b$ with condition (4.2) and with the given value

$$U^0(\pm a) = \frac{1}{\sqrt{\varepsilon}} \sin k\sqrt{\varepsilon}a. \tag{4.23}$$

This allows us to find the function U^0 in the remaining part of the interval. Namely, we have

$$U^0(x) = \frac{1}{\sqrt{\varepsilon}}\frac{\sin k\sqrt{\varepsilon}a}{\sin k(b-a)} \sin k(b-|x|), \quad a \leq |x| \leq b. \tag{4.22b}$$

Finally, the solution of the problem has the form

$$\begin{aligned} U(x) &= U^0(x) \\ &+ \frac{k\rho_1[\sqrt{\varepsilon}\cos k\sqrt{\varepsilon}a \sin k(b-a) + \sin k\sqrt{\varepsilon}a \cos k(b-a)]}{\sqrt{\varepsilon}\cos k\sqrt{\varepsilon}a \sin k(b-a)} u_1(x). \end{aligned} \tag{4.24}$$

Obviously, if we take into account (4.22), (4.20), and (4.21), then formula (4.24) again coincides with (4.3).

In contrast with the w-method, in the ρ-method it may happen that knowing ρ as a function of k is not sufficient for describing the resonant properties of the system. The resonance occurs if $\rho_1 = \infty$. The denominator of ρ_1 coincides with the left-hand side of Eq. (4.5); therefore, the resonance occurs only at eigenfrequencies k_n. If $\rho_1 = \infty$, then the resonance always takes place. Furthermore, the resonance can also occur for $\rho_1 = 0$. If $\rho_1 = 0$ since there is only one zero factor in the numerator in (4.21), then no resonance occurs, the indeterminacies of the form $\frac{0}{0}$ or $\frac{\infty}{\infty}$ arising in (4.24) can readily be evaluated, and the resulting expression is finite. But if both factors vanish simultaneously, that is, if $\cos k\sqrt{\varepsilon}a = \sin k(b-a) = 0$, then it follows from (4.3) that the total field grows infinitely. Indeed, by (4.24), the coefficient of u_1 is infinite in this case, while u_1 is finite. Moreover, U^0 also becomes infinite for $a < |x| < b$. This resonance cannot be found on the curve $\rho_1(k)$, since the ρ-method can be used for solving the following two problems of different nature: instead of the boundary conditions used in the ρ-method, the original problem can contain either the condition that the field and its derivative are continuous (just as in our problem) or the condition $U = 0$ (in this case the resonance occurs for $\rho = 0$). In the last case, another two versions of the problem are possible; this depends on whether the interior or the exterior region is considered. The eigenvalue provides information on resonances in all these problems. If $\cos k\sqrt{\varepsilon}a = 0$, then the resonance occurs in the interior region; if $\sin k(b-a) = 0$, then the resonance occurs in the exterior region. If the denominator in (4.21) vanishes, then we have a resonance in our problem. If the resonances in the interior and exterior regions coincide, then the resonance in our problem corresponds to $\rho_1 = 0$.

We must point out that the solutions of the homogeneous problem can also be used for solving the following two actual problems by the w-method: one with condition (4.2) and the other with the condition $\left.\dfrac{du}{dx}\right|_{|x|=b} = 0$. However, in these two problems the resonance cannot occur simultaneously, since the numerator and the denominator in (4.16) cannot be zero at the same time.

In solving a problem by the first version of the ρ-method, we must separately examine the frequencies at which $\rho = 0$. In particular, for these frequencies we can use the second version of the ρ-method.

5. In the second version of the ρ-method, the homogeneous problem differs from that in the first version by the following conditions for $x = \pm a$:

$$\left.\frac{du_n}{dx}\right|_{|x|=a-0} - \left.\frac{du_n}{dx}\right|_{|x|=a+0} = 0, \tag{4.25a}$$

$$u_n\big|_{|x|=a-0} - u_n\big|_{|x|=a+0} \mp \left.\frac{1}{\tilde{\rho}}\frac{du_n}{dx}\right|_{|x|=\pm a} = 0. \tag{4.25b}$$

The eigenfunction of this problem has the form

$$u_1(x) = \begin{cases} \cos k\sqrt{\varepsilon}x, & |x| < a, \\ \dfrac{\sqrt{\varepsilon}\sin k\sqrt{\varepsilon}a}{\cos k(b-a)} \sin k(b-|x|), & a \le |x| \le b. \end{cases} \tag{4.26}$$

The corresponding eigenvalue is

$$\tilde{\rho}_1 = \frac{k\sqrt{\varepsilon}\sin k\sqrt{\varepsilon}a \cos k(b-a)}{\sqrt{\varepsilon}\sin k\sqrt{\varepsilon}a \sin k(b-a) - \cos k\sqrt{\varepsilon}a \cos k(b-a)}. \tag{4.27}$$

Then the solution of the original problem has the form

$$U(x) = U^0(x) + \frac{\tilde{\rho}_1[\sin k\sqrt{\varepsilon}a \cos k(b-a) + \sqrt{\varepsilon}a \cos k\sqrt{\varepsilon}a \sin k(b-a)]}{k\varepsilon \sin k\sqrt{\varepsilon}a \cos k(b-a)} u_1(x), \tag{4.28}$$

where

$$U^0(x) = \begin{cases} \dfrac{1}{\sqrt{\varepsilon}} \sin k\sqrt{\varepsilon}|x|, & |x| \le a, \\ -\dfrac{\cos k\sqrt{\varepsilon}a}{\cos k(b-a)} \sin k(b-|x|), & a < |x| \le b. \end{cases} \tag{4.29}$$

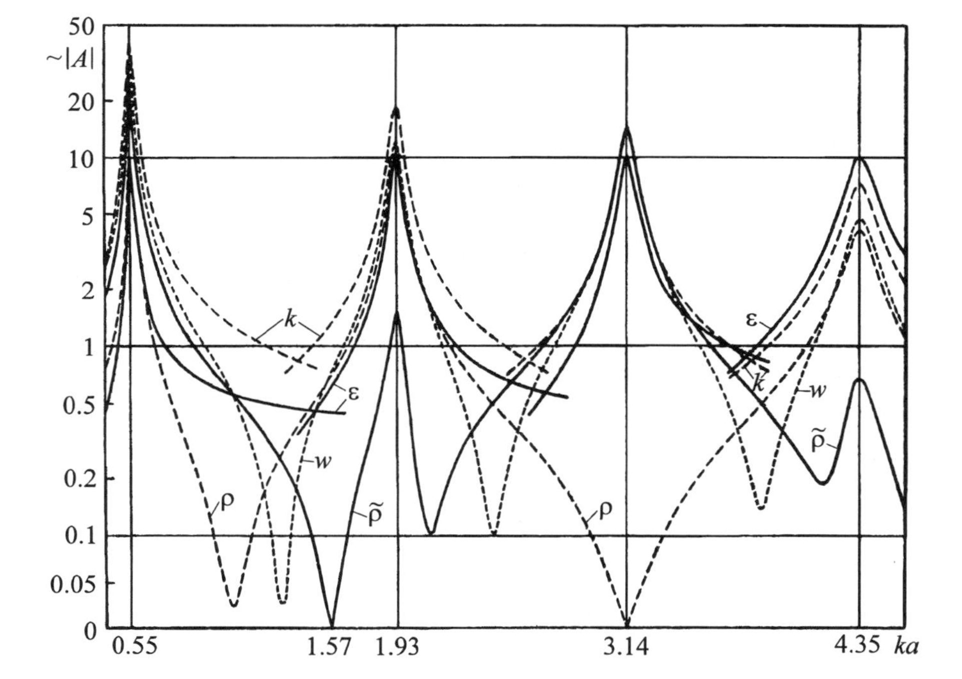

Figure 4.1

In this statement of the problem, the resonance occurs for $\tilde{\rho}_1 = \infty$, that is, also for $k = k_n$ (the denominator in $\tilde{\rho}_1$ coincides with the left-hand side of (4.5)). Just as in the first version, the resonance picture can also be distorted in the second version of the method, namely, if $\sin k\sqrt{\varepsilon}a = \cos k(b - a) = 0$. But here it is impossible for the eigenvalues $\tilde{\rho}_1$ and ρ_1 to vanish simultaneously at a resonant value of k. Therefore, by solving the problem by one of these versions of the ρ-method, we can additionally examine the vicinity of the points where an eigenvalue is zero by a different method and thus obtain complete information on the resonance properties of the system.

6. So far we have assumed that there are no losses in the system, that is, ε is a real number. Let us now consider the case of complex ε. Here we present numerical results for a resonator with losses, obtained by using various versions of the generalized method of eigenoscillations. Figure 4.1 shows how the *resonant factors* (*proportional to* $|A_n|$) *in various methods* depend on ka for $\varepsilon = 2.25 - 0.1i$ and $b = 2a$. The resonance peaks of the curves are finite. One can see that the peaks on all these curves occur for the same values of k, which are the real parts of the corresponding complex k_n. We point out that all these curves equally well describe the resonance properties of the system.

The peaks on the curves corresponding to one of the versions of the ρ-method can be weak or even disappear completely if the eigenvalue vanishes near the corresponding k. In this case the resonance is well described by an eigenvalue corresponding to another version.

The example of a system with losses shows that we must examine not only the points where ρ_n or $\tilde{\rho}_n$ vanishes but, in addition, their neighborhoods whose width is of the order of the width of the expected resonance peaks.

7. In Section 3.7 we discussed the theoretical results concerning the applicability of the Ritz method to the functionals constructed in the preceding chapter. Here and in the following subsections we illustrate the results obtained by this method by model examples in which exterior homogeneous problems are solved by the generalized method of eigenoscillations. The comparison of numerical results with exact solutions allows us to draw some practical conclusions on the convergence of the numerical procedure.

We start from the homogeneous problem of the w-method for the exterior of a disk of radius a and restrict ourselves to finding the eigenvalue w_1 that corresponds to the axisymmetric (angle-independent) eigenfunction u_1. This problem is reduced to the ordinary differential equation

$$\frac{d^2u_1}{dr^2} + \frac{1}{r}\frac{du_1}{dr} + k^2u_1 = 0, \qquad r > a, \tag{4.30}$$

with the boundary condition

$$\left(u_1 - w_1\frac{du_1}{dr}\right)\bigg|_{r=a} = 0 \tag{4.31}$$

and the radiation condition at infinity

$$u_1|_{r\to\infty} \simeq \Phi\frac{e^{-ikr}}{\sqrt{kr}}. \tag{4.32}$$

Here Φ is an arbitrary constant. The explicit solution of this problem is well known, namely,

$$u_1(r) = H_0^{(2)}(kr), \tag{4.33}$$

$$w_1 = -\frac{H_0^{(2)}(ka)}{kH_1^{(2)}(ka)}. \tag{4.34}$$

We shall employ the functional (3.35) and represent the desired eigenfunction in the form

$$u_1(r) = \frac{e^{-ikr}}{\sqrt{kr}}\sum_{j=0}^{M} B_j(kr)^{-j}, \tag{4.35}$$

thus satisfying the radiation condition. According to the Ritz method, we substitute (4.35) into the functional, differentiate with respect to B_j, equate the derivatives with zero, and arrive at a system of homogeneous algebraic equations for the B_j. The solvability condition for this system is the condition that its determinant is zero; this equation determines w_1.

Table 4.2

	ka	1	1.5	2	5	10
$k\|w_1\|$	exact	0.886	0.936	0.9592	0.9917	0.99779
	approximate	0.859	0.921	0.9502	0.9905	0.99753
arg w_1	exact	1.22	1.315	1.371	1.4852	1.5273
	approximate	1.17	1.278	1.341	1.4725	1.5210

As one might expect, this determinant depends linearly on w, that is, this equation has only one root. Here we do not write out the awkward expression for this root but give only its approximate values calculated for $M = 1$. For various ka we compare these results with the exact values (provided by (4.34)) given in Table 4.2.

For large values of ka, only two terms of the sum are required for calculating w_1 with high accuracy, since the representation (4.35) coincides in form with the asymptotics of the function $H_0^{(2)}$ for large arguments. But even for small ka, in the case where the error of the asymptotic formula is large, the Ritz method allows us to calculate the eigenvalue more precisely. This is the main advantage of the variational technique; this method allows us to calculate the eigenvalue with sufficient accuracy even if our knowledge about the eigenfunction is not complete.

8. Now let us consider a dielectric disk in a vacuum (the first polarization problem). We shall solve this problem by three versions of the generalized method of eigenoscillations: the ε-method, the ρ-method, and the s-method. In all these cases, just as previously, we restrict our consideration to axisymmetric eigenfunctions. We start from the homogeneous problem of the ε-method. In this problem the desired eigenfunctions must satisfy the equation

$$\frac{d^2u_n}{dr^2} + \frac{1}{r}\frac{du_n}{dr} + k^2\varepsilon_n u_n = 0, \qquad r < a, \tag{4.36a}$$

$$\frac{d^2u_n}{dr^2} + \frac{1}{r}\frac{du_n}{dr} + k^2 u_n = 0, \qquad r > a, \tag{4.36b}$$

be continuous together with its first-order derivatives for $r = a$, be bounded for all r, $0 \leq r < \infty$, and satisfy conditions similar to the radiation condition (4.32). Such functions can be written in the form

$$u_n(r) = \begin{cases} H_0^{(2)}(ka)J_0(k\sqrt{\varepsilon_n}r), & r \leq a, \\ J_0(k\sqrt{\varepsilon_n}a)H_0^{(2)}(kr), & r \geq a. \end{cases} \tag{4.37}$$

In this notation, all conditions of the problem are satisfied except for the condition that the derivatives must be continuous at $r = a$. The last condition yields the following transcendental equation for the eigenvalues ε_n:

$$\sqrt{\varepsilon}H_0^{(2)}(ka)J_1(k\sqrt{\varepsilon}a) - J_0(k\sqrt{\varepsilon}a)H_1^{(2)}(ka) = 0. \tag{4.38}$$

To calculate the ε_n approximately, we shall use the functional

$$L(u) = \int_0^\infty (u')^2 r dr - k^2 \varepsilon \int_0^a u^2 r dr - k^2 \int_a^\infty u^2 r dr, \tag{4.39}$$

which is obtained from (3.24). The admissible functions for this functional must be bounded, continuous at $r = a$, and satisfy the radiation condition.

Since outside the disk the equation does not contain the spectral parameter, we choose the coordinate functions so as to satisfy this equation. Inside the disk, for the coordinate functions we take a constant and $J_0(\mu_j r/a)$, where $J_0(\mu_j) = 0$. Finally, the desired functions are represented in the form

$$u_n(r) = \begin{cases} B_0 H_0^{(2)}(ka) + \sum\limits_{j=1}^{M} B_j J_0(\mu_j r a), & r \leq a, \\ B_0 H_0^{(2)}(kr), & r \geq a. \end{cases} \tag{4.40}$$

After the determinant arising in the Ritz method is calculated, we arrive at the transcendental equation

$$F(\varepsilon, k) = \varepsilon ka - \frac{2H_1^{(2)}(ka)}{H_0^{(2)}(ka)} + 4\varepsilon^2 (ka)^3 \sum_{j=1}^{M} \frac{1}{\mu_j^2[\mu_j^2 - \varepsilon(ka)^2]} = 0 \tag{4.41}$$

for the eigenvalues ε_n. For $M = \infty$ we can reduce this equation to the form (4.38) by using the following well-known representation of the Bessel function in terms of its roots [11]:

$$J_0(x) = e^{-x^2/4} \prod_{j=1}^{\infty} e^{x^2/\mu_j^2} \left(1 - \frac{x^2}{\mu_j^2}\right). \tag{4.42}$$

Using (4.41), we can readily analyze the convergence of the Ritz method. Since each term in the sum (4.41) corresponds to the coordinate function with the same number in (4.4), the rate of convergence can readily be established by preserving finitely many terms in this sum. For example, for the sequence of values $\varepsilon_n^{(M)}$ calculated by using the sum with M terms retained, we have the asymptotics

$$\varepsilon_n^{(M)} - \varepsilon_n^{(M-1)} \sim \frac{1}{M^4}. \tag{4.43}$$

Formula (4.43) was derived by using the condition $\mu_N \gg ka|\sqrt{\varepsilon}|$, and so one might expect that the number of coordinate functions for which the convergence begins increases with ka. However, it turns out that the moduli of the desired eigenvalues ε_n decrease with increasing ka, and thus the rate of convergence practically does not decrease. Numerical results that characterize the rate of convergence of the Ritz method for the first eigenvalue are given in Table 4.3.

Table 4.3

ka	N	1	2	3	4
1	$\lvert\varepsilon_1\rvert$	2.069	2.0550	2.0528	2.0521
	$\arg\varepsilon_1$	0.918	0.91493	0.91448	0.91435
2	$\lvert\varepsilon_1\rvert$	0.889	0.8781	0.8765	0.8761
	$\arg\varepsilon_1$	0.841	0.8430	0.8433	0.84346
3	$\lvert\varepsilon_1\rvert$	0.3245	0.32429	0.324277	0.324274
	$\arg\varepsilon_1$	0.515	0.5249	0.5264	0.5269
ka	N	5	10	20	Exact
1	$\lvert\varepsilon_1\rvert$	2.05187	2.05162	2.05159	2.05159
	$\arg\varepsilon_1$	0.91430	0.914256	0.914219	0.914248
2	$\lvert\varepsilon_1\rvert$	0.87589	0.87571	0.875689	0.875685
	$\arg\varepsilon_1$	0.84350	0.843546	0.843552	0.843553
3	$\lvert\varepsilon_1\rvert$	0.324273	0.324272	0.324272	0.324272
	$\arg\varepsilon_1$	0.52705	0.527218	0.527240	0.527243

9. Finally, let us consider the problem on the symmetric eigenfunctions for the ρ- and s-methods for a dielectric disk in vacuum. In the ρ-method the function must satisfy Eqs. (4.36) in the first of which the eigenvalue ε_n is replaced by a given value of ε, the continuity condition at $r = a$, and the radiation condition (4.32). Instead of the continuity condition for the derivative at $r = a$, we have the condition

$$\left.\frac{du_1}{dr}\right|_{r=a+0} - \left.\frac{du_1}{dr}\right|_{r=a-0} - \frac{1}{\rho_1}u_1(a) = 0. \tag{4.44}$$

The eigenvalue ρ_1 of this problem can be written in the explicit form

$$\rho_1 = \frac{H_0^{(2)}(ka)J_0(k\sqrt{\varepsilon}a)}{k[\sqrt{\varepsilon}H_0^{(2)}(ka)J_1(k\sqrt{\varepsilon}a) - J_0(k\sqrt{\varepsilon}a)H_1^{(2)}(ka)]}. \tag{4.45}$$

The functional that is stationary on the eigenfunction of this problem can be obtained from (4.39) by adding the term

$$\frac{a}{\rho}\,u^2(a). \tag{4.46}$$

By choosing the coordinate functions in the same form (4.40) and by carrying out all necessary calculations, we arrive at the following explicit expression for the eigenvalue:

$$\rho_1 = \frac{2}{kF(\varepsilon, k)}. \tag{4.47}$$

Numerical results obtained by this formula for various values of M are given in Table 4.4. The same estimate (4.43) holds for the rate of convergence of $1/\rho_1^{(M)}$.

Table 4.4

ε	ka	N	1	2	3	4	5
2	4	$\|k\rho_1\|$	0.89	0.0949	0.09633	0.09665	0.09676
		$\arg \rho_1$	1.11	3.0459	3.0444	3.04409	3.04398
	8	$\|k\rho_1\|$	0.42	0.079	0.96	0.348	0.359
		$\arg \rho_1$	0.44	0.92	1.82	0.395	0.368
4	4	$\|k\rho_1\|$	0.49	0.672	0.398	0.371	0.364
		$\arg \rho_1$	0.51	1.78	0.41	0.383	0.375
	8	$\|k\rho_1\|$	0.21	0.41	0.70	1.00	0.41
		$\arg \rho_1$	0.21	0.42	0.77	1.56	2.71

ε	ka	N	10	20	50	Exact
2	4	$\|k\rho_1\|$	0.096864	0.096877	0.096879	0.096879
		$\arg \rho_1$	3.043875	3.043861	3.043859	3.043859
	8	$\|k\rho_1\|$	0.3454	0.34398	0.343806	0.343794
		$\arg \rho_1$	0.3533	0.35186	0.351666	0.351653
4	4	$\|k\rho_1\|$	0.3582	0.35748	0.357390	0.357383
		$\arg \rho_1$	0.3691	0.36834	0.368234	0.368227
	8	$\|k\rho_1\|$	0.653	0.6711	0.67329	0.673438
		$\arg \rho_1$	2.427	2.4042	2.40120	2.401002

In the s-method, we have the same equations (4.30) (with ε_n replaced by ε) and the condition that the function and its first-order derivative must be continuous at $r = a$.

Instead of the radiation condition, we have the condition

$$u|_{r\to\infty} \simeq \frac{B_1 \cos kr + C_1 \sin kr}{\sqrt{kr}} \tag{4.48}$$

with some real coefficients B_1 and C_1. The eigenvalue λ_1 can be expressed via these coefficients as

$$\lambda_1 = \frac{B_1 + C_1}{B_1 - C_1} \,. \tag{4.49}$$

This problem can be solved explicitly. Furthermore, the eigenvalue has the form

$$\lambda_1 = \frac{J_0(k\sqrt{\varepsilon}a)N_1(ka) - \sqrt{\varepsilon}J_1(k\sqrt{\varepsilon}a)N_0(ka)}{J_0(k\sqrt{\varepsilon}a)J_1(ka) - \sqrt{\varepsilon}J_1(k\sqrt{\varepsilon}a)J_0(ka)} \,. \tag{4.50}$$

The approximate values of λ_0 can be calculated by the Ritz method with the use of the functional

$$\begin{aligned} L(u) \;=\; & \int_0^a \left(\frac{d^2u}{dr^2} + \frac{1}{r}\frac{du}{dr} + k^2\varepsilon u\right) ur\,dr + \int_a^\infty \left(\frac{d^2u}{dr^2} + \frac{1}{r}\frac{du}{dr} + k^2 u\right) ur\,dr \\ & + \mathrm{Re}[\lambda(B_1 - iC_1 + (i\lambda - 1)(B_1 + iC_1)]. \end{aligned} \tag{4.51}$$

Table 4.5

M	$ka = 2$	$ka = 5$	$ka = 10$
1	0.869	-1.986	-0.27539
2	0.87832	-1.990921	-0.27543
Exact	0.87830	-1.9909	-0.27543

The desired eigenfunction can be represented in the form

$$u_0(r) = \begin{cases} B_0 J_0(k\sqrt{\varepsilon} r), & r < a, \\ \sum_{j=1}^{M} (B_j \cos kr + C_j \sin kr) r^{-(j-1/2)}, & r > a. \end{cases} \tag{4.52}$$

The results of calculating λ_1 numerically by using different numbers of coordinate functions are given in Table 4.5 for $\varepsilon = 2.25$ and for several values of ka. The convergence is fast even for small ka: for $M = 2$ (that is, for four coordinate functions in V^-) we have at least four valid digits.

4.2 Scattering on a Quasistationary Level

In the present section we briefly discuss some results obtained by applying the s-method to the quantum-mechanical scattering problem stated in Section 1.5. Here, in contrast with Section 2.5, we replace i by $-i$ and moreover change the sign of eigenvalues so that they coincide with the entries of the quantum-mechanical scattering matrix.

For methodical reasons, we solve the same problem by the s-method and by the ε-method (see the technique developed in Section 1.5). The comparison of numerical results allows us to find the error arising if we take into account only the first term in formulas like (1.164) and (1.165) and one more term in the resonance conditions. We show that a two-term formula describes the scattering sufficiently accurately in a wide range of energy values. We also consider several special distributions of the potential.

1. Suppose that we must solve the Schrödinger equation with potential $U(\mathbf{r})$ so that the solution asymptotically depends on r as follows (see (1.155), (1.156)):

$$\Psi|_{r\to\infty} \simeq e^{ikz} + \frac{e^{ikr}}{kr}\Phi(\varphi, \theta). \tag{4.53}$$

Let us find a system of functions $\psi_l(\mathbf{r})$ with respect to which we can expand the solution of the problem in question. These functions must satisfy the same Schrödinger equation

$$\Delta\psi_l + [k^2 - U(\mathbf{r})]\psi_l = 0 \tag{4.54}$$

and the following conditions at infinity:

$$\psi_l|_{r\to\infty} \simeq \frac{1}{1-s_l}\left[\Phi_l^*(\varphi,\theta)\frac{e^{-ikr}}{kr} - s_l\Phi_l(\varphi,\theta)\frac{e^{ikr}}{kr}\right], \tag{4.55}$$

$$\Phi_l^*(\varphi,\theta) = \Phi_l(\pi+\varphi,\pi-\theta). \tag{4.56}$$

Here Φ_l and s_l are not given; these functions and numbers are to be determined in the course of the solution. One can readily verify that $|s_l| = 1$ for real k and $U(\mathbf{r})$. The asymptotic condition (4.55) means that ψ_l is a superposition of converging and diverging waves as $r \to \infty$ with coefficients being complex-valued functions with equal absolute values of the angles.

In the standard way we can establish the orthogonality condition

$$\int_\Omega \Phi_l \Phi_m^* \, d\Omega = \delta_{lm}. \tag{4.57}$$

We seek the solution of the original scattering problem in the form

$$\Psi = \sum_l A_l \psi_l. \tag{4.58}$$

In this representation, the Schrödinger equation is satisfied by the function $\Psi(\mathbf{r})$ term by term. We seek the coefficients A_l from the requirement that $\Psi(\mathbf{r})$ satisfies the asymptotic condition (4.53). By taking into account (4.55) and by substituting (4.58) into (4.53), we obtain

$$\frac{e^{-ikr}}{kr}\sum_l \frac{A_l\Phi_l^*}{1-s_l} - \frac{e^{ikr}}{kr}\sum_l \frac{A_l s_l \Phi_l}{1-s_l} = e^{ikr\cos\theta} + \frac{e^{ikr}}{kr}\Phi. \tag{4.59}$$

Then we multiply both sides of this relation by Φ_m, integrate over Ω, apply the orthogonality condition (4.57), equate the coefficients of e^{-ikr}/kr with zero, and finally arrive at the following explicit expression for the A_l:

$$A_l = 2\pi i(1-s_l)\Phi_l(\pi). \tag{4.60}$$

Here $\Phi_l(\pi)$ is the value of the eigenpattern in the direction toward the incident wave.

Let us show that the pattern Φ in (4.53) can be expanded with respect to eigenpatterns Φ_l with the same coefficients A_l. To this end, we multiply (4.59) by Φ_m^*, integrate, and equate the coefficients of e^{ikr}/kr with zero. With regard to (4.57), we obtain

$$\int_\Omega \Phi\Phi_l^* \, d\Omega = A_l, \tag{4.61}$$

that is, the A_l are the coefficients of the Fourier expansion of Φ with respect to Φ_l.

Thus we obtain

$$\Phi(\varphi,\theta) = 2\pi i \sum_l (l-s_l)\Phi_l(\pi)\Phi_l(\varphi,\theta) \tag{4.62}$$

and find the representation of the desired scattering pattern as an expansion with respect to the eigenpatterns Φ_l. It follows from the basic assumptions of the s-method (see Section 2.5) that the resonance energy in the scattering on a barrier (a *real quasistationary level*) is determined as the root of the equation $s_l(k) = -1$.

2. Suppose now that $U(\mathbf{r}) = U(r)$ in the homogeneous problem (4.54)–(4.56) of the s-method, that is, the potential is independent of the angular coordinates. By separating the variables in the spherical system and by restricting ourselves to solutions independent of φ, that is, by setting

$$\psi_l(r,\varphi,\theta) = \frac{\chi^{(l)}(r)}{r} P_l(\cos\theta), \tag{4.63}$$

where the $P_l(\cos\theta)$ are the Legendre polynomials, we obtain the equation

$$\chi^{(l)''} + \left\{k^2 - \left[U + \frac{l(l+1)}{r^2}\right]\right\}\chi^{(l)} = 0. \tag{4.64}$$

The condition that ψ_l is bounded at zero yields

$$\chi^{(l)}(0) = 0. \tag{4.65}$$

Since $P_l(\cos(\pi-\theta)) = (-1)^l P_l(\cos\theta)$, it follows from the asymptotic conditions (4.55) and (4.56) that

$$\chi^{(l)}|_{r\to\infty} \simeq \frac{\Phi_l^*(\theta)}{1-s_l}[e^{-ikr} - (-1)^l s_l e^{ikr}]. \tag{4.66}$$

For each l, the one-dimensional problem (4.64)–(4.66) has one eigenvalue s_l and one eigenfunction $\chi^{(l)}$. In fact, this problem coincides with the problem of finding partial waves (and the scattering matrix), which is well known in quantum mechanics; the only distinction is that in the latter problem the amplitude of the converging wave is assumed to be given in advance.

Problem (4.64)–(4.66) can be solved directly by numerical methods, for instance, by integrating the Cauchy problem for Eq. (4.64) with the initial conditions $\chi^{(l)}(0) = 0$ and $\chi^{(l)'}(0) = 1$. If U is a finite-range potential, that is, $U(r) = 0$ for $r > b$, then the general solution of Eq. (4.64) is known in the exterior domain and the eigenvalue s_l is determined from the condition that $\chi^{(l)}$ and $\chi^{(l)'}$ are continuous at the boundary of the potential. If the range of the potential is infinite, then the eigenvalue can be calculated from the limit relation

$$\lambda_l = \lim_{r\to\infty}\left[\frac{k\chi^{(l)}(r)\sin kr + \chi^{(l)'}(r)\cos kr}{k\chi^{(l)}(r)\cos kr - \chi^{(l)'}(r)\sin kr}\right]^{(-1)^l}. \tag{4.67}$$

Here $\chi^{(l)}(r)$ is the solution of Eq. (4.64) obtained by solving the Cauchy problem, $s_l = e^{2i\delta_l}$, and $\operatorname{cotan}\delta_l = \lambda_l$.

3. The problem of elastic scattering on a symmetric barrier can also be solved by the ε-method (see Section 1.5). To compare the results conveniently, we apply this

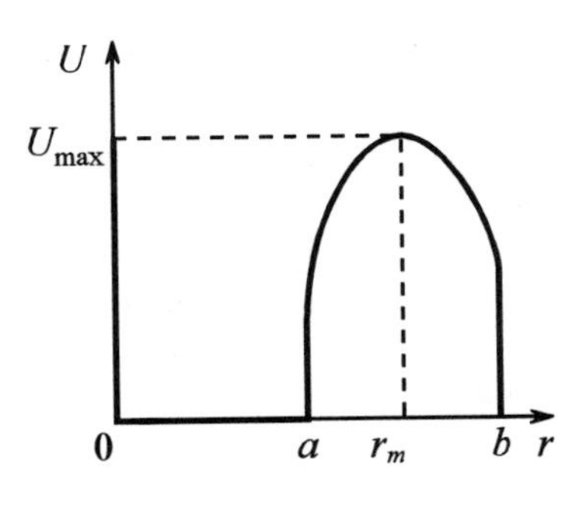

Figure 4.2

method to problem (4.64)–(4.66) rather than the original scattering problem. Thus we shall solve the homogeneous problem of the s-method, which has an explicit solution, approximately by another version of the generalized method.

For simplicity, we restrict ourselves to the case $l = 0$ and omit this index; that is, we write $\chi^{(l)} = \chi$, $s_l = s$, and $\delta_l = \delta$.

If the potential is of a barrier shape (see Fig. 4.2) and $U_{\max} > k^2$, then for almost all k (except for neighborhoods of the quasistationary levels) the function $\chi(r)$ is small in the interior of the barrier domain and the scattering occurs as if at the outer side of the barrier. Near the quasistationary levels, the particles penetrate into the barrier for a long time, and we must supplement the reflection wave by an additional term that takes this effect into account.

Just as in Section 1.5, we introduce a system of functions χ_n by the equation

$$\chi_n'' + [k^2 - U_n(r)]\chi_n = 0 \tag{4.68}$$

with the conditions

$$\chi_n(0) = 0, \tag{4.69}$$
$$\chi_n|_{r\to\infty} \simeq e^{ikr} \tag{4.70}$$

and with the auxiliary potential

$$U_n(r) = \sigma_n U(r) + (1 - \sigma_n)\overline{U}(r), \tag{4.71}$$

where

$$\overline{U}(r) = \begin{cases} U_{\max}, & r \le r_m, \\ U(r), & r > r_m. \end{cases} \tag{4.72}$$

In this homogeneous problem, σ_n plays the role of an eigenvalue. The eigenfunctions are orthogonal with the weight $U - \overline{U}$:

$$\int_0^{r_m} (U - \overline{U})\chi_n\chi_m \, dr = 0, \quad n \ne m. \tag{4.73}$$

We seek the solution of problem (4.64)–(4.66) in the form

$$\chi(r) = \bar{\chi}(r) + \sum_n A_n \chi_n(r), \tag{4.74}$$

where $\bar{\chi}$ satisfies the same problem (4.64)–(4.66) with U replaced by $\overline{U}$. We denote the value of s corresponding to this solution by $\bar{s}$. Then we have

$$s = \bar{s} + 2i \sum_n A_n. \tag{4.75}$$

The coefficients A_n can be determined from the condition that $\chi(r)$ must satisfy Eq. (4.64). By using an identity similar to (1.162), we obtain

$$A_n = \frac{\sigma_n''}{1 - \sigma_n} \frac{\int_0^{r_m} (U - \overline{U}) |\chi_n|^2 \, dr}{\sigma_n \int_0^{r_m} (U - \overline{U}) \chi_n^2 \, dr}. \tag{4.76}$$

Far from the quasistationary level, one has $|\sigma_n''| \ll |1 - \sigma_n'|$, all the coefficients A_n are small, and

$$s \approx \bar{s}. \tag{4.77}$$

Near a quasistationary level, one of the coefficients A_n becomes finite. In the sum (4.75) we preserve only this term. The second factor in (4.76) is equal to $\bar{s}$ with high accuracy. This follows from the fact that in the interior of the barrier (where the factor $(U - \overline{U})$ does not vanish) the function χ_n has resembles a standing wave, that is, its phase is almost constant and the absolute value of the ratio of integrals is close to 1. The σ_n is also close to 1, and therefore, we have

$$A_n \approx \frac{\sigma_n''}{1 - \sigma_n} \bar{s}. \tag{4.78}$$

Thus

$$s \approx \frac{1 - \sigma_n^*}{1 - \sigma_n} \bar{s}. \tag{4.79}$$

By passing to the scattering phases, we can rewrite (4.79) in the form $\delta \approx \bar{\delta} + \delta_n$, where $\delta_n = \arctan\left[\sigma_n''/(1 - \sigma_n')\right]$.

To find the eigenvalues σ_n, we can use the variational technique based on the functional

$$R(\chi) = \frac{\int_0^{\infty} \left[(k^2 - U)\chi^2 - (\chi')^2\right] dr}{\int_0^{r_m} (U - \overline{U}) \chi^2 \, dr}. \tag{4.80}$$

Table 4.6

N	3	4	5	6	Exact
$k_* a$	2.499	2.499	2.498	2.4961	2.4952
$2\hat{\delta} ka$	0.224	0.2276	0.2275	0.2286	0.2289

The admissible functions of this functional must be everywhere continuous and satisfy conditions (4.69) and (4.70).

4. For several special distributions of the potential, we solved problem (4.64)–(4.66) exactly (as the Cauchy problem) and approximately by formulas (4.77) and (4.79) with the help of the variational technique for determining the eigenvalues σ_n. Prior to solving this problem, we verified whether the Ritz method with the coordinate functions

$$v_j = \begin{cases} \left(\frac{r}{b}\right)^{2j-1} e^{ikb}, & r \leq b, \\ e^{ikr}, & r > b, \end{cases} \qquad j = 1, 2, \ldots, N, \tag{4.81}$$

converges for the functional (4.80) in the case of a rectangular barrier for which problem (4.68)–(4.71) admits an explicit solution (a transcendental equation for σ_n). It follows from Table 4.6 that a good accuracy (about 0.2% for the resonance energy k_* and about 2.5% for the resonance width $\hat{\delta} k$) is achieved even if we take into account only three coordinate functions in the Ritz method. The value of $\hat{\delta} k$ is determined by the condition $\delta = \frac{\pi}{2} \pm \frac{\pi}{4}$, the barrier width is equal to $a/4$, $U_{\max} a^2 = 20$, and $b = 3a/2$.

In Fig. 4.3, exact resonance curves (the scattering phase δ versus energy) are plotted by solid curves, the same variables calculated by formulas (4.77) are plotted by dashed curves 1, and those calculated by formulas (4.79), by dashed curves 2. The dot-and-dash curves show the phase δ_n corresponding to the eigenfunction $\chi_n(r)$.

This figure shows that the error in the scattering phase produced by formula (4.77) does not exceed 0.06π if the distance between k and the quasistationary level does not exceed $5\hat{\delta} k$. Closer to the quasistationary level, a good accuracy (not worse than 0.025π in the phase) is provided by the two-term formula (4.79). Although this formula was derived under the assumptions that hold only at resonance, it remains valid (with the same accuracy) in the entire energy range. This suggests that there apparently exists another method for deriving formula (4.79), and this method is independent of the properties of the standing wave in the interior of the barrier.

We must point out that the value of the quasistationary level k_*, which is determined as the root of the equation $\sigma_n'(k) = 1$, is somewhat different from the result obtained by the s-method (as a root of the equation $\hat{\delta} k = \pi/2$). However, this distinction is less than the half-width of the resonance curve and therefore unimportant.

Figure 4.4 shows how the resonance characteristics (k_* and δk) depend on the barrier height for the same potential distributions as in Fig. 4.3. With an increase in the barrier height, the resonance occurs for larger energies of incident particles but the

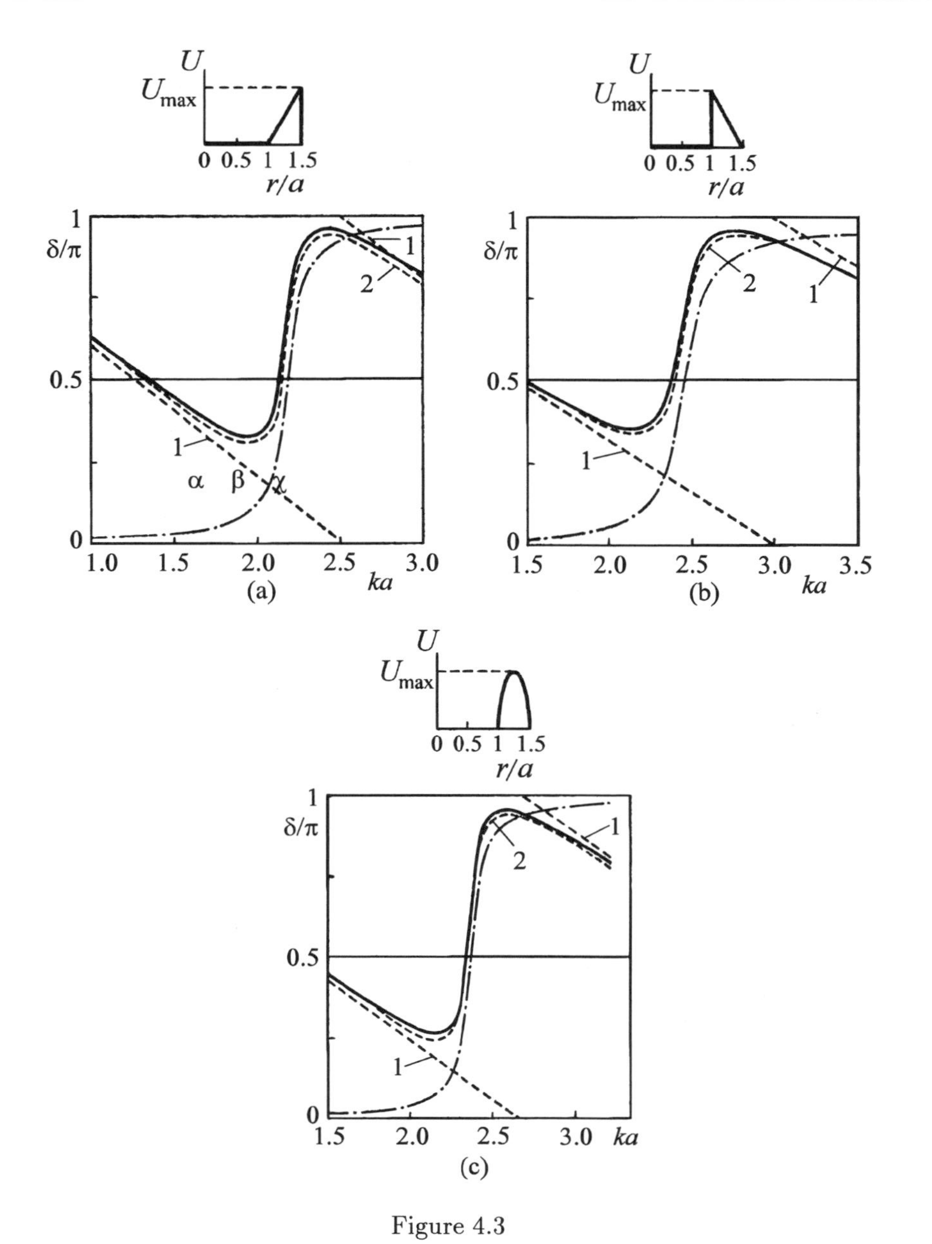

U
U_max
0 0.5 1 1.5
r/a
1
δ/π
0.5
0
1.0 1.5 2.0 2.5 3.0
ka
1
2
α β χ
(a)
1.5 2.0 2.5 3.0 3.5
(b)
1.5 2.0 2.5 3.0
(c)

Figure 4.3

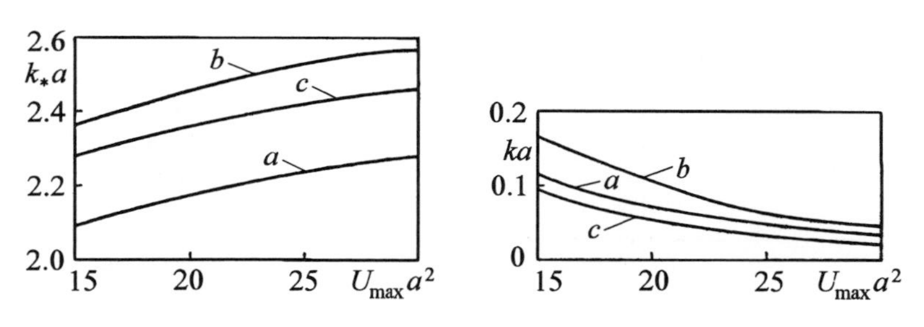

Figure 4.4

relative frequency $k_*^2/U_{\max}$ of the quasistationary level decreases. Therefore, the barrier becomes less transparent for the resonance energy, and the resonance becomes narrower.

4.3 Dielectric Body in an Open or Closed Resonator. Dielectric Waveguides (the ε-Method)

In Subsections 1–4, we find the first eigenvalue of the ε-method for three systems, namely, for a dielectric body in a resonator with semitransparent walls (Subsection 1), in a closed resonator of a special form (Subsection 2), and in a gauge resonator (Subsections 3 and 4). In the first two subsections, we apply the variational technique and the Ritz method. In Subsection 3, we use a method similar to that developed in Sections 1.3 and 1.6 for the diffraction problem. In Subsection 4, we apply two versions of the k-method to the problem of finding the eigenvalue of the ε-method. In Subsection 5, we describe a method for finding the propagation constant of the eigenwave in a dielectric waveguide; this method is logically very close to the ε-method.

1. Let us consider an open resonator formed by a dielectric body of elliptic shape covered by a circular semitransparent film (Fig. 4.5). The eigenfunctions of the homogeneous problem must satisfy Eq. (1.4a) in the domain V_1, Eq. (1.4b) in V_2 and V_3, the condition that u_n and $\dfrac{\partial u_n}{\partial N}$ are continuous on S_1, the conditions

$$\left\{\begin{array}{l} u_n^+ - u_n^- = 0, \\ \dfrac{\partial u_n^+}{\partial N} - \dfrac{\partial u_n^-}{\partial N} - \dfrac{1}{\rho} u_n = 0 \end{array}\right. \tag{4.82}$$

(with a given ρ) on S_2, and the radiation condition at infinity. We restrict consideration to oscillations that are even with respect to the angle θ.

We choose the following coordinate functions, which satisfy all boundary conditions in our problem and Eq. (1.4b) in V_3:

$$v_{pq} = \left\{\begin{array}{ll} J_p(\nu_{p0}) H_p^{(2)}(kr)\cos p\theta, & r \geq a, \\ \left[\alpha_{pq} J_p\left(\mu_{pq}\dfrac{r}{a}\right) + H_p^{(2)}(ka) J_p\left(\nu_{p0}\dfrac{r}{a}\right)\right]\cos p\theta, & r \leq a, \end{array}\right. \tag{4.83}$$

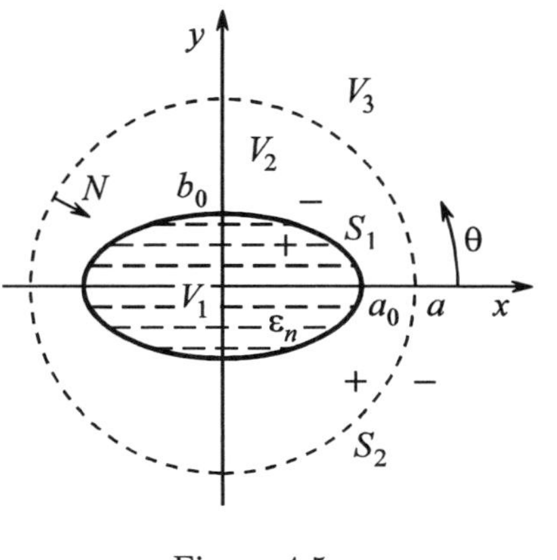

Figure 4.5

where

$$\begin{aligned}\alpha_{pq} &= \frac{kaJ_p(\nu_{p0})}{\mu_{pq}J_p'(\mu_{pq})}\left[H_p^{(2)\prime}(ka) - \frac{1}{k\rho}H_p^{(2)}(ka)\right], \\ J_p(\mu_{pq}) &= 0, \quad J_p'(\nu_{p0}) = 0.\end{aligned} \tag{4.84}$$

The functional that is stationary in the class of such functions does not contain integrals over the boundaries and over the domain V_3 and has the form

$$L(u) = \int_{V_1}(\Delta u + k^2\varepsilon u)u\,dV + \int_{V_2}(\Delta u + k^2 u)u\,dV. \tag{4.85}$$

By applying the Ritz method to (4.85), we calculate the first eigenvalue ε_1. Recall that the eigenvalues ε_n completely determine the resonance properties of a system with any permittivity $\varepsilon = \varepsilon' + i\varepsilon''$. The resonance occurs at the frequency for which $\varepsilon_n' = \varepsilon'$, and the Q-factor is determined by the imaginary part of the eigenvalue.

Numerical results are shown in Figs. 4.6 and 4.7. The dependence of the eigenvalue ε_1 on the wall transparency is shown in Fig. 4.6 for various relations between the geometric parameters. One can readily see that for $k\rho \ll 1$ the real part of the eigenvalue is linear in ρ, and the imaginary part weakly depends on the sizes of the dielectric and is mainly determined by the transparency of the wall.

The dependence of ε_1' on the resonator sizes in shown in Fig. 4.7. For small transparency, ε_1'' is almost independent of ka, and therefore, we do not show this dependence here.

2. Let us consider a closed H-shaped resonator with metallic walls and a dielectric insert (Fig. 4.8). The eigenfunctions of the homogeneous problem arising for this resonator in the ε-method must satisfy Eq. (1.4a) in the domain V^+ filled with dielectric, Eq. (1.4b) in the remaining part V^- of the resonator, the conditions that u_n and $\dfrac{\partial u_n}{\partial N}$ are continuous on the boundary S_1 of the dielectric, and the condition $u_n = 0$ on the resonator walls. Here we shall consider only the oscillations that are even in x and y.

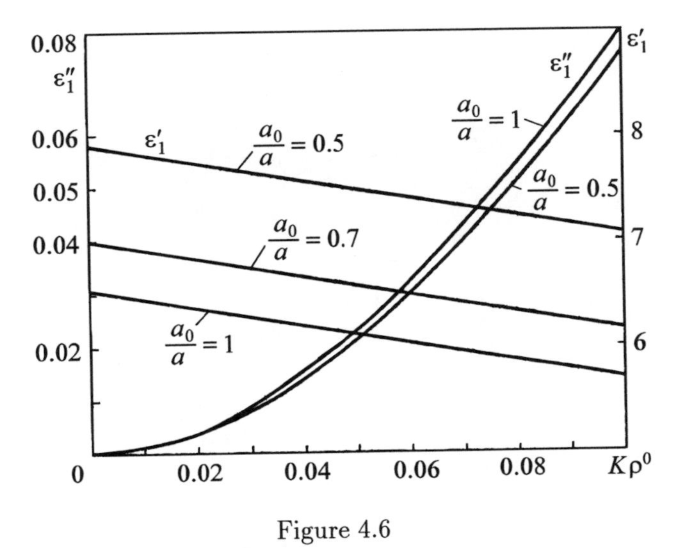

Figure 4.6

Figure 4.7

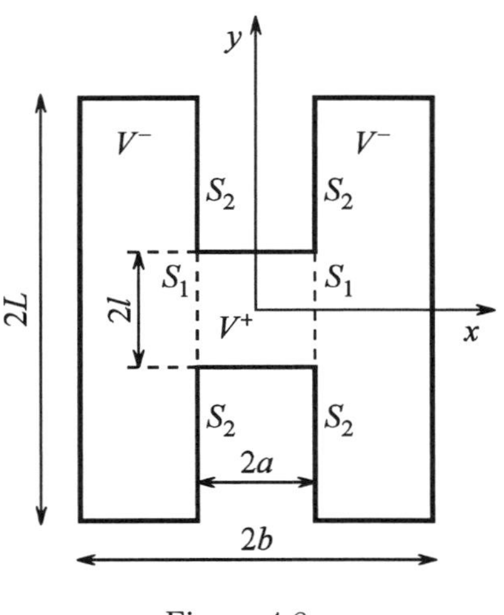

Figure 4.8

To solve this problem it is expedient first to construct a convenient system of coordinate functions and then to choose a functional that is stationary in the class of admissible functions corresponding to this system. *The boundary conditions of the homogeneous problem that are not satisfied by the coordinate functions must be made natural.*

We introduce different coordinate functions in each of the domains V^+ and V^-. Then we have two types of coordinate functions for our resonator: functions of the first type are not zero in V^+ and are identically zero in V^-; functions of the second type are not zero only in V^-. In this case, admissible functions may be discontinuous on S_1.

Since all boundaries are rectilinear, we can represent each coordinate function in the form

$$v_{pq} = f_q(x)\varphi_p(y), \tag{4.86}$$

where $f_q(x)$ and $\varphi_p(y)$ are trigonometric functions. In the domain V^+ we take a system of functions that satisfy the boundary condition on the resonator wall, that is, we set $\varphi_p(y) = \cos\left(p + \frac{1}{2}\right)\frac{\pi y}{l}$. To make this system complete in V^+, we must set $f_q(x) = \cos\frac{\pi q x}{2a}$. Obviously, any function that takes arbitrary values on the boundary S_1 can be expanded with respect to this system.

In a similar way, by satisfying the boundary condition for $|y| = L$, we can set $\varphi_p(y) = \cos\left(p + \frac{1}{2}\right)\frac{\pi y}{L}$ in V^-. By choosing the coordinate functions in this way, we can however satisfy only the condition at $|x| = b$, while the condition on the resonator walls at $|x| = a$ (this part of the wall is denoted by S_2) cannot be satisfied. Indeed, functions of the form (4.86) that vanish on S_2 are also automatically zero on S_1, while the actual solution does not vanish on S_1. To make the system complete, we must set

$$f_q(x) = \sin \frac{\pi(q+1)|b-x|}{b-a}$$

in V^-.

Thus we have constructed a system of functions that satisfy all boundary conditions except for the conditions for $|x| = a$. These conditions (a function is zero on S_2 and a function and its derivative are continuous on S_1) must be made natural for the functional. The functional with required properties can be obtained as a special case of (3.34). For our problem this functional has the form

$$\begin{aligned} L(u) &= \int_{V^+} [(\Delta u)^2 - k^2 \varepsilon u^2]\, dV + \int_{V^-} [(\Delta u)^2 - k^2 u^2]\, dV \\ &\quad + \int_{S_1} (u^+ - u^-) \left(\frac{\partial u^+}{\partial N} + \frac{\partial u^-}{\partial N} \right) dS - 2 \int_{S_2} u \frac{\partial u}{\partial N}\, dS, \end{aligned} \tag{4.87}$$

where N is the outward normal to V^-.

One can readily verify that the functional (4.87) is not positive definite, since it contains integrals over S_1 and S_2. We shall verify the convergence of the Ritz method numerically. The comparison of the results obtained for different numbers of coordinate functions shows that the least (in the absolute value) eigenvalue ε_1 can be calculated with an error about 1% if the total number of coordinate functions (in V^+ and V^-) is 6–7. It is impossible to improve the accuracy by increasing the number of coordinate functions without some special operations that improve the stability of numerical calculations. Nevertheless, an increase in the number of coordinate functions significantly (approximately to 1–5%) improves the accuracy of the calculation of higher eigenvalues.

Figure 4.9 shows how the first three eigenvalues ε_n depend on the thickness of the dielectric insert for $ka = \sqrt{2}$, $L/a = 3$, and $b/a = 2$.

3. In this and in the following section, the problem on a dielectric body placed in a closed resonator is studied in the following statement: we know the resonator shape, the body shape and position, and the first resonant frequency; it is required to find the permittivity ε of the body. The permittivity ε of various materials can be determined just in this way for centimeter waves. In fact, we seek the first eigenvalue (in terms of the ε-method) of a system consisting of a dielectric body (a sample) in a resonator.

Usually, the sample shape is chosen so that its effect on the field structure is not too destructive. For instance, in the case of a cylindrical resonator the sample is a cylinder of the same radius. Then ε can be calculated by simple formulas. However, it is not always possible to use a sample of a required shape, for instance, if nondestructive monitoring is performed or the sample material is difficult to process. In the method for measuring ε via the resonant frequency of a resonator with a sample, the shape of the sample may be arbitrary, but then the formulas obtained for calculating ε may be very complicated. In the method proposed here for solving this problem, the field of the eigenoscillations can be expanded in a series with respect to any orthogonal function system. Then we seek the value of ε for which this series satisfies the homogeneous Helmholtz equation

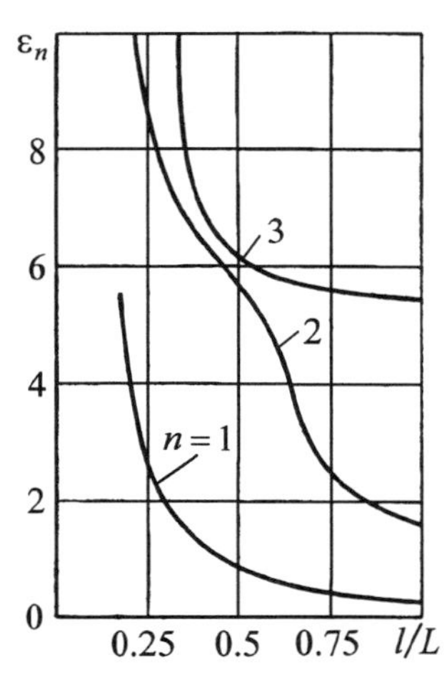

Figure 4.9

or the homogeneous Maxwell system. The equation for ε is obtained by equating an infinite determinant with zero.

In this subsection we use the system of eigenfunctions of the ε-method which corresponds to the same resonator with *another* inserted ("auxiliary") *dielectric body.* If the shape of this auxiliary body is adapted to the resonator shape, then these functions can readily be found. Moreover, if this shape is close to that of the sample, then the above-mentioned infinite determinant converges rapidly.

Let us illustrate this method for a scalar problem. Just as in Section 1.3, we first consider the permittivity ε of the sample and the permittivity $\varepsilon_{\mathrm{aux}}$ of the auxiliary body as continuous functions of the coordinates: $\varepsilon = \varepsilon(\mathbf{r})$ and $\varepsilon_{\mathrm{aux}} = \varepsilon_{\mathrm{aux}}(\mathbf{r})$. Discontinuous functions appear later, after formula (4.94). This allows us not to verify whether the boundary conditions are satisfied on the surfaces of the bodies.

We introduce the functions $u_n(\mathbf{r})$ as the solution of Eq. (1.65) with the condition (1.74), that is, of the equation

$$\Delta u_n + k^2 \varepsilon_n(\mathbf{r}) u_n = 0, \tag{4.88a}$$

$$\varepsilon_n(\mathbf{r}) = 1 + \sigma_n[\varepsilon_{\mathrm{aux}}(\mathbf{r}) - 1], \tag{4.88b}$$

where k is the measured resonant frequency. On the function $\varepsilon_{\mathrm{aux}}(\mathbf{r})$ we impose the only condition that the system "an auxiliary body in the resonator" has the same first eigenfrequency k. Then $\varepsilon_1(\mathbf{r}) = \varepsilon_{\mathrm{aux}}(\mathbf{r})$, that is, $\sigma_1 = 1$. On the resonator surface the functions $u_n(\mathbf{r})$ satisfy the conditions $u_n|_S = 0$ or $\left.\dfrac{\partial u_n}{\partial N}\right|_S = 0$. It follows from (1.87) that they are orthogonal with weight:

$$\int_{V_{\mathrm{res}}} [\varepsilon_{\mathrm{aux}}(\mathbf{r}) - 1] u_n u_m \, dV = \delta_{nm}. \tag{4.89}$$

The integral is taken over the entire volume of the resonator, in fact, over the part where $\varepsilon_{\text{aux}}(\mathbf{r}) - 1 \neq 0$.

The field $u(\mathbf{r})$ of the eigenoscillations of the resonator with a sample satisfies the equation

$$\Delta u + k^2 \varepsilon(\mathbf{r}) u = 0 \tag{4.90}$$

with the same boundary conditions on the resonator walls as $u_n(\mathbf{r})$.

We represent $u(\mathbf{r})$ by the series

$$u(\mathbf{r}) = \sum_n C_n u_n(\mathbf{r}). \tag{4.91}$$

Substituting this series into (4.90), differentiating term by term, and using (4.88a), we obtain

$$\sum_n C_n [\varepsilon(\mathbf{r}) - \varepsilon_n(\mathbf{r})] u_n = 0. \tag{4.92}$$

To be able to use (4.89), we rewrite this functional equation in the form

$$\sum_n C_n [\varepsilon(\mathbf{r}) - \varepsilon_{\text{aux}}(\mathbf{r})] u_n = \sum_n C_n [\varepsilon_n(\mathbf{r}) - \varepsilon_{\text{aux}}(\mathbf{r})] u_n \tag{4.93}$$

and then multiply it by u_m and integrate over the resonator volume with regard to (4.88b) and (4.89). We thus obtain the system of linear equations

$$\sum_n C_n \int_{V_{\text{res}}} [\varepsilon(\mathbf{r}) - \varepsilon_{\text{aux}}(\mathbf{r})] u_n u_m \, dV = C_m(\sigma_m - 1). \tag{4.94}$$

Now we assume that the sample occupies the volume V_{samp} and its permittivity is ε, while the auxiliary body occupies the volume V_{aux} and its permittivity is ε_{aux}. Then we have $\varepsilon(\mathbf{r}) = \varepsilon$ inside V_{samp} and $\varepsilon(\mathbf{r}) = 1$ outside V_{samp}, while $\varepsilon_{\text{aux}}(\mathbf{r}) = \varepsilon_{\text{aux}}$ inside V_{aux} and $\varepsilon_{\text{aux}} = 1$ outside V_{aux}. Hence the integrals in (4.94) are taken over the volume V_{aux} if, for simplicity, we assume that V_{samp} lies completely in V_{aux}. In the volume V_{samp} we have $\varepsilon(\mathbf{r}) - \varepsilon_{\text{aux}}(\mathbf{r}) = \varepsilon - \varepsilon_{\text{aux}}$, and in the volume V_{-} we have $\varepsilon(\mathbf{r}) - \varepsilon_{\text{aux}}(\mathbf{r}) = 1 - \varepsilon_{\text{aux}}$, where V_{-} is the volume of the auxiliary body that lies outside the sample. The normalization condition (4.89) and the expression for $\sigma_n - 1$ acquire the form

$$\int_{V_{\text{aux}}} u_n u_m \, dV = \frac{1}{\varepsilon_{\text{aux}} - 1} \delta_{nm}, \tag{4.95a}$$

$$\sigma_n - 1 = \frac{\varepsilon_n - \varepsilon_{\text{aux}}}{\varepsilon_{\text{aux}} - 1}. \tag{4.95b}$$

By S_{nm} we denote the real coefficients

$$S_{nm} = (\varepsilon_{\text{aux}} - 1) \int_{V_{-}} u_n u_m \, dV. \tag{4.96}$$

By using the normalization condition (4.95a), we can rewrite (4.96) in the form homogeneous in u_n and u_m:

$$S_{nm} = \frac{1}{N_n N_m} \int_{V_-} u_n u_m \, dV, \qquad N_n^2 = \int_{V_{\text{samp}}} u_n^2 \, dV. \tag{4.97}$$

Obviously, we have $S_{nm} = S_{mn}$, $S_{nm} \geq 0$, and $S_{nm}^2 \leq 1$.

In this notation, the homogeneous system (4.94) has the form

$$\sum_n C_n \left(S_{nm} - \delta_{nm} \frac{\varepsilon - \varepsilon_n}{\varepsilon - 1} \right) = 0. \tag{4.98}$$

Thus the unknown number ε satisfies the equation

$$\det \left| S_{nm} - \delta_{nm} \frac{\varepsilon - \varepsilon_n}{\varepsilon - 1} \right| = 0. \tag{4.99}$$

To find ε, we must know V_{aux} and ε_{aux} and then find the eigenelements σ_n and $u_n(\mathbf{r})$ of Eq. (4.98), calculate the integrals in (4.103), and solve Eq. (4.99). The desired variable ε is contained only in the diagonal entries of the determinant.

If the volume V_- is small in contrast with V_{aux}, then ε is close to ε_{aux} and, replacing ε by ε_{aux} everywhere in (4.99) except for the first term, we obtain an approximate value of ε. If we retain only one element ($n = 1$, $m = 1$) in the determinant (4.99), then we obtain a *coarser approximation*. In this case, we have the following explicit expression for ε:

$$\varepsilon = \frac{\varepsilon_{\text{aux}} - S_{11}}{1 - S_{11}}, \qquad S_{11} = \frac{\displaystyle\int_{V_-} u_1^2 \, dV}{\displaystyle\int_{V_{\text{aux}}} u_1^2 \, dV}. \tag{4.100}$$

Let us compare this expression with the formula

$$\varepsilon = \frac{\varepsilon_{\text{aux}} - \tilde{S}}{1 - \tilde{S}}, \tag{4.101a}$$

$$\tilde{S} = \frac{\displaystyle\int_{V_1} u u_1 \, dV}{\displaystyle\int_{V_{\text{aux}}} u u_1 \, dV}, \tag{4.101b}$$

which readily follows from (4.88a) with $n = 1$ and (4.90). Furthermore, we can replace u by u_1 in the denominator of the expression for $\tilde{S}$ in (4.101b), since, according to (4.95a), all terms in (4.91) except for the first are orthogonal to u_1. Thus the difference between the approximate solution (4.100) and the exact formula (4.101) is that the field u is replaced by u_1 in the numerator of the expression for $\tilde{S}$. In other words, replacing ε by ε_{aux} in (4.100), we do not take into account the changes occurring in the fields in V_-.

The solution of the scalar problem is based on the technique developed in Section 1.3. The same vector problem can be solved, that is, reduced to an equation of the form (4.99), by using the technique developed in Section 1.6 for the Maxwell system. The series (1.189) (without the terms $\mathbf{E}^0$ and $\mathbf{H}^0$) for the field in the resonator with a sample are substituted into Eq. (1.197) (without right-hand sides). The vector functions $\mathbf{e}_n$ and $\mathbf{h}_n$ satisfy Eqs. (1.198) and the corresponding conditions on the resonator walls. The functions $\varepsilon_n(\mathbf{r})$ are given by Eq. (4.88b), and again we have $\varepsilon_1(\mathbf{r}) = \varepsilon_{\text{aux}}(\mathbf{r})$. The fields $\mathbf{e}_n$ are orthogonal and normalized by the condition

$$\int_{V_{\text{res}}} [\varepsilon_{\text{aux}}(\mathbf{r}) - 1]\mathbf{e}_n \mathbf{e}_m \, dV = \delta_{nm} \tag{4.102}$$

similar to (4.89). Applying the operation **rot** to the series (1.189) and performing the same transformations as in the scalar case, we obtain a homogeneous system of equations for the coefficients of the series. This system differs from (4.98) by the replacement of $u_n u_m$ by $\mathbf{e}_n \mathbf{e}_m$ in the expression for S_{nm}. The desired permittivity ε of the sample satisfies the same equation (4.99) in which we set (by analogy with (4.97))

$$S_{nm} = \frac{1}{N_n N_m} \int_{V_-} \mathbf{e}_n \mathbf{e}_m \, dV, \qquad N_n^2 = \int_{V_{\text{aux}}} (\mathbf{e}_n)^2 \, dV. \tag{4.103}$$

Explicit approximate solutions of the form (4.100) can be found according to the same scheme as in the scalar problem. If V_- is small, then these solutions are close to exact formulas of the form (4.101).

4. In this subsection we solve the same problem of calculating the permittivity ε of the sample in a resonator via the measured resonant frequency by representing the functions $u(\mathbf{r})$ in the form of the series (4.91) with respect to the eigenfunctions of the k-method (see Section 0.2).

We begin with the following *simple version*: we assume that the functions $u_n(\mathbf{r})$ in (4.91) are the eigenfunctions of the empty resonator corresponding to the eigenvalues k_n, that is, they satisfy Eqs. (0.15). Obviously, they are not the functions denoted in this way in Subsection 4.3.2. They are mutually orthogonal in integrals over V_{res} (0.19), and we normalize them by the conditions

$$\int_{V_{\text{res}}} u_n u_m \, dV = \delta_{nm} \tag{4.104}$$

Substituting the series in u_n into the equation for u, differentiating term by term, using (0.15), and passing to the discontinuous distribution of $\varepsilon(\mathbf{r})$, we arrive at the system of homogeneous equations for the coefficients of the series. The existence condition for a nontrivial solution of this system is just the desired equation for ε:

$$\det \left| \int_{V_{\text{samp}}} u_n u_m \, dV - \frac{1}{\varepsilon - 1} \frac{k_n^2 - k^2}{k^2} \delta_{nm} \right| = 0, \tag{4.105}$$

which is similar to (4.99). It is expedient to use this method only for samples whose volume is small compared with that of the resonator.

Apparently, it is more efficient to use a *modified version* of the k-method. Namely, we take the functions $u_n(\mathbf{r})$ to be the eigenfunctions of the k-method for the system consisting of a resonator with an auxiliary body whose permittivity is $\varepsilon_{\text{aux}}(\mathbf{r})$. These functions satisfy the equation

$$\Delta u_n + k_n^2 \varepsilon_{\text{aux}}(\mathbf{r}) u_n = 0 \tag{4.106}$$

and are orthogonal with respect to (0.28). We normalize these functions by

$$\int_{V_{\text{res}}} \dot{\varepsilon}_{\text{aux}}(\mathbf{r}) u_n u_m \, dV = \delta_{nm}. \tag{4.107}$$

By the same transformations as previously, we obtain the following equation for ε:

$$\det \left| \int_{V_{\text{samp}}} u_n u_m \, dV + \frac{1-\varepsilon_{\text{aux}}}{\varepsilon - \varepsilon_{\text{aux}}} \int_{V_-} u_n u_m \, dV - \frac{1}{\varepsilon - \varepsilon_{\text{aux}}} \frac{k_n^2 - k^2}{k^2} \delta_{nm} \right| = 0. \tag{4.108}$$

Obviously, the determinant of this equation is also infinite.

5.[1] Since generalized methods were developed for bodies of finite volume, it is impossible to use them directly in problems on waves propagating along a dielectric waveguide. Indeed, assume that there is an auxiliary dielectric waveguide that is simple and close to the (main) waveguide that we must study. We wish to represent the waves of the main waveguide as the series in the eigenwaves of this auxiliary waveguide. Just this program was realized in Subsection 4.3.3 for a dielectric sample in a resonator by choosing an auxiliary sample and by using the system of its eigenoscillations in the ε-method for representing the field of the main dielectric body. Note that in this case our method would not require any modification if the samples were in free space.

In the problem on an open waveguide, two difficulties arise in this way. First, the system of waves of any auxiliary waveguide is finite, and the fields of these waves do not form a complete system. Second, outside the waveguide and, in particular, at infinity with respect to the radial coordinate, the dependence of the fields of the nth wave on the transverse coordinates is determined by the transverse wave number α_n, $\alpha_n^2 = h_n^2 - k^2$, where h_n is the longitudinal wave number. For different waves, the wave numbers h_n are different. Therefore, the α_n are also different and differ from α of the wave in the main waveguide. The convergence of the series becomes worse with increasing radial coordinate. Recall that in all versions of the generalized method, if the bodies are finite, then the dependence of the field on the coordinates outside the body is determined by the wave number k, which is the same for all oscillations.

The so-called *translation formula method* described below eliminates both difficulties by determining the eigenwaves of the auxiliary waveguide in an unusual way. The fields of these waves are usually determined as the eigenfunctions of a problem in which h is an eigenvalue and k is given. We suggest to determine all these waves as the eigenfunctions of the same problem in which the eigenvalue is k but not h, and moreover, *α is given in advance.* In this case the system of waves becomes infinite, all eigenvalues k_n become real, and the behavior of the fields of all waves is the same at the radial infinity. The situation arising here is typical of generalized methods. Formally, this method can be called the k-method, although logically it is close to the ε-method.

[1]The results described in this subsection were obtained by V. V. Shevchenko.

We shall illustrate this method by its scalar version. In the main waveguide, the permittivity is a function of the transverse coordinates: $\varepsilon = \varepsilon(\rho)$, where ρ is a two-dimensional vector in the plane $z = \text{const}$. This function is equal to 1 outside a cylinder with the axis z. The field of its wave has the form $u(\rho)e^{-ihz}$. It is required to find the function $u(\rho)$ and the propagation constant h. The function $u(\rho)$ satisfies the equation

$$\nabla^2 u + [k^2\varepsilon(\rho) - h^2]u = 0, \qquad \nabla^2 \equiv \partial^2/\partial x^2 + \partial^2/\partial y^2 \tag{4.109}$$

and vanishes at the radial infinity, that is, as $\rho \to \infty$. Outside the waveguide and, in particular, as $\rho \to \infty$, the equation acquires the form

$$\nabla^2 u - \alpha^2 u = 0, \qquad \alpha^2 = h^2 - k^2. \tag{4.110}$$

It is significant that the equation does not contain k in this region. We rewrite Eq. (4.109) in the form

$$\nabla^2 u + \{k^2[\varepsilon(\rho) - 1] - \alpha^2\}u = 0. \tag{4.111}$$

The permittivity of the *auxiliary waveguide* is $\varepsilon_{\text{aux}}(\rho)$, which also vanishes outside some cylinder. Its field satisfies the same equation with $\varepsilon(\rho)$ replaced by $\varepsilon_{\text{aux}}(\rho)$. We determine the fields of the eigenwaves of the auxiliary waveguide as the eigenfunctions of the equation

$$\nabla^2 u_n + \{k_n^2[\varepsilon_{\text{aux}}(\rho) - 1] - \alpha^2\}u_n = 0 \tag{4.112}$$

corresponding to the eigenvalues k_n. These fields and the numbers k_n depend on α. If $\varepsilon_{\text{aux}}(\rho)$ is a sufficiently simple function, for example, if it is equal to a number ε_{aux} for $\rho < a$ and to 1 for $\rho > a$, then the problem of finding $k_n(\alpha)$ and $u_n(\rho, \alpha)$ is reduced to solving an elementary problem on the field inside a cylinder with permittivity ε_{aux} on whose boundary the ratio of the field to its normal derivative is given. One can readily verify that the fields u_n are orthogonal with weight $\varepsilon_{\text{aux}}(\rho) - 1$ and, under an appropriate normalization, we have

$$\int [\varepsilon_1(\rho) - 1]u_n u_m \, dS = \delta_{nm}. \tag{4.113}$$

In fact, there is integration over a finite region where $\varepsilon_{\text{aux}}(\rho) \neq 1$.

The subsequent procedures reproduce those performed in Subsection 4.33. Expanding the solution of Eq. (4.84) in a series in the functions u_n,

$$u = \sum_n A_n u_n, \tag{4.114}$$

differentiating term by term, and using condition (4.113), we obtain the following system of equations for A_n:

$$\sum_n A_n \left\{ \left(1 - \frac{k_n^1}{k^2}\right) \delta_{nm} + \int [\varepsilon(\rho) - \varepsilon_{\text{aux}}(\rho)] u_n u_m \, dS \right\} = 0, \quad m = 1, 2, \ldots \tag{4.115}$$

The conditions for this equation to have a nontrivial solution reads

$$\det\left|\left(1-\frac{k_n^1}{k^2}\right)\delta_{nm}+\int[\varepsilon(\rho)-\varepsilon_{\text{aux}}(\rho)]u_n u_m\,dS\right|=0. \tag{4.116}$$

In this equation, the numbers k_n, as well as the functions u_n, depend on α. Therefore, its solution is also a function of α, $k=k(\alpha)$. In particular, if $\alpha=0$, then k is the critical frequency. The desired propagation constant is also a function of α, $h^2(\alpha)=l^2(\alpha)+\alpha^2$. The usual dependence $h=h(k)$ has the parametric form $k=k(\alpha)$, $h=h(\alpha)$.

The integrals in (4.116) are taken over the region where $\varepsilon_1(\rho)\neq\varepsilon(\rho)$. If both waveguides are homogeneous, then, just as in Subsection 4.3.3, we have

$$\int(\varepsilon-\varepsilon_{\text{aux}})u_n u_m\,dS=(\varepsilon-\varepsilon_{\text{aux}})\int_S u_n u_m\,dS+(1-\varepsilon_{\text{aux}})\int_{S_-}u_n u_m\,dS, \tag{4.117}$$

where S is the cross-section of the main waveguide and S_- is the region between the contours of the auxiliary and principal waveguides.

The scheme presented here was successfully used in some special problems. Homogeneous waveguides with complicated cross-section contours, many-layered waveguides, coupled waveguides, waveguides with filling inhomogeneous in the cross-section (isotropic and nonisotropic), etc. were considered. After some modification, this method was used for solving a two-dimensional problem on a plane waveguide between two different media.

4.4 Open Resonator Formed by a Pair of Metallic Plates (the ρ-Method)

In this section we consider a two-dimensional problem on an open resonator formed by a pair of metallic plates. Employing the technique of integral equations, we solve the homogeneous problem of the first version of the ρ-method and study the resonance properties of the system. The results obtained are compared with a known asymptotics and thus *the accuracy of the asymptotics results is estimated* for various values of geometric parameters. In the last Subsection, we briefly discuss a new approach to the solution of homogeneous integral equations.

1. Let us consider an open resonator formed by two plane-parallel metallic planes of width $2a$ arranged at the distance L from each other (Fig. 4.10).

The homogeneous problem of the ρ-method consists of the equation

$$\Delta u_n+k^2u_n=0 \tag{4.118}$$

continued to the entire plane (x,z), the boundary conditions (only for $|x|\leq a$)

$$u_n\Big|_{|x|=\frac{L}{2}+0}-u_n\Big|_{|x|=\frac{L}{2}-0}=0, \tag{4.119a}$$

$$\left.\frac{\partial u_n}{\partial z}\right|_{|x|=\frac{L}{2}+0}-\left.\frac{\partial u_n}{\partial z}\right|_{|x|=\frac{L}{2}-0}\pm\left.\frac{1}{\rho}u_n\right|_{x=\pm\frac{L}{2}}=0, \tag{4.119b}$$

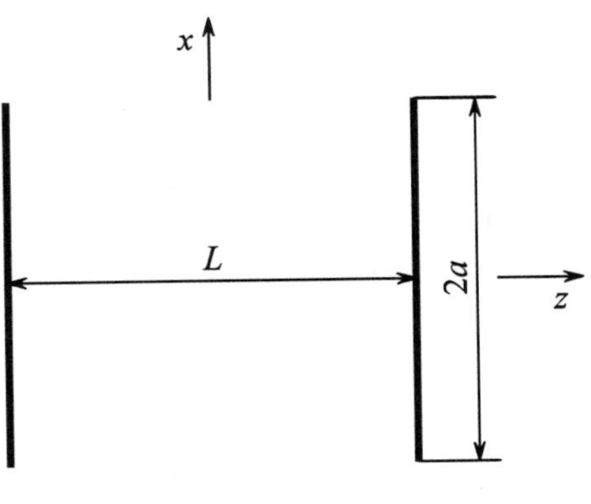

Figure 4.10

the Meiksner conditions at the plate edges, and the radiation conditions at infinity. As it follows from Subsection 2.2.2, to this problem there corresponds a homogeneous Fredholm integral equation of the second kind, namely,

$$\rho_n v_n(P) = \frac{i}{4} \int_S H_0^{(2)}(kR) v_n(P')\, dP', \tag{4.120}$$

where the points P and P' range over the surfaces S of both plates and R denotes the distance between these points.

Since this problem is symmetric, only even or odd (in x, as well as in z) solutions are possible. A function $v(P)$ is said to be even in z if the values (with sign) of this function on the two plates coincide. We shall study only solutions that are even with respect to x.

By introducing the dimensionless coordinate

$$\xi = x/a, \tag{4.121}$$

we can rewrite our equation in the form

$$\begin{aligned} \frac{\rho_n}{a} v_n(\xi) \; &= \; \frac{i}{4} \int_{-1}^{1} \Big[H_0^{(2)}(ka|\xi - \xi'|) \\ &\pm H_0^{(2)}\Big(ka\sqrt{L^2 \pm (\xi - \xi')^2}\Big)\Big] v_n(\xi')\, d\xi'. \end{aligned} \tag{4.122}$$

Here the plus sign in the square brackets corresponds to solutions that are even in z, and the minus sign, to those odd in z. To solve Eq. (4.122) numerically, we represent the unknown function $v_n(\xi)$ by the series

$$v(\xi) = \sum_{m=1}^{\infty} B_m f_m(\xi) \tag{4.123}$$

in the functions

$$f_m(\xi) = \cos \frac{(2m-1)\pi\xi}{2 + 0,824(1-i)/c}, \tag{4.124}$$

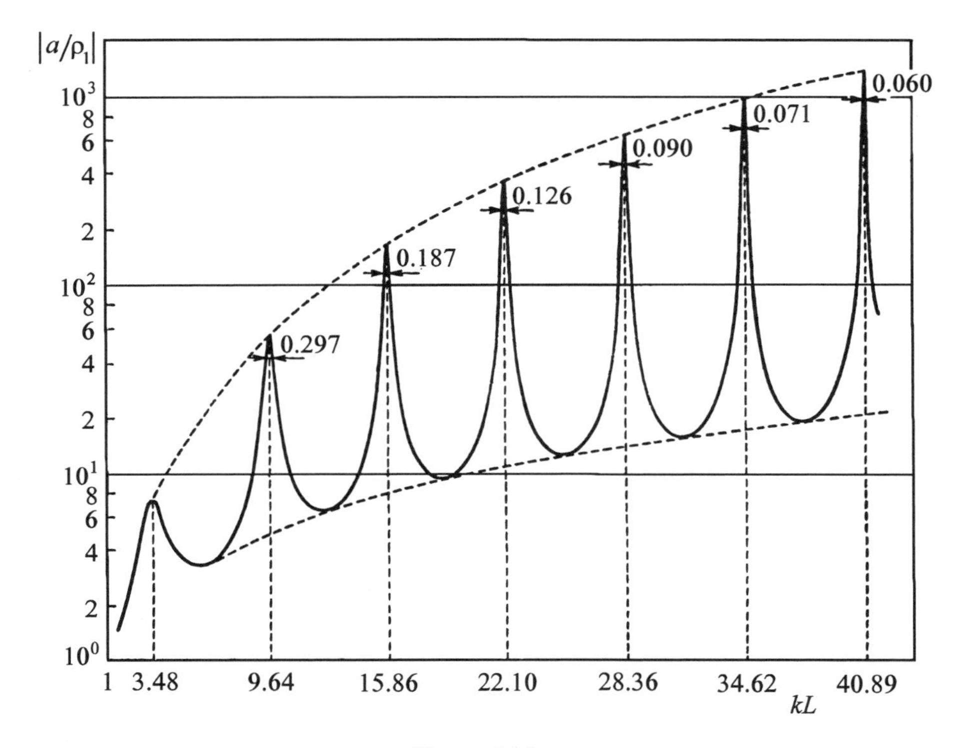

Figure 4.11

which describe the eigenoscillations of the resonator in question in the well-known asymptotic approximation [17]. Here

$$c = \frac{ka^2}{L} \tag{4.125}$$

is the quasioptical parameter of the resonator. We substitute (4.123) into (4.122), multiply on the left and on the right by $f_n(\xi)$ and integrate. This leads to an infinite system of homogeneous linear equations. We obtain the eigenvalues ρ from the condition that the determinant of this system is equal to zero.

The curve in Fig. 4.11 illustrates the resonance properties of the resonator if the actual boundary condition is $U|_S = 0$. This curve corresponds to the first eigenvalue ρ_1 of Eq. (4.122) for $L/a = 2$. There are radiation losses in the system considered, and therefore, the resonance peaks on the curve are of finite height. If the width $2\delta kL$ of the peak (corresponding to the level $1/\sqrt{2}$) is known, then one can determine the Q-factor of the resonator by the formula

$$Q = \frac{kL}{2\delta kL}. \tag{4.126}$$

For instance, we have $Q \approx 680$ for $kL = 40.89$.

Table 4.7

L/a	1.29	1.58	1.83	2.04	2.24	2.42
kL	6.43	9.58	12.72	15.86	19.01	22.15
$2\delta kL$	0.235	0.210	0.198	0.192	0.189	0.186
L/a	2.58	2.74	2.89	3.03	3.98	4.92
kL	25.29	28.43	31.58	34.70	59.85	91.27
$2\delta kL$	0.181	0.179	0.178	0.175	0.164	0.160

The curve corresponds to an eigenfunction even in z; the distance between neighboring maxima of the curve is approximately equal to 2π. The resonant frequencies of oscillations that are odd in z lie halfway between these maxima.

2. In the quasioptical approximation, c is the only parameter that is important for studying such resonators. Resonators with different values of ka and a/L have equal losses and similar field distributions if the product of these variables takes the same value. Such an approximation can be used for $ka \gg 1$ and $L \gg a$.

Let us vary ka together with L/a so that the parameter c remains constant. This allows us to find out for what relations between the parameters the quasioptical approximation is valid.

Generally speaking, the eigenoscillations obtained in the quasioptical approach and the fields described by the eigenfunctions of the ρ-method are different. However, for sufficiently large values of the Q-factor, a leading term appears in both methods if we solve the problem of excitation at the resonant frequency. If for some parameter values both methods realistically describe the physics of the process, then the corresponding eigenfunctions of both methods must be close to each other. Here we shall compare only the resonant frequencies and widths of resonance peaks rather than the eigenfunctions themselves.

The calculations were performed for $c = 1.2\pi$. It follows from the quasioptical approximation that for this value of c, the width of the resonance peak for the fundamental oscillation must be equal to $2\delta kL = 0.147$. The resonant frequencies and widths of resonance peaks obtained by the ρ-method are given in Table 4.7. If the distance between the mirrors is at least two times larger than the mirror sizes $\left(\frac{L}{a} \approx 4\right)$, then the width of the peaks can be calculated by this asymptotic method with accuracy of the order of 10%. Within the limits of the accuracy attained in numerical calculations (to the second decimal place), the resonant frequencies completely coincide with those calculated by asymptotic formulas. The eigenfunctions (both even and odd in z) of the ρ-method corresponding to the same transverse type of oscillations in quasioptics are used in Table 4.7.

3. A similar problem was studied for a *resonator with curved mirrors*. Each mirror was an arc of the circle whose center coincides with that of the other mirror. In quasioptics such resonators are said to be *confocal*. As is well known, the eigenoscillations of

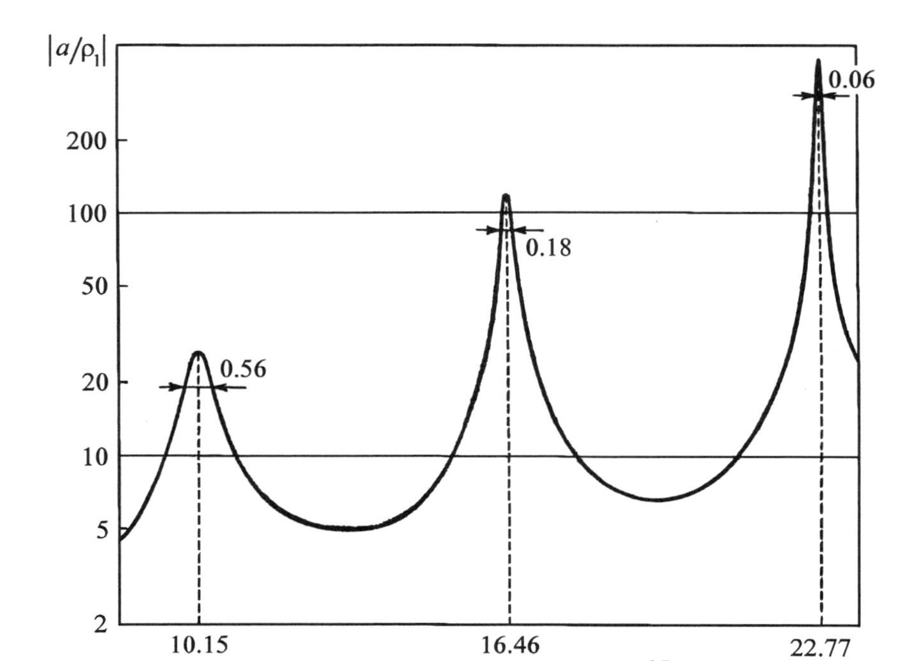

Figure 4.12

such resonators are described by spheroidal functions. These functions were chosen as coordinate functions for solving the integral equation by the ρ-method. The numerical results obtained are shown in Fig. 4.12 (for $L/a = 3$) and in Table 4.8 (for $c = 2.2$). The quasioptical approximation gives $2\delta kL = 0,110$ for the value of c corresponding to Table 4.8.

Table 4.8

L/a	2.15	2.74	3.22	3.63	4.67	5.76
kL	10.21	16.49	22.77	29.05	47.90	73.03
$2\delta kL$	0.122	0.122	0.110	0.122	0.115	0.117

It has already been pointed out that the generalized method of eigenoscillations is convenient for analysis of resonance systems with large Q-factors. Namely, in this case a leading term appears if the field is represented as a series with respect to eigenfunctions, and this leading term describes the complete solution with sufficient accuracy. On the other hand, if the Q-factor is very large, some difficulties arise in the numerical solution of the homogeneous problem. For instance, in the integral equation of the ρ-method, the desired eigenvalue becomes very small (in absolute value) near the resonance, and therefore, can be found only with low relative accuracy. However, as the above examples show, these methods allow us to study a fairly wide range of parameters.

The results of this section can readily be used in the case of resonators with semi-transparent mirrors. Such mirrors are characterized by the transparency parameter ρ in the actual boundary condition (e.g., see (2.62a)), and the resonance curve for such mirrors is given by the formula $1/|\rho - \rho_n|$, where ρ_n is the above-mentioned eigenvalue of the integral equation (4.122).

4. The method exploited here for solving the integral equation gives reliable results only for one or two eigenvalues. For the case in which one needs to calculate several eigenvalues and corresponding eigenfunctions with sufficient accuracy (e.g., if the Q-factor of the resonator is small and one needs several terms of the series for solving the diffraction problem), we propose another modification of the well-known iteration method. In this case we shall use not only the last iterations (which, as is well-known, give the eigenvalue with the largest absolute value) but also all preceding iterations (just these iterations give information on higher eigenvalues). Moreover, the total number of iterations turns out to be much less than usually.

Here we write out the basic formulas of this approach but do not derive them (see [59]). Let the initial approximation $v^{(0)}$ be a function on S, and let

$$v^{(m)} = \int_S G v^{(m-1)} \, dS \tag{4.127}$$

be the mth iteration by the kernel of the integral equation (2.49). Let us construct the function $f_N = \sum_{j=0}^{N} (-1)^j c_j v^{(j)}$ with indeterminate coefficients c_j. We assume that one of them, say, c_N, is equal to 1 and the other can be obtained from the requirement that the function f_N vanishes at N distinct points (say, uniformly spaced) on the boundary S. Then the eigenvalues ρ_n of Eq. (2.49) can be calculated as the zeros of the polynomial $P_N = \sum_{j=0}^{N} c_j \rho^j$. Then the eigenfunctions can be found by the formula

$$v_n = \sum_{j=0}^{N} \rho_n^j \sum_{m=0}^{j} (-1)^m c_{j-m} f_{N-m}. \tag{4.128}$$

The number of eigenelements that can be calculated in this way depends on the initial approximation and the density of the initial part of the spectrum of eigenvalues ρ_n. New iterations cease to improve the results as $v^{(m)}$ becomes close to the first eigenfunction. Therefore, in contrast with the usual iteration method, it is expedient to choose $v^{(0)}$ so that its expansion will contain the largest possible number of eigenfunctions with sufficiently large weights.

4.5 Open Resonator with Closed Semitransparent Walls. Leaky Waves in a Waveguide (the ρ-Method)

1. In this and in the following section, we consider open resonators with large Q-factor and weak coupling with the exterior medium. Such resonators are some kinds of traps. For an exterior excitation, the interior field in these resonators is small at almost all frequencies, while the field outside the resonator is the same as it were for the zero coupling. Only in narrow bands of frequencies (namely, near the eigenfrequencies of interior oscillations of a closed resonator of the same shape) the interior field becomes large, while the exterior field is subject to finite perturbations. It is required to find the frequency at which these two phenomena are most pronounced (the real resonant frequency of the open resonator) and the width of the resonance curve.

We apply the technique developed in Section 2.2. We use the fact that the only assumption inposed on the field U^0 (singled out as a separate term in front of the series in (2.34)) is that it must satisfy the wave equation with the same right-hand side as in the diffraction problem. More precisely, U^0 must be generated by the same sources but not necessarily in free space. We choose U^0 to be the field generated by the same sources in the process of diffraction in a similar resonator but with zero coupling, that is, in a "metallized" resonator. We assume that this field is known. In any case, it is not of a resonance character. The advantage gained by this choice of U^0 (in contrast with the case where U^0 is the incident field) is that in this case the perturbation arising in the above-mentioned regions is described only by one term of the series. In fact, this method was used in Section 4.2.

Just as in Section 4.2, we shall examine the solution of the scalar equation. The solution of the Maxwell equations would require more complicated formulas, although it follows the same scheme and leads to the same physical results.

2. First, let us solve the problem of finding the function U that satisfies Eq. (2.32) outside and inside some closed surface S, the radiation condition, and the transmission conditions (2.60) on S, where ρ is a characteristic of the resonator wall, namely, its transparency. We assume that ρ is constant on S, although there are no essential difficulties if ρ is a function of a point of the surface. For a nontransparent (metallic) wall, we have $\rho = 0$. For a two-dimensional problem on resonators (waveguides) with closely arranged longitudinal slots ($U = E_z$), the form of ρ was given in (2.62a). In three-dimensional problems, we encounter more complicated conditions, similar to (2.60). Nearly the same conditions can be imposed on the fields on both sides of a thin dielectric film with large ε.

If there are losses in a semitransparent wall, then $\operatorname{Im}\rho < 0$. If the losses in the walls are small or absent ($-\operatorname{Im}\rho \ll |\rho|$), and moreover,

$$|k\rho| \ll 1, \tag{4.129}$$

then the resonator is a trap, which is just the case that will be studied here.

Let U^0 be a field equal to zero in the interior of the resonator (in V^+) and satisfying Eq. (2.32), the radiation condition, and the boundary condition

$$U^0 = 0|_S \tag{4.130a}$$

outside the resonator (in V^-).

By introducing the eigenfunctions u_n and the eigenvalues ρ_n of the resonator by formulas (2.35a) and (2.36), respectively, we obtain the solution in the usual form (2.34). However, here the term U^0 is introduced in a way different from that in Section 2.2, and therefore, the coefficients A_n are now given by the formula

$$A_n = \frac{1}{\dfrac{1}{\rho_n} - \dfrac{1}{\rho}} \frac{\displaystyle\int_S u_n \frac{\partial U^0}{\partial N} \, dS}{\displaystyle\int_S u_n^2 \, dS} \tag{4.131}$$

rather than by (2.63a). Both formulas were obtained with the help of the orthogonality condition (2.39a). An important distinction from formula (2.63a) (that is valid for another simpler choice of U^0) is the fact that all coefficients (4.131) (possibly, except for one) are small for small $|\rho|$, while all terms in (2.63a) are not small and are of the same order outside the resonance region.

One of the coefficients in (4.131) is not small only at the frequencies for which $1/|\rho_n|$ is so large that the difference between two large variables in the denominator of A_n is finite rather than large. As we shall see, after an explicit expression for ρ_n is obtained, this effect takes place in narrow frequency ranges, since ρ_n strongly depends on frequency. The denominator in (4.131) contains the difference of two large values, namely, $(1/\rho)$, which depends on the coupling, and $(1/\rho_n)$, which is inversely proportional (in the higher-order term) to the difference between the operating frequency and the resonant frequency of a closed resonator (see (4.139) in what follows). If the real parts of these variables cancel each other out, then the resonance occurs and U differs from U^0 by one term in the series (2.34); this term is finite in V^- and large in V^+. This structure of formulas (2.34) and (4.131) and the mechanism producing an essential term in the series (2.34) are also preserved in problems considered in Sections 4.6 and 4.7.

3. Let us express the desired values $1/\rho_n$ in terms of the eigenfrequency k_m and the eigenfunction χ_m of the interior of the closed resonator, that is, in terms of eigenelements of the following selfadjoint problem in V^+:

$$\Delta\chi_m + k_m^2\chi_m = 0, \tag{4.132a}$$

$$\chi_m|_S = 0. \tag{4.132b}$$

The eigenfrequencies k_m are real. To simplify the subsequent formulas, we assume that the eigenfunctions χ_m are also real. The dimension of the index m in (4.121) is larger by one than that of the index n. For example, in three-dimensional problems, m is a set of three numbers and n is a set of two numbers.

Let us apply the Green formula to both wave equations (2.32) and (4.132a) in V^+. In the resulting surface integral $\int_S u_n \frac{\partial \chi_m}{\partial N}\, dS$, we replace the first factor according to the second formula in (2.35a). Thus we obtain the following exact expression for ρ_n:

$$\rho_n = (k_m^2 - k^2)\frac{\int_{V+} u_n \chi_m \, dV}{\int_S \frac{\partial \chi_m}{\partial N}\left(\frac{\partial u_n^+}{\partial N} - \frac{\partial u_n^-}{\partial N}\right) dS}. \tag{4.133}$$

For each n, this formula is valid for infinitely many values of m.

Let us analyze (4.133) in the region of frequencies close to the interior resonance, that is, for k close to k_m.

It follows from (4.133) that the eigenvalue ρ_n is zero for $k = k_m$. This fact is also a straightforward consequence of (4.132b) and (2.35a); namely, for $k = k_m$ there exists an eigenfunction u_n that vanishes on S. In this case, if the corresponding factor is chosen equal to 1, then for a single value of m we have

$$u_n^+ = \chi_m, \tag{4.134a}$$

$$u_n^- = 0. \tag{4.134b}$$

For any k close to k_m, the value of $|\rho_n|$ is small. The factor in (4.133) is also small, and in the higher-order terms in the integrals we can replace u_n by χ_m and $\frac{\partial u_n^+}{\partial N}$ by $\frac{\partial \chi_m}{\partial N}$ and omit $\frac{\partial u_n^-}{\partial N}$. The latter follows from the fact that $\frac{\partial u_n^-}{\partial N}$ can be calculated by using $u_n^-|_S$ in the boundary value problem in V^-, while $u_n^-|_S$ is zero in the zero-order term with respect to $(k - k_m)$. According to (2.35a), $\rho_n \frac{\partial u_n^-}{\partial N}\Big|_S$ is small and in the higher-order term we have

$$u_n^-|_S = \rho_n \frac{\partial \chi_m}{\partial N}\Big|_S. \tag{4.135a}$$

Thus ρ_n is real in the higher-order term and can be expressed in terms of the solution of problem (4.132) as

$$\rho_n \simeq -(k - k_m)A, \quad A = \frac{2k_m \int_{V+} \chi_m^2 \, dV}{\int_S \left(\frac{\partial \chi_m}{\partial N}\right)^2 dS}. \tag{4.136}$$

We also need the next (imaginary) term in ρ_n. We can obtain this term by writing out the following formula, which is similar to (2.37a) (but here we only apply the Green formula to u_n and u_n^* in V^+ and then replace $\dfrac{\partial u_n^+}{\partial N}$ and $\dfrac{\partial u_n^{+*}}{\partial N}$ according to (2.35a)):

$$\operatorname{Im}\frac{1}{\rho_n} = -\frac{\operatorname{Im}\displaystyle\int_S u_n^* \frac{\partial u_n^-}{\partial N}\, dS}{\displaystyle\int_S |u_n|^2\, dS}. \tag{4.137}$$

To find the right-hand side of (4.137) actually, we need to solve the above-mentioned first boundary problem in V^-, that is, find u_n^- in the entire exterior volume V^- by using $u_n^-|_S$ (4.135a) and the radiation condition. Then we need to find the derivative $\dfrac{\partial u_n^-}{\partial N}$ on S. To solve this problem, we need to know the Green function in the domain V^- with the condition

$$G^-|_S = 0, \tag{4.138a}$$

that is, the Green function of the metallized shell. Here we do not write out the corresponding formulas, since G^- is known only for some simple bodies. We only note that, in contrast with A determined by (4.136), the value $\operatorname{Im}(1/\rho_n)$ depends on the bodies placed near the resonator (this determines G^-). The value of $\operatorname{Im}(1/\rho_n)$ is finite, since the small value ρ_n is canceled if we substitute (4.135a) into (4.137). Thus, if by $-B$ we denote the higher-order term (finite as $k \to k_m$) term in the expansion of $\operatorname{Im}(1/\rho_n)$ in powers of $k - k_m$, then $B > 0$ (cf. (2.38)). Thus, retaining only the higher-order terms in $\operatorname{Re}(1/\rho_n)$ and $\operatorname{Im}(1/\rho_n)$, we obtain the desired formula

$$\frac{1}{\rho_n} = \frac{-1}{(k - k_m)A} - iB. \tag{4.139}$$

In fact, it is much more convenient to calculate B directly from (4.139) rather than (4.137). To this end, we need to find $\rho_n(k)$ either from the simple integral equation (2.49) or from the integral equation with real kernel (2.147), which is studied in detail in Section 4.9. For bodies of complicated shape, for which it is difficult to solve the interior problem (4.132), it may be expedient to find not only B but also the coefficient A from these equations.

Formula (4.139) contains only the variables (k_m, χ_m, G^-), which correspond to the metallized resonator.

4. Formulas (4.139) and (4.131) allow us to describe the resonator for any (small) coupling and any excitation. In particular, they allow us to find the traditional characteristic of the resonator, namely, the resonant frequency curve, that is, the dependence of the factor

$$\frac{1}{1/\rho_n - 1/\rho} \tag{4.140a}$$

in (4.131) on k. Let us find the position of the maximum of this curve and its half-width.

For simplicity, we assume that ρ is real. The denominator in (4.140a) attains its minimum at $\operatorname{Re}\dfrac{1}{\rho_n} = \dfrac{1}{\rho}$. Hence the center k_* of the resonance curve is displaced from the interior resonant frequency by

$$\Delta k = -\rho/A \quad (\Delta k = k_* - k_m). \tag{4.141}$$

Thus for resonators of the type considered here, that is, for resonators whose coupling is determined by the transmission conditions (2.60), the frequency shift is independent of the resonator position with respect to external bodies. The shift is of the order of the coupling value and its sign is opposite to the sign of ρ.

Formula (4.141) is of the same structure as the formula for the frequency shift in closed resonators, provided that the boundary condition (4.132b) is replaced by the condition

$$\chi_m = w\frac{\partial \chi_m}{\partial N}\bigg|_S . \tag{4.142}$$

It is well known that the frequency shift in such closed resonators is equal to

$$\Delta k = -\operatorname{Re} w/A, \tag{4.143}$$

where A has the same value as in (4.141).

Since V^+ and V^- are related to each other, it follows that the half-width of the resonance curve, which by (4.140a) and (4.139) is equal to

$$\delta k = \rho^2 \cdot B/A, \tag{4.144}$$

depends on the bodies placed near the resonator. The value δk is proportional to the coefficient B, which characterizes the energy flux through the surface on which there is a field proportional to $\dfrac{\partial \chi_m}{\partial N}$. The width δk is less than the shift Δk by an order of magnitude, and the amplitude A_n is not large but of the order of 1 already at the interior resonant frequency, that is, for $k = k_m$.

One can readily write out the general expression for the field at any frequency. Far from the resonant frequencies, the series in (2.34) is small (of the order of $k\rho$), and the field is close to U^0 in V^- and small in V^+. As k approaches k_*, one term in the series becomes not small.

At the center of the resonance curve, that is, for $k = k_*$, the field is equal (in the higher-order term) to

$$U = U^0 + \frac{i}{B}\,\frac{\displaystyle\int_S \frac{\partial U^0}{\partial N}\frac{\partial \chi_m}{\partial N}\, dS}{\displaystyle\int_S \left(\frac{\partial \chi_m}{\partial N}\right)^2 dS}\,\frac{u_n}{\rho}. \tag{4.145}$$

Outside (in V^-), the eigenfunction u_n is of the order of ρ (according to formula (4.135a) with ρ_n replaced by ρ) and the term added to U^0 in (4.145) is finite. Inside the resonator,

u_n is finite (see (4.134a)) and the field is large, that is, of the order of $1/\rho$. If the coupling decreases, then it follows from (4.144) that the frequency region where the field is large also decreases and, according to (4.141), approached the interior resonant frequency. In this case, the field grows for $k = k_*$.

5. Let us use formula (4.139) for calculating the wave numbers of leaky waves in a waveguide formed by bands parallel to the z-axis. The transverse section of such a waveguide is a resonator with $U = E_z$ of the type considered above.

The complex wave numbers $\hat{h}_m$ are eigenvalues of the two-dimensional equation

$$\Delta u_m + (k^2 - \hat{h}_m^2)u_m = 0 \tag{4.146}$$

with the boundary conditions (2.60) (where ρ is given in (2.62a)) and the condition of at most exponential growth at infinity. By $\hat{k}_m$ we denote the complex eigenfrequency of the two-dimensional open resonator, that is, the eigenvalue of the homogeneous problem

$$\Delta u_m + \hat{k}_m^2 u_m = 0 \tag{4.147}$$

with the same transmission conditions. Obviously,

$$\hat{h}_m^2 = k^2 - \hat{k}_m^2 \tag{4.148}$$

and the $\hat{k}_m$ are the roots of the equation

$$\rho_n(\hat{k}_m) = \rho. \tag{4.149}$$

This equation does not have real roots, since $\operatorname{Im}\rho \leq 0$, while, according to (2.38), we must have $\operatorname{Im}\rho_n > 0$ for real k.

With the help of (4.139) one can readily find the leading terms (with respect to $k\rho$) in the expansion of $\hat{k}_m$ and thus in that of $\hat{h}_m$. For simplicity, we again assume that ρ is real. Then we have

$$\hat{k}_m = k_m - \frac{\rho}{A} + i\rho^2\frac{B}{A}, \tag{4.150}$$

$$\hat{h}_m = h_m + \frac{\rho}{A}\frac{k_m}{h_m} - i\rho^2\frac{B}{A}\frac{k_m}{h_m}, \tag{4.151}$$

where $h_m = \sqrt{k^2 - k_m^2}$ is the wave number of the metallized waveguide. If $|h_m| \ll k$ (that is, near the critical frequencies), then one cannot use (4.151); in this case, we must calculate $\hat{h}_m$ directly from (4.148).

If the transparency is finite ($\rho \neq 0$), then some changes arise already in the terms that are of the first order with respect to ρ; the decay is of the order of ρ^2. Since $A > 0$ and $\rho > 0$ for the polarization in question, it follows from the fact that transparency is finite that the phase velocity decreases, that is, $\operatorname{Re}\hat{h}_m > h_m$.

6. Let us briefly summarize the results obtained for the second polarization, that is, for a resonator for which the boundary conditions (2.61) with some given $\tilde{\rho}$ are satisfied.

The field U^0 that is separated from the complete solution of (2.34) must be the field that corresponds to the diffraction problem for the same surface S with the boundary condition

$$\left.\frac{\partial U^0}{\partial N}\right|_S = 0. \tag{4.130b}$$

The system of eigenfunctions is determined by the wave equation, the radiation condition, and the transmission conditions (2.35b) on S; the $\tilde{\rho}_n$ are the eigenvalues that satisfy the second condition in (2.38).

The coefficients of the series (2.34) are of the same structure as in (4.131); their frequency dependence is mainly determined by the same factor

$$\frac{1}{1/\tilde{\rho}_n - 1/\tilde{\rho}}. \tag{4.140b}$$

The resonances for which one of A_n is not small and U significantly differs from U^0 occur near the frequencies that are the eigenvalues of the same problem (4.132), where the boundary condition (4.132b) is replaced by the boundary condition

$$\left.\frac{\partial \tilde{\chi}_m}{\partial N}\right|_S = 0. \tag{4.152}$$

Near the resonance, the function $\tilde{u}_n^+$ is close to $\tilde{\chi}_m$, and the boundary value $\left.\frac{\partial \tilde{u}_n}{\partial N}\right|_S$ is small and equal to

$$\left.\frac{\partial \tilde{u}_n}{\partial N}\right|_S = -\tilde{\rho}_n \tilde{\chi}_m|_S. \tag{4.135b}$$

The functions $\tilde{u}_n^-$ are determined by their normal derivatives on S, which, according to (4.135b), are proportional to the value of $\tilde{\chi}_m$ on the surface. In practice, we need to know the Green function G^- of the exterior domain satisfying the condition

$$\left.\frac{\partial G^-}{\partial N}\right|_S = 0. \tag{4.138b}$$

By solving this second boundary value problem, we also can find $\tilde{u}_n^-|_S$.

The main result of this study (formula (4.139)) remains valid for $\tilde{\rho}_n$. The only difference is that in the expression (4.136) for the coefficient A, $\frac{\partial \chi_m}{\partial N}$ is replaced by $\tilde{\chi}_m$ in the denominator; the expression obtained for B is similar to (4.137).

The subsequent analysis is the same as for the first polarization. Formulas (4.141) and (4.144) for the shift of the maximum of the resonance curve and for its half-width remain valid (with ρ replaced by $\tilde{\rho}$); formula (4.151) for the complex phase velocity of leaky waves in the band waveguide also remains valid; an analogy between formula (4.141) and the formula for the frequency shift arising in passing from ideal to

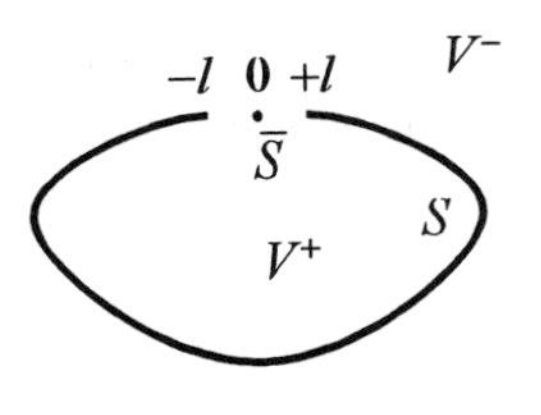

Figure 4.13

impedance walls (with $\tilde{\rho}$ replaced by $-\mathrm{Re}\,\frac{1}{w}$ in (4.141)) and all considerations on the order of magnitude of the fields in V^- and V^+ inside and outside the resonance regions are preserved. Note that since $\tilde{\rho}$ determined by (2.62b) is negative, the phase velocity of the leaky H-waves is larger in this case than in a closed waveguide.

4.6 Two-Dimensional Problem for a Metallic Resonator of Arbitrary Cross-Section with Small Hole and Two Coupled Resonators (H-Polarization; the ρ-Method)

1. In this and in the following section, we use the ρ-method to study the two-dimensional (all variables are independent of z) excitation problem for an open resonator. The resonator is a cylinder of an arbitrary cross-section with a longitudinal slot (Fig. 4.13). We assume that the slot width $2l$ is small compared with the wavelength ($2l \ll \lambda$) and the wall is infinitely thin.

The mathematical statement of the problem is as follows: find the solution of the equation

$$\Delta U + k^2 U = f \tag{4.153}$$

satisfying the condition

$$\left.\frac{\partial U}{\partial N}\right|_{S-\bar{S}} = 0, \tag{4.154}$$

on the walls of the resonator in the case of the H-polarization ($U = H_z$) or the condition

$$U|_{S-\bar{S}} = 0 \tag{4.155}$$

in the case of the E-polarization ($U = E_z$) (this case will be studied in Section 4.7). By S we denote the total contour of the cross-section of the cylinder and by $\bar{S}$ we denote the slot. Moreover, the solution must satisfy the corresponding conditions on the sharp edges and the radiation condition. If we consider the case of excitation by a plane wave, then the radiation condition is imposed on the scattered field. On the slot $\bar{S}$, the total field and its normal derivative must satisfy the continuity conditions

$$\left.(U^+ - U^-)\right|_{\bar{S}} = 0, \tag{4.156a}$$

$$\left.\left(\frac{\partial U^+}{\partial N} - \frac{\partial U^-}{\partial N}\right)\right|_{\bar{S}} = 0. \tag{4.156b}$$

We solve this problem by using the technique developed in Section 2.2 with a slight generalization of the statement of homogeneous problems.

2. Let us introduce the eigenfunctions u_n that satisfy the same equation

$$\Delta u_n + k^2 u_n = 0 \tag{4.157}$$

as the total field, the boundary condition (4.154) on the walls, and, instead of (4.156), the following transmission conditions on the slot (cf. (2.35b)):

$$\left.\left(\frac{\partial u_n^+}{\partial N} - \frac{\partial u_n^-}{\partial N}\right)\right|_{\bar{S}} = 0, \tag{4.158a}$$

$$\left.\left(u_n^+ - u_n^- + \frac{\alpha}{\tilde{\rho}_n}\frac{\partial u_n}{\partial N}\right)\right|_{\bar{S}} = 0, \tag{4.158b}$$

where α is a function on $\bar{S}$, which will be defined later. Here, in contrast with Section 2.2, we generalize the statement of the homogeneous problem by introducing the function α. Moreover, each eigenfunction must satisfy the radiation condition and some conditions on the edges. The numbers $\tilde{\rho}_n$ play the role of eigenvalues of the above-stated homogeneous problem.

By applying the Green formula for two eigenfunctions with different numbers to the exterior (V^-) and interior (V^+) regions of the resonator and by using Eq. (4.157), the boundary conditions on the walls, conditions (4.158) and the radiation conditions, we can readily show that the normal derivatives of the eigenfunctions are orthogonal on the slot $\bar{S}$ with the weight α:

$$\int_{\bar{S}} \alpha \frac{\partial u_n}{\partial N}\frac{\partial u_m}{\partial N}\, dS = 0, \quad n \neq m. \tag{4.159}$$

The formal solution of the excitation problem (4.153), (4.154), (4.156) with sources f disposed, for example, only in the exterior region is given by a single series with respect to the eigenfunctions:

$$U = U^0 + \sum_n A_n u_n \quad \text{in} \quad V^-,$$

$$U = \sum_n A_n u_n \quad \text{in} \quad V^+, \tag{4.160}$$

where U^0 is the solution of the problem of diffraction of the same sources f on the cylinder without a slot (a resonator with zero coupling):

$$\Delta U^0 + k^2 U^0 = f, \tag{4.161a}$$

$$\left.\frac{\partial U^0}{\partial N}\right|_S = 0. \tag{4.161b}$$

Obviously, the representation (4.160) with any coefficients A_n satisfies all assumptions of the original problem except for (4.156a). Since the latter condition must be satisfied, we can use the orthogonality condition (4.159) to find the coefficients

$$A_n = -\tilde{\rho}_n \frac{\displaystyle\int_{\bar{S}} U^0 \frac{\partial u_n}{\partial N}\, dS}{\displaystyle\int_{\bar{S}} \alpha \left(\frac{\partial u_n}{\partial N}\right)^2 dS}. \tag{4.162}$$

In what follows, after solving the homogeneous problem and finding the coefficients A_n, we shall verify that the sums in (4.160) contain only one term in the leading term with respect to kl.

3. Now let us solve the homogeneous problem (4.157), (4.158). By using the Green formula, we can readily reduce this problem to the following homogeneous integral equation of the second kind on the slot for the unknown functions $\varphi_n = \left.\frac{\partial u_n}{\partial N}\right|_{\bar{S}}$:

$$-\frac{\alpha}{\tilde{\rho}_n}\varphi_n = \int_{\bar{S}} \varphi_n (G^+ + G^-)\, dS. \tag{4.163}$$

Here G^+ and G^- are the Green functions for the Helmholtz equation on two disjoint domains V^+ and V^- with the boundary condition

$$\left.\frac{\partial G^+}{\partial N}\right|_S = \left.\frac{\partial G^-}{\partial N}\right|_S = 0.$$

To solve Eq. (4.163), in the kernel of this equation we need to separate the logarithmic singularity and the terms containing the resonances of the isolated domains V^+ and V^-. To this end, it is convenient to expand each term in the kernel, that is, G^+ and G^-, with respect to the eigenfunctions of the homogeneous Steklov problems (in our terminology, with respect to the eigenfunctions of the w-method, see Section 2.1) for the corresponding domains:

$$\Delta\psi_n + k^2\psi_n = 0, \tag{4.164a}$$

$$\left.\left(\psi_n = w_n \frac{\partial \psi_n}{\partial N}\right)\right|_S = 0. \tag{4.164b}$$

In V^- the functions ψ_n must also satisfy the radiation conditions.[2]

Since the further arguments are formally the same for V^+ and V^-, we can temporarily omit the signs "+" and "−."

As was already noted in Section 2.1, the eigenvalues w_n, treated as functions of the frequency, contain all information on the resonance properties of the corresponding

[2]Note that we may impose condition (4.164b) only on a part of the contour S and set, say, $\partial\psi_n/\partial N = 0$ on the complementing part. Sometimes, this statement simplifies the calculations (see the text preceding (4.181a)).

isolated domains. Recall that, say, for the boundary condition (4.154) the resonance occurs in the interior domain when kw_n becomes infinite; in the exterior domain $|kw_n|$ becomes large in the resonance conditions.

Problems (4.164) are equivalent to the integral equations

$$-\, w_n \psi_n = \int_S \psi_n G \, dS, \tag{4.165}$$

where $G = G^+$ or $G = G^-$. We assume that in both the exterior and the interior domain (according to the Hilbert–Schmidt theorem) we have the following bilinear formula, which gives the expansion of the kernel (4.165) with respect to the eigenfunctions ψ_n on S:

$$G(s, s') = -\sum_{n=0} \frac{w_n}{L(1+\delta_{0n})} \psi_n(s)\psi_n(s'). \tag{4.166}$$

Here we have used the normalization $\int_S \psi_n^2 \, dS = (1+\delta_{0n})L$, and $2L$ is the length of the complete contour S. To be definite, we restrict ourselves to problems with symmetry axis and consider only the solutions that are even in s. We also assume that for large numbers the asymptotics of the eigenvalues w_n and the eigenfunctions ψ_n has the form

$$\frac{w_n}{L} \simeq \frac{1}{\pi n} + \frac{(kL)^2}{2(\pi n)^3}, \tag{4.167a}$$

$$\psi_n(s) \simeq \cos \frac{\pi n s}{L}. \tag{4.167b}$$

At least, this assumption is valid for the exterior and interior regions of the circle and for the interior region of the rectangle. There are heuristic considerations showing that this asymptotics remains valid in the general case. In some sense, (4.167) is justified by the results on the asymptotics of the eigenelements of the Steklov problem for the Laplace equation if the contour S is of arbitrary shape.

The asymptotic expansion (4.167) allows us to separate the terms in (4.166) that become infinite at resonances and the terms with logarithmic singularity:

$$\begin{aligned} G(s, s') &= -\frac{w_0 \psi_0(s)\psi_0(s')}{2L} - \sum_{n=1} \left(\frac{w_n}{L} - \frac{1}{\pi n} \right) \psi_n(s)\psi_n(s') \\ &\quad - \sum_{n=1} \frac{\psi_n(s)\psi_n(s')}{\pi n}. \end{aligned} \tag{4.168}$$

The integral in (4.163) is taken over the narrow slot ($|s| \ll L$). Therefore, in the first sum in (4.168), which converges in view of (4.167a), we can approximately set $s = s' = 0$. Finally, by separating the logarithmic singularity in the second sum in (4.168) with the help of (4.167b) and the well-known identity

$$\sum_{n=1} \frac{\cos n\alpha \cos n\beta}{n} = -\frac{1}{2} \ln 2|\cos\alpha - \cos\beta|, \tag{4.169}$$

we see that, up to an unessential constant term (which does not contain resonances), the Green function on the slot can be represented in the form

$$G(s,s')|_{\bar{S}} \simeq C + \frac{1}{2\pi}\ln\left|\left(\frac{\pi s}{L}\right)^2 - \left(\frac{\pi s'}{L}\right)^2\right|, \tag{4.170}$$

where

$$C = -\frac{w_0\psi_0^2(0)}{2L} - \sum_{n=1}\left(\frac{w_n}{L} - \frac{1}{\pi n}\right)\psi_n^2(0), \tag{4.171}$$

and only the higher-order terms with respect to $\frac{s}{L}$ are retained in the logarithm. The last two formulas will play the key role in solving Eq. (4.163).

Let us eliminate the small parameter l in the limits of integration in (4.163). For this, instead of s, we introduce a new variable ξ by the formula

$$s = l\cos\xi. \tag{4.172}$$

After simple transformations with the help of Eq. (4.169), the integral equation (4.163) can be rewritten in the form

$$\begin{aligned} -\frac{\alpha(\xi)\varphi_n(\xi)}{\tilde{\rho}_n} &= l\int_0^\pi \varphi_n(\xi')\sin\xi'\Bigg(C^+ + C^- - \frac{2}{\pi}\ln\frac{\pi l}{2L} \\ &\quad - \frac{2}{\pi}\sum_{n=1}\frac{\cos 2n\xi\cos 2n\xi'}{n}\Bigg)\,d\xi'. \end{aligned} \tag{4.173}$$

Finally, if we set $\alpha(\xi)$ equal to

$$\alpha = l\sin\xi = \sqrt{l^2 - s^2}, \tag{4.174}$$

then (4.173) becomes the simplest equation for the function $\varphi_n(\xi)\sin\xi$. By solving this equation, we find

$$-\frac{1}{\tilde{\rho}_0} = \pi(C^+ + C^-) + 2\ln\frac{\pi l}{2L}, \tag{4.175a}$$

$$\varphi_0 = \frac{1}{l\sin\xi} = \frac{1}{\sqrt{l^2 - s^2}}, \tag{4.175b}$$

$$\tilde{\rho}_n = n \quad (n > 0), \tag{4.175c}$$

$$\varphi_n = \frac{\cos 2n\xi}{l\sin\xi}. \tag{4.175d}$$

It follows from formulas (4.175) that all resonances are contained only in the first eigenvalue $\tilde{\rho}_0$, since only this value depends on the frequency.

4. Let us return to the excitation problem. Note that, according to (4.175b) and (4.175d), the solution (4.160) contains the desired singularity on the edges and the choice of α according to (4.174) ensures, in particular, that the integral in the denominator in (4.162) converges. By substituting (4.175b) and (4.175d) into (4.162), we find the coefficients

$$A_n = -\tilde{\rho}_n \frac{2}{\pi(1+\delta_{0n})} \int_0^{\pi} U^0(l\cos\xi)\cos 2n\xi \, d\xi; \tag{4.176}$$

that is, up to terms of the order of $(kl)^2$, we have

$$A_0 = -\tilde{\rho}_0 U^0(0), \quad A_n = 0 \quad \text{for } n > 0. \tag{4.177}$$

For example, the current on the resonator wall far from the slot is described by the formulas

$$U^-(s) = U^0(s) - \frac{U^0(0)G^-(0,s)}{C^+ + C^- + \frac{2}{\pi}\ln\frac{\pi l}{2L}}, \tag{4.178a}$$

$$U^+(s) = \frac{U^0(0)G^+(0,s)}{C^+ + C^- + \frac{2}{\pi}\ln\frac{\pi l}{2L}}. \tag{4.178b}$$

Resonance phenomena are most instructive in the case of a logarithmically narrow slot. As we shall see, these effects are of the same character as those for the resonator with semitransparent walls studied in the preceding section. Indeed, according to the formulas obtained there, $\tilde{\rho}_0$ and the corresponding field in the interior of the resonator are almost always small (proportional to $1/\ln\frac{\pi l}{2L}$). The fact that the resonance occurs means that at some frequencies close to the eigenfrequencies of a given region the large value $\ln\frac{\pi l}{2L}$ in the denominator $\tilde{\rho}_0$ is compensated for by one of the terms $C^+ + C^-$, for example, by the term $\frac{w_n^+}{L}\psi_n^2(0)$ if the resonance occurs in V^+. In this case, $\tilde{\rho}_0$ and the field complementary to U^0 in V^- are of the order of 1 (that is, the exterior field is under the action of a finite perturbation), and the field in the interior of the resonator becomes large, namely, proportional to $\ln\frac{\pi l}{2L}$. Formula (4.178b) also implies that at the eigenfrequency of the closed domain V^+, the field in the interior of the resonator is already not large (of the order of 1) and the field in the exterior domain is equal to U^0, since in this case the second term in (4.178a) is zero.

Thus the resonance condition for the nth oscillation can be stated in the form

$$\frac{w_n}{L(1+\delta_{0n})}\psi_n^2(0) = \frac{2}{\pi}\ln\frac{\pi l}{2L}. \tag{4.179}$$

This condition allows us to find the resonant frequency shift caused by the slot. For example, for a circular cross-section of radius a we have $w_n^+ = \frac{J_n(ka)}{kJ_n'(ka)}$, and the frequency shift, say, for $n = 0$ is given by

$$\Delta k \approx -\frac{1}{4k_m a^2 \ln\delta/2}, \tag{4.180}$$

where 2δ is the angular size of the slot and $J_1(k_m a) = 0$.

For a rectangular cross-section, we can also explicitly write out w_n^+ if in the auxiliary problem (4.164) only on one wall of the resonator (where the slot is cut) we have condition (4.164b), and on the other three walls we have the condition $\frac{\partial \psi_n}{\partial N} = 0$. Then

$$w_n^+ = -\frac{1}{\beta_n \tan \beta_n b}. \tag{4.181a}$$

Here $\beta_n = \sqrt{k^2 - \left(\frac{\pi n}{a}\right)^2}$, $2a$ is the size of the walls one of which contains the slot, and b is the size of the other walls. In this case, for example, we obtain

$$\Delta k = -\frac{1}{4ma \ln \frac{\pi l}{2a}} \tag{4.181b}$$

for the oscillation for which $n = 0$.

In the general case of an arbitrary contour S, for the frequency shift we have the following formula, similar to (4.141):

$$\Delta k = -\frac{\tilde{\rho}}{\overline{A}}, \tag{4.182}$$

where

$$\tilde{\rho} = \frac{\pi}{4L} \frac{1}{\ln \frac{\pi l}{2L}}, \tag{4.183a}$$

$$\overline{A} = \frac{k_m \int_{V+} \chi_m^2 \, dV}{L \chi_m^2(0)}, \tag{4.183b}$$

and k_m and χ_m are the eigenfrequencies and the eigenfunctions of the interior problem with the condition $\left.\frac{\partial \chi_m}{\partial N}\right|_S = 0$.

This formula can readily be obtained if we apply the Green formula to the functions ψ_n and χ_m in V^+ and use the boundary condition (4.164b), the resonance condition (4.179), and the fact that one of the functions χ_m is close to ψ_n for k close to k_m.

5. The above technique can also be used in full strength for the analysis of closed resonators coupled via a small slot (Fig. 4.14) with the following obvious change in the statement of the problem: there is no radiation condition. In this case the only change in all formulas is caused by the difference in the lengths ($2L^+$ and $2L^-$) of the contours S^+ and S^-. It turnes out that this difference results in replacing $\ln \frac{\pi l}{2L}$ by $\ln \frac{\pi l}{2\sqrt{L^+ L^-}}$ in all formulas.

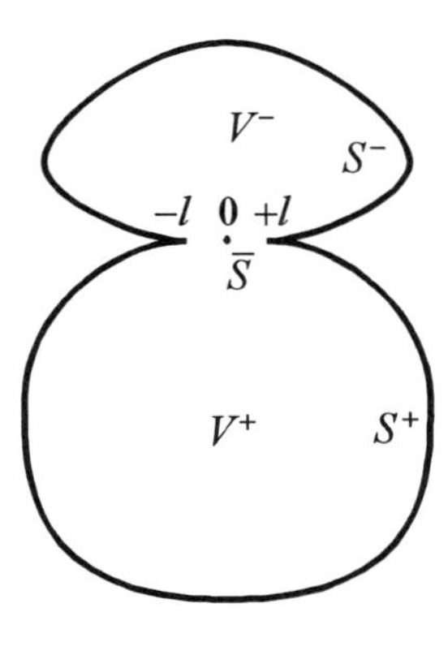

Figure 4.14

Obviously, here w_n^-, as well as w_n^+, is real and contains only infinite resonances. Therefore, in the coupled system there exist only total resonances ($\tilde{\rho}_0 = \infty$) for which

$$C^+ + C^- + \frac{2}{\pi} \ln \frac{\pi l}{2\sqrt{L^+ L^-}} = 0. \tag{4.184}$$

Another difference from the open system is that both terms in (4.178a) become infinite at the eigenfrequency of V^-. However, if we evaluate this indeterminacy, then we see that the field in V^- is large but finite and is of the order of $\ln[\pi l(2\sqrt{L^+L^-})^{-1}]$.

There are no essential difficulties if the eigenfrequencies of the domains V^+ and V^- are close to each other or coincide.

Let us write out the following formulas for the shift Δk of the resonant frequencies from the eigenfrequency k in one of the regions for three specific resonance systems:

$$\Delta k = -\left(2a^2 k_{0m} \ln \frac{\delta_+ \delta_-}{4}\right)^{-1}, \tag{4.185a}$$

$$\Delta k = -\left(2a_- \ln \frac{\pi^2 l^2}{4a_+ a_-}\right)^{-1}, \tag{4.185b}$$

$$\Delta k = -\left(2a_-^2 k_{0m} \ln \frac{\pi l^2}{4a_+ a_-}\right)^{-1}, \tag{4.185c}$$

$$\Delta k = -\left(2a_+ \ln \frac{\pi l^2}{4a_+ a_-}\right)^{-1}. \tag{4.185d}$$

Formula (4.185a) corresponds to the case of two coupled circular regions. Here $2\delta_-$ is the angular size, a is the radius of the circle with respect to the eigenfrequency k_{0m} ($J_1(k_{0m}a) = 0$) of which the above shift is calculated, and $2\delta_+$ is the angular size of the slot in the other circle. Formula (4.185b) gives the shift in the case of two coupled rectangular domains. Here $2a_-$ and b_- are the dimensions of the rectangle with respect to the eigenfrequency $k = \pi/b_-$ of which the above shift is calculated, and $2a_+$ and b_+ are the dimensions of the other rectangle. The slot is cut in adjacent walls with dimensions $2a_-$ and $2a_+$. Formulas (4.185c) and (4.185d) correspond to a circle of radius a_- coupled with a rectangle with wall dimensions $2a_+$ and b. The slot is cut in the wall of width $2a_+$. Formula (4.185c) gives the shift with respect to the eigenfrequency k_{0m} ($J_1(k_{0m}a_-) = 0$) of the circle, and formula (4.185d), with respect to the eigenfrequency $k = \pi/b$ of the rectangle.

Finally, note that the phase velocities of the eigenwaves of the corresponding waveguides can be expressed in an elementary way in terms of the roots of Eq. (4.184).

4.7 Two-Dimensional Problem for a Metallic Resonator of Arbitrary Cross-Section with Small Hole (E-Polarization; the ρ-Method)

1. For the E-polarization, we introduce the eigenfunctions u_n which must satisfy Eq. (4.157), the boundary condition (4.155) on the walls, and the following transmission conditions on the slot (cf. (2.35a)):

$$(u_n^+ - u_n^-)|_{\overline{S}} = 0, \tag{4.186a}$$

$$\left(\frac{\partial u_n^+}{\partial N} - \frac{\partial u_n^-}{\partial N} - \frac{u_n}{\alpha \rho_n}\right)\Big|_{\overline{S}} = 0, \tag{4.186b}$$

where α is a function of the coordinates on $\overline{S}$. Each eigenfunction must also satisfy the radiation condition and the corresponding conditions near the sharp edges. The eigenfunctions thus introduced are orthogonal with weight $1/\alpha$ on the slot:

$$\int_{\overline{S}} \frac{u_n u_m}{\alpha}\, dS = 0, \quad n \neq m. \tag{4.187}$$

The formal solution of the diffraction problem (4.153), (4.155) is given by the series (4.160), where U^0 is the solution of Eq. (4.161a) with the condition $U^0|_S = 0$ and the coefficients A_n have the form

$$A_n = \rho_n \frac{\displaystyle\int_{\overline{S}} u_n \frac{\partial U^0}{\partial N}\, dS}{\displaystyle\int_{\overline{S}} u_n^2/\alpha\, dS}. \tag{4.188}$$

2. Let us reduce the homogeneous problem to an equation on the slot. For the polarization considered, this equation is integro-differential rather than integral. There arises a difficulty due to the fact that we must differentiate the field given by the Green formula along the normal to $\overline{S}$ and substitute this derivative into the integrand. If in this formula we use the Green functions for V^+ and V^- with the condition $G|_S = 0$, then, in general, we cannot perform differentiation in the integrand, since the kernel has a strong singularity if the arguments coincide. However, this procedure becomes possible if we use the Green function of the impedance body, that is, the Green function with the condition $\left(G - w\dfrac{\partial G}{\partial N}\right)\Big|_S = 0$. In this case, obviously, $\dfrac{\partial G}{\partial N}\Big|_S$ has the same singularity as the function G itself (that is, a logarithmic singularity), and thus one more differentiation can be performed in the integrand. To have only the integral over the slot in the desired equation, we must temporarily impose the same impedance boundary condition (instead of (4.155)) on the eigenfunctions on the boundary S. Then, after appropiate transformations, we can equate w with zero and thus return to the original eigenfunctions.

We can rewrite the terms of the kernel $\dfrac{\partial^2 G(s,s')}{\partial N_s \partial N_{s'}}$ as series in the eigenfunctions ψ_n of the Steklov problem (4.164) by using the following integral equation, which is similar to (4.165) but is obtained with the help of the Green function of the impedance body:

$$\psi_n(s) = (w_n - w) \int_S \psi_n(s') \frac{\partial^2 G(s,s')}{\partial N_s \partial N_{s'}}\, ds'. \tag{4.189}$$

From this we obtain the desired bilinear formula giving the decomposition of the kernel $\dfrac{\partial^2 G}{\partial N_s \partial N_{s'}}$ on S. On the slot $\overline{S}$, after $\psi_n(s)$ is (approximately) replaced by $\psi_n(0)\cos\dfrac{\pi n s}{L}$ and after several formal transformations are performed to improve the convergence, this formula becomes

$$\begin{aligned}\frac{\partial^2 G}{\partial N_s \partial N_{s'}}(s,s') \;&\simeq\; \frac{\psi_0^2(0)}{2L(w_0 - w)} \\ &\quad - \frac{L^2}{\pi^2}\frac{\partial^2}{\partial s^2}\sum_{n=1} \frac{\psi_n^2(0)\cos\dfrac{\pi n s}{L}\cos\dfrac{\pi n s'}{L}}{L n^2 (w_n - w)}.\end{aligned} \tag{4.190}$$

By substituting this representation into the expression for the normal derivative of the eigenfunction on $\overline{S}$ (given by the Green function), by carrying out the double differentiation with respect to s before the integral sign, by passing to the limit as $w \to 0$, and by using the boundary condition (4.186b), we arrive at the desired equation

$$\begin{aligned}\frac{u_n(s)}{\alpha(s)\rho_n} &= \frac{1}{2L}\left(\frac{\psi_0^{+2}(0)}{w_0^+} + \frac{\psi_0^{-2}(0)}{w_0^-}\right)\int_{\overline{S}} u_n\, ds' \\ &\quad - \frac{1}{\pi^2}\frac{d^2}{ds^2}\int_{\overline{S}} u_n(s') \sum_{m=1}\left(\frac{L\psi_m^{+2}(0)}{w_m^+} + \frac{L\psi_m^{-2}(0)}{w_m^-}\right)\frac{\cos\dfrac{\pi m s}{L}\cos\dfrac{\pi m s'}{L}}{m^2}\, ds'.\end{aligned} \tag{4.191}$$

If we substitute $\dfrac{L}{w_m}$ into (4.191) and use the asymptotics (4.167a), then we can, in fact, differentiate any of the arising terms with respect to s and then set $s = s' = 0$ where this does not lead to divergence. By omitting the unimportant summands that do not contain resonance terms, we replace (4.191) by the equation

$$\begin{aligned}\frac{u_n(s)}{\alpha(s)\rho_n} &= \frac{C^+ + C^-}{L^2}\int_{-l}^{+l} u_n(s')\,ds' \\ &+ \frac{1}{\pi}\frac{d^2}{ds^2}\int_{-l}^{+l} u_n(s')\ln\left|\left(\frac{\pi s'}{L}\right)^2 - \left(\frac{\pi s}{L}\right)^2\right| ds', \end{aligned} \tag{4.192}$$

where C^+ and C^- have the following form for the domains V^+ and V^-:

$$C = \frac{L\psi_0^2(0)}{2w_0} + \sum_{n=1}\left(\frac{L}{w_n} - \pi n + \frac{(kL)^2}{2\pi n}\right)\psi_n^2(0). \tag{4.193}$$

Finally, if we change the variables by Eq. (4.172) and choose $\alpha(s)$ according to (4.174), then Eq. (4.192) can be solved explicitly, and its eigenelements are given by

$$\frac{1}{\rho_0} = \frac{\pi}{2}\left(\frac{l}{L}\right)^2 (C^+ + C^-) + 2, \tag{4.194a}$$

$$u_0 = \sqrt{1 - \left(\frac{s}{l}\right)^2}, \tag{4.194b}$$

$$\frac{1}{\rho_n} = 2(2n+1), \quad n > 0, \tag{4.194c}$$

$$u_n = \sin\left[(2n+1)\arccos\frac{s}{l}\right]. \tag{4.194d}$$

It follows from these formulas that, just as in the case of the H-polarization, all resonances are contained only in the first eigenvalue ρ_0, and the eigenfunctions satisfy the desired condition near the sharp edges.

3. By substituting (4.194b) and (4.194d) into (4.188), we find the coefficients

$$A_n = \rho_n \frac{2l}{\pi}\int_0^{\pi} \frac{\partial U^0}{\partial N}(l\cos\xi)\sin(2n+1)\xi\sin\xi\,d\xi. \tag{4.195}$$

It follows from this formula that with accuracy up to the terms of the order of $(kl)^3$ we have

$$A_0 = \rho_0 l\frac{\partial U^0}{\partial N}(0), \quad A_n = 0 \quad \text{for } n > 0. \tag{4.196}$$

Furthermore, the field on the slot is equal to

$$U(s) = \frac{\sqrt{l^2 - s^2}\dfrac{\partial U^0}{\partial N}(0)}{\dfrac{\pi}{2}\left(\dfrac{l}{L}\right)^2 (C^+ + C^-) + 2}, \tag{4.197}$$

and far from the resonator (at the point q), we have

$$U(q) = U^0(q) - \frac{l^2 \dfrac{\partial U^0}{\partial N}(0) \dfrac{\partial G^-}{\partial N}(0, q)}{\left(\dfrac{l}{L}\right)^2 (C^+ + C^-) + \dfrac{4}{\pi}}. \tag{4.198a}$$

Finally, in the interior of the resonator, far from the walls, we have

$$U(q) = \frac{l^2 \dfrac{\partial U^0}{\partial N}(0) \dfrac{\partial G^-}{\partial N}(0, q)}{\left(\dfrac{l}{L}\right)^2 (C^+ + C^-) + \dfrac{4}{\pi}}. \tag{4.198b}$$

It follows from the analysis of (4.198) and the comparison of these formulas with (4.178) that the solutions of both polarizations possess the same resonance properties with the only difference that for the E-polarization the role of a small parameter is played by $\left(\dfrac{l}{L}\right)^2$ rather than by $\dfrac{1}{\ln \dfrac{\pi l}{2L}}$. Here the resonance condition has the form

$$\frac{w_n}{L} = -\frac{\pi}{4(1 + \delta_{0n})} \left(\frac{l}{L}\right)^2 \psi_n^2(0), \tag{4.199}$$

and the frequency shift in the general case of an arbitrary contour is given by formula (4.182)

$$\Delta k = -\rho / \overline{A}, \tag{4.200a}$$

where

$$\rho = \frac{L}{2\pi} \left(\frac{\pi l}{2L}\right)^2, \quad \overline{A} = \frac{k_m \int_{V+} \chi_m^2 \, dV}{L \left(\dfrac{\partial \chi_m}{\partial N}(0)\right)^2}, \tag{4.200b}$$

and k_m and χ_m are, respectively, the eigenfrequencies and the eigenfunctions of the "closed" problem with the condition $\chi_m|_S = 0$. These formulas can be derived by the scheme described in Section 4.6, where the resonance condition (4.199) was used.

In the special case of a circular contour of radius a and $n = 0$, the resonant frequency shift is

$$\Delta k = -k_m \delta^2 / 8, \quad J_0(k_m a) = 0, \tag{4.201}$$

where 2δ is the angular size of the slot. This formula is in a good agreement with the rigorous results obtained by the s-method (see Section 4.10, Fig. 4.19) provided that δ does not exceed $\simeq \pi/8$. In this case, the error in (4.201) does not exceed 13%.

The technique developed in this section, just as in the case of the H-polarization, can trivially be extended to the problem on closed resonators coupled via a small hole and to the problem on the eigenwaves of the corresponding waveguides.

4. The results of this and the preceding section allow us to write out the propagation constants $\hat{h}_m$ for the leaky waves in a waveguide with a narrow slot (just as this was done in Section 4.5 for closed semitransparent surfaces). The transverse wave numbers $\hat{k}_m$ for the H- and E-waves in such a waveguide are, respectively, the roots of the equations

$$\frac{1}{\tilde{\rho}_0(k)} = 0, \tag{4.202a}$$

$$\frac{1}{\rho_0(k)} = 0, \tag{4.202b}$$

or, which is the same, the zeros of the resonant denominators in (4.178) and (4.198). One can readily that near the eigenfrequencies ($k \approx k_m$) of the corresponding interior problems, the values $1/\tilde{\rho}_0(k)$ and $1/\rho_0(k)$ are of the same structure, namely, are proportional to the expression (cf. (4.139))

$$\frac{1}{\rho} + \frac{1}{(k - k_m)\overline{A}} + i\overline{B}. \tag{4.203}$$

By equating (4.203) with zero, we obtain the following formulas for $\hat{k}_m$ and $\hat{h}_m$ (similar to formulas (4.150) and (4.151) obtained in Section 4.5):

$$\hat{k}_m = k_m - \frac{\rho}{\overline{A}} + i\rho^2 \frac{\overline{B}}{\overline{A}}, \tag{4.204}$$

$$\hat{h}_m = h_m + \frac{\rho}{\overline{A}} \frac{k_m}{h_m} - i\rho^2 \frac{\overline{B}}{\overline{A}} \frac{k_m}{h_m}. \tag{4.205}$$

For the H-waves, in these formulas ρ must be replaced by $\tilde{\rho}$ given by (4.183a), $\overline{A}$ is given in (4.183b), $\overline{B} = 2L \operatorname{Im} C^-$ in (4.171), and $\operatorname{Im} C^- = \operatorname{Im} G^-(0,0)$. For the E-waves, ρ is given in the first formula in (4.200b), $\overline{A}$ in the second formula in (4.200b), and $\overline{B} = \frac{2}{L} \operatorname{Im} C^-$ in (4.193). Since $\tilde{\rho} < 0$, it follows from (4.205) that $\operatorname{Re} \hat{h}_m < \operatorname{Re} h_m$ for the H-waves, that is, the slot increases the phase velocity of H-waves, just as in the case of a semitransparent wall (Section 4.5). Since $\rho > 0$, we have $\operatorname{Re} \hat{h}_m > \operatorname{Re} h_m$ for the E-waves, that is, the slot decreases the phase velocity of E-waves (cf. Section 4.5).

To find the imaginary parts of $\hat{k}_m$ and $\hat{h}_m$ by these formulas, we need to know the Green function for the exterior region of the metallized waveguide. In fact, this function is known only for a circular cylinder. In fact, in problems with slots (Section 4.7) for E-waves, one can find $\overline{B}$ directly from the integral equation (2.49), but this would require calculations in the complex domain. In Section 4.10 these values will be found with the help of the solution of the real integral equation corresponding to the s-method.

All main results of the last two sections were essentially obtained on the basis of formulas (4.170), (4.171), and (4.190). Obviously, these formulas can be used not only for the analysis of eigenoscillations of resonators and leaky waves in waveguides with longitudinal slots, that is, for solving the homogeneous equations (4.163) and (4.191). The same kernels are contained in the nonhomogeneous equations of the diffraction problems.

These expansions can also be used for solving the diffraction problems with the help of the technique of conformal mappings. The method developed in this section (based on the theory of Section 2.2) provides a shorter way to the result and, in principle, is not subject to the restriction that the slot width must be small. By using computers, we can find the eigenelements (which are, by definition, independent of the excitation) of a resonator or a waveguide with a slot of finite width.

4.8 Open Dielectric Resonator with Large Permittivity (E-Polarization; the ρ-Method)

1. In this section we study an open resonator, that is, a body with large permittivity ε, $|\varepsilon| \gg 1$. To solve the diffraction problem means to find the field U that satisfies the equations

$$\Delta U + k^2 U = f, \tag{4.206a}$$
$$\Delta U + k^2 \varepsilon U = 0 \tag{4.206b}$$

outside the body (in V^-) and inside the body (in V^+), respectively. On the boundary S of the body, the field U and its first-order derivative must be continuous:

$$(U^+ - U^-)|_S = 0, \quad \left(\frac{\partial U^+}{\partial N} - \frac{\partial U^-}{\partial N} \right) \Big|_S = 0. \tag{4.207}$$

Moreover, the field U, as well as the fields U^0 and u_n introduced later, must satisfy the radiation conditions.

We define the field U^0, just as in Section 4.5, as the result of diffraction of the same sources on a body with the same surface S on which, however, the field U^0 must be zero:

$$\Delta U^0 + k^2 U^0 = f \quad \text{in } V^-, \tag{4.208a}$$
$$U^0|_S = 0. \tag{4.208b}$$

The eigenfunctions u_n must satisfy the homogeneous equations

$$\Delta u_n + k^2 u_n = 0 \text{ in } V^-, \qquad \Delta u_n + k^2 \varepsilon u_n = 0 \text{ in } V^+ \tag{4.209}$$

and the following boundary conditions on S, which contain the eigenvalue ρ_n:

$$(u_n^+ - u_n^-)|_S = 0, \tag{4.210a}$$
$$\left(\frac{\partial u_n^+}{\partial N} - \frac{\partial u_n^-}{\partial N} - \frac{u_n}{\rho_n} \right) \Big|_S = 0. \tag{4.210b}$$

The solution of problem (4.206), (4.207) can be represented by the series (2.34) with

$$A_n = -\rho_n \frac{\displaystyle\int_S u_n \frac{\partial U^0}{\partial N}\, dS}{\displaystyle\int_S u_n^2\, dS}. \tag{4.211}$$

The problem is again reduced to finding ρ_n, that is, the basic characteristic of the resonator.

2. Let us introduce the eigenfrequencies k_m and the eigenfunctions χ_m for the interior domain of a metallized resonator as the eigenelements of the following homogeneous problem in V^+:

$$\Delta\chi_m + k^2\varepsilon\chi_m = 0, \tag{4.212a}$$

$$\left.\frac{\partial \chi_m}{\partial N}\right|_S = 0. \tag{4.212b}$$

In this case, the boundary conditions (4.208b) in V^-, that is, for U^0, and (4.212b) in V^+, that is, for χ_m, are different.

Using the equations for u_n and χ_m, we can readily see that the exact expression for ρ_n has the form

$$\rho_n = \frac{\displaystyle\int_S \chi_m u_n\, dS}{\displaystyle\varepsilon(k^2 - k_m^2)\int_{V+} \chi_m u_n\, dV - \int_S \chi_m \frac{\partial u_n^-}{\partial N}\, dS}. \tag{4.213}$$

The frequencies k_m are close to each other; the distance between neighboring frequencies is of the order of $1/\sqrt{|\varepsilon|}$. Therefore, although formula (4.213) holds for any m, in this formula we must use the frequency k_m nearest to the operating frequency. Only in this case it is possible to calculate the resonance shift with respect to k_m and the resonance width. It follows from (4.216) and (4.217) that both these variables are, at least, $1/\sqrt{|\varepsilon|}$ times less than the distance between the frequencies k_m.

If the frequency k is not close to k_m, then ρ_n is small (of the order of $1/|\varepsilon|$) and all terms in the series (2.34) are small. For $k = k_m$, we have $\left.\frac{\partial u_n^+}{\partial N}\right|_S = 0$, so that the function u_n coincides with χ_m in V^+. For k close to k_m, we can replace $u_n|_S$ by $\chi_m|_S$. To find the function $\left.\frac{\partial u_n^-}{\partial N}\right|_S$ via the function $u_n|_S$ occurring in (4.193) is a boundary value problem in V^-. We shall denote the solution of this problem by the symbol L. Then the desired expression for the eigenvalue ρ_n near the resonance is given by

$$\rho_n = \frac{\displaystyle\int_S \chi_m\, dS}{\displaystyle\varepsilon(k - k_m)2k_m \int_{V+} \chi_m^2\, dV - \int_S \chi_m L(\chi_m)\, dS}. \tag{4.214}$$

3. The second factor in (4.211) is finite. In this factor we can replace u_n by χ_m. Then the frequency dependence of the field complementary to U^0 is determined by the frequency dependence of ρ_n in (4.214). We write

$$\int_S \chi_m L(\chi_m)\, dS = \alpha + i\beta. \tag{4.215}$$

In practice, to find α and β, we need to solve the above-mentioned boundary value problem in V^-. For example, we have $L = -\dfrac{kH_n^{(2)\prime}(ka)}{H_n^{(2)}(ka)} \cos n\theta$ for a disk and $L = ik$ for the one-dimensional problem. Therefore, $\alpha = 0$ and the frequency shift in (4.216) is zero, etc.

To simplify the formulas, we consider only real values of ε. In this case, the shift of the resonance curve center with respect to the interior resonance k_m is of the order of $1/\varepsilon$ and is given by

$$\Delta k_m = \frac{1}{\varepsilon} \frac{\alpha}{2k_m \int_{V+} \chi_m^2\, dV}. \tag{4.216}$$

The width of the resonance curve is of the same order:

$$\delta k = \frac{1}{\varepsilon} \frac{\beta}{2k_m \int_{V+} \chi_m^2\, dV}. \tag{4.217}$$

The maximum value of the factor ρ_n (that is, its value at the center of the resonance curve) is

$$\max |\rho_n| = \frac{\int_S \chi_m^2\, dS}{|\beta|}. \tag{4.218}$$

The maximum absolute value of the term '$A_n u_n$ complementary to U^0 (which is the only term that we must take into account near the resonance) is equal to

$$\max |A_n u_n| = \left| \frac{\int_S \chi_m \dfrac{\partial U^0}{\partial N}\, dS \; u_n}{\beta} \right|, \tag{4.219}$$

where u_n outside the resonator can be obtained from the boundary value problem in V^- in terms of the value $u_n|_S = \chi_m|_S$, and inside the resonator (in V^+) we have $u_n \approx \chi_m$.

There is a special case in which the resonator in question is itself inside another loss-free resonator or a beyond-limit waveguide, that is, if there are no radiation losses. In this case the denominator in (4.219) or in (4.218) is zero. Obviously, in this case we must take into account the imaginary part of ε, since the problem on a loss-free resonator at the resonant frequency displaced by (4.216) with given sources has no solution and no physical meaning.

Inside the resonator (in V^+), the field is not large. It is small far from the resonant frequency and becomes finite at resonance. To this finite field, there corresponds a large amount of stored energy ($|\varepsilon| \gg 1$).

We also note that there are two distinctions between the dielectric resonator and the resonator studied in Section 4.5. The frequency shift depends on the resonator position with respect to the exterior bodies, and the resonator width is of the same order as the shift. In this sense, the jump behavior of ε forms a less powerful barrier at the boundary of the body than a fine grating.

4.9 Open Resonator with Closed Semitransparent Walls (*E*-Polarization; the *s*-Method)

In Section 4.5 we studied resonators with semitransparent walls and large Q-factor by the ρ-method for both polarizations. We described the structure of eigenvalues near the spectrum of the closed problem and performed a qualitative investigation of all characteristics of such resonators. It turned out that the resonators studied in Sections 4.6 and 4.7 possess the same properties. However, to find all numerical parameters (first of all, the resonance width and the decay of leaky waves) without solving the interior and exterior problems for a metallized waveguide in advance, we must perform calculations in the complex domain.

In this and in the following section, we solve some of these problems by the method developed in Section 2.5. This allows us to simplify the calculations significantly, since, as was already noted, the kernel of the integral equation of the s-method is real.

1. To study resonators with large Q-factor, it is expedient to define U^0 in the representation (2.151) of the total field as the diffraction field of the same sources on metallized resonators of the same shape (just as was done in the preceding sections). In other words, U^0 must satisfy Eq. (2.152) in V^- with the boundary condition $U^0|_S = 0$ and the radiation condition, whereas $U^0 = 0$ in V^+. Obviously, in this case the coefficients in (2.151) differ from those in (2.155b) and have the form

$$A_n = -\frac{\rho}{i + \lambda_n} \frac{\displaystyle\int_S u_n \frac{\partial U^0}{\partial N}\, dS}{\displaystyle\int_S u_n \operatorname{Im} u_n^s \, dS}. \tag{4.220}$$

In what follows, we show that the coefficients (4.220) are not small only at resonances.

To study the resonance effects, we need to know the structure of the eigenvalues λ_n of the s-method and of the coefficients (4.220) near the eigenfrequencies k_m of the corresponding closed problems, since in the resonators in question the resonance occurs near these frequencies. To this end, let us match the functions (introduced in Section 2.5) u_n and u_n^s for k close to k_m and the eigenfunctions χ_m of closed problems (4.132) that correspond to the eigenvalues k_m. Applying the Green formula to the two functions χ_m and $\operatorname{Re} u_n^s$ in the interior domain V^+ and using (2.149) and the fact that $s_n = \dfrac{i - \lambda_n}{i + \lambda_n}$, we obtain the following expression for λ_n:

$$\lambda_n = \frac{\rho - \dfrac{(k^2 - k_m^2) \displaystyle\int_{V+} \operatorname{Re} u_n^s \chi_m \, dV}{\displaystyle\int_S u_n \frac{\partial \chi_m}{\partial N} \, dS}}{\dfrac{\displaystyle\int_S \frac{\partial \chi_m}{\partial N} \operatorname{Im} u_n^s \, dS}{\displaystyle\int_S u_n \frac{\partial \chi_m}{\partial N} \, dS}}. \tag{4.221}$$

It follows from the Green formula that the denominator is equal to

$$\frac{(k^2 - k_m^2) \displaystyle\int_{V+} \chi_m \operatorname{Im} u_n^s \, dV}{\displaystyle\int_S u_n \frac{\partial \chi_m}{\partial N} \, dS}. \tag{4.222}$$

Therefore, $|\lambda_n| \to \infty$ as $k \to k_m$. In this case, from Eq. (2.147b) we obtain the following limit equation for the two-dimensional problem:

$$\int_S u_n J_0(k_m R) \, dS = 0. \tag{4.223a}$$

In view of (2.148), this implies that in the entire volume (in particular, in V^+) we have

$$\operatorname{Im} u_n^s(k_m) = 0. \tag{4.223b}$$

Since (4.223a) coincides with the equation for the normal derivative

$$\int_S \frac{\partial \chi_m}{\partial N} J_0(k_m R) \, dS = 0 \tag{4.224}$$

of the eigenfunction χ_m of the closed problem, the functions $\left.\dfrac{\partial \chi_m}{\partial N}\right|_S$ and $u_n|_S$ coincide for $k = k_m$ up to a factor independent of the coordinates, that is,

$$u_n(k_m)|_S = \left.\frac{\partial \chi_m}{\partial N}\right|_S. \tag{4.225}$$

By comparing the expression (2.148) for $\operatorname{Re} u_n^s$ with the expression for the function χ_m via $\dfrac{\partial \chi_m}{\partial N}$ on S and with regard to (4.223b), we obtain the equation

$$u_n^s(k_m) = -\chi_m \tag{4.226}$$

in V^+. Thus $u_n^s(k_m) = 0$ on the contour S, and since u_n^s satisfies the radiation conditions, we have

$$u_n^s(k_m) \equiv 0 \tag{4.227}$$

in the entire region V^-. If we assume that the function u_n^s is analytic in k, then for k close to k_m we obtain

$$u_n^s(k) \approx -\chi_m + (k - k_m)\frac{\partial u_n^s}{\partial k}(k_m) \quad \text{in} \quad V^+, \tag{4.228a}$$

$$u_n^s(k) \approx (k - k_m)\frac{\partial u_n^s}{\partial k}(k_m) \quad \text{in} \quad V^-. \tag{4.228b}$$

By analogy, on S we have

$$u_n(k) \approx \frac{\partial \chi_m}{\partial N} + (k - k_m)\frac{\partial u_n}{\partial k}(k_m). \tag{4.229}$$

Thus

$$\int_S \frac{\partial \chi_m}{\partial N} \operatorname{Im} u_n^s \, dS = \int_S u_n \operatorname{Im} u_n^s \, dS \tag{4.230}$$

in the leading terms with respect to $k - k_m$. Applying the Green formula to the functions u_n and u_n^s in the regions V^+ and V^- and using the boundary conditions and the conditions at infinity for these functions, we arrive at the relation (which holds for any k)

$$\int_S u_n \operatorname{Im} u_n^s \, dS = -k \int_{S_\infty} |u_n^s|^2 \, dS. \tag{4.231}$$

It follows from (4.228b)–(4.231) that the denominator in (4.221) is proportional to $(k - k_m)^2$:

$$\int_S \frac{\partial \chi_m}{\partial N} \operatorname{Im} u_n^s \, dS \approx -(k - k_m)^2 k \int_{S_\infty} \left| \frac{\partial u_n^s}{\partial k}(k_m) \right|^2 dS. \tag{4.232}$$

By using formulas (4.225), (4.226), and (4.232) in (4.221), we find the desired structure of $\lambda_n(k)$ for k close to k_m:

$$\lambda_n(k) \approx -\frac{\rho + (k - k_m)A}{(k - k_m)^2 C}. \tag{4.233}$$

The coefficients A and C are independent of the frequency and the wall transparency ρ; moreover, A is determined by formula (4.136) and

$$C = \frac{k_m \int_{S_\infty} \left| \frac{\partial u_n^s}{\partial k}(k_m) \right|^2 dS}{\int_S \left(\frac{\partial \chi_m}{\partial N} \right)^2 dS}. \tag{4.234}$$

One can show that the coefficient C is proportional to the variable B (introduced in Section 4.5):

$$C = A^2 B. \tag{4.235}$$

Note that, in general, formula (4.233) holds for any ρ. However, we are mostly interested in resonators with small transparency of the walls.

2. Formula (4.233) for $\lambda_n(k)$ allows us to describe the behavior of the coefficients A_n (see (2.155b) and (4.220)) near the eigenfrequencies k_m and to study the resonance phenomenon. By substituting (4.233) into (2.155b) and by carrying out simple transformations, we obtain

$$A_n \approx -\frac{\dfrac{1}{\rho} D}{\dfrac{1}{\rho} + \dfrac{1}{(k-k_m)A} + iB}, \qquad D = \frac{\displaystyle\int_{V+} U^0 \chi_m \, dV}{\displaystyle\int_{V+} \chi_m^2 \, dV} \tag{4.236}$$

for the expansion (2.151), where U^0 is the field of sources in free space. Otherwise, if U^0 is the solution of the diffraction problem on the (metallized) resonator with zero coupling, then the use of (4.233) in (4.220) results in

$$A_n \approx -\frac{\dfrac{1}{k-k_m} D}{\dfrac{1}{\rho} + \dfrac{1}{(k-k_m)A} + iB}, \qquad D = \frac{\displaystyle\int_S \frac{U^0}{\partial N} \frac{\chi_m}{\partial N} \, dS}{2k_m \displaystyle\int_{V+} \chi_m^2 \, dV}. \tag{4.237}$$

It also follows from these formulas (see (4.141)) that the resonance occurs at the frequency

$$k_* = k_m - \rho/A. \tag{4.238}$$

In this case, according to (4.236) and (4.228), that is, for a simple choice of U^0, the total field (2.151) in V^+ is large, since the resonance term $A_n u_n^s$ in this field is of the order of $1/\rho$, while in V^- the term $A_n u_n^s$ is of the order of 1, and there is no leading term in the expansion corresponding to this region. Near the resonance, at the eigenfrequency of the interior problem ($k = k_m$), the corresponding term is identically zero, and the total field in V^+ is already not large but of the order of 1. Far from the resonant frequencies, the terms $A_n u_n^s$ are finite everywhere. However, one can show that the total field in the interior of the resonator is small and proportional to ρ.

For another choice of U^0, according to (4.237), one resonant term (complementary to U^0) of the order of 1 arises also in the exterior region under the resonance conditions. In this case all other terms are small and proportional to ρ. In other words, under this approach, the finite perturbation of the total field in V^- observed at resonance is described by one term. This is just the advantage gained by a more complicated choice of U^0. Far from the resonance, the field complementary to U^0 is of the order of ρ everywhere. For $k = k_m$, the corresponding term is absent only in the exterior region.

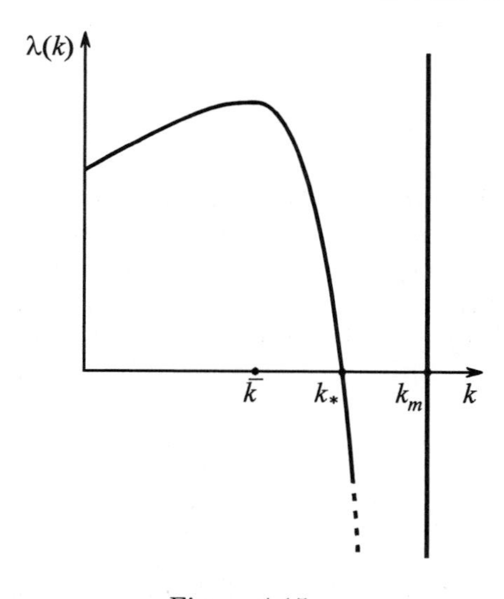

Figure 4.15

It follows from (4.233) that the dependence of λ_n on the frequency near the resonances is of the standard form depicted in Fig. 4.15. At the resonant frequency (4.238), we have $\lambda_n(k_*) = 0$ (in this case $s_n = 1$). At the eigenfrequency of the closed problem, we have $|\lambda_n(k_m)| = \infty$ (in this case $s_n = -1$). For $k = \bar{k}$, the curve $\lambda_n(k)$ attains its maximum, proportional to $1/\rho$, and we have

$$k_* - \bar{k} = k_m - k_*. \tag{4.239}$$

3. The results obtained allow us (just as in Section 4.5) to write out the propagation constants $\hat{h}_m$ of the leaky waves in waveguides of arbitrary cross-sections with semitransparent walls, as well as to calculate $\hat{h}_m$ by using the integral equation (2.147b) and formula (4.233). Indeed, the transverse wave numbers (the complex eigenfrequencies) $\hat{k}_m$ are the zeros of the resonant denominators in (4.236) and (4.237), or, which is the same, the roots of the equation

$$\lambda_n(k) = -i. \tag{4.240}$$

This yields the expressions (4.150) and (4.151) for $\hat{k}_m$ and $\hat{h}_m$. The values of B (and of A and k_m if necessary) contained in (4.150) can be found with the help of formula (4.233) on the curve $\lambda_n(k)$, which is obtained in the course of numerical solution of the integral equation (2.147b).

Shown in Figs. 4.16 and 4.17 are numerical results obtained for elliptic and rectangular resonators. The dashed lines show the shift Δka (multiplied by a) of the resonant frequency with respect to the eigenfrequency $k_m a$ of the closed problem. The solid curves show a variable proportional to the decay of $\hat{h}''_m$ in the corresponding waveguides. All variables are functions of the wall transparency. The dot-and-dash line

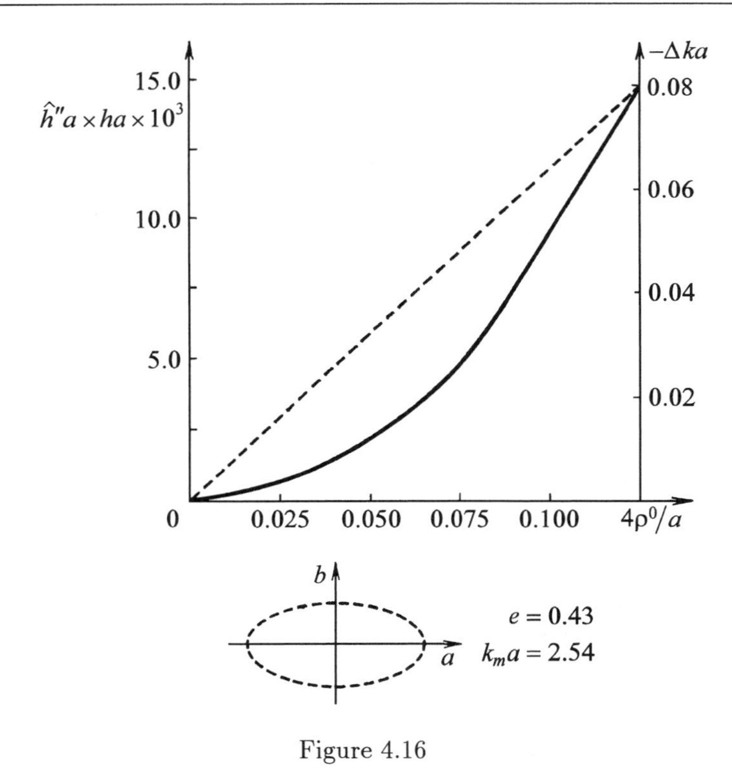

Figure 4.16

in Fig. 4.17 illustrates the variable Δka calculated by approximate formulas (4.141) and (4.136). The error of these formulas increases with increasing wall transparency.

Thus the above-developed technique allows one to find the complex propagation constants for leaky waves in waveguides with semitransparent walls and any shape of the cross-section. Obviously, formula (4.234) is not used in this case. Moreover, it is possible not to solve the interior problem in advance, that is, one can find A and B directly from (2.147b) and (4.233).

Obviously, in the same way one can also find complex eigenfrequencies of resonators; however, in problems of excitation of resonators it is very important from the physical viewpoint to know the values $\lambda_n(k)$ corresponding to the real frequencies. To find them, we again need only to find the solutions of (2.147b).

4.10 Waveguide with a Longitudinal Slot. Leaky Waves (E-Polarization; the s-Method)

1. In Section 4.7 we obtained explicit formulas for the shift of resonant frequencies which occurs due to a small cut in two-dimensional metallic resonators of arbitrary shape; we also found the structure (4.205) of the propagation constants $\hat{h}_m$ of the leaky

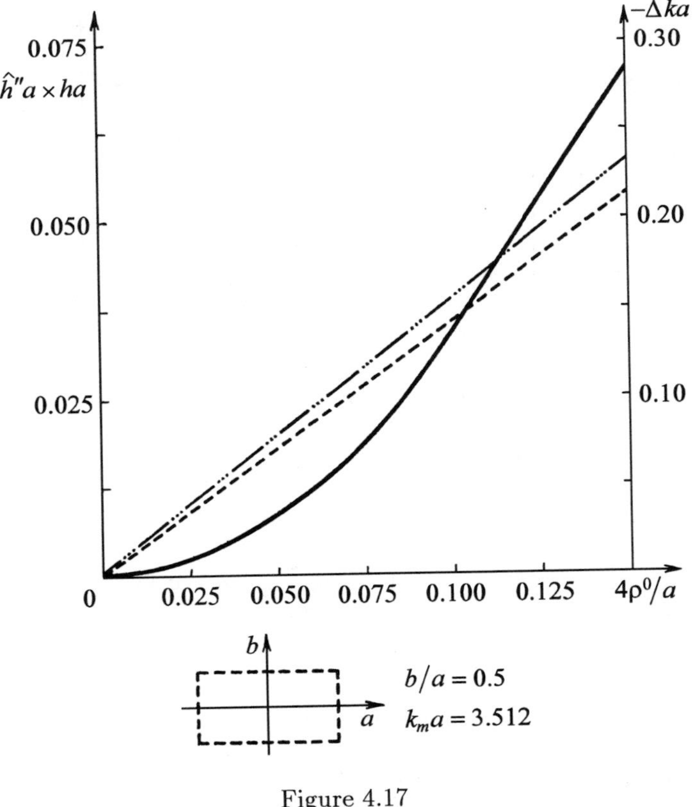

Figure 4.17

waves in the corresponding waveguides. However, as was already pointed out, significant difficulties arise in calculating $\hat{h}_m$ by formula (4.205), since it is required to solve the exterior Steklov problem for an arbitrary contour.

In the present section we show how the integral equation (2.147b) of the s-method can effectively be used for calculating the propagation constants $\hat{h}_m$. In the case of a metallized ($\rho = 0$) boundary S (if there are slots, then S is nonclosed), this equation has the form

$$\int_S u_n(N_0(kR) - \lambda_n J_0(kR))\, dS = 0. \tag{4.241}$$

Here k is the frequency, R is the distance between the points of observation and integration, and $\lambda_n = i\dfrac{1 - s_n}{1 + s_n}$ is an eigenvalue.

The eigenelements of this equation, just as functions of the frequency, characterize the resonance properties of two-dimensional resonators and in the excitation problems play the role similar to that of eigenfrequencies and the corresponding eigenoscillations in closed systems. The function $\lambda_n(k)$ allows one to find the resonant frequencies for resonators with slots of any dimensions. If the slots are small (just as we assume), then these frequencies are close to the eigenfrequencies k_m of the corresponding closed resonators. Hence it is important to know the formula (similar to (4.233)) that describes the behavior of the eigenfrequencies λ_n near the frequencies k_m. This formula allows us to calculate $\hat{h}_m$ in a waveguide with a longitudinal slot. To derive this formula, we rewrite λ_n in terms of the first eigenvalue ρ_0 (4.194a) (obtained in Section 4.7). Comparing the statement of the homogeneous problem in Section 4.7 with that in Section 2.5 (formulas (2.137), (2.138), and (2.140)) and using the Green formula, the boundary conditions, and the conditions at infinity for both statements of the problem, we obtain

$$\bar{s}_n \approx \frac{\rho_0}{\rho_0^*} s_n, \tag{4.242a}$$

whence

$$\bar{\lambda}_n \approx \frac{\mathrm{Re}\,[(i+\lambda_n)/\rho_0]}{\mathrm{Im}\,[(i+\lambda_n)/\rho_0]}. \tag{4.242b}$$

The bar indicates variables related to the problem with the slot, and s_n and λ_n correspond to the problem for the completely metallized surface (S is closed, the slot is absent). The eigenvalue $\lambda_n(k)$ can be calculated for $k \approx k_m$ from (4.233) with $\rho = 0$ as

$$\lambda_n \approx -\frac{1}{(k-k_m)AB}. \tag{4.243}$$

By substituting (4.243) and the structure (4.203) of ρ_0 near k_m (obtained with the help of (4.194a)) into (4.242b), we arrive at the desired formula for $\bar{\lambda}_n$ with k close to k_m:

$$\bar{\lambda}_n(k) \approx -\frac{\rho + (k-k_m)\overline{A} + (k-k_m)^2 \rho \overline{A}\,\overline{B} AB}{\rho(k-k_m)(AB - \overline{A}\,\overline{B}) + (k-k_m)^2 \overline{A} AB}. \tag{4.244}$$

The coefficients in (4.244) were calculated in the preceding sections: ρ, $\overline{A}$, and $\overline{B}$ in Section 4.7, and A and B in Section 4.5. Recall that $\overline{A}$, $\overline{B}$, A, and B are independent of the frequency and of ρ.

2. The roots of the equation

$$\bar{\lambda}_n(k) = -i \tag{4.245}$$

are the transverse wave numbers $\hat{k}_m$ (the complex eigenfrequencies) in terms of which the desired propagation constants $\hat{h}_m$ are expressed. This fact allows us to find formulas for $\hat{k}_m$ and $\hat{h}_m$, which are obviously the same as in Section 4.7 ((4.204) and (4.205)).

The variable $\overline{B}$ contained in these formulas (as well as $\overline{A}$ and k for more complicated cross-sections of waveguides) can be found with the help of formula (4.244) from the curve $\bar{\lambda}_n(k)$ obtained by solving Eq. (4.241) numerically for a nonclosed boundary S. In this case, it is expedient to calculate the product AB in advance by formula (4.243) along the curve $\lambda_n(k)$ of the same equation but with a closed boundary S.

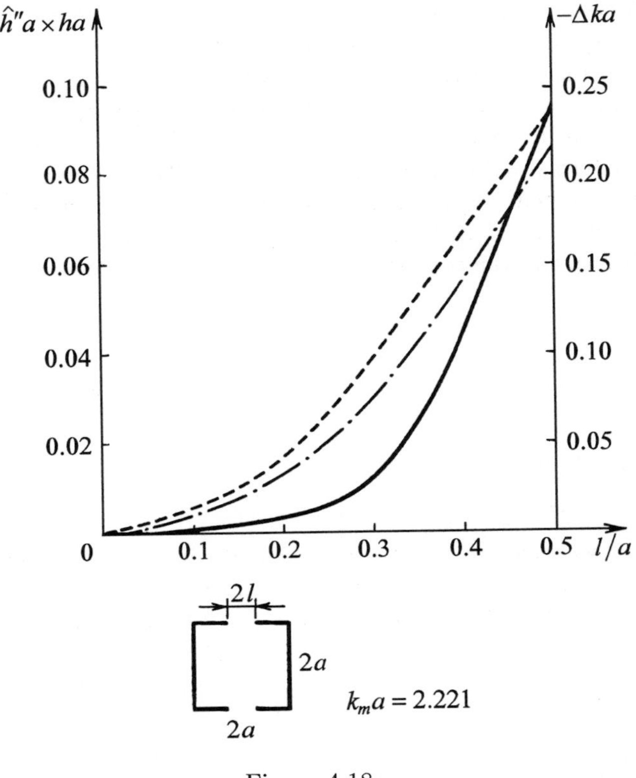

Figure 4.18

The results obtained by this method for cylindrical resonators of rectangular and circular cross-section are shown in Figs. 4.18 and 4.19. The dotted curves show the shift Δka of the resonant frequency from the eigenfrequency $k_m a$ of a closed resonator calculated with the help of Eq. (4.241). The dot-and-dash curves show the same variable calculated by the approximate formulas (4.200). The solid curves show a variable proportional to the decay of $\hat{h}_m$ in the corresponding waveguides. All these variables are functions of slot dimensions.

In conclusion, we note that for the complex eigenfrequency we have the following formula, which is more general than (4.150) and (4.204):

$$\hat{k}_m \simeq k_* - \frac{i}{\dfrac{d\bar{\lambda}_n}{dk}(k_*)}. \tag{4.246}$$

This formula remains valid for any resonators with large Q-factor, including resonators that are not close to closed resonators. Obviously, this formula gives an approximate solution of Eq. (4.245) and of the similar equation (4.240) (with $\bar{\lambda}_n$ replaced by λ_n)

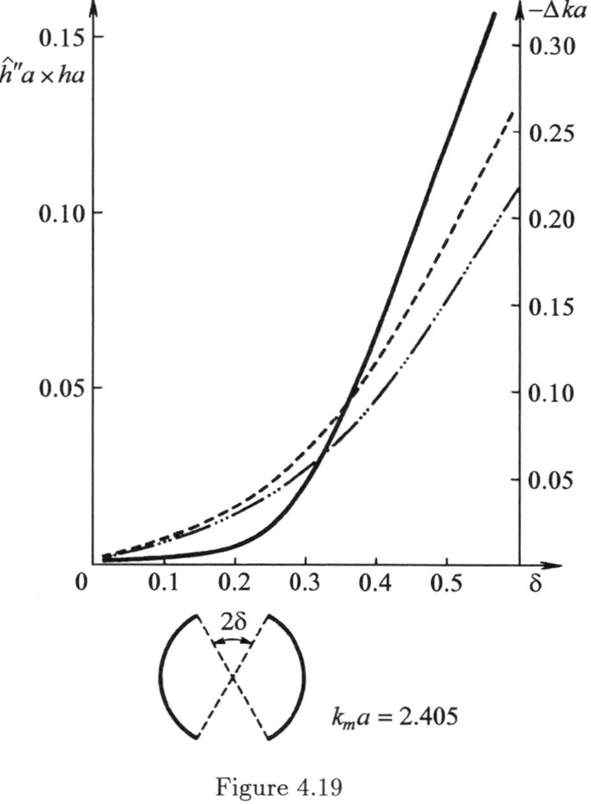

Figure 4.19

provided that the Q-factor is large, that is, if

$$\left| k_* \overline{\frac{d\lambda_n}{dk}}(k_*) \right| \gg 1. \tag{4.247}$$

However, for resonators close to closed ones, from the computational viewpoint, it is more convenient to use special cases of formula (4.246) (that is, Eqs. (4.150) and (4.204)), based on the above-obtained structures of eigenvalues near the eigenfrequency of the closed problem.

4.11 Bibliographical Remarks

In the main text of Chapters 1–4, we usually gave no bibliographical references; nor did we compare our method with other methods. In this section, we list the main papers in which other authors suggested similar ideas (Subsection 1; we indicate only one or two papers in each direction), compare some aspects of the generalized method of eigenoscillations with other directions in diffraction theory (Subsection 2), and state some possible generalizations of the method (Subsection 3).

Throughout the book we use the monographs and textbooks [48, 17, 35, 46, 6, 14] on diffraction theory.

1. The method of eigenfrequencies for closed resonators can be found in numerous textbooks (e.g., see [17]). The usual quantum-mechanical scattering theory [46, 2] is an example of application of this method to open systems in which the eigenfunctions (corresponding to the poles of the scattering matrix) grow at infinity and the field expansion contains also the integral over the continuous spectrum. The method of eigenfrequencies was applied to open electromagnetic systems in [48]. The difficulties encountered in stationary diffraction problems because of the growth of the eigenfunctions are discussed, e.g., in [2].

The method of Chapter 1 is close to the so-called method of Sturm functions. The paper [37] was one of the first papers in which the representation of the field in the form of a series with respect to these functions without resorting to the continuous spectrum was considered in detail. In that paper, the solution of the Schrödinger equation is expanded in a series in the eigenfunctions of an auxiliary homogeneous problem, in which $k^2 = 0$ is assumed, so that the auxiliary problem is selfadjoint. The spectral parameter is the coefficient of the potential energy (the coupling constant). The eigenfunctions decay exponentially at infinity, and so the expansion is valid only in a finite region.

In [33], k is taken real in the auxiliary problem, and the eigenfunctions satisfy the radiation condition. This idea, as applied to a one-dimensional quantum-mechanical problem, was suggested long ago, and formulas (1.99)–(1.101) of the version described in Subsection 1.3.4 with $\varepsilon\mathbf{r}$ replaced by $1 - U(\mathbf{r})/k^2$ coincide with the formulas given in the beginning of Section 3 in the survey [5].

The generalized method of eigenoscillations differs from [37] mainly in that k in the homogeneous equation is taken equal to k in the diffraction problem. In contrast with [33], in [37] only the diffracted field is expanded in a series, and hence the series is convergent everywhere. Apparently, the term corresponding to the scattering on the auxiliary potential $U(\mathbf{r})$ has never been isolated in this series earlier. It is the isolation of this term that permits one to obtain an explicit single-term expression in the resonance conditions. In the terminology adopted in quantum mechanics, this means that the eigenfunctions are generated by an equation different from the usual Lippmann–Schwinger equation (1.85), namely, by the *Lippmann–Schwinger equation of the distorted wave method* [46], that is, by Eq. (1.106) or, in other notation, Eq. (1.173). The series used in Sections 4.5–4.10 also contain an analog of this term, namely, the term corresponding to the diffraction on a metallized resonator. Apparently, the corresponding technique has never been applied to Eqs. (1.121) and (1.123).

The method of Section 2.1 is related to Steklov's work concerning the scalar wave equation for a closed volume with spectral parameter appearing as a coefficient in the boundary condition of the third kind. The generalization of these results to exterior problems and to two-sided boundary conditions (transmission conditions) has apparently never been carried out. Section 4.6 uses the results obtained in [39] for the asymptotics of the eigenelements of the Steklov problem for the Laplace equation.

The s-method is close to the method which was apparently suggested for the first time (for radio location and antenna problems) in [9]. In electromagnetic theory, the scattering operator was introduced in [9, 48] (without the additional condition (2.139b) or (2.140b)). The notion of eigenoscillations in an open domain, which depend on the shape of the body and are described by the eigenfunctions of the scattering operator, with a discrete spectrum consisting of eigenvalues ($-s_n$ in our notation; see Section 2.5) was introduced in [9]. The ideas of [9] were developed in [13], where it was shown that the currents corresponding to these eigenoscillations coincide with the system of eigenfunctions of some real operator on the surface of the body (in the scalar notation, this corresponds to Eq. (2.147) with $\rho = 0$) and where a generalization to nonideally conductive and dielectric bodies was given. The eigenfunctions obtained in [9] were used, in particular, in [16].

In the present monograph, the corresponding technique is developed as a special case of a more general method (Chapter 2) in which the spectral parameter is introduced into the boundary condition. The isolation of U^0 permits one to study high-Q systems (open resonators and waveguides, in particular, resonators with weak coupling) effectively. The technique is also developed for a surface with a transmission condition on the boundary (small values of $|\rho|$ correspond to weak coupling) and for a resonator with narrow slots. The structure of the eigenvalues near the spectrum of the closed resonator, as well as the character of the field inside and outside the resonator, is found, and a number of concrete problems are solved.

Stationary Rayleigh type functionals for the eigenvalues in closed domains are considered in detail, e.g., in [35], where one can also find several examples showing how to make some boundary conditions natural. The general methods of indeterminate coefficients for constructing functionals for which given boundary conditions are natural (Section 3.3) has not been used earlier. The variational technique has not apparently been used in the calculation of other eigenvalues of electromagnetic problems. In the construction of stationary functionals in an infinite domain, the fact that k is real is essential.

The examples given in Chapter 4 (except in Section 4.1) contain solutions of new problems or more complete solutions of known problems. Scattering on a quasistationary level (Section 4.2) was studied long ago by the eigenfrequency method, which leads to much more laborious computations. The asymptotic theory ($L \gg a \gg \lambda$) of open resonators was originally developed in [7]. The analytic form of this theory was found in [48] (in particular, our formula (4.124) has been borrowed from this book). The theory of waveguides with narrow slots (but necessarily with a flange) is given, e.g., in [8], where one can also find bibliographical references on the topic. The representation of the field in a waveguide by a sum of leaky waves is given in [40]. A method for determining the propagation constants for these waves is given in Section 4.10.

The material of Subsection 4.3.5 is taken from [41]. This method has been later developed in [25, 24, 45, 42, 43, 38].

The present book is mainly comprised by the authors' results, partly published in 1968–1975 in [18, 19, 21, 23, 26, 50, 51, 52, 53, 54, 55, 60]. A review of these papers can be found in [56, 57]. Some new problems are considered in [58].

2. In this subsection we indicate some remote analogies between various aspects of the generalized method of eigenoscillations and some new trends in diffraction theory.

- The diffraction problem for two bodies is not additive: the field is not just the sum of the fields occurring in the diffraction by each of the bodies separately. It is often expedient to single out a solution of a simpler problem from the complete solution. For example, in series like (1.3), U^0 can be understood as the solution of the diffraction problem on one of the two bodies, or, more generally, on a body that is in a sense close to the diffracting system. The field is represented in the same "pseudoadditive" form in the Schwarzschild successive diffraction method, in the "Fredholm denominator" method (e.g., see [34, 28]), and in various versions of the operator semi-inversion method (e.g., see [30]).

- The *discrete representation* of the field in scattering problems is used in several recent methods (e.g., see Section IV in the survey [31]). In some of these methods, the fields are expanded in spheroidal functions outside a sufficiently large sphere. In other methods, the expansion is carried out with respect to functions that, just as in the generalized method of eigenoscillations, are determined by the body on which the diffraction occurs.

- A discrete representation of the field can also be achieved by introducing *complex coordinates* [62]. There is an analogy between this procedure, the introduction of Sturm functions [37] (the eigenfunctions of a potential well surrounded by an infinitely high barrier), and the introduction of an auxiliary barrier waveguide [19] in the theory of irregular waveguides. The farther is the observation point from the body, the worse the series provided by these methods converge. There is also a close method for expanding the field near an open waveguide, which uses the *Mittag–Leffler expansion in simple fractions in the complex plane* [40, 61].

- There are various forms of *integral equations* to which one can reduce the diffraction problem. To each of these nonhomogeneous equations we can assign a homogeneous equation and then use the eigenfunctions of the latter in the expansion of the solution. If the equation has a complicated kernel (like in [22], where this method is suggested), then the numerical implementation may be difficult. The spectral parameter need not occur in the kernel as a factor.

3. The method presented in this book can be developed in several directions. Some of these directions are as follows.

- Applications to specific problems, development of numerical algorithms, and obtaining new numerical and analytical results.
- A generalization of the technique by introducing other spectral parameters (say, describing the geometry), if this will be suggested by new problems.
- Applications of the method to linear equations other than those of electromagnetic theory and acoustics, say, to elasticity equations. Even for systems with finitely many degrees of freedom, the choice of a spectral parameter other than the frequency can be expedient (for example, a system with losses can be studied without passing into the complex domain).
- Generalizations to bodies whose boundaries stretch to infinity or to bodies for which $\varepsilon \to 1$ as $r \to \infty$.
- Generalizations to nonstationary problems. Complex eigenfrequencies, which seem to be an inconvenient tool in the study of steady-state phenomena, pretty well describe "extremely nonstationary" processes like light fading-out in the absence of sources. However, for intermediate transient problems in which the field exists for $-\infty < t < \infty$ but is created by sources of variable amplitude or frequency, the application of complex eigenfrequencies will necessitate the introduction of the continuous spectrum. One can try to avoid this by introducing a different spectral parameter. A similar situation occurs for bodies that move or change their shape if the frequency is too low for the adiabatic approximation to apply. Then the field will be described by a system of differential equations for the coefficients in expansions like (1.3) similar to systems of waveguide equations in the method of cross-sections.

Bibliography for Chapters 1–4

[1] N. P. Abovskii, N. P. Andreev, and A. P. Deruga. *Variational Principles in the Theory of Shells.* Nauka, Moscow 1978 (Russian).

[2] A.I. Baz', Ya. B. Zel'dovich, and A. M. Perelomov. *Scattering, Reactions, and Disintegration in Nonrelativistic Quantum Mechanics.* Nauka, Moscow 1971 (Russian). Engl. transl.: Israel Program for Scientific Translations, Jerusalem 1969.

[3] R. Courant and D. Hilbert. *Methoden der Mathematischen Physik I.* Heidelberger Taschenbücher. Springer-Verlag, Berlin, Heidelberg, New York 1968.

[4] V. A. Dikarev. The Ritz method in the theory of generalized eigenoscillations. *Radiotekhn. i Èlektron.* **26**(1)(1981), 18–26. Engl. transl. in: *Radio Engrg. Electron. Phys.* **26**(1)(1981), 16–23.

[5] L. Dolph. Recent developments in some non-self-adjoint problems of mathematical physics. *Bull. Amer. Math. Soc.* **67**(1)1961, 1–69.

[6] V. A. Fok. *Problems of Diffraction and Propagation of Electromagnetic Waves.* Sovetskoe Radio, Moscow 1970 (Russian). Engl. edition: *Electromagnetic Diffraction and Propagation Problems.* Pergamon Press, Oxford 1965.

[7] A. G. Fox and T. Li. Resonant modes in a laser interferometer. *Bell Sys. Tech. J.* **40**(1961), 453.

[8] Kh. L. Garb and P. Sh. Fridberg. Dispersion properties of a rectangular waveguide loaded by a half-space via a small slot in a wall of finite thickness. *Radiotekhn. i Èlektron.* **19**(1)(1974), 1–3.

[9] R. J. Garbacz. Modal expansion for resonance scattering phenomena. *Proc. IEEE* **53**(8)(1965), 856.

[10] I. Ts. Gokhberg and M. G. Krein. *Introduction into the Theory of Linear Nonselfadjoint Operators in Hilbert Space.* Nauka, Moscow 1965 (Russian). Engl. transl.: Amer. Math. Soc., Providence 1969.

[11] I. S. Gradshtein and I. M. Ryzhik. *Tables of Integrals, Sums, Series, and Products.* Fizmatgiz, Moscow 1963 (Russian). Engl. transl.: Academic Press, New York–London 1965.

[12] A. D. Grigor'ev. Analysis of azimuth-inhomogeneous oscillation modes in axisymmetric resonators. *Radiotekhn. i Èlektron.* **24**(6)(1979), 1211–1213.

[13] R. F. Harrington. Characteristic modes for antennas and scatterers. In: R. Mittra (ed.), *Numerical and Asymptotic Techniques in Electromagnetics*, Topics in Applied Physics, Vol. 3, 1975, 51–87.

[14] H.Hönl, A. V. Maue, and K. Westpfahl. *Theorie der Beugung, Handbuch der Physik*, Vol. XXV/1, Springer-Verlag, Berlin, New York 1961.

[15] D. Hilbert. *Theorie der linearen Integralgleichungen.* Teubner, Leipzig 1912.

[16] V. V. Karnishin, V. V. Akindinov, and V. V. Vishin. A method for studying electromagnetic wave scattering in a resonance region. *Radiotekhn. i Èlektron.* **15**(1)(1970), 14–20. Engl. transl. in: *Radio Engrg. Electron. Phys.* **15**(1)(1970).

[17] B. Z. Katsenelenbaum. *High-Frequency Electromagnetic Theory.* Nauka, Moscow 1966 (Russian).

[18] B. Z. Katsenelenbaum. Forced electromagnetic oscillations of dielectric bodies in an infinite domain and eigenfunctions of the discrete spectrum. *Radiotekhn. i Èlektron.* **13**(4)(1968), 568–590.

[19] B. Z. Katsenelenbaum. The expansion of forced oscillations of nonclosed systems in the discrete spectrum eigenfunctions. *Radiotekhn. i Èlektron.* **14**(1)(1969), 25–30.

[20] B. Z. Katsenelenbaum. On the convergence rate of the series obtained by spectral methods for Green's function of diffraction problems. *Radiotekhn. i Èlektron.* **27**(4)(1982), 696–698. Engl. transl. in: *Radio Engrg. Electron. Phys.* **27**(4)(1982), 53–55.

[21] B. Z. Katsenelenbaum, E. N. Korshunova, and A. N. Sivov. Computation of propagation constants for leaky waves in waveguides of arbitrary cross-section with a semitransparent wall or with a longitudinal slot. *Akusticheskii Zhurnal* **21**(3)(1975), 493.

[22] B. Z. Katsenelenbaum and A. N. Sivov. The rigorous statement of the problem on free and forced oscillations of an open resonator. *Radiotekhn. i Èlektron.* **12**(7)(1967), 1184–1191.

[23] B. Z. Katsenelenbaum and A. N. Sivov. To the excitation theory of high-Q open resonators. *Radiotekhn. i Èlektron.* **19**(12)(1974), 2449–2457.

[24] Yu. V. Kolesnichenko and V. V. Shevchenko. Dispersion and polarization characteristics of modes in fibre lightguides with noncircular cores. *Telecommunications and Radio Engrg.* **43**(9)(1988), 116–124.

[25] Yu. V. Kolesnichenko and V. V. Shevchenko. Fiber-optic dielectric waveguides with a noncircular core. *Sov. J. of Communications Technology and Electronics* **33**(11)(1988), 127–136.

[26] E. N. Korshunova and A. N. Sivov. An application of the generalized eigenoscillation method to the computation of propagation constants for leaky E-waves in waveguides of arbitrary cross-section with longitudinal slots. *Radiotekhn. i Èlektron.* **20**(6)(1975), 1038–1041.

[27] M. A. Krasnosel'skii, G. M.Vainikko, P. P. Zabreiko, Ya. B. Rutitskii, and V. Ya. Stetsenko. *Approximate Solution of Operator Equations.* Nauka, Moscow 1969 (Russian). Engl. transl.: Wolters-Northhoff Publishing, Groningen 1972.

[28] E. P. Kurushin, E. I. Nefedov, and A. T. Fialkovskii. *Diffraction of Electromagnetic Waves on Anisotropic Structures.* Nauka, Moscow 1975 (Russian).

[29] Ismo V. Lindel. Variational methods for nonstandard eigenvalue problems in waveguide and resonator analysis. *IEEE Transactions on Microwave Theory and Techniques* **MTT-30**(8)(1982), 1194–1204.

[30] L. N. Litvinenko. The successive refinement method for the solution of infinite systems of linear algebraic equations. *Doklady Akad. Nauk SSSR* **203**(1)(1972), 64–67. Engl. transl. in: *Soviet Phys. Dokl.* **17**(3)(1972), 205–207.

[31] G. D. Malushkov. Methods for solving electromagnetic excitation problems for bodies of revolution (a review). *Izvestiya Vuzov, Radiofizika* **18**(11)(1975), 1563–1590.

[32] S. G. Mikhlin. *Variational Methods in Mathematical Physics.* Nauka, Moscow 1967 (Russian). Engl. transl.: Macmillan, New York 1964.

[33] I. M. Narodetskii. The Hilbert–Schmidt expansion of the two-particle potential scattering amplitude. *Yadernaya Fizika* **9**(5)(1969), 1086. Preprint No. 621, Institute for Theoretical and Experimental Physics of the Academy of Sciences of the USSR, Moscow 1968.

[34] E. I. Nefedov and A. T. Fialkovskii. *Asymptotic Theory of Diffraction of Electromagnetic Waves on Finite Structures.* Nauka, Moscow 1972 (Russian).

[35] V. V. Nikol'skii. *Variational Methods for Interior Problems of Electromagnetic Theory.* Nauka, Moscow 1967 (Russian).

[36] V. V. Nikol'skii and T. I. Lavrova. The solution of eigenwave problems by the method of minimal autonomous blocks. *Radiotekhn. i Èlektron.* **24**(8)(1979), 1518–1527.

[37] M. Rotenberg. Application of Sturmian functions to the Schrödinger three-body problem: Elastic E^+-H scattering. *Ann. Phys.* **19**(2)(1962), 1962.

[38] A. D. Scaballanovich and V. V. Shevchenko. Mode properties of a double-core lightguide with nonidentical circular cores. *J. of Communications Technology and Electronics* **40**(3)(1995), 75–80.

[39] S. E. Shamma. Asymptotic behavior of Stekloff eigenfunctions. *J. Appl. Math. Phys.* **20**(3)(1971).

[40] A. D. Shatrov. Discrete field representations in the excitation problem for a dielectric plate. *Radiotekhn. i Èlektron.* **15**(9)(1970), 1808–1815. Engl. transl. in: *Radio Engrg. Electron. Phys.* **15**(1970), 1564-1573.

[41] V. V. Shevchenko. Shift-formula method in the theory of dielectric waveguides and fiber lightguides. *Sov. J. of Communications Technology and Electronics* **31**(9)(1986), 28–42.

[42] V. V. Shevchenko. On modal theory for coupled optical dielectric waveguides. *Sov. Lightwave Communications* **3**(4)(1993), 199–211.

[43] V. V. Shevchenko and N. Espinoza-Ortiz. Shift formula method for asymmetrical planar waveguides. *J. of Communications Technology and Electronics* **38**(16)(1993), 121–128.

[44] V. A. Steklov. General Methods for Solving the Main Problems of Mathematical Physics. Doctor of Sciences Thesis, St. Petersburg 1901.

[45] S. V. Stoyanov and V. V. Shevchenko. Modes of inhomogeneous anisotropic fibre lightguide. *Telecommunications and Radio Engineering* **45**(8)(1990), 114–122.

[46] J. R. Taylor. *Scattering Theory. The Quantum Theory on Nonrelativistic Collisions.* Wiley, New York 1972.

[47] G. M. Vainikko. *Regular Convergence of Operators and Approximate Solution of Equations, Itogi Nauki i tekhniki. Matematicheskii Analiz*, Vol. 16. VINITI, Moscow 1979 (Russian).

[48] L. A. Vainshtein. *Open Resonators and Open Waveguides.* Sovetskoe Radio, Moscow 1966 (Russian).

[49] G. I. Veselov and V. M. Temnov. The method of partial domains for diffraction problems with noncoordinate boundaries. *Izvestiya Vuzov. Radiofizika* **28**(7)(1984), 912–928.

[50] N. N. Voitovich, B. Z. Katsenelenbaum, E. N. Korshunova, and A. N. Sivov. The solution of exterior diffraction problems and the computation of propagation constants for open waveguides with the help of a real integral equation. *Radiotekhn. i Èlektron.* **20**(6)(1975), 1129–1137.

[51] N. N. Voitovich, B. Z. Katsenelenbaum, and A. N. Sivov. The surface current method for the construction of discrete spectrum eigenfunctions in diffraction problems. *Radiotekhn. i Èlektron.* **15**(4)(1970), 685–696.

[52] N. N. Voitovich, B. Z. Katsenelenbaum, and A. N. Sivov. Stationary functionals for the generalized eigenfunction method in diffraction theory. *Radiotekhn. i Èlektron.* **17**(2)(1972), 268–275.

[53] N. N. Voitovich, B. Z. Katsenelenbaum, and A. N. Sivov. To the theory of scattering by a quasistationary level. Preprint No. 28 (143), Institute for Radiotechnics and Electronics of the Academy of Sciences of the USSR, Moscow 1973.

[54] N. N. Voitovich, B. Z. Katsenelenbaum, and A. N. Sivov. The s-method for problems of diffraction on bounded bodies. Preprint No. 29 (144), Institute for Radiotechnics and Electronics of the Academy of Sciences of the USSR, Moscow 1973.

[55] N. N. Voitovich, B. Z. Katsenelenbaum, and A. N. Sivov. The excitation of a two-dimensional metallic resonator with a small hole (a cylinder with a slot). *Radiotekhn. i Èlektron.* **19**(12)(1974), 2458–2469.

[56] N. N. Voitovich, B. Z. Katsenelenbaum, and A. N. Sivov. The generalized eigenoscillation method in diffraction theory. In: *Studies in Radiotechnics and Electronics in 1954–1974*, Vol. 1(1974), 241–262. Institute for Radiotechnics and Electronics of the Academy of Sciences of the USSR, Moscow.

[57] N. N. Voitovich, B. Z. Katsenelenbaum, and A. N. Sivov. The generalized eigenoscillation method in diffraction theory. *Uspekhi Fizicheskikh Nauk* **118**(4)(1976), 709–736. Engl. transl. in: *Soviet Phys. Uspekhi* **19**(4)(1976), 337–352.

[58] N. N. Voitovich, A. I. Kidisyuk, and A. I. Rovenchak. A numerical implementation of the generalized eigenoscillation method for problems on two-dimensional complex-shaped resonators. In: *Diffraction Theory and Wave Propagation*, Vol. II(1977), 221–224. Moscow.

[59] N. N. Voitovich and A. I. Rovenchak. Modification of the method of successive approximations for homogeneous problems. *Zhurn. Vyshisl. Matem. i Matem. Fiziki* **22**(2)(1982), 348–357.

[60] N. N. Voitovich and N. P. Santalov. Some applications of the generalized eigenoscillation method. *Radiotekhn. i Èlektron.* **19**(12)(1974), 2625–2629.

[61] N. N. Voitovich and A. D. Shatrov. Excitation of an open waveguide with dielectric walls. *Radiotekhn. i Èlektron.* **18**(4)(1973), 687–694.

[62] V. Yu. Zavadskii. *Computation of Wave Fields in Open Regions and Waveguides.* Nauka, Moscow 1972 (Russian).

Chapter 5

Spectral Properties of Diffraction Problems

5.1 Introduction

1. The preceding chapters of this book, as well as the earlier work of B. Z. Katsenelenbaum, A. N. Sivov, and N. N. Voitovich, put forward a number of significant questions addressed to mathematicians. Apparently, the most important of these questions is as follows. Throughout the preceding chapters it was assumed that functions belonging to sufficiently wide function classes can be expanded in convergent series with respect to the eigenfunctions of the spectral problems considered in the book. Are these expansions actually possible? The present chapter mostly deals with this question. It is by no means so easy as one might expect at first glance, and the answer requires an insight into some areas of functional analysis and operator theory. In this section we use an example to explain the difficulties encountered and to give a qualitative description of the results. Precise definitions and complete statements will be given in the subsequent sections; here we only outline their contents.[1]

Consider the Helmholtz equation

$$\Delta u + k^2 u = 0 \tag{5.1}$$

in the three-dimensional space $\mathbf{R}^3$ outside a smooth connected closed surface S. We subject the solution to the conditions

$$u^+ = u^-, \quad u^\pm - \lambda(\partial_\nu u^- - \partial_\nu u^+) = g \quad \text{on } S \tag{5.2}$$

and the radiation condition at infinity (see Eq. (0.35) and Eq. (5.137) below)[2]. We assume that the nonzero number k may be complex:

$$k = k_1 + ik_2 \neq 0, \quad \text{where} \quad k_2 \leq 0. \tag{5.3}$$

---

[1]See also "Concluding Remarks".

[2]By ν we denote the unit normal vector on S pointing into the infinite domain, and ∂_ν stands for the derivative along the direction of this vector. The superscripts $^+$ and $^-$, just as in the preceding chapters, refer to the exterior and the interior side of the surface, respectively. For convenience, we use the same notation, λ or μ, for the spectral parameter in all the problems (in Eq. (2.60), the spectral

If k is not real, then the radiation condition is understood to imply exponential decay of the solution at infinity.

The substitution

$$u(x) = -\frac{1}{4\pi}\int_S \frac{e^{-ik|x-y|}}{|x-y|}\varphi(y)\,dS_y \tag{5.4}$$

readily reduces the problem to a Fredholm integral equation of the second kind, namely to the equation

$$A(k)\varphi - \lambda\varphi = g \tag{5.5}$$

on S, where

$$A(k)\varphi(x) = -\frac{1}{4\pi}\int_S \frac{e^{-ik|x-y|}}{|x-y|}\varphi(y)\,dS_y, \qquad x \in S. \tag{5.6}$$

Under the assumptions that all the functions involved are smooth and λ is not zero and does not coincide with any of the eigenvalues of the operator $A(k)$, Eq. (5.5) is equivalent to the original problem: $u(x)$ is determined by $\varphi(x)$ according to (5.4), and $\varphi(x)$ is determined by $u(x)$ using the formula $\varphi = \partial_\nu u^- - \partial_\nu u^+$ (see Subsection 5.7.1).

From now on we shall consider Eq. (5.5). For this it is convenient to introduce the Hilbert space $L_2(S)$ of square integrable complex-valued functions on S and assume that g and φ are elements of $L_2(S)$. The inner product in this space is defined by the well-known formula

$$(\varphi, \psi)_S = \int_S \varphi(x)\,\overline{\psi(x)}\,dS, \tag{5.7}$$

where $\overline{\psi}$ is the complex conjugate of ψ.[3]

First, let $k_1 = \operatorname{Re} k = 0$. Then one can readily verify that $A(k)$ is a selfadjoint operator in $L_2(S)$, that is,

$$(A(k)\varphi, \psi)_S = (\varphi, A(k)\psi)_S \tag{5.8}$$

for all $\varphi, \psi \in L_2(S)$. Since, moreover, $A(k)$ is compact (that is, Eq. (5.5) is a Fredholm equation), it follows by a well-known theorem of functional analysis that in $L_2(S)$

parameter was denoted by ρ), and a point in space will be denoted by x or y (instead of r or r'). The other differences in the notation (there are only few) will also be indicated in footnotes.

In Section 2.2, the nonhomogeneous equation $\Delta u + k^2 u = f$ with homogeneous boundary condition $g = 0$ was considered. That problem can be reduced to the one considered here by subtracting the integral

$$-\frac{1}{4\pi}\int_{\mathbf{R}^3} \frac{e^{-ik|x-y|}}{|x-y|} f(y)\,dy$$

(the incident field) from the solution.

[3]The notation $\overline{\psi}$ is used here instead of ψ^*. In the preceding chapters, the bar has a different meaning.

there exists an orthonormal basis $\{\varphi_j\}$, $j = 1, 2, \ldots$, consisting of eigenfunctions of the operator $A(k)$ (e.g., see [21, Section 61]). Let $A(k)\varphi_j = \lambda_j \varphi_j$; the eigenvalues λ_j are real. Suppose that the φ_j and λ_j are known (this is a very strong assumption).

An arbitrary function $g \in L_2(S)$ can be expanded in the series

$$g = \sum_{j=1}^{\infty} c_j \varphi_j, \tag{5.9}$$

which is convergent in the norm of $L_2(S)$ (that is, in the mean). Here $c_j = (g, \varphi_j)_S$ are the Fourier coefficients of g with respect to the system $\{\varphi_j\}$.

Under the above assumptions on λ, Eq. (5.5) has a unique solution φ (e.g., see [21, Section 58]). The solution can be sought in the form of a series, $\varphi = \sum d_j \varphi_j$. By substituting this series into (5.5), we readily find the d_j and arrive at the formula

$$\varphi = \sum_{j=1}^{\infty} \frac{c_j}{\lambda_j - \lambda} \varphi_j. \tag{5.10}$$

This is one of the classical methods in the theory of Fredholm equations; it can be called the *spectral method*.

2. Now let $k_1 \neq 0$. Then the operator $A(k)$ is no longer selfadjoint:

$$(A(k)\varphi, \psi)_S = (\varphi, A^*(k)\psi)_S, \tag{5.11}$$

where the operator $A^*(k)$, the *adjoint* of $A(k)$, is obtained by replacing $-ik$ in (5.6) by $i\bar{k}$ (in view of the symmetry of the kernel of the integral operator $A(k)$). Now the eigenvalues λ are not real (cf. Eq. (2.38)). In Section 2.2, it was suggested to apply the spectral method just in this case (more precisely, for $k_2 = 0$). What is the difference between this situation and the one considered in Subsection 1? Let us consider some technical aspects, postponing the discussion of a more essential problem until Subsection 3.

First, we must take into account the fact that, along with eigenfunctions, there can be associated functions, and then the use of the latter cannot be avoided. We can draw an analogy with linear algebra. Let A be a linear operator in the n-dimensional complex space $\mathbf{C}^n$, and let J be the Jordan normal form of the matrix of this operator. If J is a diagonal matrix, then there is a basis of eigenvectors of A in $\mathbf{C}^n$. However, if J has at least one Jordan block of size > 1, then there is no basis of eigenvectors of A in $\mathbf{C}^n$, but there is a basis consisting of eigenvectors and associated vectors.

We have to note that it is not easy to indicate associated functions explicitly. In the sequel, we shall find them explicitly only in the one-dimensional versions of some problems considered in the book, see Subsections 5.7.8 and 5.10.5.

Suppose that the space $L_2(S)$ possesses a basis $\{\varphi_j\}$, $j = 1, 2, \ldots$, consisting of *root functions* (that is, eigenfunctions and associated functions) of the operator $A(k)$, that is, a system such that each function $g \in L_2(S)$ can be expanded uniquely in a series (5.9) convergent to g in the mean. (Later on, we shall see that this is also a very strong assumption.) In general, the elements φ_j of this basis are not pairwise orthogonal, and to find the Fourier coefficients c_j, we need the system $\{\psi_l\}$ *biorthogonal* to $\{\varphi_j\}$, that is, the system such that $(\varphi_j, \psi_l)_S = \delta_{jl}$, $j, l = 1, 2, \ldots$. Then $c_j = (g, \psi_j)_S$.

One can show by using the symmetry of the kernel of the operator $A(k)$ that if all its eigenvalues are simple, then the function ψ_j is proportional to $\overline{\varphi}_j$; after an appropriate normalization, ψ_j coincides with $\overline{\varphi}_j$. In other words, in this case we can directly use the "real orthogonality" of the eigenfunctions in the manner adopted in the preceding chapters. If there are also eigenvalues of multiplicity > 1, then one can prove that for a special choice of the eigenfunctions corresponding to these eigenvalues we again have $\psi_j = \overline{\varphi}_j$. The corresponding terms in (5.10) become somewhat more complicated. We discuss this in more detail in Subsection 5.2.6.

Thus if the space $L_2(S)$ possesses a basis consisting of root functions of the operator $A(k)$, then the spectral method can be applied (though with some technical complications).

3. However, the question of the existence of such a basis is by no means trivial. There exist nonselfadjoint compact operators that do not have root vectors at all. Examples are given by Volterra integral operators, say, the operator $\int_0^x \varphi(t)\,dt$ in the space L_2 on the interval $[0, 1]$.

The existence of a basis consisting of root functions cannot be proved in general for operators studied in this book. For the operator $A(k)$, our main goal will be to establish that its system of root functions possesses a property close to the property of being a basis and sufficient for the justification of the spectral method. Namely, we shall show in Section 5.7 that the Fourier series (5.9) with respect to this system of an arbitrary function $g \in L_2(S)$ converges to g in the mean after inserting parentheses (in a way independent of g). Every pair of parentheses singles out a group of terms $c_l\varphi_l + \cdots + c_m\varphi_m$ corresponding to very close eigenvalues. In some cases the parentheses can be removed, but this may slow down the convergence of the series. If g is infinitely smooth, then the series with parentheses converges to g uniformly on S and can be differentiated term by term as many times as desired; moreover, the convergence is very rapid. Formula (5.4) with φ_j substituted for φ defines the root functions u_j of the original problem. One can use these root functions to construct series expansions for functions given outside S.

A similar situation (with some simplifications and additional details) occurs in the two-dimensional case, where S is a closed curve on the plane.

We shall also obtain some information about the distribution of eigenvalues of the operator $A(k)$; in particular, we shall indicate the asymptotics of the eigenvalues.

As can be seen from the contents, in Sections 5.8–5.11 we analyze other problems. In a close form, these problems were considered in Sections 0.2, 1.1–1.3, and 2.1–2.6. The results of our Sections 5.8 and 5.11 are essentially the same in nature as those of Section 5.7. The results of Section 5.9 are weaker but still sufficient for the justification of the spectral method. For the problems considered in Section 5.10, under some assumptions we show that there are no associated functions and that for an appropriate choice of an (indefinite) inner product the eigenfunctions form an orthonormal basis.

4. The proofs are mainly based on the tools of two areas of mathematics, which were mainly formed in the third fourth of the 20th century, namely, the theory of nonselfadjoint operators in an abstract Hilbert space and the theory of elliptic differential and pseudodifferential operators and general elliptic boundary value problems in Sobolev L_2-spaces. We first give, mainly in Sections 5.2–5.6, the necessary definitions and statements of theorems in a form convenient to us, with brief explanations and bibliographical references, and then apply these theorems to specific problems and operators.

The operator (5.6) is an especially transparent model for the application of these theorems. Let us indicate the properties of this operator which play the most important role here. To be definite, suppose that $k_2 = 0$. We write $A(k)$ in the form $A_R(k)+iA_I(k)$, where $A_R(k)$ and $A_I(k)$ are the real and imaginary parts of the operator A, respectively; these are the integral operators with kernels

$$-\frac{\cos k|x-y|}{4\pi|x-y|} \quad \text{and} \quad \frac{\sin k|x-y|}{4\pi|x-y|}. \tag{5.12}$$

The second of these functions is infinitely smooth in (x, y), that is, has continuous derivatives of all orders, whereas the first function has the same singularity at $x - y = 0$ as the kernel of the operator $A(k)$. Furthermore, we shall see that there exists an (unbounded) inverse $A_R^{-1}(k)$ (possibly after an unessential modification of the terms in the sum $A_R(k)+iA_I(k)$). Taken together, these two facts mean that $A(k)$ is a very weak perturbation of the selfadjoint operator $A_R(k)$. For weak perturbations of selfadjoint operators, we can use theorems on the basis property with parentheses (formulated in Section 5.6).

The operators $A(k)$ and $A_R(k)$ belong to the class of *pseudodifferential operators*, which contains usual differential operators as well as the main integral and integro-differential operators of mathematical physics. The calculus of pseudodifferential operators is outlined in Section 5.4. To each pseudodifferential operator, this calculus assigns its *symbol*, which has a certain *order*. For example, the Laplace operator on the plane has the symbol $-(\xi_1^2 + \xi_2^2)$ of order 2. In this example, the symbol is independent of x and homogeneous in ξ. In the general case, the symbol depends on x and ξ and can be expanded, at least locally in x, in an asymptotic sum of terms positively homogeneous in ξ of decreasing orders; in the case of differential operators, this is a usual finite sum. The leading part of this sum is called the *principal symbol*; it has the maximal order of homogeneity. By definition, this order is assigned to the complete symbol. The operators $A(k)$ and $A_R(k)$ for $k_2 = 0$ have the same symbol of order -1 (the principal symbol is given in (5.158)). The operator $A_I(k)$ has zero symbol; its order is $-\infty$ by convention. The principal symbol of the product of two pseudodifferential operators is the product of principal symbols of the factors; the order of the product is the sum of orders of the factors unless the product of the principal symbols is zero identically. We use these properties many times in Sections 5.7–5.11.

The class of pseudodifferential operators contains the important subclass of *elliptic* pseudodifferential operators. For example, the Laplace operator, $A(k)$, and $A_R(k)$ are elliptic. The Maxwell system is not elliptic, but the problems for this system considered in Section 5.11 will be reduced to equations with elliptic pseudodifferential operators on S.

For elliptic equations, there is a well-developed theory of solvability in the Sobolev spaces $H^s = W_2^{(s)}$ (see Section 5.5) and in other function spaces. The results of this theory will also be used. For example, it follows from the ellipticity of the operator $A_R(k)$ and from the fact that its principal symbol is real that the inverse $A_R^{-1}(k)$ exists (possibly, after a modification of $A_R(k)$ on some finite-dimensional subspace of functions) and has a symbol of order $+1$. The use of the spaces H^s permits one to give the following definition of the *order of an operator*: an operator of finite order m is an operator that acts as a bounded operator from H^{s+m} into H^s for each s. Then a pseudodifferential operator with symbol of order m proves to be an operator of order m. The space H^s coincides with L_2 for $s = 0$, becomes narrower as s increases, and consists of smooth functions for large s. Thus, an operator of negative order increases the smoothness of functions, whereas an operator of positive order in general decreases the smoothness. Operators of positive order can also be considered as unbounded operators in H^s with fixed s. The H^s are Hilbert spaces, and therefore they are especially convenient for our aims.

In Section 5.10 we touch upon the theory of selfadjoint operators in spaces with indefinite metric.

5. It is obvious from the previous subsection that we use a wide variety of mathematical tools. The difficulties that will be encountered by a reader who has not got acquainted with these tools in advance are due to the large amount of mathematical concepts involved rather than awkward mathematical proofs (for these, we refer the readers to special literature) or complicated calculations (there are virtually none). However, the notions and theorems included in Sections 5.2–5.6 are undoubtedly significant and interesting in themselves, and the problems considered in Sections 5.7–5.11 permit one to understand the essence of these notions and theorems via important specific examples.

This chapter can also be of interest to a mathematically well-prepared reader, who will see how the numerous techniques and results of the modern elliptic theory and spectral theory of operators are amplified, combined, and applied to the analysis of virtually classical integral or integro-differential operators related to the Helmholtz equation and the Maxwell system and of (nonstandard) spectral problems for them. Our analysis of these operators is undoubtedly deeper than in existing textbooks (e.g., see [76] and [31]).

To understand the subsequent text, the reader must be acquainted with the foundations (or, better to say, elements) of functional analysis and operator theory. For references, one cane use some sections in [21, Chapters 1–5], [54, Chapters 1, 3], or [61, Chapters 3–5, 7–9]. We assume that the reader is familiar with the notion of a distribution (generalized function). The necessary information about distributions and their Fourier transforms can be found in some sections of [40, Chapters 1, 2]. We also give numerous references to mathematical monographs and journal articles, where the reader can find the proofs omitted here or additional details.

Among all the problems considered in Chapters 1 and 2, we consider the most typical and probably the most important problems admitting the reduction to Fredholm equations in $L_2(S)$ or $L_2(V^+)$, where V^+ is the domain bounded by the surface (or curve) S. The detailed analysis of all versions of these problems would require too much place. In the main part of this chapter, we eliminate all secondary mathematical difficulties (which are especially abundant in specific applied problems analyzed in Chapter 4) by assuming that the surface S is smooth and closed (that is, has no edges and holes) and the permittivity $\varepsilon(x)$ changes by a jump across S. Unfortunately, little has been found out for the problems considered in Sections 1.2 and 1.4, and we do not touch these problems here. As to the variational approach to evaluating the eigenvalues of nonselfadjoint problems considered in the book, it was already discussed in Chapter 3.

6. Up to this point, the text of the present introduction mainly reproduces the corresponding text in the Russian book published in 1977. Let us now say a few words about what new was included in this chapter. First, there are many local improvements and refinements in the exposition, sometimes accompanied by a certain strengthening of the results. Here we do not try to list them. Second, the new text reflects the progress achieved since 1977 in the general spectral theory of elliptic equations. In particular, we present deeper results concerning the spectral asymptotics for general nonselfadjoint operators (see Subsection 5.6.6) and for pseudodifferential operators on one-dimensional smooth boundaries of two-dimensional domains (see Subsections 5.5.4 and 5.7.5 and Remark 5.8.17). Third, in Section 5.12 we describe the recently obtained analogs of the results of Sections 5.7 and 5.8 for domains with Lipschitz boundaries (such boundaries may have corners, edges, and conical points). Here the mathematical technique is quite different; in particular, there is no calculus of pseudodifferential operators on Lipschitz surfaces.

Finally, the bibliography was substantially renewed.

The spectral approach to a specific physical problem probably always gives rise to the hope for a successful insight into the properties of the corresponding physical system. On the other hand, the analysis, by mathematical tools, of the spectral approach to diffraction problems that is proposed in the preceding chapters has already led to some progress in some directions of the spectral theory of elliptic operators. It is this analysis that is the subject of the present chapter.

5.2 Vector Systems and Nonselfadjoint Operators in Hilbert Spaces

1. Vector systems. Let $\mathfrak{H}$ be a complex infinite-dimensional separable Hilbert space with inner product (f,g) and norm $\|f\|^2 = (f,f)^{1/2}$. We introduce several notions pertaining to a system of vectors $\{f_j\}$, $j = 1,2,\dots$, in this space. In particular, we explain when this system is called a complete system, an Abel–Lidskii system, a basis with parentheses, a Riesz basis with parentheses, or a Bari basis with parentheses. In this sequence, each property is a strengthening of the preceding one. We shall give a few assertions. Some of them are obvious; the proof of the other assertions, as well as further details, can be found in [43, Sections 6.1–6.3].

The system $\{f_j\}$ is said to be *complete* if for any vector $f \in \mathfrak{H}$ and any number $\varepsilon > 0$ there exist a positive integer m and coefficients $a_1, \dots, a_m$ such that

$$\left\| f - \sum_{j=1}^{m} a_j f_j \right\| < \varepsilon. \tag{5.13}$$

To decrease ε (that is, to increase the accuracy), in general we have not only to increase m but also to change the already found coefficients $a_1, \dots, a_m$.

The system $\{f_j\}$ is said to be *minimal* if none of the vectors f_j is contained in the closed linear span of the other vectors of this system.

Let $\{f_j\}$ be a minimal system. Then in $\mathfrak{H}$ there exists a system $\{g_l\}$, $l = 1, 2, \dots$, biorthogonal to $\{f_j\}$: $(f_j, g_l) = \delta_{jl}$, $j, l = 1, 2, \dots$. If, moreover, $\{f_j\}$ is a complete system, then $\{g_l\}$ is uniquely determined and is also a minimal system. If $\{f_j\}$ is a complete orthonormal system, then $g_j = f_j$ for all j.

Let $\{f_j\}$ be a minimal system, and let $\{g_l\}$ be a system biorthogonal to $\{f_j\}$. Suppose that some vector $f \in \mathfrak{H}$ is the sum of a series

$$\sum_{j=1}^{\infty} c_j f_j. \tag{5.14}$$

Then, obviously,

$$c_j = (f, g_j). \tag{5.15}$$

Now if f is an arbitrary vector from $\mathfrak{H}$, then the numbers (5.15) are called the *Fourier coefficients*, and the formal series (5.14) is called the *Fourier series of the vector f with respect to the system* $\{f_j\}$. However, this series may be divergent, and even if it is convergent, it may converge to some other vector instead of f.[4] The system $\{f_j\}$ is called a *basis* in $\mathfrak{H}$ if the Fourier series (5.14) of each vector $f \in \mathfrak{H}$ converges to f. In this case $\{g_l\}$ is also a basis (Banach's theorem, see [43, Section 6.1]). Obviously, any basis is a complete system.

The partial sum $c_1 f_1 + \dots + c_m f_m$ of the Fourier series of f need not be the best approximation of f by linear combinations of the vectors $f_1, \dots, f_m$ even if $\{f_j\}$ is a basis. (The best approximation is obviously given by $b_1 h_1 + \dots + b_m h_m$, where $h_1, \dots, h_m$ is an orthonormal basis in the linear span of $f_1, \dots, f_m$ and the coefficients are given by $b_j = (f, h_j)$.) Nevertheless, the basis property of the system $\{f_j\}$ permits one to increase m infinitely without changing the coefficients c_j already found according to formula (5.15).

Suppose that $\{e_j\}$, $j = 1, 2, \dots$, is an orthonormal basis in $\mathfrak{H}$, B is a bounded linear operator in $\mathfrak{H}$ with bounded inverse B^{-1}, and $f_j = Be_j$. Then $\{f_j\}$ is also a basis in $\mathfrak{H}$.

[4]If the system $\{g_j\}$ is complete, then the series can converge only to f.

Such a basis $\{f_j\}$ is called a *basis equivalent to an orthonormal basis*, or a *Riesz basis*. The following properties of a Riesz basis can readily be derived from the definition.

1° The space $\mathfrak{H}$ can be equipped with the new inner product $\langle f, g\rangle = (B^{-1}f, B^{-1}g)$ such that the basis $\{f_j\}$ is orthonormal with respect to it and the norms $\langle f, f\rangle^{1/2}$ and $\|f\|$ are *equivalent*, that is, their ratio is bounded above and below by positive constants.

2° $C_1\|f\|^2 \le \sum_{j=1}^{\infty} |c_j|^2 \le C_2\|f\|^2$ for each f with some positive constants C_1 and C_2.

These two inequalities replace the Parseval identity for an orthonormal system and can be used to obtain two-sided estimates of the norm of the remainder of the Fourier series.

3° The system $\{f_j\}$ is an *unconditional basis*, that is, remains a basis after an arbitrary permutation of the vectors forming the system.

In fact, each of these three properties is equivalent to the assertion that $\{f_j\}$ is a Riesz basis if we supplement 2° by the requirement that the system is complete and minimal and supplement 3° by the requirement that the system is *almost normalized* in the sense that $0 < C_1' \le \|f_j\| \le C_2'$ (the Bari and Lorch theorems; see [43, Section 6.2]).

Two systems $\{f_j\}$ and $\{e_j\}$, $j = 1, 2, \ldots$, are said to be *quadratically close* if

$$\sum \|f_j - e_j\|^2 < \infty. \tag{5.16}$$

A complete minimal system quadratically close to an orthonormal basis is necessarily a Riesz basis (the Bari theorem; see [43, Section 6.3]). Such a basis is called a *basis quadratically close to an orthonormal basis*, or a *Bari basis*.

A system $\{f_j\}$ is called a *basis with parentheses* in $\mathfrak{H}$ if there exists an increasing sequence of indices $\{m_l\}$ such that for each vector $f \in \mathfrak{H}$ the sequence of partial sums with numbers m_l of the Fourier series (5.14) of this vector converges to f, that is,

$$f = \sum_{l=0}^{\infty} P_l f, \quad \text{where } P_l f = \sum_{j=m_l+1}^{m_{l+1}} c_j f_j \tag{5.17}$$

and $m_0 = 0$. Here $\{m_l\}$ is independent of f. This property also implies the completeness.

Let $\mathfrak{M}_l$ be the linear span of the vectors $f_{m_l+1}, \ldots, f_{m_{l+1}}$. In (5.17) we defined the projection operator P_l that transforms each vector f into the part of its Fourier series lying in $\mathfrak{M}_l$. If $\{f_j\}$ is a complete minimal system, then P_l is the projection of $\mathfrak{H}$ onto $\mathfrak{M}_l$ along the closed linear span of the other $\mathfrak{M}_k$. If $\{f_j\}$ is an orthonormal basis in $\mathfrak{H}$, then P_l is the orthogonal projection onto $\mathfrak{M}_l$.

It obviously follows from (5.17) that

$$f = c_1 f_1 + \cdots + c_{m_l} f_{m_l} + P_{l+1} f + P_{l+2} f + \cdots \tag{5.18}$$

for each l, that is, the parentheses can be removed up to an arbitrarily large number. However, it may happen that the partial sums of the series (5.18) with numbers lying between m_k, $k = 0, \ldots, l$, approximate f worse than those with the numbers m_k.

We also make the following remark. Let $h_{m_l+1}, \ldots, h_{m_{l+1}}$ be an orthonormal basis in $\mathfrak{M}_l$ for each l. Then it follows from (5.17) that $\{h_j\}$, $j = 1, 2, \ldots$, is a basis in $\mathfrak{H}$

(without parentheses). Indeed, for $m_l < j \le m_{l+1}$ one has

$$\|b_{m_l+1}h_{m_l+1} + \cdots + b_j h_j\|^2 = |b_{m_l+1}|^2 + \cdots + |b_j|^2 \le \|P_l f\|^2,$$

and $\|P_l f\| \to 0$ as $l \to \infty$. The passage from the series $\sum c_j f_j$ to the series $\sum b_l h_l$ can be viewed as some averaging of the former series.

In the next subsection we define some properties of a vector system stronger than that of being a basis with parentheses.

2. Riesz and Bari bases with parentheses. Suppose that $\{f_j\}$, $j = 1, 2, \ldots$, is a minimal system, $\mathfrak{M}_l$ is the linear span of the vectors $f_{m_l+1}, \ldots, f_{m_{l+1}}$, $\{h_j\}$, $j = 1, 2, \ldots$, is an orthonormal basis in $\mathfrak{H}$, and $\mathfrak{N}_l$ is the linear span of the vectors $h_{m_l+1}, \ldots, h_{m_{l+1}}$. If there exists a bounded boundedly invertible operator B in $\mathfrak{H}$ such that $\mathfrak{M}_l = B\mathfrak{N}_l$ for all l, then $\{f_j\}$ is a basis with parentheses in $\mathfrak{H}$. In this case, $\{f_j\}$ is called a *Riesz basis with parentheses* in $\mathfrak{H}$.

One can show (cf. [43, Section 6.5]) that the following properties are equivalent to the assertion that $\{f_j\}$ is a Riesz basis with parentheses.

1° The space $\mathfrak{H}$ can be equipped with a new inner product $\langle f, g\rangle$ such that the system of orthonormal bases in $\mathfrak{M}_l$, $l = 0, 1, \ldots$, with respect to this inner product is an orthonormal basis in $\mathfrak{H}$ and the norm $\langle f, f\rangle^{1/2}$ is equivalent to $\|f\|$.

2° $C_1\|f\|^2 \le \sum \|P_l f\|^2 \le C_2\|f\|^2$ for all f with some positive constants C_1 and C_2.

3° One can arbitrarily rearrange terms in the series $f = P_0 f + P_1 f + \ldots$. The converse is also true, and therefore a Riesz basis with parentheses is also called an *unconditional basis with parentheses.*

Let $\{h_j\}$, $j = 1, 2, \ldots$, be some orthonormal basis in $\mathfrak{H}$, let $0 = \widetilde{m}_0 < \widetilde{m}_1 < \ldots$, and let Q_l be the orthogonal projection on the linear span of the vectors $h_{\widetilde{m}_l+1}, \ldots, h_{\widetilde{m}_{l+1}}$. Suppose that $\{f_j\}$ is a complete minimal system satisfying the condition

$$\sum_{l=0}^{\infty} \|P_l - Q_l\|^2 < \infty. \tag{5.19}$$

Then $\{f_j\}$ is a Riesz basis with parentheses (this follows from a theorem due to Markus; see [43, Section 6.5]). Such a system will be called a *Bari basis with parentheses.*[5] Condition (5.19) says that the series $\sum P_l f$ and $\sum Q_l f$ are in a sense "equiconvergent." In the following we shall encounter systems $\{f_j\}$ satisfying the much stronger condition

$$l^N \|P_l - Q_l\| \le C_N \tag{5.20}$$

for each positive integer N. This condition implies "rapid equiconvergence" of the series $\sum P_l f$ and $\sum Q_l f$ for any f. We also note that (5.19) implies $\{m_l\} = \{\widetilde{m}_l\}$ starting from some number.

3. Spectral properties of compact operators. If A is a linear operator in $\mathfrak{H}$, then by $\Sigma(A)$ we denote its *spectrum*, that is, the set of all complex numbers λ such that

[5]Here we slightly deviate from the terminology used in [69] and [43, Section 6.5]

the operator $A - \lambda I$ does not have a bounded inverse. Here and in the following I is the *identity operator*. If A is bounded, then $\Sigma(A)$ is contained in the disk $\{\lambda : |\lambda| \leq \|A\|\}$. The complement of the spectrum of A is called the *resolvent set* of A. The spectrum is always closed, and the resolvent set is open.

Let A be a *compact* (or *completely continuous*) operator. Then $\Sigma(A)$ consists of 0 and at most countably many eigenvalues that can accumulate only at 0 (e.g., see [21, Section 58]). To each eigenvalue $\lambda \neq 0$ there corresponds a finite-dimensional *root subspace* $\mathfrak{L}(\lambda) = \mathfrak{L}_A(\lambda)$ consisting of all vectors f such that $(A - \lambda I)^m f = 0$ for some positive integer m. If $f \neq 0$, then f is called a *root vector* and the least $m = m(f)$ is called the *order* of f. Root vectors of order 1 are called *eigenvectors*, and root vectors of order > 1 are called *associated vectors*. If 0 is also an eigenvalue, then we shall usually assume that the corresponding root vectors and the zero vector form a finite-dimensional subspace.[6] The dimension $d(\lambda) = \dim \mathfrak{L}(\lambda)$ is called the (algebraic) *multiplicity* of the eigenvalue λ. If it is greater than 1, then either $\mathfrak{L}(\lambda)$ consists of 0 and eigenvectors, or there are also associated vectors. We set $m(\lambda) = \max m(f)$, where the maximum is taken over all $f \neq 0$ in $\mathfrak{L}(\lambda)$; this is the maximal size of Jordan blocks of the matrix of A in $\mathfrak{L}(\lambda)$.

Usually we arrange all the eigenvalues of A in a sequence $\{\lambda_j\}$, $j = 1, 2, \ldots$, as follows: if 0 is not an eigenvalue, then $|\lambda_j| \geq |\lambda_{j+1}|$ for all j; if 0 is an eigenvalue, then $\lambda_1 = \cdots = \lambda_{d(0)} = 0$ and $|\lambda_j| \geq |\lambda_{j+1}|$ for $j > d(0)$; in this sequence, each eigenvalue λ is repeated $d(\lambda)$ times. In each $\mathfrak{L}(\lambda)$ we choose a basis and combine these bases into a system $\{f_j\}$, $j = 1, 2, \ldots$, in such a way that $f_j \in \mathfrak{L}(\lambda_j)$ for all j. We fix the system $\{f_j\}$ and call it the *system of root vectors* of the operator A. This system is minimal. For operators that will be considered in the sequel, the system is always infinite. Note also that the space $\mathfrak{M}_l$ (see Subsection 1) will always be a root subspace or a finite sum of root subspaces corresponding to close eigenvalues.

The numbers $\mu_j = \lambda_j^{-1}$ for nonzero λ_j are called the *characteristic numbers* of the operator A.

4. Operators with discrete spectrum. Consider an *unbounded* closed operator L in $\mathfrak{H}$ with dense domain $\mathfrak{D}(L)$. One says that L is an *operator with discrete spectrum*, or an *operator with compact resolvent*, if the *resolvent* $R_L(\mu) = (L - \mu I)^{-1}$ exists and is a compact operator for at least one $\mu = \mu_0$. In this case it can be proved (e.g., see [54, Section 3.6]) that the spectrum $\Sigma(L)$ consists of at most countably many eigenvalues with the unique possible limit point ∞. To each eigenvalue there corresponds a finite-dimensional root subspace $\mathfrak{L}(\mu) = \mathfrak{L}_L(\mu)$. The resolvent $R_L(\mu)$ is compact for each $\mu \notin \Sigma(L)$.

Any operator with discrete spectrum is necessarily unbounded, since the inverse of a bounded operator cannot be compact.

We shall usually arrange the eigenvalues of L in a sequence $\{\mu_j\}$, $j = 1, 2, \ldots$, such that $|\mu_1| \leq |\mu_2| \leq \ldots$ and each eigenvalue μ is repeated $d(\mu)$ times, where the multiplicity $d(\mu)$ is defined in the same way as for compact operators, that is, as the dimension of the corresponding root subspace. We combine bases in all $\mathfrak{L}(\mu)$ into a

[6]This assumption does not pertain to the operators that will be studied in Section 5.10.

system of root vectors $\{f_j\}$ of the operator L in such a way that $f_j \in \mathfrak{L}(\mu_j)$ for all j. This is again a minimal system.

If the inverse $L^{-1} = A$ exists (this can always be ensured by replacing L by $L - \mu_0 I$ with an appropriate μ_0), then our notation is consistent in the following manner: $\mu_j = \lambda_j^{-1}$, $\mathfrak{L}_L(\mu_j) = \mathfrak{L}_A(\lambda_j)$, and $\{f_j\}$ is a common system of root vectors of the operators L and A.

The resolvent $R_L(\mu)$ is a holomorphic operator-valued function outside $\Sigma(L)$ and has poles of order $d(\mu_j)$ at each $\mu_j \in \Sigma(L)$.

Here, as well as in the previous section, we of course have in mind nonselfadjoint operators. A selfadjoint operator (compact or with discrete spectrum) has an orthonormal basis of eigenvectors. The same is true of any *normal* operator, that is, an operator that commutes with its adjoint. Conversely, if an operator has an orthonormal basis of eigenvectors, then it is normal. A trivial example of a normal operator is given by the product of a selfadjoint operator by a complex number. The resolvent of a normal operator with discrete spectrum has only simple poles.

5. Abel–Lidskii systems. Here we give the definition due to Lidskii [65] of a property related to a system $\{f_j\}$ of root vectors of an operator L with discrete spectrum (or the compact operator $A = L^{-1}$). This property is intermediate between the completeness and the property of being a basis with parentheses and consists in the assumption that the Fourier series (5.14) of each vector f with respect to the system $\{f_j\}$ of root vectors of a given operator is summable to f by some special method. In [65], this method was called the Abel method of order α. We shall call it the *Abel–Lidskii method of order* α, and the system $\{f_j\}$ itself will be called an *Abel–Lidskii system of order* α.

Suppose that all except finitely many eigenvalues μ_j of the operator L (the characteristic numbers of A) are contained in some sector $\Theta = \{\lambda : |\arg\lambda| < \theta\}$. Let α be a positive number such that $\alpha\theta < \pi/2$. We set $\lambda^\alpha = |\lambda|^\alpha e^{i\alpha\arg\lambda}$ in this sector, so that $|\exp(-\lambda^\alpha t)|$ rapidly decays for $t = \text{const} > 0$ and $\lambda \in \Theta$ as $\lambda \to \infty$.

Suppose first that the system $\{f_j\}$ does not contain associated vectors corresponding to eigenvalues $\mu_j \in \Theta$. In this case we say that $\{f_j\}$ is an *Abel–Lidskii system of order* α if there exists an increasing sequence of indices $0 = m_0 < m_1 < \cdots < m_l < \ldots$ such that for each $f \in \mathfrak{H}$ the series

$$\sum_{l=0}^{\infty} P_l(t)f, \quad \text{where } P_l(t)f = \sum_{j=m_l+1}^{m_{l+1}} c_j e_j(t) f_j, \tag{5.21}$$

converges for $t > 0$ and the sum $f(t)$ of this series tends to f in $\mathfrak{H}$ as $t \to +0$. Here the function $e_j(t) = e_j(t;\alpha)$ is equal to $\exp(-\mu_j^\alpha t)$ if $\mu_j \in \Theta$, and moreover, all terms corresponding to the same eigenvalue are contained in the same term of the sum over l. For j such that $\mu_j \notin \Theta$ (say, for $\mu_j = 0$ if 0 is an eigenvalue) we set $e_j(t) \equiv 1$.

In the general case, when there exist associated vectors, the definition is generalized as follows. Let $f_p, \ldots, f_q$ be a basis in the root subspace $\mathfrak{L}_L(\mu)$, $\mu \in \Theta$. Then the sum

$c_p e_p(t) f_p + \cdots + c_q e_q(t) f_q$ is replaced by the integral

$$-\frac{1}{2\pi i}\int_{|\lambda-\mu|=\varepsilon} \exp(-\lambda^\alpha t) R_L(\lambda) f \, d\lambda, \tag{5.22}$$

where the integration contour lies in Θ and surrounds the single eigenvalue μ counterclockwise. At $t = 0$ this integral is equal to the projection of f on $\mathfrak{L}_L(\mu)$, that is, to $c_p f_p + \cdots + c_q f_q$ (e.g., see [54, Section 3.5]). If μ is a simple pole of the resolvent, that is, $m(\mu) = 1$, then the integral coincides with $c_p e_p(t) f_p + \cdots + c_q e_q(t) f_q$. Each term $P_l(t) f$ in (5.21) is a finite sum of integrals of the form (5.22).

Note that the inequality $\alpha\theta < \pi/2$ for α is in fact unessential if θ can be chosen arbitrarily small. However, to enlarge α makes no sense.

If we consider a completely continuous operator A that does not have an unbounded inverse L (say, if $Af = 0$ for some $f \neq 0$), then $R_L(\lambda)$ in (5.22) must be replaced by the *modified resolvent*

$$R_A^{(m)}(\lambda) = A(I - \lambda A)^{-1} \tag{5.23}$$

of the operator A. In this case the μ_j are the characteristic numbers of A.

We can also assume that $\Theta = \{\lambda : |\arg(\lambda/\lambda_0)| < \theta\}$, where $|\lambda_0| = 1$. Then everywhere in the above formulas we replace λ^α by $|\lambda|^\alpha e^{i\alpha \arg(\lambda/\lambda_0)} = (\lambda/\lambda_0)^\alpha$.

When speaking on the basis property with parentheses in Subsections 1–2, we essentially also considered a summability method for the series (5.14). Namely, the method was to pass from the sequence of all partial sums of the series to some subsequence. In contrast with that method, the Abel–Lidskii method is closely related to the operator in question, since the functions $e_j(t; \alpha)$ depend on the operator. We continue the discussion of the Abel–Lidskii method in Subsections 5.6.1 and 5.6.3.

6. Construction of the system biorthogonal to the system of root vectors. The use of symmetry. Let A be a compact operator in $\mathfrak{H}$, and let A^* be the adjoint of A. The following three assertions are well known (e.g., see [21, Chapter 5] or [54, Section 3.6]).

1° *The operator A^* is compact.*

2° *A nonzero number λ is an eigenvalue of A if and only if $\overline{\lambda}$ is an eigenvalue of A^*.*

3° *If λ is a nonzero eigenvalue of A, then* $\dim \mathfrak{L}_L(\lambda) = \dim \mathfrak{L}_{A^*}(\overline{\lambda})$.

The operator A is said to be *dissipative* if $\operatorname{Im}(Af, f) \geq 0$ for all f. This definition makes sense for an arbitrary bounded operator.

4° *Suppose that A is a dissipative operator and λ is a real eigenvalue of A. Then $\mathfrak{L}_A(\lambda) = \mathfrak{L}_{A^*}(\lambda)$, and $A = A^*$ on this subspace. In particular, this subspace does not contain associated vectors* (see [43, Section 5.1]).

In particular, this assertion is true of $\mathfrak{L}_A(0)$ if A is a dissipative operator such that 0 is an eigenvalue of A.

Note that if A is dissipative, then A^* is not dissipative (unless $A = A^*$), but the operator $-A^*$ is dissipative.

An unbounded operator L is said to be dissipative if $\operatorname{Im}(Lf, f) \geq 0$ for $f \in \mathfrak{D}(L)$. If L has a bounded inverse $L^{-1} = A$, then $-A$ is dissipative.

Specific operators considered in Sections 5.7–5.9 will usually be dissipative.

It is easy to verify the following two assertions (cf. [21, Chapter 5] and [54, Section 3.6]); the second of these assertion is a refinement of 3°.

5° *If λ and λ' are distinct eigenvalues of the operator A, then $\mathfrak{L}_A(\lambda)$ and $\mathfrak{L}_{A^*}(\overline{\lambda'})$ are orthogonal.*

6° *If λ is a nonzero eigenvalue of the operator A, then* $\dim \mathfrak{L}_A^m(\lambda) = \dim \mathfrak{L}_{A^*}^m(\overline{\lambda})$, *where $\mathfrak{L}_A^m(\lambda)$ is the subspace of $\mathfrak{L}_A(\lambda)$ consisting of zero and root vectors of order $\le m$.*

For $m = 1$, this is shown in [21, 54]. The case $m > 1$ can be reduced to $m = 1$, since $(A - \lambda I)^m = T - \lambda_1 I$ and $(A^* - \overline{\lambda} I)^m = T^* - \overline{\lambda}_1 I$, where $\lambda_1 = \lambda^m$ and T is a compact operator.

7° *Let $f_p, \dots, f_q$ be a basis in $\mathfrak{L}_A^m(\lambda)$, $\lambda \ne 0$, and let $g_p, \dots, g_q$ be a basis in $\mathfrak{L}_{A^*}^m(\overline{\lambda})$. The matrix $\{(f_j, g_l)\}$, $j, l = p, \dots, q$, is nondegenerate if and only if $m = m(\lambda)$, that is, if $\mathfrak{L}_A(\lambda)$ (and $\mathfrak{L}_{A^*}(\overline{\lambda})$) does not contain root vectors of order $> m$.*

For $m = 1$, it suffices to note that the solvability of the equation $(A - \lambda I)f = f_0$, where $f_0 = a_p f_p + \dots + a_q f_q \ne 0$, is equivalent to the orthogonality of f_0 to $g_p, \dots, g_q$ according to the theory of Fredholm equations. The case $m > 1$ can be reduced to $m = 1$ by the same trick as above.

Assertion 7° with $m = 1$ can be viewed as a criterion (though not very useful) for the absence of associated vectors.

In the remaining part of Section 5.2, as well as in Section 5.6, we shall *assume* that if $\lambda = 0$ is an eigenvalue of A, then $\mathfrak{L}_A(0)$ and $\mathfrak{L}_{A^*}(0)$ have the *same* finite dimension, and moreover, if $f_1, \dots, f_{d(0)}$ and $g_1, \dots, g_{d(0)}$ are bases in $\mathfrak{L}_A(0)$ and $\mathfrak{L}_{A^*}(0)$, respectively, then the matrix $\{(f_j, g_l)\}$, $j, l = 1, \dots, d(0)$, is *nondegenerate*. This condition is obviously satisfied if A is a dissipative operator with finite-dimensional $\mathfrak{L}_A(0)$ (see 4°).

Let $\{f_j\}$ and $\{h_j\}$ be systems of root vectors of the operators A and A^* such that $f_j \in \mathfrak{L}_A(\lambda)$ if and only if $h_j \in \mathfrak{L}_{A^*}(\overline{\lambda})$. Assertions 5° and 7° show that *the system $\{g_l\}$ biorthogonal to $\{f_j\}$ can be constructed as follows.* Let $h_p, \dots, h_q$ be a basis in some $\mathfrak{L}_{A^*}(\overline{\lambda})$. Then for $l = p, \dots, q$ we set $g_l = b_{pl} h_p + \dots + b_{ql} h_q$. The conditions $(f_j, g_l) = \delta_{jl}$, $j, l = p, \dots, q$, imply a system of linear algebraic equations with nonzero determinant for the coefficients b_{jl}, and so these coefficients are uniquely determined from this system.

In other words, let $\Gamma = (\Gamma_{jl})$ be the matrix of inner products $\Gamma_{jl} = (f_j, g_l)$, $j, l = p, \dots, q$; it is nondegenerate by virtue of 7°. We set $(g_p, \dots, g_q) = (h_p, \dots, h_q)B$. The matrix B is determined by the equation $\Gamma \overline{B} = E$, where E is the unit matrix, whence we obtain $B = \overline{\Gamma}^{-1}$.

Now suppose that the space $\mathfrak{H}$ is equipped with an operation of passing from a vector f to a "complex conjugate" vector $\overline{f}$ such that $(\overline{f}, \overline{g}) = \overline{(f, g)}$. (For example, $\mathfrak{H}$ consists of complex-valued functions defined on some set.) Let M be a linear operator in $\mathfrak{H}$ with domain $\mathfrak{D}(M)$. We introduce the *complex conjugate* operator $\overline{M}$ by setting

$$\overline{M} f = \overline{M \overline{f}} \quad \text{for } f \in \mathfrak{D}(\overline{M}) = \overline{\mathfrak{D}(M)} = \{f : \overline{f} \in \mathfrak{D}(M)\}. \tag{5.24}$$

Many of the operators related to problems considered in this book possess the *symmetry property* expressed by the formula

$$M^* = \overline{M}; \tag{5.25a}$$

this property is often encountered in mathematical physics (cf. [36, Section 3]). For example, the integral operator (5.6) with complex-valued symmetric kernel (which is of course not Hermitian symmetric) possesses this property. Note in addition that property (5.25a) can be rewritten in the form

$$M = M', \tag{5.25b}$$

where

$$M' = \overline{M}^* \tag{5.26}$$

is the so-called *transpose* of M. If $\mathfrak{H} = L_2$ with the usual inner product $\int f(x) \cdot \overline{g(x)}\, dx$, then property (5.25a) or (5.25b) means that

$$\int Mf(x) \cdot g(x)\, dx = \int f(x) \cdot Mg(x)\, dx \tag{5.27}$$

for all $f, g \in L_2$.

It is obvious that if $\{f_j\}$ is a system of root vectors of A, then $\{\overline{f}_j\}$ is a system of root vectors of $\overline{A}$. Suppose that $A^* = \overline{A}$. *Then, together with* $\{f_j\}$, *the system* $\{\overline{f}_j\}$ *of root vectors of the operator* A^* *is also known to us, and in the above notation we must set* $(g_p, \ldots, g_q) = (\overline{f}_p, \ldots, \overline{f}_q)\overline{\Gamma}^{-1}$, *where* Γ *is the matrix of inner products* $\Gamma_{jl} = (f_j, \overline{f}_l)$.

Moreover, the following assertion holds.

8° *Under condition* (5.25), *one can choose a basis* $(f_p, \ldots, f_q)$ *in each root subspace of the operator* A *in a manner such that the systems* $\{f_j\}$ *and* $\{\overline{f}_j\}$, $j = 1, 2, \ldots$, *will be biorthogonal.*

If the inner product is given by $\int f(x)\overline{g(x)}\, dx$, then this means that

$$\int f_j(x) f_l(x)\, dx = \delta_{jl}. \tag{5.28}$$

Indeed, let $\{f'_p, \ldots, f'_q\}$ be some basis in $\mathfrak{L}_A(\lambda)$, and let Γ be the matrix of inner products $(f'_j, \overline{f}'_l)$, $j, l = p, \ldots, q$. We set

$$\begin{pmatrix} f_p \\ \vdots \\ f_q \end{pmatrix} = T \begin{pmatrix} f'_p \\ \vdots \\ f'_q \end{pmatrix}.$$

Then $(\overline{f}_p, \ldots, \overline{f}_q) = (\overline{f}'_p, \ldots, \overline{f}'_q)T^*$, and for the matrix T we obtain the equation $T\Gamma T' = E$. This equation for T is solvable, since the matrix Γ is nondegenerate and symmetric, $\Gamma = \Gamma'$ (e.g., see [68, Chapter 6, Subsection 23.3, Theorem 4]).

In particular, if all the eigenvalues are simple, then $(f_j, \overline{f}_j) \neq 0$ for any root vector f_j, and we obtain $(f_j, \overline{f}_j) = 1$ for an appropriate normalization of these vectors. However, if $\dim \mathfrak{L}_A(\lambda) > 1$, then the inner products $(f, \overline{f})$ for nonzero vectors $f \in \mathfrak{L}_A(\lambda)$ are not necessarily nonzero even if $\mathfrak{L}_A(\lambda)$ does not contain associated vectors.

Thus, condition (5.25) simplifies the construction of the system biorthogonal to $\{f_j\}$. However, condition (5.25) has yet another consequence which will play a very essential role. Recall that for each bounded operator A, the *real* and *imaginary parts* of A are the selfadjoint operators given by the formulas

$$A_R = \operatorname{Re} A = \frac{A + A^*}{2}, \qquad A_I = \operatorname{Im} A = \frac{A - A^*}{2i}. \tag{5.29}$$

For an unbounded operator A, this definition makes sense only if $\mathfrak{D}(A) = \mathfrak{D}(A^*)$.

If $A^* = \overline{A}$, these formulas acquire the form

$$A_R = \frac{A + \overline{A}}{2}, \qquad A_I = \frac{A - \overline{A}}{2i}. \tag{5.30}$$

For easy reference, we state the following assertion:

9° *If* $A^* = \overline{A}$, *then* $A = A_R + iA_I$, *where* A_R *and* A_I *are the selfadjoint operators defined by formulas* (5.30).

The inner product (Af, f) is called the *quadratic form* of the operator A. One can readily verify that if A_R and A_I are the real and imaginary parts of A, respectively, then

$$\operatorname{Re}(Af, f) = (A_R f, f), \qquad \operatorname{Im}(Af, f) = (A_I f, f). \tag{5.31}$$

Hence the dissipativity of A is equivalent to the *nonnegativity* of A_I: $(A_I f, f) \geq 0$ for all f. In turn, this is equivalent to the nonnegativity of the eigenvalues of A_I. Note that in this case $A_I f \neq 0$ implies $(A_I f, f) > 0$. These assertions can easily be verified by using the orthonormal basis of eigenvectors of A_I.

7. Approximate solutions of Fredholm equations. Consider the equation

$$Af - \lambda f = g \tag{5.32}$$

with a compact operator A. Suppose that λ does not belong to the spectrum $\Sigma(A)$. Then the resolvent $R_A(\lambda) = (A - \lambda I)^{-1}$ is a bounded operator, and hence

$$\|f\| \leq C_\lambda \|g\| \tag{5.33}$$

with some constant C_λ independent of f. If f' is a solution for a right-hand side g', then

$$\|f - f'\| \leq C_\lambda \|g - g'\|. \tag{5.34}$$

Suppose that the system $\{f_j\}$ of root vectors of A is at least complete. Then for each $g \in \mathfrak{H}$ there exists a linear combination $g' = a_1 f_1 + \cdots + a_m f_m$ arbitrarily close to g in the norm. The estimate (5.33) shows that the corresponding solution f' is close to f. Moreover, f' can be constructed from g' essentially by the tools of linear algebra. Indeed, it suffices to indicate the solution for a vector g'' of the form $a_p f_p + \cdots + a_q f_q$, where $f_p, \ldots, f_q$ is a basis in some $\mathfrak{L}_A(\lambda_j)$. This solution has the form $f'' = b_p f_p + \cdots + b_q f_q$. Let a and b be column vectors with coordinates $a_p, \ldots, a_q$ and $b_p, \ldots, b_q$, respectively, and let A_{λ_j} be the matrix of the operator A in $\mathfrak{L}_A(\lambda_j)$ in the basis $f_p, \ldots, f_q$. Then $(A_{\lambda_j} - \lambda E)b = a$, where E is the identity matrix; it follows that $b = (A_{\lambda_j} - \lambda E)^{-1}a$ for $\lambda \neq \lambda_j$. For $a \neq 0$, the right-hand side, viewed as a function of λ, has a pole at λ_j. The maximal possible order of this pole is $m(\lambda_j)$. If in $\mathfrak{L}_A(\lambda_j)$ there are no associated vectors (and only in this case), then the pole is always simple.

If $a_p, \ldots, a_q$ are the Fourier coefficients of the vector g with respect to the system $\{f_j\}$, then $b_p, \ldots, b_q$ will be the Fourier coefficients of the solution.

Thus, the knowledge of the system $\{f_j\}$ of root vectors and of the eigenvalues of the operator A permits us to construct approximate solutions of Eq. (5.32) provided that the system $\{f_j\}$ is at least complete. The "better" this system is (see the series of definitions in Subsections 1, 2, and 5), the more convenient this spectral method is for solving Eq. (5.32) (cf. (5.10)).

The results of the last two subsections can easily be transferred to the case of an operator L with discrete spectrum (instead of a compact operator A).

5.3 Sobolev Spaces

Here we give a list of definitions and theorems to be used in the sequel. The proofs can be found, e.g., in [94] and [99].

1. Definitions. Let s be an arbitrary real number. The *space* $H^s(\mathbf{R}^n)$, $n \geq 1$, is defined as the completion of the space $C_0^\infty(\mathbf{R}^n)$ of compactly supported smooth complex-valued functions on $\mathbf{R}^n$ with respect to the norm

$$\|u(x)\|_{\mathbf{R}^n,s} = \left(\int_{\mathbf{R}^n} (1 + |\xi|^2)^s |(Fu)(\xi)|^2 \, d\xi \right)^{1/2}, \tag{5.35}$$

where F is the *Fourier transform*

$$(Fu)(\xi) = \int_{\mathbf{R}^n} e^{-ix\cdot\xi} u(x) \, dx, \tag{5.36}$$

$x \cdot \xi = x_1\xi_1 + \cdots + x_n\xi_n$. (Note that in Chapter 3 Sobolev spaces were denoted by H_s rather than H^s.) For nonnegative integer s, the norm (5.35) is equivalent to the norm

$$\left(\sum_{|\alpha|\leq s} \int |D^\alpha u(x)|^2 \, dx \right)^{1/2}, \tag{5.37}$$

that is, the ratio of these two norms is bounded above and below by positive constants. Here and in the following we write $D^\alpha = D_1^{\alpha_1} \cdots D_n^{\alpha_n}$ and $D_j = -i\partial/\partial x_j$, $|\alpha| =$

$\alpha_1 + \cdots + \alpha_n$. The integral is taken over $\mathbf{R}^n$. For fractional $s > 0$, the norm (5.35) is equivalent to the norm

$$\Big(\sum_{|\alpha| \le [s]} \int |D^\alpha u(x)|^2 \, dx + \sum_{|\alpha| = [s]} \iint \frac{|D^\alpha u(x) - D^\alpha u(y)|^2}{|x-y|^{n+2s-2|s|}} \, dx \, dy \Big)^{1/2}, \tag{5.38}$$

where $[s]$ is the integral part of s and the integrals are also taken over $\mathbf{R}^n$. We take the liberty of preserving the same notation $\|u\|_{\mathbf{R}^n,s}$ for the norms (5.37) and (5.38).

Let V be a domain in $\mathbf{R}^n$. The *space* $H^s(V)$ is defined for $s \ge 0$ as the completion of the space $C_0^\infty(\overline{V})$ of infinitely smooth compactly supported functions in the closure $\overline{V}$ of V with respect to the norm $\|u(x)\|_{V,s}$ defined by (5.37) for integer s and by (5.38) for fractional s, where the integrals are taken over V.

Now let us consider an n-dimensional compact infinitely smooth surface $\mathfrak{S}$ in $\mathbf{R}^{n+1}$ (here $n \ge 1$; for $n = 1$ this is a closed curve without self-intersections). Locally, $\mathfrak{S}$ is specified by an equation of the form $x_j = \Phi(x_1, \dots, \hat{x}_j, \dots, x_n)$, where Φ is a C^∞ function and the hat ˆ over a coordinate indicates that this coordinate is omitted. Each point of $\mathfrak{S}$ has a neighborhood in $\mathfrak{S}$ in which there are local coordinates, say, $y = (y_1, \dots, y_n)$. On the intersection of two such *coordinate neighborhoods*, the corresponding local coordinate systems y and $\tilde{y}$ are related by a nondegenerate infinitely smooth transformation. For the above local representation of $\mathfrak{S}$ by the equation $x_j = \Phi$, one of infinitely many possible local coordinate systems is $(x_1, \dots, \hat{x}_j, \dots, x_{n+1})$.

In numerous ways, one can construct a finite system of C^∞ functions $\varepsilon_1, \dots, \varepsilon_K$ on $\mathfrak{S}$ such that the support of each ε_j is contained in a single coordinate neighborhood and the functions $\kappa_j = \varepsilon_j^2$, $j = 1, \dots, K$, form a (C^∞) *partition of unity on* $\mathfrak{S}$: $\sum \kappa_j \equiv 1$ on $\mathfrak{S}$.

For each real s, we introduce a norm on the space $C^\infty(\mathfrak{S})$ of infinitely smooth functions on $\mathfrak{S}$ by setting

$$\|u\|_{\mathfrak{S},s} = \Big(\sum_{j=1}^{K} \|\varepsilon_j u\|^2_{\mathbf{R}^n,s} \Big)^{1/2}. \tag{5.39}$$

Here we mean that the function $\varepsilon_j u$ is represented via the local coordinates on the support of ε_j and is continued by zero to the corresponding $\mathbf{R}^n$; the norm $\|\varepsilon_j u\|_{\mathbf{R}^n,s}$ is calculated in these coordinates. The dependence of the norm (5.39) on the specific choice of the functions ε_j and the local coordinate systems is unessential: various norms of this kind are pairwise equivalent. (It even suffices to assume that $\varepsilon_j \ge 0$ for all j and $\sum \varepsilon_j > 0$ on $\mathfrak{S}$.) Note also that the norm $\|u\|_{\mathfrak{S},0}$ is equivalent to the L_2 norm $\left(\int_{\mathfrak{S}} |u|^2 \, d\mathfrak{S} \right)^{1/2}$, where $d\mathfrak{S}$ is the n-dimensional surface area element on $\mathfrak{S}$ determined by the metric induced on $\mathfrak{S}$ by the Euclidean metric on $\mathbf{R}^{n+1}$. The *space* $H^s(\mathfrak{S})$ is defined as the completion of $C^\infty(\mathfrak{S})$ with respect to the norm (5.39). By virtue of this definition, $C^\infty(\mathfrak{S})$ is dense in $H^s(\mathfrak{S})$.

2. Properties of the spaces H^s. The assertions given in this subsection equally pertain to Sobolev spaces on $\mathbf{R}^n$, V, or $\mathfrak{S}$, unless otherwise is specified. Some of these assertions are obvious.

1° *The space* H^0 *coincides with the space* L_2 *of Lebesgue square integrable functions.*

2° $H^{s_1} \supset H^{s_2}$ *for* $s_1 < s_2$, *and the embedding operator (which takes each function* $u \in H^{s_2}$ *to the same function in* H^{s_1}*) is bounded. For the case of a bounded domain* V *or a compact surface* $\mathfrak{S}$, *this operator is compact. The space* H^{s_2} *is dense in* H^{s_1}.

3° *For positive integer* s, *the space* H^s *consists of functions* u *belonging to* L_2 *together with the derivatives* $D^\alpha u$ *of order* $|\alpha| \leq s$ *in the sense of distributions.* Here $H^s = H^s(\mathbf{R}^n)$ or $H^s = H^s(V)$. On $\mathfrak{S}$, the derivatives $D^\alpha u$ are defined only locally (in local coordinates), and if $u \in H^s(\mathfrak{S})$ has a sufficiently small support, then $D^\alpha u \in L_2(\mathfrak{S})$ for $|\alpha| \leq s$.

The spaces $H^s(\mathbf{R}^n)$ and $H^s(\mathfrak{S})$ with negative s contain distributions (generalized functions; further on, we sometimes omit the word "generalized"). For example, the delta function $\delta(x)$ in $\mathbf{R}^n$ belongs to $H^s(\mathbf{R}^n)$ for $s < -n/2$.

4° *The expressions* (5.35) *and* (5.37)–(5.39) *are well defined for elements of the corresponding spaces* H^s *and define norms on these spaces.*

5° *The operator of multiplication by a function that is continuous and bounded together with the derivatives of order* $|\alpha| \leq m$ *is a bounded operator in* H^s *for* $|s| \leq m$. *(For the mth-order derivatives, it suffices to assume that they are bounded and measurable.) On* $\mathfrak{S}$, *local derivatives are meant.*

6° *The operator* D^α *(the derivative in the sense of distributions) is bounded from* $H^{s+\alpha}$ *into* H^s. *Here* $H^s = H^s(\mathbf{R}^n)$ *or* $H^s(V)$.

Now we assume for simplicity that V has a compact connected $(n-1)$-dimensional sufficiently regular boundary S.

7° *The space* $H^s(V)$ *consists of the restrictions to* V *of functions from* $H^s(\mathbf{R}^n)$; *the restriction operator acts boundedly from* $H^s(\mathbf{R}^n)$ *into* $H^s(V)$. *This operator has a bounded right inverse, which is an operator of continuation of functions from* $H^s(V)$ *to functions from* $H^s(\mathbf{R}^n)$.

We do not describe the construction of the continuation operator. It exists even if S is a Lipschitz surface (e.g., see [96, Chapter 6]; we recall the definition of a Lipschitz surface in Section 5.12).

8° *Suppose that an* $(n-1)$*-dimensional sufficiently smooth surface* S *(a curve for* $n = 2$; *a two-point set for* $n = 1$*) divides the space* $\mathbf{R}^n$ *into the interior and exterior domains* V^+ *and* V^-, s *is a positive integer, and* $u(x)$ *is a function belonging to* $H^s(V^+)$ *and* $H^s(V^-)$. *Then* $u \in H^s(\mathbf{R}^n)$ *if and only if* $\partial_\nu^l u^+ = \partial_\nu^l u^-$, $l = 0, \ldots, s-1$, *on* S. Here, just as previously, ∂_ν is the outward normal derivative on S.

A similar assertion is valid, say, for the ball

$$E_R = \{x \in \mathbf{R}^n : |x| < R\} \tag{5.40}$$

of large radius R in $\mathbf{R}^n$ instead of $\mathbf{R}^n$ (a disk for $n = 2$) and for the intersection

$$V_R^- = V^- \cap E_R \tag{5.41}$$

instead of V^-. Here we assume that S is contained in the interior of the ball.

9° *For any function from* $H^s(V)$ *with* $s > 1/2$ *and* $n \geq 2$, *its boundary value (or trace) on* S *is well defined as an element of* $H^{s-1/2}(S)$; *it coincides with the usual boundary value if the function is continuous in* $\overline{V}$. *The trace operator that takes each function to its boundary value on* S *is continuous from* $H^s(V)$ *to* $H^{s-1/2}(S)$. *Likewise, any function from* $H^s(\mathbf{R}^n)$ *for* $s > 1/2$ *has a well-defined trace on* S*; the trace belongs to* $H^{s-1/2}(S)$*, and the trace operator is bounded from* $H^s(\mathbf{R}^n)$ *to* $H^{s-1/2}(S)$.

10° *For* $s > n/2$*, each function belonging to the space* H^s *in* $\mathbf{R}^n$*,* V*, or* $\mathfrak{S}$ *is bounded and continuous* (possibly, after correction on a set of Lebesgue measure zero).

The last assertion is a special case of the famous *Sobolev embedding theorem.* To state this theorem in a more general form, we introduce the spaces $C^{(s)}$. The *space* $C^{(s)}(\overline{V})$ for integer $s \geq 0$ consists of functions $u(x)$ continuous in V up to the boundary together with the derivatives of order $\leq s$ and such that the norm

$$\sup_{x\in V, |\alpha|\leq s} |D^\alpha u(x)| \tag{5.42}$$

is finite. For fractional $s > 0$, this space consists of functions $u(x)$ continuous in V up to the boundary together with the derivatives of order $\leq [s]$ and such that the norm

$$\sup_{x\in V, |\alpha|\leq [s]} |D^\alpha u(x)| + \sup_{x,y\in V, |\alpha|=[s]} \frac{|D^\alpha u(x) - D^\alpha u(y)|}{|x-y|^{s-[s]}} \tag{5.43}$$

is finite. In particular, the same definition applies to the *space* $C^{(s)}(\mathbf{R}^n)$ and the corresponding norm. The convergence $u_n(x) \to u(x)$ of a sequence of functions in $C^{(s)}(\overline{V})$ for integer s is just the uniform convergence $u_n(x) \to u(x)$ together with the derivatives of order $|\alpha| \leq s$. For $s' > s$ we have $C^{(s')}(\overline{V}) \subset C^{(s)}(\overline{V})$, and the embedding operator is bounded.

Finally, the *space* $C^{(s)}(\mathfrak{S})$ consists of all functions u on $\mathfrak{S}$ such that all $\kappa_j u$, $j = 1, \ldots, K$, belong to $C^{(s)}(\mathbf{R}^n)$ if we express them via the local coordinates and continue by zero to the corresponding $\mathbf{R}^n$. The norm of a function u in $C^{(s)}(\mathfrak{S})$ is defined as the maximum of the norms of the functions $\kappa_j u$ in the corresponding $C^{(s)}(\mathbf{R}^n)$; here again the choice of the partition of unity and the corresponding local coordinates is unessential.

Obviously, $C^{(s)} \subset H^s$ and the embedding operator is bounded in the cases of a bounded domain V and a compact surface $\mathfrak{S}$, $s \geq 0$.

11° *For* $s > n/2 + t$*, where* $t \geq 0$*, the space* H^s *is contained in* $C^{(t)}$*, and the embedding operator is bounded.*

This is the above-mentioned generalization of Assertion 10°. Here again we mean that the functions may be modified on subsets of measure zero.

12° *All* H^s *are Hilbert spaces.* The corresponding inner products can readily be written out. For example, if the norm is defined by (5.37), then the inner product is given by

$$\sum_{|\alpha|\leq s} \int D^\alpha u \cdot \overline{D^\alpha v}\, dx. \tag{5.44}$$

13° *The inner product* (u, v) *in* $H^0 = H^0(\mathbf{R}^n)$ *or* $H^0(\mathfrak{S})$ *can be extended by continuity to* $H^s \times H^{-s}$ *for any* $s \in \mathbf{R}$. *Moreover, the generalized Schwartz inequality*

$|(u,v)| \leq C_s\|u\|_s\|v\|_{-s}$ *holds. The dual space of* H^s *can be identified with* H^{-s}*: each linear continuous functional on* H^s *has the form* (u,v) *for some* $v \in H^{-s}$.

The proof for the case of $\mathbf{R}^n$ uses the Fourier transform and the operators

$$\Lambda^s = F^{-1}(1+|\xi|^2)^{s/2}F, \tag{5.45}$$

which define continuous isomorphisms of $H^s(\mathbf{R}^n)$ onto $H^0(\mathbf{R}^n)$. These are elliptic pseudodifferential operators (see Subsection 5.5.1); their analogs can also be constructed on $\mathfrak{S}$ (see Subsection 5.5.2).

14° *For integer* $m > 0$ *and for arbitrary* s*, each element* $u \in H^{s-m}(\mathbf{R}^n)$ *can be represented in the form*

$$u = \sum_{|\alpha|\leq m} D^\alpha u_\alpha,$$

where

$$u_\alpha \in H^s(\mathbf{R}^n) \quad \text{and} \quad \|u_\alpha\|_{\mathbf{R}^n,s} \leq C_{ms}\|u\|_{\mathbf{R}^n,s-m}.$$

In particular, the elements of $H^{-m}(\mathbf{R}^n)$ are derivatives (in the sense of distributions) of order $\leq m$ of functions belonging to $L_2(\mathbf{R}^n)$ and finite sums of such derivatives.

15° *Let* $K(x,y)$ *be a function from* $C_0^{(2m)}(\mathbf{R}^n \times \mathbf{R}^n)$ *with positive integer* m*. Then the operator defined by the formula*

$$Au(x) = \int_{\mathbf{R}^n} K(x,y)u(y)\,dy, \qquad u \in H^0(\mathbf{R}^n), \tag{5.46}$$

can be extended to a bounded operator from $H^{-m}(\mathbf{R}^n)$ *to* $H^m(\mathbf{R}^n)$.

Let us derive this assertion from 14°. For this we substitute $u = \sum D^\alpha u_\alpha$ into (5.46), where $u_\alpha \in H^0(\mathbf{R}^n)$, $|\alpha| \leq m$. By transferring the operators D^α to K, we obtain

$$Au(x) = \sum_{|\alpha|\leq m} \int_{\mathbf{R}^n} K_\alpha(x,y)u_\alpha(y)\,dy,$$

where the kernels $K_\alpha(x,y) = (-1)^{|\alpha|}D_y^\alpha K(x,y)$ are compactly supported and continuous in (x,y) together with the x-derivatives of order $\leq m$. Using the norm (5.37), we see that

$$Au \in H^m(\mathbf{R}^n) \quad \text{and} \quad \|Au\|_{\mathbf{R}^n,m} \leq C\sum \|u_\alpha\|_{\mathbf{R}^n,0}.$$

It remains to use the inequality in 14° with $s = 0$.

16° *The following interpolation inequalities are true for Sobolev norms in* $\mathbf{R}^n$*, in a bounded domain with sufficiently regular (Lipschitz) boundary, and on a sufficiently smooth surface. Let* $0 < t < s$*. Then for any* $\varepsilon > 0$ *one has*

$$\|u\|_t \leq \varepsilon\|u\|_s + C_\varepsilon\|u\|_0$$

with some constant C_ε.

An operator acting on functions defined in $\mathbf{R}^n$ or on $\mathfrak{S}$ is called an *operator of order* (not higher than) γ if it is a continuous operator from H^s to $H^{s-\gamma}$ for each s. If an operator A of order γ is not an operator of order $\gamma-\varepsilon$ for any $\varepsilon > 0$, then γ is called the *exact order* of A. Operators of negative order are also called *smoothing operators*, and operators that have order $-n$ for every positive integer n are called *operators of order* $-\infty$, or *infinitely smoothing operators.* Smoothing operators on $\mathfrak{S}$ are compact in each $H^s(\mathfrak{S})$ by virtue of 2°. Infinitely smoothing operators on $\mathfrak{S}$ are integral operators with infinitely differentiable kernels. An integral operator in $\mathbf{R}^n$ with compactly supported infinitely differentiable kernel is infinitely smoothing in $\mathbf{R}^n$.

We shall also use some generalizations of the above definitions and assertions to the case of vector-valued functions; these will be given in Subsection 5.11.2.

3. Sobolev spaces on a closed curve. Suppose that S is an infinitely smooth closed curve in $\mathbf{R}^n$, $n \geq 2$, without self-intersections, σ is the arc length on S measured from some given point, and $|S|$ is the total length of the curve. We set

$$x = \frac{2\pi\sigma}{|S|}. \tag{5.47}$$

Now functions on S can be identified with 2π-periodic functions $u(x)$ on $\mathbf{R}$, $u(x+2\pi) \equiv u(x)$, that is, with functions on the circle. The space of such functions possesses the natural basis $\{e^{imx}\}$, $m = 0, \pm 1, \dots$, and to each function $u(x)$ we can assign the sequence

$$c_m = \frac{1}{2\pi}\int_0^{2\pi} u(x)e^{-imx}\,dx \tag{5.48}$$

of its Fourier coefficients. One can show that the Sobolev norm $\|u\|_{S,s}$ is equivalent to the norm

$$\left\{ \sum_{m=-\infty}^{\infty} (1 + |m|^2)^s |c_m|^2 \right\}^{1/2}. \tag{5.49}$$

This remark remains valid for negative s as well; in this case, formula (5.48) can be interpreted as the action of the distribution $u(x)$ on the circle on the test function $(2\pi)^{-1}e^{-imx}$.

5.4 Pseudodifferential Operators

Here we use the definitions and notation introduced in Section 5.3. The proofs and further details can be found, e.g., in [97] and [93]; see also the survey [12].

1. Pseudodifferential operators in $\mathbf{R}^n$. Let $a(x, \xi)$ be a function defined for $x \in \mathbf{R}^n$ and $\xi \in \mathbf{R}^n \setminus \{0\}$. We call it a *homogeneous symbol of order* γ if the following conditions are satisfied: $a(x, \xi)$ is positively homogeneous in ξ of order γ (that is, $a(x, \rho\xi) = \rho^\gamma a(x, \xi)$ for $\rho > 0$), infinitely differentiable for $\xi \neq 0$, and (for simplicity) independent of x for sufficiently large $|x|$, that is, for $|x| \geq C = C(a)$. Here γ is a real number.

If $\gamma < 0$, then $a(x,\xi)$ has a singularity at $\xi = 0$; if $\gamma \geq 0$ and $a(x,\xi)$ is not a polynomial, it is not infinitely smooth at that point. This is inconvenient, and to avoid this inconvenience, we proceed as follows. Let $\theta(\xi)$ be an arbitrary C^∞ function on $\mathbf{R}^n$ that is equal to 0 in a neighborhood of the origin and to 1 outside a larger neighborhood. We set $a_\theta(x,\xi) = a(x,\xi)\theta(\xi)$.

Now we define the *operator* $A = a(x,D)$ on functions $u \in C_0^\infty(\mathbf{R}^n)$ by setting

$$\begin{aligned} a(x,D)u(x) &= F^{-1}_{\xi\to x} a_\theta(x,\xi)(Fu)(\xi) \\ &= \frac{1}{(2\pi)^n}\int_{\mathbf{R}^n_\xi} e^{ix\cdot\xi} a_\theta(x,\xi)\int_{\mathbf{R}^n_y} e^{-iy\cdot\xi}u(y)\,dy\,d\xi. \end{aligned} \tag{5.50}$$

Here F^{-1} is the inverse Fourier transform

$$(F^{-1}w)(x) = \frac{1}{(2\pi)^n}\int_{\mathbf{R}^n} e^{ix\cdot\xi} w(\xi)\,d\xi.$$

The operator $A = a(x,\xi)$ is called a *homogeneous pseudodifferential operator with symbol* $a(x,\xi)$. The dependence of A on the choice of the function $\theta(\xi)$ is unessential; namely, the difference of two pseudodifferential operators with the same symbol $a(x,\xi)$ and with distinct $\theta(\xi)$ is an infinitely smoothing operator.

If $a(x,\xi) = \sum_{|\alpha|=\gamma} a_\alpha(x)\xi^\alpha$ is a (homogeneous) polynomial (here $\xi^\alpha = \xi_1^{\alpha_1}\cdots\xi_n^{\alpha_n}$), then there is no need to multiply it by $\theta(\xi)$. In this case $a(x,D) = \sum_{|\alpha|=\gamma} a_\alpha(x)D^\alpha$ is a homogeneous differential operator of (nonnegative integer) order γ. It can be represented in the form (5.50) with $a(x,\xi)$ instead of $a_\theta(x,\xi)$.

The following assertion holds (it is obvious if the symbol is independent of x, by virtue of (5.35)).

1° $\|a(x,D)u(x)\|_{\mathbf{R}^n,s-\gamma} \leq C_s\|u(x)\|_{\mathbf{R}^n,s}$ *for every s, where the constant is independent of $u(x)$.*

This inequality permits one to extend the homogeneous pseudodifferential operator $a(x,D)$ from the dense subset $C_0^\infty(\mathbf{R}^n) \subset H^s(\mathbf{R}^n)$ to the entire $H^s(\mathbf{R}^n)$ for every s with this inequality being preserved, so that the resulting operator is bounded from $H^s(\mathbf{R}^n)$ to $H^{s-\gamma}(\mathbf{R}^n)$. In particular, this is obviously true of a homogeneous differential operator $a(x,D)$. Here we everywhere assume that the symbol is homogeneous of order γ.

Now we can define a *pseudodifferential operator with principal symbol* $a(x,\xi)$ of order γ by the formula $A = a(x,D) + T$, where T is an arbitrary operator of order $\gamma - 1$ in the Sobolev scale $H^s(\mathbf{R}^n)$ (see the definition of order of an operator in the end of Subsection 5.3.2). Clearly, the following assertion holds.

2° *A pseudodifferential operator A with principal symbol a_γ of order γ is an operator of order γ.*

One can show that if $a_\gamma(x,\xi) \not\equiv 0$, then γ is the exact order of this operator.

In the following, we say "a pseudodifferential operator of order γ" instead of "a pseudodifferential operator with principal symbol of order γ".

Pseudodifferential operators that will be considered in the following admit the expansion $A = A_\gamma + \cdots + A_{\gamma-q} + T_{\gamma-q-1}$ for every positive integer q, where A_j is a homogeneous pseudodifferential operator, independent of q, with symbol $a_{\gamma-j}(x,\xi)$ of order $\gamma - j$ and $T_{\gamma-q-1}$ is an operator of order $\gamma - q - 1$ (it is also a pseudodifferential operator, since we can increase q). Such pseudodifferential operators are said to be *polyhomogeneous*. For a polyhomogeneous pseudodifferential operator in $\mathbf{R}^n$ we can introduce the *complete symbol*. The complete symbol is an infinitely differentiable function $a(x,\xi)$ that is independent of x for large x, satisfies for all α and β the estimates

$$|\partial_x^\alpha \partial_\xi^\beta a(x,\xi)| \leq C_{\alpha\beta}(1 + |\xi|)^{\gamma-|\beta|} \tag{5.51}$$

(here and in the following we write, say, $\partial_x^\alpha = \partial_1^{\alpha_1} \ldots \partial_n^{\alpha_n}$, $\partial_j = \partial/\partial x_j$), and gives the representation of the pseudodifferential operator $A = a(x,D)$ in the form

$$Au(x) = F^{-1}_{\xi\to x} a(x,\xi)(Fu)(\xi) = \frac{1}{(2\pi)^n} \int_{\mathbf{R}^n} \int_{\mathbf{R}^n} e^{i(x-y)\cdot\xi} a(x,\xi) u(y)\, dy\, d\xi. \tag{5.52}$$

The complete symbol admits the asymptotic expansion in the symbols of the operators $A_\gamma, A_{\gamma-1}, \ldots$, namely,

$$a(x,\xi) = a_\gamma(x,\xi) + \cdots + a_{\gamma-q}(x,\xi) + O(|\xi|^{\gamma-q-1}), \qquad \xi \to \infty. \tag{5.53}$$

This expansion can be differentiated as many times as desired with respect to x and ξ for $\xi \neq 0$; the differentiation in ξ reduces the order in ξ of each term of the expansion by 1.[7]

Any differential operator

$$Au(x) = \sum_{|\alpha|\leq\gamma} a_\alpha(x) D^\alpha \tag{5.54a}$$

with coefficients $a_\alpha(x) \in C^\infty(\mathbf{R}^n)$ independent of x for large $|x|$ is a (polyhomogeneous) pseudodifferential operator. The principal and complete symbols of A are, respectively,

$$\sum_{|\alpha|=\gamma} a_\alpha(x)\xi^\alpha \quad \text{and} \quad \sum_{|\alpha|\leq\gamma} a_\alpha(x)\xi^\alpha. \tag{5.54b}$$

The operator of multiplication by a function $a_0(x)$ is a still more special case of a pseudodifferential operator; this is a homogeneous pseudodifferential operator with symbol $a_0(x)$ of order zero. A more general class of homogeneous pseudodifferential operators of order zero is formed by *singular integral operators*. A singular integral operator admits the representation

$$Au(x) = a(x)u(x) + \int K(x, x-y) u(y)\, dy + A_1 u(x), \tag{5.55a}$$

[7]One can also consider more general polyhomogeneous pseudodifferential operators, for which $a_{\gamma-q}(x,\xi)$ is positively homogeneous of order m_q in ξ and $\gamma = m_0 > m_1 > \ldots$, $m_q \to -\infty$ $(q \to \infty)$. In particular, the operator $A - \lambda I$ belongs to this class if γ is not an integer.

where the kernel $K(x,z)$ is infinitely smooth for $z \neq 0$ and positively homogeneous in z of order $-n$, and A_1 is a pseudodifferential operator of order -1. Although the kernel $K(x,z)$ has a critical singularity at $z=0$, it possesses the property

$$\int_{|z|=1} K(x,z)\,dS_z = 0,$$

which permits one to define the integral in (5.55a) in the sense of the Cauchy principal value:

$$\text{p.v.} \int K(x, x-y)u(y)\,dy = \lim_{\varepsilon_j \to 0} \int_{\varepsilon_1 < |x-y| < \varepsilon_2^{-1}} K(x, x-y)u(y)\,dy. \tag{5.56}$$

The kernel and the difference (symbol)$-a(x)$ are related by the Fourier transform (in the sense of distributions); see [88] or the survey [12, Subsection 1.7].

Pseudodifferential operators of negative order play a more essential role in the following; they will be considered in Subsection 2.

The following three assertions of the *calculus of pseudodifferential operators* will be important to us. The first two of them are obvious for symbols independent of x, and the third can readily be verified for differential operators.

3° (The Composition Theorem.) *Let A and B be pseudodifferential operators with principal symbols a_γ and b_δ, respectively. Then AB is a pseudodifferential operator with principal symbol $a_\gamma b_\delta$.*

It follows that the (scalar) pseudodifferential operators AB and BA have the same principal symbol $a_\gamma b_\delta$. Consequently, if A and B are pseudodifferential operators of order γ and δ, respectively, then AB and BA are pseudodifferential operators of order $\gamma+\delta$, while the commutator $AB-BA$ is a pseudodifferential operator of order $\gamma+\delta-1$.

4° *Let A be a pseudodifferential operator with principal symbol a_γ. Then the formal adjoint A^* of the operator A with respect to the inner product in $H^0(\mathbf{R}^n)$ is a pseudodifferential operator with principal symbol $\overline{a}_\gamma$.*

The formal adjointness is understood in the following sense:

$$(Au, v)_{\mathbf{R}^n} = (u, A^*v)_{\mathbf{R}^n}, \quad \text{where } (u,v)_{\mathbf{R}^n} = \int_{\mathbf{R}^n} u\overline{v}\,dx,$$

for $u, v \in C_0^\infty(\mathbf{R}^n)$. With regard to 2° and 4°, the first of these equalities is valid for $u, v \in H^\beta(\mathbf{R}^n)$, where $\beta = \max(0,\gamma)$ and γ is the order of the operators A and A^*. Moreover, with regard to assertion 13° of Subsection 5.3.2, this equality remains true for $u \in H^{s+\gamma}(\mathbf{R}^n)$ and $v \in H^{-s}(\mathbf{R}^n)$ for any s.

5° *Under an infinitely differentiable nondegenerate coordinate transformation $x = x(\tilde{x})$, a pseudodifferential operator A in $\mathbf{R}^n_x$ with principal symbol $a(x,\xi)$ becomes a pseudodifferential operator $\tilde{A}$ in $\mathbf{R}^n_{\tilde{x}}$ with principal symbol*

$$\tilde{a}(\tilde{x},\tilde{\xi}) = a(x(\tilde{x}), (\partial x/\partial\tilde{x})'^{-1}\tilde{\xi}), \tag{5.57}$$

where $\partial x/\partial\tilde{x}$ is the Jacobi matrix and the prime stands for the matrix transposition. More precisely, if $\varphi(x), u(x) \in C_0^\infty(\mathbf{R}^n)$ vanish outside the domain where the transformation is defined and nondegenerate, then the calculation of $\varphi(x)(Au)(x)$ followed by the substitution $x = x(\tilde{x})$ gives $\varphi(x(\tilde{x}))\widetilde{A}u(x(\tilde{x}))$, where $\widetilde{A}$ is a pseudodifferential operator with principal symbol (5.57).

So far, we have given simplified statements. However, there exists a *complete symbolic calculus* in $\mathbf{R}^n$ for polyhomogeneous (and even more general) pseudodifferential operators.

First, one can prove that *if a sequence* $\{a_{\gamma-j}(x,\xi)\}$ *of homogeneous symbols of orders* $\gamma - j$, $j = 0, 1, \ldots$, *is given, then there exists a complete symbol* $a(x,\xi)$ *with the asymptotic expansion* (5.53) (*and one can construct the corresponding pseudodifferential operator* (5.52)).

If A *and* B *are polyhomogeneous pseudodifferential operators with complete symbols* $a(x,\xi)$ *and* $b(x,\xi)$, *then the complete symbol* $c(x,\xi)$ *of the product* AB *has the asymptotic expansion*

$$c(x,\xi) \sim \sum_{\alpha} \frac{1}{\alpha!}\, \partial_\xi^\alpha a(x,\xi) D_x^\alpha b(x,\xi), \tag{5.58}$$

where $\alpha! = \alpha_1! \ldots \alpha_n!$. One can substitute the asymptotic expansions of the complete symbols $a(x,\xi)$ and $b(x,\xi)$ into this formula and then rearrange the terms in a descending order of degrees of homogeneity.

Let A *be a pseudodifferential operator with complete symbol* $a(x,\xi)$. *The complete symbol* $a^{(*)}(x,\xi)$ *of the formally adjoint pseudodifferential operator* A^* *possesses the asymptotic expansion*

$$a^{(*)}(x,\xi) \sim \sum_{a} \frac{1}{\alpha!}\, \partial_\xi^\alpha D^\alpha \overline{a(x,\xi)}. \tag{5.59}$$

There is also a somewhat awkward rule expressing the complete symbol of the pseudodifferential operator arising under a change of variables in terms of the complete symbol, and its derivatives, of the original pseudodifferential operator.

Note that sometimes it is more convenient to consider $a(x,\xi)$ as the complete symbol of A if instead of (5.52) we only have

$$Au(x) = F^{-1}_{\xi\to x} a(x,\xi)(Fu)(\xi) + Tu(x),$$

where T is an operator of order $-\infty$.

For further details, e.g., see [97], [93], or the survey [12].

2. Integral operators with weakly singular kernels. In the study of diffraction problems, the following fact will be important. The integral operators generated by single or double layer potentials on a surface $S \subset \mathbf{R}^3$ or a curve $S \subset \mathbf{R}^2$ for the Helmholtz equation (in particular, the operator (5.6)) are pseudodifferential operators of order -1. (The definitions of pseudodifferential operators on a closed surface or curve will be given below in Subsections 3 and 4.) This will be shown in Sections 5.7 and 5.8, where the principal symbol of the operator corresponding to the single layer potential will also be calculated. As a preliminary step, let us first consider general integral operators in $\mathbf{R}^n$ with kernels *homogeneous in the generalized sense of order* $\delta > -n$:

$$Au(x) = \int_{\mathbf{R}^n} K(x, x-y)\vartheta(x-y)u(y)\,dy. \tag{5.60}$$

Here $\vartheta(z) \in C_0^\infty(\mathbf{R}^n)$ is equal to 1 in a neighborhood of the origin, and we impose the following conditions on the function $K(x,z)$:

(i) If δ is not a nonnegative integer, then $K(x,z)$ is positively homogeneous in z of order δ in the ordinary sense: $K(x,\rho z) = \rho^\delta K(x,z)$ for $\rho > 0$. Furthermore, $K(x,z)$ is infinitely differentiable for $z \neq 0$ and is zero for $|x| \geq C = C_K$.

(ii) If δ is a nonnegative integer, then

$$K(x,z) = K_1(x,z) + \sum_{|\alpha|=\delta} b_\alpha(x) z^\alpha \log|z|. \tag{5.61}$$

Here K_1 has the same properties as K in (i), and $b_\alpha(x) \in C_0^\infty(\mathbf{R}^n)$.

It is shown in [88] that under these conditions the operator (5.60) is a pseudodifferential operator of order $\gamma = -n - \delta$ with principal symbol $a(x,\xi)$ equal to the Fourier transform of the kernel $K(x,z)$ with respect to z. More precisely, the principal symbol coincides with this Fourier transform, understood in the sense of distributions, for $\xi \neq 0$. (This means that

$$\int_{\mathbf{R}^n} K(x,z)\overline{\psi(z)}\,dz = \frac{1}{(2\pi)^n}\int_{\mathbf{R}^n} a(x,\xi)\overline{\varphi(\xi)}\,d\xi$$

provided that $\varphi(\xi)$ belongs to $C_0^\infty(\mathbf{R}^n)$ and vanishes in a neighborhood of the origin, and $\psi(y)$ is the inverse Fourier transform of $\varphi(\xi)$.) Moreover, this is a complete symbol of the operator (5.60) in the following sense: the operator (5.60) is a homogeneous pseudodifferential operator up to an infinitely smoothing term. The complete symbol may well be zero identically; this happens in case (ii) if $K(x,z) = K_1(x,z) = \sum_{|\alpha|=\delta} c_\alpha(x) z^\alpha$, where $c_\alpha(x) \in C_0^\infty(\mathbf{R}^n)$. Obviously, in this case the operator (5.60) is infinitely smoothing.

The following examples are basic for the subsequent exposition:

$$\begin{aligned} K(x,z) &= b(x)|z|^l, && n = 2, \\ K(x,z) &= c(x)|z|^m \log|z|, && n = 1, \end{aligned} \tag{5.62}$$

where $l \geq -1$ is odd and $m \geq 0$ is even, and b and c are compactly supported smooth functions in $\mathbf{R}^2$ and $\mathbf{R}^1$, respectively. For even l, the first of the functions (5.62) gives an infinitely smoothing operator by (5.60).

To continue the preparation for Sections 5.7 and 5.8, let us now consider integral operators with inhomogeneous kernels of a rather special form. Such kernels appear when one considers the restrictions of the above-mentioned potentials to S. Four auxiliary assertions will be established.

To make the notation closer to that used in Section 5.7 and further on, we now write $x = x(t)$ and $y = x(\tilde{t})$ (where t or $\tilde{t}$ are local coordinates on S); here $\dim t = \dim \tilde{t} = n-1$, $n \geq 2$. Consider the operator

$$Au(t) = \int_{\mathbf{R}^{n-1}} g(t,\tilde{t})|x(t) - x(\tilde{t})|^{-q} u(\tilde{t})\, d\tilde{t} \tag{5.63}$$

with $g(t,\tilde{t}) \in C_0^\infty(\mathbf{R}^{n-1} \times \mathbf{R}^{n-1})$. It will be convenient to represent this function in the form $g(t,\tilde{t}) = g(t,t+\tau) = g_1(t,\tau)$, where $\tau = \tilde{t}-t$. Then $g_1(t,\tau) \in C_0^\infty(\mathbf{R}^{n-1}\times\mathbf{R}^{n-1})$; let $g_1 = 0$ outside the set $\{(t,\tau) : |t| < C_1,\ |\tau| < C_2\}$. Let p be the maximum nonnegative integer such that $D_\tau^\alpha g_1(t,\tau) = 0$ at $\tau = 0$ for $|\alpha| < p$. If $p \geq 1$, then *$g_1(t,\tau)$ has a zero of order p at $\tau = 0$, and if $p = 0$, then $g_1(t,0) \not\equiv 0$*. Next, let $x(t) = (x_1(t), \ldots, x_n(t))$, where all $x_j(t)$ belong to $C^\infty(\mathbf{R}^{n-1})$. (In Section 5.7 and further on, $x = x(t)$ will be a parametric representation of a part of an $(n-1)$-dimensional surface in $\mathbf{R}^n$.) We assume that the vectors $\partial x(t)/\partial t_j$, $j = 1, \ldots, n-1$, are pairwise orthogonal and everywhere nonvanishing. Finally, let q be a *positive integer such that $p - q > -(n-1)$*. This inequality guarantees the existence of the integral (5.63). We can restrict ourselves to the cases $n = 2$ and $n = 3$, but the following assertion holds for every $n \geq 2$.

Proposition 5.4.1 *Under the above assumptions, A is a pseudodifferential operator of order $\gamma = -(n-1) - (p-q)$ in $\mathbf{R}^{n-1}$.*

Proof. Let us study the kernel of the operator (5.63). To be definite, suppose that $n = 3$. By Taylor's formula,

$$x_j(\tilde{t}) - x_j(t) = \frac{\partial x_j(t)}{\partial t_1}\tau_1 + \frac{\partial x_j(t)}{\partial t_2}\tau_2 + r_j(t,\tau), \tag{5.64}$$

where the function $r_j(t,\tau)$ is infinitely differentiable (as well as the other terms in this formula) and satisfies the inequality $|r_j(t,\tau)| \leq C_3|\tau|^2$ for $|t| \leq C_1$ and $|\tau| \leq C_2$, $j = 1,2,3$. By assumption, the vectors $\partial x(t)/\partial t_1$ and $\partial x(t)/\partial t_2$ are orthogonal. Let $h_1(t)$ and $h_2(t)$ be the (nonzero) lengths of these vectors. It follows from (5.64) that

$$|x(t) - x(\tilde{t})|^2 = h_1^2(t)\tau_1^2 + h_2^2(t)\tau_2^2 + R(t,\tau), \tag{5.65}$$

where $R(t,\tau)$ is a C^∞ function satisfying the inequality $|R(t,\tau)| \leq C_4|\tau|^3$ for $|t| \leq C_1$ and $|\tau| \leq C_2$. Let $H(t,\tau)$ be the sum of the first two terms on the right-hand side in (5.65). This function does not vanish for $|t| \leq C_1$ and $\tau \neq 0$. It follows from (5.65) that

$$|x(t) - x(\tilde{t})|^{-q} = H^{-q/2}(t,\tau)(1 + R_1(t,\tau))^{-q/2},$$

where $R_1(t,\tau) = R(t,\tau)H^{-1}(t,\tau)$ and $|R_1(t,\tau)| \leq C_5|\tau|$ for $|t| \leq C_1$ and $0 < |\tau| < C_2$. Under the additional assumption that $C_5|\tau| \leq C_6 < 1$, we can expand $(1 + R_1(t,\tau))^{-q/2}$ in the binomial series. Let $b(\tau)$ be a C^∞ function equal to 1 in a neighborhood of the origin and vanishing for $C_5|\tau| \geq C_6$. Then

$$b(\tau)|x(t) - x(\tilde{t})|^{-q} = b(\tau)\Big(H^{-q/2}(t,\tau) + \sum_{j=1}^{\infty} c_j R^j(t,\tau) H^{-j-q/2}(t,\tau)\Big),$$

where the c_j are some numbers. For $j > q$, the function $R^j(t,\tau)H^{-j-q/2}(t,\tau)$ will be continuous everywhere, including the points at which $\tau = 0$, if we define it to be zero for $\tau = 0$. The larger j, the smoother this function is. On the other hand, we can use Taylor's formula to expand the infinitely smooth function $R^j(t,\tau)$ in powers of τ to an arbitrary given order. This expansion starts from terms of order $3j$. A similar expansion is possible for $g_1(t,\tau)$; it starts from terms of order p. This permits one to obtain the following formula for each positive integer $m > p - q$:

$$b(\tau)g_1(t,\tau)|x(t) - x(\tilde{t})|^{-q} = b(\tau)\Big(\sum_{l=p-q}^{m} K_l(t,\tau) + R^{(m)}(t,\tau)\Big), \tag{5.66}$$

where the $K_l(t,\tau)$ are functions infinitely differentiable for $\tau \neq 0$ and positively homogeneous of order l in τ, and $R^{(m)}(t,\tau) \in C^{(m)}(\mathbf{R}^{n-1} \times \mathbf{R}^{n-1})$; all these functions vanish for $|t| \geq C_1$. Clearly, if in (5.66) we replace $b(\tau)$ by $1 - b(\tau)$, then the resulting function will be infinitely smooth. According to what was said in the beginning of this subsection, we obtain an expansion

$$A = A_\gamma + A_{\gamma-1} + \cdots + A_{-m-(n-1)} + T^{(m)},$$

where A_j is a pseudodifferential operator of order j and $T^{(m)}$ is an integral operator with kernel belonging to $C_0^{(m)}(\mathbf{R}^{n-1} \times \mathbf{R}^{n-1})$. Moreover, the A_j are independent of m. Let us take an arbitrary s. For sufficiently large m, we find, with regard to assertions 2° from Subsection 1 and 15° from Subsection 5.3.2, that $A - A_\gamma$ is a bounded operator from $H^s(\mathbf{R}^{n-1})$ to $H^{s-\gamma-1}(\mathbf{R}^{n-1})$. Consequently, $A - A_\gamma$ is an operator of order $\gamma - 1$. □

The complete symbol of A is the asymptotic sum of the symbols of the operators $A_\gamma, A_{\gamma-1}, \ldots$. The principal symbol of A coincides with the symbol of A_γ; as shown by the proof, the latter operator is an integral operator with kernel

$$b(\tau) \sum_{\alpha_1+\alpha_2=p} \frac{\tau_1^{\alpha_1}\tau_2^{\alpha_2}}{\alpha_1!\alpha_2!} \frac{\partial^{\alpha_1+\alpha_2}}{\partial\tau_1^{\alpha_1}\partial\tau_2^{\alpha_2}} g_1(t,0)H^{-q/2}(t,\tau), \tag{5.67}$$

where $\tau = \tilde{t} - t$.

Proposition 5.4.2 *Let $n = 3$, $q = 1$, and $g(t,t) \not\equiv 0$ (i.e., $p = 0$). Then the operator* (5.67) *has the principal symbol*

$$2\pi g(t,t)h_1^{-1}(t)h_2^{-1}(t)(h_1^{-2}(t)\xi_1^2 + h_2^{-2}(t)\xi_2^2)^{-1/2}. \tag{5.68}$$

Proof. The Fourier transform in the sense of distributions of the function $|t|^{-1}$ on the plane is $2\pi|\xi|^{-1}$ (see [40, Section II.3]). The desired result follows from this by a change of variables. □

Proposition 5.4.3 *For $n = 2$, the operator*

$$Au(t) = \int_{\mathbf{R}^1} g(t,\tilde{t}) \log|x(t) - x(\tilde{t})| u(\tilde{t})\, d\tilde{t} \tag{5.69}$$

is a pseudodifferential operator of order $-p-1$.

Here $t, \tilde{t} \in \mathbf{R}^1$, $x(t) = (x_1(t), x_2(t))$, and the assumptions about $g(t, \tilde{t})$, p, and $x(t)$ are the same as before the statement of Proposition 5.4.1. The proof is similar to that of Proposition 5.4.1 with the only difference that $H^{1/2}(t, \tau) = h(t)|\tau|$, where $h(t) = |x'(t)|$, and the Taylor series for $\log(1 + R_1(t, \tau))$ is used instead of the binomial series.

The principal symbol of the operator (5.69) coincides with the symbol of the integral operator A_γ, $\gamma = -p - 1$, which has the kernel

$$b(\tau) \frac{\tau^p}{p!} \frac{\partial^p}{\partial \tau^p} g_1(t, 0) \log |\tau|, \quad \text{where } \tau = \tilde{t} - t. \tag{5.70}$$

Proposition 5.4.4 *For $p = 0$, the principal symbol of the operator* (5.69) *is equal to*

$$-\pi g(t, t)|\xi|^{-1}. \tag{5.71}$$

This follows from the fact that the Fourier transform of the distribution $\log |\tau|$ on the real line is equal to $-\pi|\xi|^{-1}$ outside the origin (see [40, Sections II.2 and I.4]).

3. Pseudodifferential operators on a surface $\mathfrak{S}$. Let $\mathfrak{S}$ have the same meaning as in Subsection 5.3.1. A *pseudodifferential operator A of order γ on $\mathfrak{S}$* is a linear operator defined originally on $C^\infty(\mathfrak{S})$ and satisfying the following two conditions. Let $\omega_1, \omega_2 \in C^\infty(\mathfrak{S})$.

(a) If the supports of ω_1 and ω_2 are disjoint, then $\omega_1 A(\omega_2 u)$ is an operator of order $-\infty$.

(b) If the supports of ω_1 and ω_2 lie in the same coordinate neighborhood, then $\omega_1 A(\omega_2 u) = \omega_1 \tilde{A}(\omega_2 u)$, where the functions on the right-hand side are expressed via the corresponding local coordinates and $\tilde{A}$ is a pseudodifferential operator of order γ in $\mathbf{R}^n$.

Using a partition of unity $\{\kappa_j\}$ (see Subsection 5.3.1), we can represent A in the form

$$Au = \sum_{j,l} \kappa_j A(\kappa_l u). \tag{5.72}$$

Without loss of generality we can assume that the supports of any two functions κ_j and κ_l either are disjoint or lie in the same coordinate neighborhood; then each term on the right-hand side has the form indicated in (a) or (b). We can readily conclude that A acts in $C^\infty(\mathfrak{S})$ and can be extended to an operator continuous from $H^s(\mathfrak{S})$ to $H^{s-\gamma}(\mathfrak{S})$ for all s, that is, an operator of order γ on $\mathfrak{S}$.

Assertion 5° in Subsection 1 shows that the *principal symbol* of a pseudodifferential operator A of order γ is well defined as a function on the *cotangent bundle* $T^*\mathfrak{S}$ of the manifold $\mathfrak{S}$ (e.g., see [37, Part II, Chapter 1]); the cotangent bundle consists of *cotangent vectors.* (More precisely, the symbol is defined on nonzero cotangent vectors.) Let us explain these notions and simultaneously consider the *tangent bundle* $T\mathfrak{S}$, which will be needed in Section 5.11 (see [37, Part II, Chapter 1]). The tangent bundle consists of all *tangent vectors* issuing from points of $\mathfrak{S}$ and tangent to $\mathfrak{S}$. The tangent vectors at a given point $x \in \mathfrak{S}$ form the *tangent space* $T_x\mathfrak{S}$ of $\mathfrak{S}$ at that point. This is a linear space, and we can consider linear homogeneous functions of elements of this space. These functions are called *cotangent vectors*; they form the *cotangent space* $T_x^*\mathfrak{S}$. Let $t = (t_1, \ldots, t_n)$ be local coordinates on $\mathfrak{S}$. Then each vector tangent to $\mathfrak{S}$ at a point on $\mathfrak{S}$ with coordinates t is determined by a $2n$-tuple $(t, \eta) = (t_1, \ldots, t_n, \eta_1, \ldots, \eta_n)$ of numbers which can be viewed as local coordinates in $T\mathfrak{S} = \bigcup T_x\mathfrak{S}$. (The sets $T_x\mathfrak{S}$ and $T_y\mathfrak{S}$ are treated as disjoint for $x \neq y$.) Each linear homogeneous function of η (for a given t) has the form $\xi_1\eta_1 + \ldots + \xi_n\eta_n$. Hence the corresponding cotangent vector is also specified by the $2n$-tuple $(t, \xi) = (t_1, \ldots, t_n, \xi_1, \ldots, \xi_n)$ of numbers, which can be viewed as local coordinates on $T^*\mathfrak{S} = \bigcup T_x^*\mathfrak{S}$. (Here $T_x^*\mathfrak{S}$ and $T_y^*\mathfrak{S}$ are also disjoint for $x \neq y$.) However, the coordinate transformation laws on $T\mathfrak{S}$ and $T^*\mathfrak{S}$ are not the same. If, locally on $\mathfrak{S}$, we pass from coordinates t to new coordinates $\tilde{t}$, $t = t(\tilde{t})$, then (t, η) passes into $(\tilde{t}, \tilde{\eta}) = (t(\tilde{t}), (\partial t(\tilde{t})/\partial\tilde{t})\tilde{\eta})$ (where $\partial t/\partial\tilde{t}$ is the Jacobi matrix), whereas (t, ξ) passes into $(\tilde{t}, \tilde{\xi}) = (t(\tilde{t}), (\partial t(\tilde{t})/\partial\tilde{t})'^{-1}\tilde{\xi})$. This, in conjunction with (5.57), implies that the principal symbol of a pseudodifferential operator is a function on $T^*\mathfrak{S}$.

Locally, the principal symbol is represented by a function $a(t, \xi)$ that is defined for $\xi \neq 0$ and is infinitely differentiable and positively homogeneous of order γ. Within a coordinate neighborhood, one can also define the complete symbol (cf. Subsection 1).

Assertions 2°, 3°, and 4° of Subsection 1 remain valid for pseudodifferential operators on $\mathfrak{S}$ if we define the inner product by

$$(u, v)_{\mathfrak{S}} = \int_{\mathfrak{S}} u\bar{v}\, d\mathfrak{S} \tag{5.73}$$

(see also the end of Subsection 5.5.2). We could transform the examples of pseudodifferential operators given in Subsection 2 to obtain examples of pseudodifferential operators on a surface, but it is more convenient to postpone this until Sections 5.7 and 5.8. Note that the notation there will be slightly different: we shall consider a surface S in $\mathbf{R}^3$ or a closed curve in $\mathbf{R}^2$. In Section 5.11 we shall consider matrix pseudodifferential operators.

4. Pseudodifferential operators on a closed curve. Let S be an infinitely smooth closed curve without self-intersections in $\mathbf{R}^n$, $n \geq 2$. As was already noted in Subsection 5.3.3, each function on S can be identified with a 2π-periodic function $u(t)$ on $\mathbf{R}$. Now we point out that a polyhomogeneous pseudodifferential operator on S admits the "discrete" representation

$$Au(t) = a(t, D)u(t) = \sum_{-\infty}^{\infty} e^{imt}a(t, m)c_m(u), \tag{5.74}$$

where the $c_m(u)$ are the Fourier coefficients (5.48). (The proof can be found in [10].) Here $a(t, \xi)$ is the complete symbol, 2π-periodic in t; thus it is globally defined, just as for pseudodifferential operators in $\mathbf{R}^n$. The symbol is used in (5.74) only at the integer points $\xi = m$; it admits the asymptotic expansion

$$\begin{aligned} a(t, m) &= a_\gamma(t, m) + \cdots + a_{\gamma-q}(t, m) + O(|m|^{\gamma-q-1}) \\ &= a_\gamma(t, \operatorname{sgn} m)|m|^\gamma + \cdots + a_{\gamma-q}(t, \operatorname{sgn} m)|m|^{\gamma-q} + O(|m|^{\gamma-q-1}) \end{aligned} \tag{5.75}$$

as $m \to \infty$ for any q. All $a_j(t,m)$ are 2π-periodic C^∞ functions of t, and the expansion (5.75) can be differentiated with respect to t as many times as desired. Formula (5.74) is similar to (5.52), but the role of the direct and inverse Fourier transforms is played by the transitions from the function $u(t)$ to the sequence $\{c_m(u)\}$ of its Fourier coefficients and then to a Fourier type series.

For pseudodifferential operators on a C^∞ closed curve (essentially, on a circle), there is a *complete symbolic calculus* similar to that for pseudodifferential operators in $\mathbf{R}^n$. All symbols in this calculus, as well as all terms in their asymptotic expansions, are 2π-periodic.

5.5 Elliptic Pseudodifferential Operators and Boundary Value Problems

1. Elliptic pseudodifferential operators on a closed surface (e.g., see [97], [93], and the survey [12]). Let A be a pseudodifferential operator of order γ on a surface $\mathfrak{S}$. One says that A is *elliptic* if its principal symbol a_γ does not vanish (on nonzero cotangent vectors). The following assertions hold for an elliptic operator A.

1° (An *a priori estimate.*) *If* $u \in H^{s+\gamma}(\mathfrak{S})$, *then*

$$\|u\|_{\mathfrak{S},s+\gamma} \le C_s(\|Au\|_{\mathfrak{S},s} + \|u\|_{\mathfrak{S},s+\gamma-1}) \tag{5.76}$$

for any s *with a constant* C_s *independent of* u.

Here $\|u\|_{\mathfrak{S},s+\gamma-1}$ can be replaced by $\|u\|_{\mathfrak{S},s'}$ with any fixed $s' < s+\gamma$. We also note that *if* $Au = 0$ *implies* $u = 0$, *that is, the uniqueness holds for the equation* $Au = f$, *then the stronger inequality* $\|u\|_{\mathfrak{S},s+\gamma} \le C_s'\|Au\|_{\mathfrak{S},s}$ *is valid.*

2° *The space* $\operatorname{Ker} A = \mathfrak{L}^1_A(0)$ *of solutions of the equation* $Au = 0$ *has a finite dimension* l.

3° *The range of the operator* $A\colon H^{s+\gamma}(\mathfrak{S}) \to H^s(\mathfrak{S})$ *is closed and has a finite codimension* l' (that is, has a complement of dimension l').

A bounded operator from one Hilbert (or Banach) space to another with finite-dimensional kernel and finite-codimensional closed range is said to be *Fredholm.* Thus, *an elliptic pseudodifferential operator of order* γ *is Fredholm from* $H^{s+\gamma}(\mathfrak{S})$ *to* $H^s(\mathfrak{S})$ *for every* s.

The simplest way to prove the Fredholm property of an elliptic pseudodifferential operator A of order γ with principal symbol a_γ is to construct and use a *parametrix* B for A, that is, an elliptic pseudodifferential operator of order $-\gamma$ with principal symbol a_γ^{-1} such that

$$BA = I + T_1 \quad \text{and} \quad AB = I + T_2, \tag{5.77}$$

where T_1 and T_2 are smoothing operators (see the end of Subsection 5.3.2) and hence compact operators in each $H^s(\mathfrak{S})$. It is well known that an operator of the form $I+T$ in each Hilbert (or Banach) space is Fredholm if T is a compact operator. Hence, in particular, it readily follows from the first relation in (5.77) that $\operatorname{Ker} A$ is finite-dimensional, and the second relation implies that the range of A has a finite-dimensional complement. The fact that the range of A is closed can be derived from either of these relations. Assertion 1°, as well as Assertion 4° below, can also be derived from the first relation in (5.77). The ellipticity condition can be derived from 1°; thus, *the ellipticity of a pseudodifferential operator A is equivalent to the Fredholm property of A viewed as an operator from $H^{s+\gamma}(\mathfrak{S})$ to $H^s(\mathfrak{S})$.*

4° (The *smoothness theorem.*) *If $Au = f$, $u \in H^{s+\gamma}(\mathfrak{S})$, but $f \in H^{s+t}(\mathfrak{S})$, where $t > 0$, then $u \in H^{s+t+\gamma}(\mathfrak{S})$.* (If f has an extra smoothness, then so does u.) In particular, if $f \in C^\infty(\mathfrak{S})$, then $u \in C^\infty(\mathfrak{S})$ (see Assertion 11° in Subsection 5.3.2). The local version of this assertion is also valid; namely, if $u \in H^s(\mathfrak{S})$ and $f \in C^\infty$ in a neighborhood of some point of $\mathfrak{S}$, then $u \in C^\infty$ in this neighborhood.

It follows from this assertion that $\operatorname{Ker} A$ consists of infinitely smooth functions.

For $\gamma \neq 0$, the root functions of an elliptic pseudodifferential operator also belong to $C^\infty(\mathfrak{S})$. For nonzero eigenvalues, this follows from the fact that the operator $A - \lambda I$ is elliptic (of order 0 for $\gamma < 0$ and γ for $\gamma > 0$).

For any Fredholm operator A, the difference $l - l'$ is called the *index* of A. The index is invariant under homotopies (continuous deformations) of the operator in question such that the Fredholm property is preserved (in our case, such that the ellipticity is preserved; in particular, the index remains unchanged if an arbitrary lower-order operator is added to A). The index of the operator $I + T$, where T is a compact operator, is zero. For a Fredholm operator A of index zero, the inverse operator A^{-1} exists if and only if $l = 0$.

The formal adjoint A^* of a pseudodifferential operator A (see Assertion 4° in Subsection 5.4.1) is elliptic together with A, so that similar assertions hold for A^*. One can prove that $l' = \dim \operatorname{Ker} A^*$. Moreover, if $v_1, \ldots, v_{l'}$ is a basis in $\operatorname{Ker} A^*$, then the range of A can be described as follows.

5° *Let $f \in H^s(\mathfrak{S})$, $s \geq 0$. It is necessary and sufficient for the solvability of the equation $Au = f$ in $H^{s+\gamma}(\mathfrak{S})$ that the conditions $(f, v_j)_{\mathfrak{S}} = 0$ hold $(j = 1, \ldots, l')$.*

The index of an elliptic pseudodifferential operator is independent of s, since $\operatorname{Ker} A$ and $\operatorname{Ker} A^*$ consist of infinitely smooth functions and hence are independent of s. In cases of interest to us, we shall see that the index is zero. In particular, the index of a selfadjoint pseudodifferential operator A is zero and remains equal to zero if a lower-order pseudodifferential operator is added to A.

6° *If $l = l' = 0$, then A^{-1} exists and is a pseudodifferential operator of order $-\gamma$ with principal symbol a_γ^{-1}. If $a_\gamma = \overline{a}_\gamma$, then the index of A is zero.*

Here the first assertion can be derived from (5.77). The second assertion can readily be obtained by a deformation of A to the selfadjoint pseudodifferential operator $\operatorname{Re} A$ without changing the principal symbol.

Note also that the existence of A^{-1} implies the existence of A^{*-1} (and vice versa). This can readily be derived from 5° and 6°.

Remark An elliptic pseudodifferential operator A of *positive* order γ can be viewed not only as a bounded operator from $H^{s+\gamma}(\mathfrak{S})$ to $H^s(\mathfrak{S})$, but also as an *unbounded* *operator in* $H^s(\mathfrak{S})$ *with domain* $\mathfrak{D}_s(A) = H^{s+\gamma}(\mathfrak{S})$. The domain is dense in $H^s(\mathfrak{S})$. If the resolvent set of A is nonempty, then A is an operator with discrete spectrum (i.e. the resolvent is compact). In this case, the index of A is zero. The spectrum is independent of s since the root functions belong to $C^\infty(\mathfrak{S})$.

Furthermore, *A is a closed operator* in $H^s(\mathfrak{S})$: if $u_n \in \mathfrak{D}_s(A)$, $u_n \to u$, and $Au_n \to f$ in $H^s(\mathfrak{S})$, then $u \in \mathfrak{D}_s(A)$ and $Au = f$. Indeed, $Au_n \to Au$ in $H^{s-\gamma}(\mathfrak{S})$ (see Assertion 2° in Subsection 5.4.1); consequently, $Au = f$ in $H^{s-\gamma}(\mathfrak{S})$, $u \in H^{s+\gamma}(\mathfrak{S})$ by virtue of 4°, and $Au = f$ in $H^s(\mathfrak{S})$.

One can also show that the operator A and the formal adjoint A^* (see Subsections 5.4.1 and 5.4.3), viewed as operators in $H^0(\mathfrak{S})$ with domain $H^\gamma(\mathfrak{S})$, are adjoints of each other in the sense of the theory of operator in Hilbert spaces (e.g., see [21, Section 44]).

A pseudodifferential operator A of negative order can be viewed as a compact operator in any $H^s(\mathfrak{S})$. Its spectrum is independent of s, since it consists of nonzero eigenvalues and zero, even if A is not elliptic or elliptic with nonzero index. The root functions corresponding to the eigenvalue zero belong to $C^\infty(\mathfrak{S})$ if A is elliptic.

The ellipticity of a pseudodifferential operator in $\mathbf{R}^n$ is also defined by the condition $a_\gamma(x,\xi) \neq 0$ (for $\xi \neq 0$), where a_γ is the principal symbol. For elliptic operators in $\mathbf{R}^n$, assertions similar to 1° and 4° are valid.

2. Selfadjoint elliptic pseudodifferential operators. In this subsection we assume that $A_0 = A_0^*$ is an elliptic pseudodifferential operator of order $\gamma > 0$ on $\mathfrak{S}$. Then A_0 is selfadjoint as an operator in $H^0(\mathfrak{S})$ with domain $\mathfrak{D}(A_0) = H^\gamma(\mathfrak{S})$ (see the end of the preceding subsection); in particular, the spectrum $\Sigma(A_0)$ lies on the real axis.

1° *The operator* A_0*, viewed as an operator in* $H^s(\mathfrak{S})$ *with domain* $H^{s+\gamma}(\mathfrak{S})$*, has a discrete spectrum independent of* s*,* $-\infty < s < \infty$.

Indeed, let $\operatorname{Im}\mu \neq 0$. Then μ does not belong to the spectrum of A_0, and the operator $(A_0 - \mu I)^{-1}$ exists, is a pseudodifferential operator of order $-\gamma$, and hence is compact in each $H^s(\mathfrak{S})$ (see the end of Subsection 5.3.2). It follows that A_0 is an operator with discrete spectrum (see Subsection 5.2.4).

By virtue of the ellipticity, the principal symbol $a_\gamma = \overline{a}_\gamma$ is nonzero (on nonzero cotangent vectors). One can show that if $a_\gamma > 0$, then the spectrum $\Sigma(A_0)$ lies on some ray $\{\mu : \mu \geq \mu_0\}$; if $a_\gamma < 0$, then the spectrum lies on some ray $\{\mu : \mu \leq \mu_0\}$. A selfadjoint operator with such a spectrum (not necessarily a pseudodifferential operator) is said to be *bounded below* or *above*, respectively. In both cases the operator is called *semibounded.* If $\mu_0 = 0$, then the operator is said to be *nonnegative* (resp., *nonpositive*). If, moreover, 0 does not belong to the spectrum, then the operator is said to be *positive* (resp., *negative*). Equivalent definitions can be given in terms of the quadratic form of the operator. In particular, A is bounded below if $(Af, f) \geq \mu(f, f)$, nonnegative if $(Af, f) \geq 0$ (cf. the end of Subsection 5.2.6), and positive if $(Af, f) > 0$ $(0 \neq f \in \mathfrak{D}(A))$.

For simplicity, assume that 0 is not an eigenvalue of A_0. We can define the powers A_0^s $(-\infty < s < \infty$; the definition will be recalled in the next subsection in (5.81)).

Seeley [89] proved that $A_0^{s/\gamma}$ is a pseudodifferential operator of order s with principal symbol $a^{s/\gamma}$. This operator defines an isomorphism of $H^s(\mathfrak{S})$ onto $H^0(\mathfrak{S})$: $\|u\|_{\mathfrak{S},s} \leq C_1\|A^{s/\gamma}u\|_{\mathfrak{S},0} \leq C_2\|u\|_{\mathfrak{S},s}$ (cf. the operator (5.45) in $\mathbf{R}^n$). Let

$$\langle u, v\rangle_{\mathfrak{S},s} = (A_0^{s/\gamma}u, A_0^{s/\gamma}v)_{\mathfrak{S}}. \tag{5.78}$$

This Hermitian form is a new inner product on $H^s(\mathfrak{S})$. Obviously, $\langle A_0u, v\rangle_{\mathfrak{S},s} = \langle u, A_0v\rangle_{\mathfrak{S},s}$ for $u, v \in H^{s+\gamma}(\mathfrak{S})$. The following assertion holds.

2° *The operator A_0 viewed as an operator in $H^s(\mathfrak{S})$ with domain $H^{s+\gamma}(\mathfrak{S})$ is selfadjoint with respect to the inner product (5.78). The corresponding norm $\langle u, u\rangle_{\mathfrak{S},s}^{1/2}$ is equivalent to the norm $\|u\|_{\mathfrak{S},s}$.*

For simplicity, suppose that the principal symbol a_γ is positive. Let ν_j be the eigenvalues of A_0 ($\mu_0 = \nu_1 \leq \nu_2 \leq \ldots$). By $N(t)$ we denote the *distribution function for the eigenvalues of A*, defined as the number of eigenvalues on the interval $[\nu_1, t)$ (counted with their multiplicities). The following important asymptotic formula holds:

$$N(t) = Ct^{n/\gamma} + O(t^{(n-1)/\gamma}), \qquad t \to +\infty, \tag{5.79}$$

where $C = C(n) = (2\pi)^{-n}V$ and V is the volume of the domain in $T^*\mathfrak{S}$ determined by the inequality $a_\gamma(x, \xi) < 1$ (see [50] and the surveys [12, 86]). It readily follows from (5.79) that

$$\nu_j = C'j^{\gamma/n} + O(j^{(\gamma-1)/n}), \qquad j \to \infty, \tag{5.80}$$

where $C' = C'(n) = C^{-\gamma/n}$. The remainder estimates in these formulas cannot be sharpened in general.

A similar result holds for the characteristic numbers of a selfadjoint elliptic pseudodifferential operator of negative order $-\gamma$ with positive principal symbol.

Let us also note that, for given A_0, the following meaning can be assigned to the form $(f, g)_{\mathfrak{S}}$, where $f \in H^s(\mathfrak{S})$ and $g \in H^{-s}(\mathfrak{S})$, for any s:

$$(f, g)_{\mathfrak{S}} = (A_0^{s/\gamma}f, A_0^{-s/\gamma}g)_{\mathfrak{S}}.$$

Assertion 5° of Subsection 1 remains valid for $s < 0$. Moreover, if B and B^* are formally adjoint pseudodifferential operators of order γ, then

$$(Bf, g)_{\mathfrak{S}} = (f, B^*g)_{\mathfrak{S}}, \qquad f \in H^{s+\gamma}(\mathfrak{S}),\ g \in H^{-s}(\mathfrak{S}),$$

for any s.

3. An abstract analog of the scale $H^s(\mathfrak{S})$. Let $\mathfrak{H}$ be a separable Hilbert space, and let L_0 be a selfadjoint operator with discrete spectrum in $\mathfrak{H}$. Next, let $\{h_j\}$, $j = 1, 2, \ldots$, be an orthonormal basis in $\mathfrak{H}$ consisting of eigenvectors of L_0; $L_0h_j = \nu_jh_j$. We assume that 0 is not an eigenvalue of L_0 and the eigenvalues are numbered in a nondecreasing order of their absolute values. Let us introduce a scale of Hilbert spaces $\mathfrak{H}^s$, related to L_0. This scale, which will be used in Section 5.6, is more convenient than the single Hilbert space $\mathfrak{H}$ in applications to pseudodifferential operators on a closed surface.

Recall that the *real powers* L_0^s of the operator L_0 are defined as follows:

$$\text{if } f = \sum_{j=1}^{\infty} b_j h_j, \text{ then } L_0^s f = \sum_{j=1}^{\infty} b_j \nu_j^s h_j; \tag{5.81}$$

moreover, for $s > 0$ the domain $\mathfrak{D}(L_0^s)$ consists of vectors f for which the latter series converges in $\mathfrak{H}$. Here we must in advance choose some way of defining powers of complex numbers, since the ν_j are not necessarily positive. For example, we set

$$z^s = |z|^s e^{is \arg z}, \quad \text{where } -\pi < |\arg z| \leq \pi.$$

One must have in mind that if some of the eigenvalues of L_0^s are not real, then this operator is no longer selfadjoint; it is only normal.

The powers of a compact selfadjoint operator with nonzero eigenvalues are defined in a similar manner.

Let us fix a positive number γ (which will be the *order* of L_0 in the scale of spaces to be defined). We set $\mathfrak{H}^s = \mathfrak{D}(L_0^{s/\gamma})$ for $s \geq 0$. Then $\mathfrak{H}^0 = \mathfrak{H}$; for $s > 0$, the subspace $\mathfrak{H}^s \subset \mathfrak{H}$ is a Hilbert space with respect to the inner product

$$(f, g)_s = (L_0^{s/\gamma} f, L_0^{s/\gamma} g). \tag{5.82a}$$

If $g = \sum c_j h_j$, then this formula can be rewritten as

$$(f, g)_s = \sum_{j=1}^{\infty} |\nu_j|^{2s/\gamma} b_j \bar{c}_j. \tag{5.82b}$$

The norm $(f, f)_s^{1/2}$ will be denoted by $\|f\|_s$.

For $s < 0$, we define $\mathfrak{H}^s$ as the space of formal series $f \sim \sum b_j h_j$ (not necessarily convergent in $\mathfrak{H}$) with finite norm $\|f\|_s = (\sum |\nu_j|^{2s/\gamma} |b_j|^2)^{1/2}$. This is also a Hilbert space with respect to the inner product (5.82).

By $\mathfrak{H}^\infty$ we denote the intersection of all $\mathfrak{H}^s$; this is the space of "infinitely smooth" vectors.

If $L_0 = A_0$ is a selfadjoint pseudodifferential operator of order $\gamma > 0$ on a surface $\mathfrak{S}$, then $\mathfrak{H}^s = H^s(\mathfrak{S})$. In general, the scale $\mathfrak{H}^s$ depends on L_0. However, this scale and the operator L_0 considered in this scale preserve some of the properties of the Sobolev scale $H^s(\mathfrak{S})$ and the operator A_0 in that scale (see Subsection 2). Let us list some of these properties.

1° $\mathfrak{H}^{s_1} \supset \mathfrak{H}^{s_2}$ *for* $s_1 < s_2$, *and the embedding operator is compact. The space* $\mathfrak{H}^{s_2}$ *is dense in* $\mathfrak{H}^{s_1}$.

2° *For any real* s, *the operator* L_0 *is a one-to-one continuous mapping of* $\mathfrak{H}^{s+\gamma}$ *onto* $\mathfrak{H}^s$.

3° *If* $L_0 f = g$, $f \in \mathfrak{H}^{s+\gamma}$, *and* $g \in \mathfrak{H}^{s+t}$, $t > 0$, *then* $f \in \mathfrak{H}^{s+t+\gamma}$. *In particular, the eigenvectors of* L_0 *belong to* $\mathfrak{H}^\infty$.

4° *In each* $\mathfrak{H}^s$, *the operator* L_0 *with* $\mathfrak{D}(L_0) = \mathfrak{H}^{s+\gamma}$ *is a selfadjoint operator with discrete spectrum. The eigenvalues and eigenvectors of this operator are independent of* s.

All these assertions can readily be derived from the definition of $\mathfrak{H}^s$. To prove the compactness in 1°, one verifies that for each $\varepsilon > 0$ the image of a bounded set from $\mathfrak{H}^{s_2}$ possesses a finite ε-net in $\mathfrak{H}^{s_1}$.

Let L_1 be a linear operator acting in $\mathfrak{H}^\infty$. We say that it is an *operator of order* δ *in the scale* $\mathfrak{H}^s$ if $\|L_1 f\|_s \le C_s \|f\|_{s+\delta}$ for $f \in \mathfrak{H}^\infty$ and for each s. An operator L_1 with this property can be extended to a bounded operator from $\mathfrak{H}^{s+\delta}$ to $\mathfrak{H}^s$ for each s. If $\delta > 0$, then L_1 can also be viewed as an unbounded operator in $\mathfrak{H}^s$ with domain $\mathfrak{H}^{s+\delta}$.

In Section 5.6 we shall consider operators L of the form $L_0 + L_1$, where L_1 is an operator of some order $\delta < \gamma$. Obviously, they are operators of order γ. Assertion 3° can be generalized to such operators L as follows.

5° *If* $Lf = g$, $f \in \mathfrak{H}^{s+\gamma}$, *and* $g \in \mathfrak{H}^{s+t}$, $t > 0$, *then* $f \in \mathfrak{H}^{s+t+\gamma}$. *In particular, the root vectors of* L *lie in* $\mathfrak{H}^\infty$.

Indeed, $f = L_0^{-1} g - L_0^{-1} L_1 f$. For simplicity, let $g \in \mathfrak{H}^\infty$. We first find that $f \in \mathfrak{H}^{s+\gamma+(\gamma-\delta)}$, then, by induction, $f \in \mathfrak{H}^{s+\gamma+2(\gamma-\delta)}$, etc., so that $f \in \mathfrak{H}^\infty$.

Assertions 3° and 5° are analogs of the smoothness theorem 4° from Subsection 1 for elliptic pseudodifferential operators. Let us develop this analogy further.

6° L *is a Fredholm operator from* $\mathfrak{H}^{s+\gamma}$ *to* $\mathfrak{H}^s$ *with index zero.*

Indeed, L_0^{-1} can be used as a parametrix for L:

$$L_0^{-1} L = I + L_0^{-1} L_1 \quad \text{and} \quad L L_0^{-1} = I + L_1 L_0^{-1},$$

and here $L_0^{-1} L_1$ and $L_1 L_0^{-1}$ are compact operators in each $\mathfrak{H}^s$. The deformation $\{L_0 + tL_1\}$, $0 \le t \le 1$, shows that L and L_0 have the same index (zero).

4. Elliptic pseudodifferential operators on a closed curve. Let us return to pseudodifferential operators on a closed curve (see Subsection 5.4.4). In (5.75), $a_\gamma(t, m)$ is the principal symbol. It was proved in [85, 10] that if an elliptic pseudodifferential operator A of nonzero order has a real principal symbol, then one can perform a similarity transformation

$$A \mapsto B = \Psi^{-1} A \Psi \tag{5.83}$$

such that the resulting operator B will be a pseudodifferential operator whose complete symbol is independent of t:

$$Bu(t) = \sum_{-\infty}^{\infty} e^{imt} b(m) c_m(u) + Tu(t), \tag{5.84}$$

where T is an operator of order $-\infty$. Here Ψ is a so-called *Fourier integral operator.* A Fourier integral operator on the circle has the form

$$\Psi u(t) = \sum_{-\infty}^{+\infty} e^{i\psi(t,m)} h(t, m) c_m(u) + T_1 u(t), \tag{5.85}$$

where $h(t,\xi)$ is the complete symbol of an elliptic pseudodifferential operator of order zero and T_1 is an operator of order $-\infty$. This formula resembles (5.74), but the essential difference from pseudodifferential operators is that in the general case the product mt in the exponent is replaced by $\psi(t,m)$, where the *phase function* $\psi(t,\xi)$ has the following properties: it is real, positively homogeneous of order 1 in ξ, 2π-periodic in t, and infinitely smooth for $\xi \neq 0$. If the principal symbol of A is already independent of t, then the phase function of Ψ has the simplest form $\psi(t,\xi) = t\xi$, that is, Ψ is an elliptic pseudodifferential operator of order zero. In this case, the terms of the expansion of the symbol $h(t,m)$ of this pseudodifferential operator and the respective terms in the expansion of the symbol $b(m)$ of B are successively calculated with the help of the composition theorem for pseudodifferential operators (see Assertion 3° in Subsection 5.4.1) from the requirement that the terms in the expansion of $b(m)$ must be independent of t. (A nontrivial phase function $\psi(t,m)$ is needed only at the initial step to reduce the principal symbol of A to a symbol independent of t.) Unfortunately, here we cannot give further details (see [85] and [10]).

The operator $B - T$ has the orthonormal basis in $H^0(S)$ consisting of the eigenfunctions e^{imt} and hence is a normal operator. The corresponding eigenvalues are $b(m)$. Thus B is an infinitely weak perturbation of a normal operator. One can show that the asymptotic expansion

$$b(m) = b_\gamma(\operatorname{sgn} m)|m|^\gamma + \ldots + b_{\gamma-q}(\operatorname{sgn} m)|m|^{\gamma-q} + O(|m|^{\gamma-q-1}) \qquad (5.86)$$

of the symbol $b(m)$ of the operator B as $m \to \infty$ is at the same time the asymptotic expansion of the eigenvalues of B in powers of m^{-1}, where m is the number of the eigenvalue. Since similar operators have the same spectrum, the same asymptotic expansion holds for the eigenvalues of A. The existence of such an expansion is specific to elliptic pseudodifferential operators of nonzero order on a (one-dimensional) closed curve.

Remark The asymptotics (5.86) permits one to show that to ensure the summability of the Fourier series in the root functions, parentheses must enclose groups of at most two terms, starting from some number (see [10]; cf. Subsection 5.6.2 below). The corresponding A-invariant subspaces are one- or two-dimensional.

One can also obtain an expansion similar to (5.86) for the root functions:[8]

$$u_n(t) = e^{i\varphi(t,m)}[h_0(t,\operatorname{sgn} m) + h_{-1}(t,\operatorname{sgn} m)|m|^{-1} + \ldots], \qquad (5.87)$$

where the expression in square brackets is the asymptotic expansion of the symbol $h(t,m)$ occurring in (5.85).

These expansions resemble similar expansions of eigenvalues and eigen- or root functions of one-dimensional boundary value problems for differential equations with infinitely smooth coefficients on an interval (e.g., see [77]).

5. Elliptic boundary value problems. For elliptic equations, there is a general theory of elliptic boundary value problems with differential or pseudodifferential

[8]More precisely, for the eigenfunctions in the corresponding one-dimensional A-invariant subspaces and for functions forming bases in two-dimensional invariant subspaces.

boundary conditions. The results of this theory are similar to those of the theory of elliptic pseudodifferential operators on a closed surface (e.g., see [66, Chapter II], [49, Chapter X], [4], [83], [46], the survey [14], and the references given therein). Here we restrict ourselves to considering examples. Namely, we indicate the consequences of the general theory for several boundary value problems for the equation

$$Lu(x) = -\Delta u(x) + \varepsilon(x)u(x) = f(x) \tag{5.88}$$

in a bounded domain $V \subset \mathbf{R}^n$ with infinitely smooth $(n-1)$-dimensional boundary S. (In the subsequent exposition, somewhat more general equations will be encountered.) For simplicity, let $\varepsilon \in C^\infty(\overline{V})$. We assume that the boundary condition has the form

$$Bu(x) = \partial_\nu u(x) + Au(x) = g(x) \quad \text{on } S, \tag{5.89}$$

where ∂_ν is the outward normal derivative and A is a first-order pseudodifferential operator on S.

Let us first state the *ellipticity conditions* for this problem. The first of these conditions says that the operator L must be elliptic, that is, the principal symbol of L must be nonzero for $x \in \overline{V}$ and real $\xi \neq 0$. This condition is obviously satisfied, since the symbol is equal to $|\xi|^2$. Now let x_0 be an arbitrary point of the boundary S. For simplicity, we shift the origin to the point x_0 and rotate the coordinate axes so that the tangent plane to the boundary will have the equation $x_n = 0$ and the x_n-axis will be directed along the inward normal to the boundary. In a neighborhood of x_0, the surface S is then given by the equation $x_n = \Phi(x')$ with an infinitely smooth function Φ of the variables $x' = (x_1, \dots, x_{n-1})$, so that x' can be used as a system of local coordinates on S. Let (x', ξ') be the corresponding local coordinates in T^*S, and let $a(x', \xi')$ be the principal symbol of the operator A in these coordinates. (Here we slightly change the notation; earlier $a(x, \xi)$ was used to denote the complete symbol.) Consider the following problem on the half-line $t = x_n > 0$:

$$v''(t) - |\xi'|^2 v(t) = 0 \qquad \text{for } t > 0, \tag{5.90a}$$
$$-v'(t) + a(0, \xi')v(t) = h \qquad \text{for } t = 0. \tag{5.90b}$$

The second condition of ellipticity of the problem, known as the *Shapiro–Lopatinskii condition*, is as follows. For any real $\xi' \neq 0$ and any number h, the auxiliary problem (5.90) must have a unique solution with $|v(t)| \to 0$ as $t \to +\infty$. This condition must be satisfied at each point of the boundary S.

Let us explain how Eqs. (5.90) were compiled. The term with $\varepsilon(x)$ in (5.88) is treated as a lower-order term and is not taken into account in the ellipticity conditions. The Laplace operator Δ' with respect to the "tangent" variables x' has the symbol $-|\xi'|^2$. The sign "minus" before v' in (5.90b) is used, since the normal ν is directed to negative values of t. The principal symbol of A is taken at the point $x_0' = 0$.

The general form of a solution of Eq. (5.90a) decaying as $t \to +\infty$ is $v(t) = Ce^{-|\xi'|t}$. Let us substitute this expression into (5.90b):

$$(|\xi'| + a(0, \xi'))C = h.$$

We see that the second ellipticity condition at the given boundary point is the inequality

$$|\xi'| + a(0,\xi') \neq 0 \quad \text{for } \xi' \neq 0. \tag{5.91}$$

Obviously, this inequality holds for the boundary condition

$$\partial_\nu u(x) + b(x)u(x) = g(x) \quad \text{on } S, \tag{5.92}$$

where $b(x)$ is an infinitely smooth function on S, since in this case the principal symbol of A is zero. In particular, the ellipticity condition holds for the Neumann problem ($b \equiv 0$). Problems with nontrivial pseudodifferential operators in the boundary conditions will be encountered in Sections 5.7–5.9. Elliptic boundary value problems that do not contain pseudodifferential operators will be referred to as *classical* ones.

If the second ellipticity condition is satisfied everywhere on S and $\varepsilon(x) \in C^\infty(\overline{V})$ (say, $\varepsilon(x) \equiv -k^2$), then the following assertions hold.

1° (An a priori estimate.)

$$\|u\|_{V,s+2} \leq C_s(\|f\|_{V,s} + \|g\|_{S,s+1/2} + \|u\|_{V,0}) \tag{5.93}$$

for every $s \geq 0$. If the uniqueness holds ($f = 0$ and $g = 0$ implies $u = 0$), then the term $\|u\|_{V,0}$ can be omitted.

Note that if $u \in H^{s+2}(V)$, then $f \in H^s(V)$ and $g \in H^{s+1/2}(S)$ by virtue of Assertions 5°, 6°, and 9° of Subsection 5.3.2. Likewise, $u \in H^{s+3/2}(S)$ on S, $\partial_\nu u \in H^{s+1/2}(S)$, and the terms $\|u\|_{S,s+3/2} + \|\partial_\nu u\|_{S,s+1/2}$ can be included in the left-hand side, since they can be estimated by $\|u\|_{V,s+2}$ (see Assertion 9° in Section 5.3).

2° *The subspace of solutions of the homogeneous problem (with $f = 0$ and $g = 0$) is finite-dimensional.*

3° *For given $f \in H^s(V)$ and $g \in H^{s+1/2}(S)$, the problem has a solution in $H^{s+2}(V)$ if and only if f and g satisfy finitely many conditions of orthogonality to some given functions in V and on S.* These conditions may be absent; then the problem is solvable for any $f \in H^s(V)$ and $g \in H^{s+1/2}(S)$.

It follows from 3° that the range of the operator

$$(L, B)\colon H^{s+2}(V) \to H^s(V) \times H^{s+1/2}(S), \qquad s \geq 0, \tag{5.94}$$

corresponding to problem (5.88), (5.89), is closed.

Thus, *ellipticity implies the Fredholm property of the operator* (5.94). *The converse is also true* (ellipticity can be derived from the a priori estimate).

4° (The smoothness theorem.) *If $u \in H^{s+2}(V)$, $f \in H^{s+t}(V)$, and $g \in H^{s+t+1/2}(S)$, where $t > 0$, then $u \in H^{s+t+2}(V)$. In particular, if $f \in C^\infty(\overline{V})$ and $g \in C^\infty(S)$, then $u \in C^\infty(\overline{V})$.*

Hence the space of solutions of the homogeneous problem (5.88), (5.89) (with $f = 0$ and $g = 0$) consists of infinitely smooth functions.

The difference $l - l'$, where l is the dimension of the space of solutions of the homogeneous problem and l' is the number of linearly independent solvability conditions for the nonhomogeneous problem, is called the *index* of the problem. One can relate l' to the dimension of the space of solutions of the "formally adjoint" homogeneous problem (e.g., see [66, Chapter II]), but we do not dwell on this. Note that the index does not change under homotopies (continuous transformations of the problem) provided that the ellipticity conditions are preserved in the course of these homotopies. In particular, we can arbitrarily vary lower-order terms in the elliptic equation and the boundary conditions. The index of the Neumann problem for the Laplace (or Helmholtz) equation is zero. We shall see that the index is zero in all cases that are of interest to us.

To the elliptic problem (5.88), (5.89) with homogeneous boundary condition ($g = 0$) we can assign an unbounded operator L_B in $H_0(V)$. This operator is defined by the formula $L_B u = Lu$ on the domain $\mathfrak{D}(L_B) = \{u \in H^2(V) : Bu = 0\}$. The equation $L_B u = \mu u$ corresponds to the spectral problem

$$Lu = \mu u \quad \text{in } V, \qquad Bu = 0 \quad \text{on } S. \tag{5.95}$$

The spectrum of the operator L_B is discrete provided that the resolvent set is nonempty. The index in this case is zero. The root functions of L_B belong to $C^\infty(\overline{V})$.

If this operator is selfadjoint, that is, the function $\varepsilon(x)$ is real and

$$(\Delta u, v)_V = (u, \Delta v)_V \quad \text{for} \quad u, v \in \mathfrak{D}(L_B),$$

and if it is bounded below, then the following asymptotic formula, similar to (5.79), holds for the distribution function of the eigenvalues:

$$N(t) = Ct^{n/2} + O(t^{(n-1)/2}), \qquad t \to \infty, \tag{5.96}$$

where $C = C(n)$ is independent of the boundary condition and is calculated as $C(n)$ in (5.79) in terms of the principal symbol of the elliptic operator in V; in our case this constant is equal to $(4\pi/3)|V|$ for $n = 3$ and $\pi|V|$ for $n = 2$ (here $|V|$ is the volume of the domain V for $n = 3$ and the area of V for $n = 2$); see [90], [52] and the surveys [14], [86]. Here the eigenvalues are numbered and the function $N(t)$ is defined in the same way as in the end of Subsection 2. The remainder estimate in (5.96) cannot be sharpened in general. Formula (5.96) is equivalent to the formula

$$\nu_j = C' j^{2/n} + O(j^{1/n}), \qquad j \to \infty, \tag{5.97}$$

for the eigenvalues, where $C' = C'(n) = C^{-2/n}$.

Let us indicate modifications of some of the above assertions. Before the operator $\Delta u(x)$ in (5.88), we can insert a coefficient $c(x) \in C^\infty(\overline{V})$. Then the principal symbol of the operator (5.88) will be equal to $c(x)|\xi|^2$, and the ellipticity condition will read

$$c(x) \neq 0 \quad \text{in } \overline{V}^+.$$

Equation (5.90a) preserves its form (after canceling the factor $c(0)$). To preserve the selfadjointness in the case of real $c(x)$, it is convenient to introduce the factor c^{-1} in the inner product, that is, define it as $\int_{V+} c^{-1} u\overline{v}\, dx$.

If the coefficients are only finitely smooth, then the range of admissible s is bounded above. For example, if $\varepsilon(x) \in C^{(0)}(\overline{V})$ and in the boundary condition (5.92) we have $b(x) \in C^{(2)}(S)$, then one can only take $s = 0$, and the smoothness theorem fails.

Condition (5.89) or (5.92) can be replaced by the Dirichlet condition

$$u(x) = g(x) \quad \text{on } S. \tag{5.98}$$

It also satisfies the ellipticity condition (in this case, the boundary condition (5.90b) is replaced by $v(0) = h$). Furthermore, $\|g\|_{S,s+1/2}$ in 1° is replaced by $\|g\|_{S,s+3/2}$; the statements of other assertions are modified in a similar way. The index of the Dirichlet problem for the Laplace or Helmholtz equation is zero. The operator corresponding to the problem with homogeneous boundary condition is selfadjoint and bounded below and the asymptotic formulas (5.96) and (5.97) are valid.

The boundary S may consist of more than one component, and the boundary conditions may be different on different components (say, we have condition (5.89) on one of the components and the Dirichlet condition on the other). Analogs of Assertions 1°–4° for this case can readily be stated.

We can also consider the *transmission problem*, in which the domain V is divided into, say, two parts V_+ and V_-, the interior and exterior parts, by a closed infinitely smooth surface S_1 that does not have common points with S. The coefficient $\varepsilon(x)$ may have a jump on S_1. Suppose that the conditions

$$\begin{aligned} &\alpha_1(x)u^+(x) + \alpha_2(x)u^-(x) = g_1(x), \\ &\beta_1(x)\partial_\nu u^+(x) + \beta_2(x)\partial_\nu u^-(x) + \beta_3(x)u^+(x) + \beta_4(x)u^-(x) = g_2(x) \end{aligned} \tag{5.99}$$

are posed on S_1, where, to be definite, we assume that the normal is directed into V_-. The second ellipticity condition, of Shapiro–Lopatinskii type, is imposed at all points of S and S_1. Let us state this condition at a point $x_0 \in S_1$. We shift the origin to x_0 and direct the x_n-axis along the normal. Then we must consider the following problem on the line:

$$\begin{aligned} &v_+''(t) - |\xi'|^2 v_+(t) = 0 \quad \text{for } t < 0, \\ &v_-''(t) - |\xi'|^2 v_-(t) = 0 \quad \text{for } t > 0, \\ &\alpha_1(0)v_+(0) + \alpha_2(0)v_-(0) = h_1, \\ &\beta_1(0)v_+'(0) + \beta_2(0)v_-'(0) = h_2. \end{aligned} \tag{5.100}$$

The second ellipticity condition at a given point says that for $\xi' \neq 0$ and for any h_1 and h_2 this problem must be uniquely solvable in the class of functions decaying at infinity. It is easy to calculate that this condition is equivalent to the inequality

$$\alpha_1(0)\beta_2(0) + \alpha_2(0)\beta_1(0) \neq 0. \tag{5.101}$$

For such elliptic problems with transmission conditions on S_1 and boundary conditions on S, analogs of Assertions 1°–4°, following from more general theorems, are valid (e.g., see [87] and [84]). We can readily state these analogs using the above propositions as a model. We shall do this in Sections 5.7–5.9 for problems considered there. Now we only note that it is convenient to assume that the solution belongs to H^2 on both sides of S_1.

As to the index, we shall only make the following remark. Suppose that we have transferred the nonhomogeneity from the transmission conditions to the elliptic equation by subtracting some function from the solution (thus, now we have $g_1 = g_2 = 0$). Next, suppose that we can continuously deform the original transmission conditions, with the preservation of the ellipticity conditions (5.101), to the conditions

$$u^+ = u^-, \quad \partial_\nu u^+ = \partial_\nu u^- \quad \text{on} \quad S_1. \tag{5.102}$$

These last conditions can be "erased," since they are equivalent to the requirement that the solution of Eq. (5.88) belongs to $H^2(V)$ provided that this solution belongs to H^2 on both sides of S_1 (See Assertion 8° in Subsection 5.3.2). After that, we can consider the problem in V ignoring S_1, and the index of the original problem coincides with the index of the resulting problem.

One can also consider problems for elliptic systems, say, for the vector Helmholtz equation. One problem of this kind is mentioned in Subsection 5.11.4. In Subsection 5.10.4 we touch the Dirichlet problem for a fourth-order elliptic equation and for a system of elliptic equations.

6. Ellipticity with parameter. There are versions of theorems given in Subsections 1 and 5 for elliptic equations and boundary value problems containing terms with an additional parameter (see [2], [20], [89], [46], and the surveys [12, 14]). These theorems deal with the following situation. The parameter varies in some sector of the complex plane with vertex at the origin. This parameter is included in the principal symbol. If instead of the usual ellipticity conditions some strengthened conditions of *ellipticity with parameter* are satisfied, then for parameter values lying in this sector and sufficiently large in modulus the equation or the boundary value problem in question is *uniquely solvable* with an a priori estimate whose left-hand side also contains the modulus of the parameter. Here we consider an example of elliptic boundary value problem with parameter; some modifications will be treated later where necessary.

Consider the equation

$$\Delta u(x) + \varepsilon(x) - \mu\sigma(x)u(x) = f(x) \tag{5.103}$$

in a bounded domain $V \subset \mathbf{R}^n$ with the homogeneous (for simplicity) boundary condition (5.89) ($g = 0$). We assume that the functions $\varepsilon(x)$ and $\sigma(x)$ are sufficiently smooth. First, suppose that $\sigma(x)$ is positive. The principal symbol of the operator on the left-hand side in (5.103) now includes the parameter μ and is equal to

$$-|\xi|^2 - \mu\sigma(x). \tag{5.104}$$

Thus, the parameter μ is considered as having the "weight" 2 with respect to differentiation. The expression (5.104) is nonzero for $\mu \notin \mathbf{R}_-$. Let us choose a closed sector Ψ

with vertex the origin on the complex plane such that Ψ does not contain points of $\mathbf{R}_-$ except the origin. The first condition of ellipticity with parameter in Ψ, saying that this principal symbol is not zero for $x \in \overline{V}$ and $|\xi| + |\mu| \neq 0$, where ξ is real and $\mu \in \Psi$, is satisfied. Let us state the second condition of ellipticity with parameter in this sector, which pertains to all boundary points. We choose an arbitrary point $x_0 \in S$, pass to the same coordinate system as at the beginning of Subsection 4, and consider the following problem on the half-line $t = x_n > 0$:

$$v''(t) - (|\xi'|^2 + \mu\sigma(0))v(t) = 0 \qquad \text{for } t > 0, \tag{5.105a}$$

$$-v'(t) + a(0, \xi')v(t) = h \qquad \text{for } t = 0. \tag{5.105b}$$

The condition of ellipticity with parameter $\mu \in \Psi$ at the given point of the boundary S says that this problem must have a unique solution decaying as $t \to \infty$ for each complex number h and for $|\xi| + |\mu| \neq 0$, where ξ' is real and $\mu \in \Psi$. The main result is as follows.

1° *If the problem is elliptic with parameter in Ψ, then for $\mu \in \Psi$ with sufficiently large modulus, for any $f \in H^0(V)$ there exists a unique solution in $H^2(V)$, and the a priori estimate*

$$\|u\|_{V,2} + |\mu|\|u\|_{V,0} \leq C_\mu \|f\|_{V,0} \tag{5.106}$$

is valid.

Here we use the same spaces as in Subsection 4. For simplicity, we restrict ourselves to the case $s = 0$ and $g = 0$.

The second condition of ellipticity with parameter in Ψ at a given point of the boundary can be rewritten in the form of the following inequality, similar to (5.91):

$$(|\xi'|^2 + \mu\sigma(0))^{1/2} + a(0, \xi') \neq 0, \tag{5.107}$$

where the square root is chosen to have a positive real part. For example, this condition is satisfied if the principal symbol a of the pseudodifferential operator A is positive.

If $\sigma(x)$ is a complex-valued function and $|\arg \sigma(x)| \leq \theta (< \pi)$, then to ensure that the expression (5.104) is not zero for $\mu \in \Psi$, we take

$$\Psi = \{\mu : |\arg \mu| \leq \pi - \theta - \varepsilon\}$$

with $\varepsilon > 0$.

Obviously, the ellipticity conditions are contained in the conditions of ellipticity with parameter and can be obtained from the latter by setting $\mu = 0$.

Note that if L_B is a selfadjoint operator, then the ellipticity with parameter along the ray $\mathbf{R}_+$ implies that the operator is bounded above.

There are analogs of Assertion 1° for pseudodifferential equations with parameter on closed surfaces (see [89], [46], and the survey [12]). Let A be a pseudodifferential operator of positive order γ on a closed surface $\mathfrak{S}$ with principal symbol $a_\gamma(x, \xi)$. Then for the left-hand side of the equation

$$(A - \mu I)u = f, \tag{5.108}$$

the principal symbol with parameter is defined as

$$a_\gamma(x,\xi) - \mu, \tag{5.109}$$

whence the parameter μ is considered as having the weight γ. The condition of *ellipticity with parameter* in the sector Ψ says that the expression (5.109) must not be zero on nonzero cotangent vectors for $\mu \in \Psi$. (It is automatically nonzero also on zero cotangent vectors $(x, 0)$ for $0 \neq \mu \in \Lambda$.) Let us state the corresponding result.

2° *If Eq.* (5.108) *is elliptic with parameter in* Ψ*, then for any given real* s *and for sufficiently large (in modulus)* $\mu \in \Psi$*, this equation with an arbitrary right-hand side in* $H^s(\mathfrak{S})$ *has a unique solution* $u \in H^{s+\gamma}(\mathfrak{S})$*, and the solution satisfies the a priori estimate*

$$\|u\|_{V,s+\gamma} + |\mu|\|u\|_{V,s} \leq C^s\|f\|_{V,s}. \tag{5.110}$$

5.6 Tests for Completeness and Summability and the Spectral Asymptotics for Nonselfadjoint Operators

Let $\mathfrak{H}$ be a separable Hilbert space. Denote by L an operator with discrete spectrum, by A a compact operator in $\mathfrak{H}$, and by $\{f_j\}$ a system of root vectors of L or A (see Subsections 5.2.3 and 5.2.4). We shall use the following notation (for definitions, see Subsections 5.2.2 and 5.2.5):

L (or A)$\in \mathbf{A}(\alpha, \mathfrak{H})$ if $\{f_j\}$ is an Abel–Lidskii system of order α in $\mathfrak{H}$ $(\alpha > 0)$;
L (or A)$\in \mathbf{R}(\mathfrak{H})$ if $\{f_j\}$ is a Riesz basis with parentheses in $\mathfrak{H}$;
L (or A)$\in \mathbf{B}(\mathfrak{H})$ if $\{f_j\}$ is a Bari basis with parentheses in $\mathfrak{H}$.
In this sequence, each property is stronger than the previous one.

1. Operators close to selfadjoint ones. Consider the operator $L = L_0 + L_1$ in $\mathfrak{H}$, where L_0 is a selfadjoint operator with discrete spectrum and with positive (for simplicity) eigenvalues ν_j and L_1 is an operator subordinate to some power L_0^q of L_0. More precisely, we impose the following two conditions on L:

(i) $\nu_j(L_0) \geq Cj^p$ for some $C > 0$ and $p > 0$;

(ii) $\|L_1 L_0^{-q}\| = b < \infty$ for some $q < 1$.

Recall that the eigenvalues $\nu_j = \nu_j(L_0)$ are numbered in a nondecreasing order taking their multiplicities into account. By $\|\cdot\|$ we denote the operator norm in $\mathfrak{H}$.

First, let us indicate a consequence of assumption (ii) relative to the localization of eigenvalues of L.

1° *All eigenvalues of* L *lie in the union of disks* $\{\lambda : |\lambda - \nu_j| \leq b\nu_j^q\}$.

This result is easy to obtain. We have

$$L - \lambda I = (I + L_1 R_{L_0}(\lambda))(L_0 - \lambda I).$$

This operator has the inverse

$$R_L(\lambda) = R_{L_0}(\lambda)(I + L_1 R_{L_0}(\lambda))^{-1} \tag{5.111}$$

if $\lambda \notin \Sigma(L_0)$ and $\|L_1 R_{L_0}(\lambda)\| < 1$. By virtue of assumption (ii), the latter inequality holds if

$$\|L_0^q R_{L_0}(\lambda)\| < b^{-1}.$$

The operator $L_0^q R_{L_0}(\lambda)$ is normal, since it has an orthonormal basis consisting of eigenvectors (the same as for L_0). It follows that the norm of this operator is equal to the lowest upper bound of the moduli of its eigenvalues:

$$\|L_0^q R_{L_0}(\lambda)\| = \sup |\nu_j^q(\nu_j - \lambda)^{-1}| \leq b^{-1}(1+\varepsilon)^{-1} \tag{5.112}$$

if $|\lambda - \nu_j| \geq (1+\varepsilon)b\nu_j^q$ for all j; here $\varepsilon > 0$. Furthermore, for these λ the resolvent (5.111) exists and satisfies the estimate

$$\|R_L(\lambda)\| \leq C_\varepsilon \|R_{L_0}(\lambda)\|, \tag{5.113}$$

where $C_\varepsilon = (1 - (1+\varepsilon)^{-1})^{-1} = (1+\varepsilon)\varepsilon^{-1}$.

One can readily derive the following assertion from 1°.

2° *L is an operator with discrete spectrum, and all its eigenvalues μ_k are contained in the union Ω_L of some neighborhood of the origin with a domain of the form $\{\lambda = \sigma + i\tau : \sigma > 0,\ |\tau| < C\sigma^q\}$.*

Here the discreteness of the spectrum, that is, the compactness of the resolvent $R_L(\lambda)$, obviously follows from Eq. (5.111), which shows that $R_L(\lambda)$ is the product of the compact operator $R_{L_0}(\lambda)$ by a bounded operator.

For $q > 0$, the domain Ω_L is often said to be *parabolic*, since the curves $\tau = \pm C\sigma^q$ resemble parabolas. For $q = 0$ this domain is a half-strip for large σ, whereas for $q < 0$ it can be called *hyperbolic*, since the curves $\tau = \pm C\sigma^q$ resemble hyperbolas with the asymptote $\mathbf{R}_+$, where $\mathbf{R}_+ = \{\lambda = \sigma : \sigma \geq 0\}$ is the positive real axis.

So far, we have not used condition (i).

3° *Under conditions* (i) *and* (ii), *the system of root vectors of the operator L is complete in $\mathfrak{H}$.*

This result goes back to [57] (see also [43, Chapter 5]). Let us state a more general test for completeness (see [39, Chapter 11]), of which 3° is a consequence. For this purpose, we need the notion of s-numbers of a compact operator T. The operator T^*T is selfadjoint and *nonnegative*:

$$(T^*Tf, g) = (Tf, Tg) = (f, T^*Tg) \quad \text{and} \quad (T^*Tf, f) = (Tf, Tf) \geq 0$$

for any $f, g \in \mathfrak{H}$, which is equivalent to the fact that all eigenvalues of T^*T are nonnegative. Hence we can define the square root $(T^*T)^{1/2}$; this is also a compact selfadjoint nonnegative operator. Let $s_j(T)$, $j = 1, 2, \ldots$, be the sequence of its nonzero eigenvalues (numbered in a nondecreasing order taking their multiplicities into account). These numbers are called the *s-numbers* of T. If T is selfadjoint or at least normal, then the s-numbers are just the moduli of nonzero eigenvalues of T.

One can show that

$$s_j(T^*) = s_j(T), \quad s_{j+k-1}(T_1 + T_2) \leq s_j(T_1) + s_k(T_2), \tag{5.114a}$$

$$s_j(BT) \leq \|B\| s_j(T), \quad \text{and} \quad s_j(TB) \leq \|B\| s_j(T) \tag{5.114b}$$

if T_1 and T_2 are compact and B is a bounded operator (e.g., see [43, Chapter 2]).

By definition, the *class* $S^{(p)}(\mathfrak{H})$ of compact operators T in $\mathfrak{H}$ consists of operators T such that $s_j(T) \leq C(T) j^{-p}$, where $C(T)$ is a positive constant depending on T.[9] By using the obvious *Hilbert identity*

$$R_L(\lambda) - R_L(\tilde{\lambda}) = (\lambda - \tilde{\lambda}) R_L(\lambda) R_L(\tilde{\lambda}) \tag{5.115}$$

in conjunction with the inequalities in (5.114), one can readily verify that if the resolvent $R_L(\lambda)$ of an operator L with discrete spectrum belongs to $S^{(p)}(\mathfrak{H})$ for some $\lambda \notin \Sigma(L)$, then it belongs to $S^{(p)}(\mathfrak{H})$ for all $\lambda \notin \Sigma(L)$. Under our condition (i), we obviously have $R_{L_0}(\lambda) \in S^{(p)}(\mathfrak{H})$; if both conditions (i) and (ii) are satisfied, then one also has $R_L(\lambda) \in S^{(p)}(\mathfrak{H})$ by virtue of (5.111) and the inequalities in (5.114).

Remark If T is an operator acting from H^0 into H^s in a bounded n-dimensional domain or on an n-dimensional surface (which is sufficiently regular), then T as an operator in H^0 belongs to $S^{(p)}$ with $p = s/n$. See [80] or [99].

The following test for completeness is valid.

Theorem 5.6.1 *Let Θ be an open sector with vertex at the origin and angle θ, $0 < \theta < 2\pi$, at the vertex, and let L be an operator with discrete spectrum whose resolvent belongs to $S^{(p)}(\mathfrak{H})$, where $\theta < \pi p$. Suppose that in the complementary sector $\Theta_1 = \mathbf{C} \setminus \Theta$ there are no points of the spectrum $\Sigma(L)$ sufficiently far from the origin and the estimate*

$$\|R_L(\lambda)\| = O(|\lambda|^N), \qquad \lambda \to \infty, \tag{5.116}$$

holds for some N.[10] *Then the system of root vectors of L is complete in $\mathfrak{H}$.*

Moreover, to ensure the completeness, it suffices to assume the existence of rays $\Gamma_1, \dots, \Gamma_{N'}$ issuing from the origin such that the angles between neighboring rays are less than πp, the rays do not contain points of the spectrum $\Sigma(L)$ sufficiently far form the origin, and the resolvent satisfies the estimate (5.116) along these rays.

The proof is based on some tools from the operator theory and the theory of functions of a complex variable, including a representation and some estimates for the resolvent on the complex plane and the Phragmen–Lindelöf theorem.

In our case, we can take Θ to be an arbitrarily small sectorial neighborhood $\{\lambda : |\arg \lambda| < \theta\}$ of $\mathbf{R}_+$ (that is, θ is arbitrarily small) and $N = -1$. The estimate $\|R_{L_0}(\lambda)\| = O(|\lambda|^{-1})$ holds in Θ_1 and can be transferred to $R_L(\lambda)$ by virtue of (5.111).

Now we shall state a deeper theorem about summability.

[9] In the literature, the class $S_\rho(\mathfrak{H})$ of compact operators T such that $\sum s_j^\rho(T) < \infty$ is used more often. Clearly, $S^{(p)}(\mathfrak{H}) \subset S_\rho(\mathfrak{H})$ for $\rho > 1/p$.

[10] The norm of the resolvent cannot decay faster than $O(|\lambda|^{-1})$, so that $N \geq -1$. See [3, Section 12].

Theorem 5.6.2 *Under conditions* (i) *and* (ii), *one has*

$$L \in \mathbf{A}(\alpha, \mathfrak{H}), \quad \textit{if } p(1-q) < 1,\ \alpha > p^{-1} - (1-q), \tag{5.117a}$$

$$L \in \mathbf{R}(\mathfrak{H}), \quad \textit{if } p(1-q) = 1, \tag{5.117b}$$

$$L \in \mathbf{B}(\mathfrak{H}), \quad \textit{if } p(1-q) > 1. \tag{5.117c}$$

The first of these results was obtained in [56], the second in [55, 56] (a constructive proof was proposed in [71]; see also [70, Chapter 1]), and the third in [69] (see also [104]). More precisely, the first two results were obtained for $q \geq 0$ in the papers just indicated. The values $q \in (-\infty, 1)$ are considered in [69], [104] for $p(1-q) > 1$ and in [8, 12]. Concerning some strengthening of the first result, see [91, 92].

Let us give some explanations. In the proof of the theorem, we can assume that the spectra of the operators L and L_0 lie in the sector $\Theta_\varepsilon = \{\lambda : |\arg\lambda| < \varepsilon\}$ for a given (arbitrarily small) $\varepsilon > 0$; this can always be achieved by shifting the spectral parameter, that is, by replacing $\lambda - \lambda_0$ by λ and $L_0 - \lambda_0 I$ by L_0 with an appropriate λ_0. The sector Θ_ε is divided into a triangle $\Theta^{(0)}$ and trapezoids $\Theta^{(1)}, \Theta^{(2)}, \ldots$ by vertical segments with abscissae α_j, $j = 0, 1, \ldots$, forming an increasing sequence of positive numbers. These segments must not contain the eigenvalues of the operators L_0 and L_1, and, which is important, certain optimal estimates for the norms of the resolvents $R_L(\lambda)$ and $R_{L_0}(\lambda)$ must be satisfied on these segments. (We do not dwell on these estimates here.) A sequence $\{\alpha_j\}$ with these properties is said to be *admissible*. The existence of an admissible sequence for $p(1-q) < 1$ can be proved relatively easily by virtue of the fact that in this case the disks $\{\lambda : |\lambda - \nu_j| < (1+\varepsilon)b\nu_j^q\}$ cannot cover $\mathbf{R}_+$ completely, and, moreover, there must be infinitely many gaps between neighboring disks. For $p(1-q) \leq 1$, there may be no gaps at all, and the proof of the existence of admissible α_j is based on deep results of the theory of functions of a complex variable and the operator theory.

Let Γ_l be the contour of the domain $\Theta^{(l)}$, going around counterclockwise. The projections P_l in (5.17) for cases (5.117b) and (5.117c) are defined by the formula

$$P_l = -\frac{1}{2\pi i}\int_{\Gamma_l} R_L(\lambda)\,d\lambda. \tag{5.118}$$

Likewise,

$$Q_l = -\frac{1}{2\pi i}\int_{\Gamma_l} R_{L_0}(\lambda)\,d\lambda. \tag{5.119}$$

These are the so-called *Riesz projections* corresponding to the operators L and L_0. The operators $P_l(t)$ in (5.21) for case (5.117a) are defined by

$$P_l(t) = -\frac{1}{2\pi i}\int_{\Gamma_l} \exp(-\lambda^\alpha t) R_L(\lambda)\,d\lambda. \tag{5.120}$$

Note that the *Hilbert identity* for the resolvents (cf. (5.115))

$$R_L(\lambda) - R_{L_0}(\lambda) = -R_L(\lambda) L_1 R_{L_0}(\lambda) \tag{5.121}$$

permits one to represent the difference $P_l - Q_l$ of projections in the form

$$P_l - Q_l = \frac{1}{2\pi i} \int_{\Gamma_l} R_L(\lambda) B L_0^q R_{L_0}(\lambda)\, d\lambda, \tag{5.122}$$

where B is a bounded operator, and when studying the series $\sum P_l f$, one can use the "good" properties of the series $\sum Q_l f$. Likewise, when studying the series $\sum P_l(t) f$, one can use the "good" properties of the series $\sum Q_l(t) f$, where the operator $Q_l(t)$ is defined by a formula of the form (5.120) with $R_{L_0}(\lambda)$ instead of $R_L(\lambda)$.

Analogs of assertions (5.117a)–(5.117c) can also be obtained when one deals with the scale $\{\mathfrak{H}^s\}$ (see Subsection 5.5.3) for all s simultaneously. Moreover, analogs of the same assertions hold for the case in which the operator L_0 is not semibounded and its eigenvalues tend to $\pm\infty$. In this case, all but finitely many eigenvalues of the operator L lie in two vertical sectors bisected by $\mathbf{R}$, and it is convenient to consider two-sided infinite series of the form $\sum_{-\infty}^{\infty} P_l f$. Finally, analogs of these assertions can be obtained for weak perturbations of selfadjoint compact operators. We shall shortly discuss these analogs in the next subsection.

2. Corollaries and strengthening of results for elliptic operators close to selfadjoint ones. Let A_0 be an elliptic operator of order $\gamma > 0$ selfadjoint in $H^0(S)$, and let A_1 be a pseudodifferential operator of order δ, $-\infty \leq \delta < \gamma$. Then conditions (i) and (ii) of the preceding subsection are satisfied with

$$p = \gamma/n \quad \text{and} \quad q = \delta/\gamma, \tag{5.123}$$

and hence the corresponding results can be applied. However, these assumptions and results were stated in a form that does not reflect the specific features of elliptic pseudodifferential operators, in particular, those which we shall be concerned with. First, for pseudodifferential operators, instead of (i), we have an asymptotic formula with an estimate for the remainder (see (5.80)). We shall explain that this permits one to have some control of the frequency of inserting the parentheses into the Fourier series with respect to the root functions. Second, it is now natural to assume that $-\infty < q < 1$ and even admit that q may be $-\infty$; we shall have to deal with such operators. For negative q, one can improve the result (5.117c) by estimating the rate of convergence of the difference $\sum_1^{\infty}(P_l - Q_l) f$ of Fourier series with parentheses with respect to the root functions of L and the eigenfunctions of L_0. Third, by considering the operators L and L_0 in a Sobolev scale (or in an abstract scale of Hilbert spaces), we can take into account the smoothness of the function to be expanded: for a smooth function, the Fourier series with parentheses with respect to the root functions of L converges rapidly. These questions will also be discussed.

Note, however, that in applications one has to distinguish between two cases. The first case is that of elliptic pseudodifferential operators on a closed surface. Such operators arise from reduction to the boundary of elliptic problems with a spectral parameter in boundary or transmission conditions. (See Sections 5.1, 5.7, 5.8, and 5.11.) The second case is that of elliptic boundary value problems with homogeneous boundary conditions and with a spectral parameter in the elliptic equation itself (see Section 5.9).

In both cases, knowing the asymptotics of the eigenvalues with the remainder permits one to have some control of how frequently the parentheses must be inserted into the series. However, negative values of q naturally arise only in the first case. In the second case, if we assume that the problem is classical, q is necessarily nonnegative, since the perturbation of a differential operator by a differential operator (of lower order) has a nonnegative order. Moreover, the scale generated by an elliptic pseudodifferential operator on a closed surface is natural; it coincides with the Sobolev scale on the surface. In the second case, the high-order spaces in the scale generated by a differential operator with homogeneous boundary conditions consist of functions that satisfy a large number of artificial boundary conditions. This scale is henceforth useless.

We intend to subject the sequence $\{\alpha_l\}$ to the inequalities

$$l^{\rho\beta} < \alpha_l < (l+1)^{\rho\beta}, \qquad l \geq l_0, \tag{5.124}$$

and the problem is to find out which values of $\rho > 0$ can be taken. With this approach, we cannot exclude the case in which the annulus $\{\mu : \alpha_l < |\mu| < \alpha_{l+1}\}$ for some l does not contain any eigenvalues of L_0 and/or L; then the corresponding operator P_l or Q_l is zero. (In the notation of Subsection 5.2.1, we now must assume that $\{m_l\}$ is a nondecreasing sequence.) In any case, by indicating the possible values of ρ, we estimate the admissible intervals between the parentheses in the series $\sum c_j f_j$ and make it possible to estimate the rate of convergence of the resulting series with parentheses.

Let us again consider the operator $L = L_0 + L_1$ (see the beginning of Subsection 1). However, now we replace (i) by the condition

(i′) $\nu_j(L_0) = cj^p + O(j^r)$, where $c > 0$ and $p - 1 \leq r < p$.

The main part of Theorem 5.6.2 can be proved for any $q < 1$ with the additional assumption (5.124) for α_l with sufficiently large ρ. More precisely, assume for simplicity that

$$\rho \geq (p-r)^{-1}. \tag{5.125}$$

Then one can verify the following:

1. Assertion (5.117a) with $\alpha > p^{-1} - (1-q)$ is valid for every $q < 1$ under condition (5.124) provided that condition (5.125) is satisfied and $\rho > [p(1-q)]^{-1}$.

2. Assertion (5.117b) is valid for every $q < 1$ under condition (5.124) provided that condition (5.125) is satisfied, $\rho > 1$, and $\rho p > 1$. In the case $q < 0$ we must impose the additional condition $\|L_0^{-q/2} L_1 L_0^{-q/2}\| < \infty$.

3. Assertion (5.117c) is valid for every $q < 1$ under condition (5.124) provided that condition (5.125) is satisfied and either $2\rho[p(1-q)-1] > 1$ for $-p^{-1} < q < 1$ or $2\rho p > 1$ for $q \leq -p^{-1}$.

These results are formulated in [12]. In [9] one can find the proof of the following two close theorems concerning the rate of convergence of series with parentheses. For simplicity, we assume that the operator L_1 acts in the scale of spaces generated by L_0. Let the orders of L_0 and L_1 in this scale be γ and δ, respectively, and let $q = \delta/\gamma$. We subject the admissible sequence $\{\alpha_l\}$ to condition (5.124) and impose condition (5.125) on ρ.

Theorem 5.6.3 *Suppose that the above assumptions are satisfied, $p(1-q) > 1$, and*

$$\eta = \rho(p(1-q) - 1) > 0.$$

Then for each $q < 1$ there exists an admissible sequence $\{\alpha_l\}$ satisfying condition (5.124) such that

$$m^{\eta+\rho q t}\Big\| \sum_{l=m}^{m'} (P_l - Q_l) f \Big\|_s \leq C_3 \|f\|_{s+t}, \tag{5.126}$$

$$m^{\rho q t}\Big\| \sum_{l=m}^{m'} P_l f \Big\|_s \leq C_4 \|f\|_{s+t} \tag{5.127}$$

for $f \in \mathfrak{H}^{s+t}$, $t \geq 0$, and $m' \geq m$, where the constants C_3 and C_4 are independent of f, m, and m'.

Let us comment on the statement and indicate some corollaries.

In (5.126) we estimate the rate of *equiconvergence* of the series $\sum P_l f$ and $\sum Q_l f$, taking the additional smoothness of f into account. Even for $t = 0$, (5.126) implies the convergence of the series $\sum (P_l - Q_l)$ with respect to the operator norm in $\mathfrak{H}^s$. By setting $m' = \infty$, we obtain an estimate for the norm of the remainder in this series. By setting $m' = m$, we obtain an estimate for the norm of the general term of the series. In particular, the following assertions hold.

4° *For $\eta > 1/2$, condition (5.19) is satisfied in each $\mathfrak{H}^s$.*

5° *If $\inf \gamma = -\infty$ (that is, L_1 is an infinitely smoothing operator), then the norm of the operator $P_l - Q_l$ in $\mathfrak{H}^s$ decays faster than l^{-N} for any positive integer N.*

Inequality (5.126) ensures the convergence of the series $\sum P_l f$ in $\mathfrak{H}^s$ to f for any $f \in \mathfrak{H}^s$, since the series $\sum Q_l$ necessarily converges to f. Inequality (5.127) shows that for $f \in \mathfrak{H}^{s+t}$, the larger t, the better is the rate of convergence of the series $\sum P_l f$ in $\mathfrak{H}^s$. By setting $m' = \infty$ in (5.127), we obtain an estimate for the norm of the remainder in the series $\sum P_j f$; by setting $m' = m$, we obtain an estimate of the norm of a general term of this series. In particular, the following assertion holds.

6° *If $f \in \mathfrak{H}^\infty$ (that is, f is an "infinitely smooth" vector), then $\|P_l f\|_s$ decays faster than l^{-N} for every positive integer N.*

Theorem 5.6.4 *Let the assumptions of the preceding theorem be satisfied, but suppose that instead of the inequality $p(1-q) > 1$ we have only the equality $p(1-q) = 1$. Then for each $q < 1$ there exists an admissible sequence $\{\alpha_l\}$ satisfying condition (5.124) such that $\sum P_l f = f$ in $\mathfrak{H}^s$ for each $f \in \mathfrak{H}^s$. Furthermore, for $f \in \mathfrak{H}^{s+t}$, $t \geq 0$, and $m' \geq m$ inequality (5.127) holds.*

As was told already, all these results can directly be applied to elliptic pseudodifferential operators of positive order of the form $A = A_0 + A_1$ on a closed surface $\mathfrak{S}$, where A_0 is a selfadjoint positive pseudodifferential operator of positive order γ and A_1 is a pseudodifferential operator of lower order δ, $\mathfrak{H}^s = H^s(\mathfrak{S})$. Moreover, A may have several eigenvalues on $(-\infty, 0]$, since we can, say, make a shift of the spectral parameter.

Now let B be an elliptic pseudodifferential operator *of negative order* γ on a surface $\mathfrak{S}$ with positive principal symbol, and let $B = B_0 + B_1$, where B_0 is a selfadjoint elliptic pseudodifferential operator of order $\gamma_1 < 0$ and B_1 is a pseudodifferential operator of order $\delta_1 < \gamma_1$. If $\operatorname{Ker} B_0$ and $\operatorname{Ker} B$ contain only zero, then there exist pseudodifferential operators B_0^{-1} and B^{-1} (see Subsection 5.5.1), and

$$B^{-1} = B_0^{-1}(I + B_1 B_0^{-1})^{-1} = A_0 + A_1, \tag{5.128}$$

where $A_0 = B_0^{-1}$ and $A_1 = -B_0^{-1}(B_1 B_0^{-1})(I + B_1 B_0^{-1})^{-1}$. These are pseudodifferential operators of orders $\gamma = -\gamma_1$ and $\delta = -2\gamma_1 + \delta_1$, respectively, and we arrive at the preceding case, since A_0 is a selfadjoint operator with positive principal symbol. Condition (i′) is satisfied with $p = \gamma / \dim \mathfrak{S}$ and $r = (\gamma - 1)/\dim \mathfrak{S}$. Thus, *we can apply the results formulated above in this section.* Note that $\gamma - \delta = \gamma_1 - \delta_1$, that is, the difference between the orders of the operators A_0 and A_1 is the same as for B_0 and B_1. In the problems considered in Sections 5.7, 5.8, and 5.11, we have $\gamma_1 = -1$ and $\delta_1 = -3$ or $-\infty$.

Finally, *let* $\operatorname{Ker} B \neq \{0\}$. In this case we adopt the assumption given in Subsection 5.2.6 after 7°; in particular, $\mathfrak{L}_B(0)$ is assumed to be finite-dimensional. This permits one to modify the Jordan matrix of the operator B in $\mathfrak{L}_B(0)$ by replacing the eigenvalue 0 by an arbitrary nonzero number distinct from all the other eigenvalues. This is equivalent to adding some finite-dimensional operator of order $-\infty$ to B and B_1 (indeed, $\mathfrak{L}_B(0) \subset C^\infty(\mathfrak{S})$; see the smoothness theorem 4° in Subsection 5.5.1). This procedure does not affect the spectral properties of B which are of interest to us. If $\operatorname{Ker} B_0 \neq \{0\}$, then $\mathfrak{L}_{B_0}(0) = \operatorname{Ker} B_0$ is necessarily finite-dimensional (see Assertion 2° in Subsection 5.5.1) and also consists of infinitely smooth functions. In a similar way, we can modify B_0 in $\mathfrak{L}_{B_0}(0)$ by adding some finite-dimensional operator of order $-\infty$ and by subtracting it from B_1. Thus, *the results given so far in this section can be applied under the above assumption about* $\mathfrak{L}_B(0)$.

Next, our results for the spaces $H^s(\mathfrak{S})$ imply some assertions for the spaces $C^{(s)}(\mathfrak{S})$ via the embedding theorem (see Subsection 5.3.2, Assertions 10° and 11°). In particular, we find that if A_1 (or B_1) is an operator of order $-\infty$, then for each function $f \in C^\infty(\mathfrak{S})$ its Fourier series with parentheses with respect to the root functions of the operator A (or B) is uniformly convergent, and moreover, is majorized by the series $C_N \sum_l l^{-N}$ with arbitrarily large N, and the rate of convergence remains the same after local termwise differentiation of the series arbitrarily many times.

Remark In the case of a closed smooth curve, parentheses can be inserted in such a way that they separate groups of at most two terms. See the Remark at the end of Subsection 5.5.4.

3. Operators with rays of minimal growth. Here we shall give a theorem that applies to operators that are far from being selfadjoint, that is, operators that fail to have a selfadjoint leading part. We advise the reader to compare this theorem with Theorem 5.6.1.

Theorem 5.6.5 *Let Θ be a closed sector with vertex at the origin and angle θ, $0 < \theta < 2\pi$, at the vertex, and let L be an operator with discrete spectrum whose resolvent belongs to $S^{(p)}(\mathfrak{H})$, where $\theta < \pi p$. Suppose that the complementary sector $\Theta_1 = \mathbf{C} \setminus \Theta$ does not contain points of the spectrum $\Sigma(L)$ sufficiently far from the origin and the estimate*

$$\|R_L(\lambda)\| = O(|\lambda|^{-1}), \qquad \lambda \to \infty, \tag{5.129}$$

holds in Θ_1. Under these assumptions, $L \in \mathbf{A}(\alpha, H)$ if $\alpha \in (p^{-1}, \pi\theta^{-1})$.

The rays $\{\lambda : \arg\lambda = \varphi\}$ with such an estimate are called the *rays of minimal growth of the resolvent* [2]. (It would be more adequate to call them rays of maximal decay, since $\|R_L(\lambda)\|$ cannot decay faster than $|\lambda|^{-1}$.) This theorem can be applied to problems for which we can establish ellipticity with parameter (see Subsection 5.5.6) in the sector Θ_1.

Let us outline the main points of the proof. By shifting the spectral parameter if necessary and by multiplying it by an appropriate number, we can ensure that the spectrum $\Sigma(L)$ lies in the sector $\Lambda_\theta = \{\lambda : |\arg\lambda| < \theta/2\}$ and outside the disk $\{\lambda : |\lambda| < 1\}$ and that in the union of this disk with the complement of Λ_θ the following inequality holds:

$$\|R_L(\lambda)\| \le C(1 + |\lambda|)^{-1}. \tag{5.130}$$

Let Γ be the boundary of the sector Λ_θ. Consider the operator

$$I(t) = -\frac{1}{2\pi i}\int_\Gamma \exp(-\lambda^\alpha t) R_L(\lambda)\, d\lambda, \tag{5.131}$$

where for noninteger α the branch of λ^α is chosen as indicated in Subsection 5.2.5 and the contour is going in the positive direction with respect to Λ_θ (Λ_θ lies to the left of the direction of motion). It suffices to prove the following three assertions:

(a) *There exists a sequence of arcs $\{\lambda : |\lambda| = R_k, |\arg\lambda| \le \theta\}$ such that $R_k \to \infty$ and $\|R_L(\lambda)\| = \exp(o(R_k^\alpha))$ on these arcs.*

(b) *$I(t)f \to f$ as $t \to +0$ if $f \in \mathfrak{D}(L)$.*

(c) *$\|I(t)\| \le$ const for $t \in (0, 1)$.*

Assertions (a) and (b) are actually proved in [65]. Theorem 5.6.5 is a version of a theorem in [65], where the conditions were imposed on the quadratic form of the operator, while we impose conditions on the resolvent. Assertion (a) can be derived from the assumptions of the theorem with the use of some theorems of the operator theory and the theory of functions of a complex variable (mainly the same as in the proof of Theorem 5.6.1). Assertion (a) permits one to reduce (5.131) to the form (5.21). It follows from (b) and (c) that $I(t)f \to f$ as $t \to +0$ for all $f \in \mathfrak{H}$.

The prove of assertion (b) is fairly easy. Let Γ_1 be the contour obtained from Γ by replacing the segments $\{\lambda : |\lambda| \le 1/2, \arg\lambda = \pm\theta\}$ by the arc $\{\lambda : |\lambda| = 1/2, |\arg\lambda| \le \theta\}$. In (5.131) we replace Γ by Γ_1 and substitute $f = L^{-1}g$ (using the assumption that $f \in \mathfrak{D}(L)$). Since

$$L^{-1} - \lambda^{-1} I = \lambda^{-1} L^{-1}(\lambda I - L),$$

we can replace L^{-1} by λ^{-1}:

$$I(t)f = -\frac{1}{2\pi i}\int_{\Gamma_1} \exp(-\lambda^a t)\lambda^{-1} R_L(\lambda) g\, d\mu.$$

The inequality $|\lambda|^{-1}\|R_L(\lambda)\| \le C|\lambda|^{-2}$ permits one to pass to the limit as $t \to +0$:

$$\lim_{t\to+0} I(t)f = -\frac{1}{2\pi i}\int_{\Gamma_1} \lambda^{-1} R_L(\lambda) g\, d\mu.$$

The right-hand side can be evaluated as the residue of the integrand at $\mu = 0$; this residue is readily seen to be equal to f.

Let us prove assertion (c) (suggested to the author of this chapter by V. I. Matsaev). First, we assume that α is an integer. Let us make the change of variables $\lambda = \zeta t^{-1/\alpha}$, which maps the contour Γ into itself. Let Γ_2 be the contour obtained from Γ by replacing the segments $\{\zeta : |\zeta| \le 1,\ \arg\zeta = \pm\theta\}$ by the arc $\{\zeta : |\zeta| = 1,\ |\arg\zeta| \ge \theta\}$. We have

$$I(t) = -\frac{1}{2\pi i}\int_{\Gamma_2} \exp(-\zeta^\alpha) R_L(\zeta t^{-1/\alpha})\, d\zeta \cdot t^{-1/\alpha}.$$

Using the estimate (5.129) and the exponential decay of $|\exp(-\zeta^\alpha)|$ on the rays entering Γ_2, we readily obtain (c).

If α is not an integer, then we have to supplement Γ_2 by the contour Γ_3 consisting of the two edges of the cut along the interval $(-1, 0)$ of the real axis on the complex plane. The integral over Γ_3 is equal to

$$-\frac{1}{2\pi i}\int_{-1}^{0} \exp(-|\zeta|^\alpha \cos\alpha\pi)(-2i)\sin(|\zeta|^\alpha \sin\alpha\pi) R_L(\zeta t^{-1/\alpha})\, d\zeta \cdot t^{-1/\alpha}.$$

It can also be estimated readily. □

Note that under the assumptions of Theorem 5.6.5, a condition of the form (5.124) (some control of the arrangement of the parentheses) is not available, in contrast with the situation in (5.117a) in Theorem 5.6.2.

Note also that the upper bound for α is unnecessary if θ can be taken arbitrarily small.

An analog of Theorem 5.6.5 is valid for a compact operator $A \in S^{(p)}(\mathfrak{H})$ under the assumption that $\|R_A^{(m)}(\lambda)\| = O(|\lambda|^{-1})$ in the sector complementary to Θ provided that A is the inverse of an operator with discrete spectrum.

4. Tests for the basis property. Two special tests for the basis property will be mentioned below in Section 5.9 (see (5.256)) and Subsection 5.10.2.

5. Eigenvalue asymptotics for nonselfadjoint operators. A very general theorem stating that the eigenvalue asymptotics with a remainder estimate is preserved under a weak perturbation of a given selfadjoint operator is proved in [72] (see also [70, Chapter 1]). An application of this theorem to elliptic pseudodifferential operators on a closed surface gives the following result.

1°. *Let L_0 be an elliptic selfadjoint pseudodifferential operator of order $\gamma > 0$ with positive principal symbol, and let $L = L_0 + L_1$, where L_1 is a pseudodifferential operator of order $\leq \gamma - 1$. Then the asymptotic formula* (5.80) *for the eigenvalues of L_0 remains valid for the eigenvalues of L.*

We recall that the eigenvalues $\mu_j(L)$ are, in general, complex numbers numbered in a nondecreasing order of their absolute values.

A similar result is valid for the characteristic numbers of an elliptic operator $A = A_0 + A_1$, where A_0 is a selfadjoint operator of negative order $-\gamma$ with positive principal symbol and A_1 is a pseudodifferential operator of order $\leq -\gamma - 1$.

Next, an assertion similar to 1° is valid for operators L_B corresponding to elliptic boundary value problems with homogeneous boundary conditions. For example, formulas of the form (5.96)–(5.97) remain valid for the eigenvalues of the operators corresponding to classical problems obtained from selfadjoint problems with nonnegative spectrum by perturbing the elliptic equation and the boundary conditions by lower-order terms [72].

For pseudodifferential operators and for operators corresponding to problems with homogeneous boundary conditions, if they are far from being selfadjoint, we can indicate a rough asymptotics for the moduli of the eigenvalues provided that there is a sector of ellipticity with parameter and the complementary sector is not too large.

Let us state the result for pseudodifferential operators [18, 11].

2° *Suppose that a pseudodifferential operator A of order $\gamma > 0$ on a closed n-dimensional surface $\mathfrak{S}$ is elliptic with parameter in the exterior of a sector with angle θ at the vertex, where $\theta < \pi\gamma/n$. Then the moduli of the eigenvalues μ_j of the operator A numbered in a nondecreasing order (taking their multiplicities into account) satisfy the relation $|\mu_j| \asymp j^{\gamma/n}$, which means that the ratio of the left- to the right-hand side is enclosed between some positive constants for sufficiently large j. If the ellipticity with parameter holds outside an arbitrarily narrow sectorial neighborhood of a given ray, then we have the usual asymptotics $\mu_j = C' j^{\gamma/n} + o(j^{\gamma/n})$, where C' is calculated as in* (5.80).

6. General weak perturbations of selfadjoint operators. Now let us consider an operator A that is the inverse of an operator L with discrete spectrum and has the form

$$A = A_0(I + T_1), \tag{5.132}$$

where A_0 is a selfadjoint operator belonging to $S^{(p)}(\mathfrak{H})$ for some p and T_1 is only known to be a compact operator. Thus, we deal with a perturbation of A_0 by an operator relatively compact with respect to A_0, which is the subordination to A_0 in a very weak sense. We shall deal with such operators A in Section 5.12. Suppose that the eigenvalues of A_0 are different from zero and are all positive (or negative) starting from some number. (The latter assumption is solely made for simplicity.) We claim that the estimate $\|R_A^{(m)}(\lambda)\| = O(|\lambda|^{-1})$ is satisfied outside an arbitrarily narrow sectorial neighborhood of the ray $\mathbf{R}_+$ (or, respectively, $\mathbf{R}_-$). This is a consequence of the following formula for $R_A^{(m)}(\lambda)$ ([43, Subsection 5.8.1]):

$$R_A^{(m)}(\lambda) = R_{A_0}^m(\lambda)B(\lambda),$$

where $B(\lambda)$ is an operator with a uniformly bounded norm. Hence we obtain the following assertion.

1° *Under these conditions,* $A \in \mathbf{A}(\alpha, \mathfrak{H})$ *provided that* $\alpha = p^{-1} + \varepsilon$, *where* $\varepsilon > 0$.

We also need the following result, which goes back to [57, 58] (see also [43, Section 5.11]).

2° *Let the characteristic numbers of the operator* A_0 *satisfy the relation*

$$\nu_j(A_0) = cj^p + o(j^p), \qquad j \to \infty, \tag{5.133}$$

with some $c \neq 0$. *Then just the same formula is valid for the characteristic numbers of* A.

5.7 Scalar Problem with Spectral Parameter in Transmission Conditions

1. Statement of the problem and the relationship with an elliptic pseudo-differential equation on S. Here we consider a problem slightly more general than in Section 5.1.[11]

Problem I We seek a solution of the Helmholtz equation in $\mathbf{R}^n$ ($n = 1$, 2, or 3)

$$\Delta u(x) + k^2 u(x) = 0 \tag{5.134}$$

outside S (here S is a closed connected infinitely smooth surface for $n = 3$, a closed curve without self-intersections for $n = 2$, and a two-point set for $n = 1$) under the conditions

$$u^+(x) = u^-(x) \tag{5.135}$$

and

$$\sigma(x)u^\pm(x) - \lambda(\partial_\nu u^-(x) - \partial_\nu u^+(x)) = g(x) \tag{5.136}$$

on S and the radiation condition at infinity. Here $\sigma(x)$ is a given positive (until Subsection 9) function from $C^\infty(S)$ and λ is the spectral parameter. Recall that $k = k_1 + ik_2 \neq 0$ is given; $k_2 \leq 0$ (see (5.3)). The radiation condition can be represented in the form (cf. Eq. (0.35))

$$\partial_r u(x) + iku(x) = e^{-ikr} o(r^{-(n-1)/2}) \quad \text{as } r = |x| \to \infty, \tag{5.137}$$

where $\partial_r = \partial/\partial r$. As we have already noted in Section 5.1, here the nonhomogeneous term is transferred from the right-hand side of the Helmholtz equation to the right-hand side of the boundary condition by subtracting the incident field from the solution. The smoothness of the function g in (5.136) is determined by the smoothness of the original

[11] Cf. Section 2.2.

right-hand side f of the Helmholtz equation near S. The reader should have in mind this remark in Sections 5.8 and 5.11; we shall not repeat it. Note also that λ may be complex-valued.

Let V^+ and V^- be, respectively, the interior and exterior domains into which $\mathbf{R}^n$ is divided by S. By $\Phi(x-y,k)$ we denote the Green function of Eq. (5.134) with the radiation condition at infinity; thus,

$$\Phi(x,k) = \begin{cases} i(2k)^{-1}e^{-ik|x|} & \text{for } n=1, \\ -(4i)^{-1}H_0^{(2)}(k|x|) & \text{for } n=2, \\ -(4\pi|x|)^{-1}e^{-ik|x|} & \text{for } n=3, \end{cases} \tag{5.138}$$

where $H_0^{(2)}(t)$ is the Hankel function of the second kind and of order zero. Until Subsection 8, we shall consider the cases $n=2$ and $n=3$. Let us introduce the single and double layer potentials

$$u(x) = \mathcal{A}(k)\varphi(x) = \int_S \Phi(x-y,k)\varphi(y)\,dS_y, \qquad x \in \mathbf{R}^n, \tag{5.139}$$

$$v(x) = \mathcal{B}(k)\psi(x) = \int_S \partial_{\nu_y}\Phi(x-y,k)\,\psi(y)\,dS_y, \qquad x \notin S, \tag{5.140}$$

with smooth densities φ and ψ; according to our convention, the normal is directed into V^-. The functions $u(x)$ and $v(x)$ satisfy the Helmholtz equation (5.134) and the radiation condition (5.137). Let us also introduce the following integral operators on S:

$$A(k)\varphi(x) = \int_S \Phi(x-y,k)\varphi(y)\,dS_y, \qquad x \in S, \tag{5.141}$$

$$B(k)\psi(x) = \int_S \partial_{\nu_y}\Phi(x-y,k)\,\psi(y)\,dS_y, \qquad x \in S \tag{5.142}$$

(the direct value of the double layer potential).

It is well known (e.g., see [31]) that the operators (5.141) and (5.142) are compact in $H^0(S) = L_2(S)$ (since their kernels have a weak singularity) and that the potentials (5.139) and (5.140) satisfy the following limit relations on S:

$$\mathcal{A}(k)\varphi^{\pm} = A(k)\varphi^{\pm}, \tag{5.143}$$

$$\partial_\nu\,\mathcal{A}(k)\varphi^{\pm} = \mp\frac{1}{2}\varphi + B'(k)\varphi, \tag{5.144}$$

$$\mathcal{B}(k)\psi^{\pm} = \pm\frac{1}{2}\psi + B(k)\psi, \tag{5.145}$$

$$\partial_\nu\,\mathcal{B}(k)\psi^{+} = \partial_\nu\mathcal{B}(k)\psi^{-}, \tag{5.146}$$

where

$$B'(k)\psi(x) = \int_S \partial_{\nu_x}\Phi(x-y,k)\psi(y)\,dS_y, \qquad x \in S, \tag{5.147}$$

is the transpose of the operator $B(k)$ (see (5.26)).

Let us recall Green's first and second formulas for the Laplace equation:

$$\int_{V+} \Delta u \cdot v \, dx = - \int_{V+} \nabla u \cdot \nabla v \, dx + \int_S \partial_\nu u^+ \cdot v^+ \, dS \tag{5.148}$$

and

$$\int_{V+} (\Delta u \cdot v - u \cdot \Delta v) \, dx = \int_S (\partial_\nu u^+ \cdot v^+ - u^+ \cdot \partial_\nu v^+) \, dS. \tag{5.149}$$

Using Green's second formula, we derive the following integral representation for solutions of the Helmholtz equation defined in V^+ and smooth up to S:

$$u(x) = \mathcal{B}(k)(u^+)(x) - \mathcal{A}(k)(\partial_\nu u^+)(x). \tag{5.150}$$

Note that the right-hand side is zero for $x \in V^-$. The solutions defined in V^-, smooth up to S, and satisfying the radiation condition can be represented in the similar form

$$u(x) = \mathcal{A}(k)(\partial_\nu u^-)(x) - \mathcal{B}(k)(u^-)(x). \tag{5.151}$$

Note that the right-hand side is zero for $x \in V^+$.

The last two relations in conjunction with Eqs. (5.143)–(5.145) imply the following relations on S for the boundary values of the solution of the Helmholtz equation in V^+ and V^-:

$$\left(B(k) - \frac{1}{2} I\right) u^+ = A(k) \partial_\nu u^+, \tag{5.152}$$

$$\left(B(k) + \frac{1}{2} I\right) u^- = A(k) \partial_\nu u^-. \tag{5.153}$$

We have recalled classical formulas that will be used in the subsequent exposition. Now let us set

$$A_\sigma(k)\varphi = \sigma A(k)\varphi. \tag{5.154}$$

If $u(x)$ is a solution of Problem I, then it follows from the first transmission condition (5.135), the representations (5.150) and (5.151), and the remarks after them that

$$u = \mathcal{A}(k)\varphi \quad \text{outside } S, \tag{5.155}$$

where

$$\varphi = \partial_\nu u^- - \partial_\nu u^+. \tag{5.156}$$

By substituting (5.155) into the second transmission condition (5.136), we arrive at the Fredholm equation

$$A_\sigma(k)\varphi - \lambda\varphi = g, \qquad x \in S. \tag{5.157}$$

For $\lambda \notin \Sigma(A_\sigma)$, this equation is equivalent to Problem I. In the spectral problem (with $g = 0$), this equivalence holds only for the eigenfunctions. We do not write out the more cumbersome equations for the associated functions. In the following, our main task is to study the spectral properties of the operator $A_\sigma(k)$.

For $n = 3$, we can introduce local orthogonal curvilinear coordinates $t = (t_1, t_2)$ on S and express the Cartesian coordinates as C^∞ functions $x_j = x_j(t)$, $j = 1, 2, 3$. Let $(t, \xi) = (t_1, t_2, \xi_1, \xi_2)$ be the corresponding coordinates on T^*S (see Subsection 5.4.3), and let $h_1(t), h_2(t)$ be the Lamé coefficients, that is, the lengths of the vectors $\partial x(t)/\partial t_1$ and $\partial x(t)/\partial t_2$. For $n = 2$, we use similar notation: t is a parameter on S, $x_j = x_j(t)$, $j = 1, 2$, (t, ξ) are coordinates on T^*S, and $h(t)$ is the length of the vector $x'(t)$. For the parametrization (5.47), $h(t)$ is independent of t and is equal to $|S|/2\pi$, since the modulus of the derivative of the position vector of a point with respect to the arc length is equal to 1.

For $n = 3$, it is possible to assume that $h_1(t) \equiv h_2(t)$ (e.g., see [95, Chapter 9, Addendum 1]). At a fixed point t_0, we may assume that $h_1(t_0) = h_2(t_0) = 1$.

Theorem 5.7.1 *$A_\sigma(k)$ is an elliptic pseudodifferential operator of order -1 on S with principal symbol*

$$-\frac{1}{2}\sigma h|\xi|^{-1} \quad \textit{and} \quad -\frac{1}{2}\sigma(h_1^{-2}\xi_1^2 + h_2^{-2}\xi_2^2)^{-1/2} \tag{5.158}$$

for $n = 2$ and $n = 3$, respectively.

Proof. To be definite, assume that $n = 3$. Let $\omega_1, \omega_2 \in C^\infty(S)$. Condition (a) of Subsection 5.4.3 is obviously satisfied: if the supports of ω_1 and ω_2 are disjoint, then $\omega_1 A_\sigma(k)(\omega_2\varphi)$ is an integral operator with C^∞ kernel and hence an infinitely smoothing operator. Suppose that the supports of ω_1 and ω_2 lie in the same coordinate neighborhood (with coordinates t). Then $\omega_1 A_\sigma(k)(\omega_2\varphi)$ is an integral operator in $\mathbf{R}^2$ whose kernel can be analyzed by expanding the exponential in (5.138) in a power series in $|x|$. The leading term of the kernel in the local coordinates has the form

$$-(4\pi)^{-1}\omega_1(t)|x(t) - x(\tilde{t})|^{-1}\sigma(x(t))\omega_2(\tilde{t})h_1(\tilde{t})h_2(\tilde{t}). \tag{5.159}$$

Now Propositions 5.4.1 and 5.4.2 ensure the desired result. For $n = 2$, we can carry out a similar proof with the use of Propositions 5.4.3 and 5.4.4. (See also Subsection 5 below.) □

Corollary 5.7.2 *The root functions of the operator $A_\sigma(k)$ belong to $C^\infty(S)$. If 0 is an eigenvalue of the operator $A_\sigma(k)$, then $\operatorname{Ker} A_\sigma(k)$ is finite-dimensional.*

This follows from the fact that the kernel of an elliptic pseudodifferential operator is finite-dimensional (Assertion 2° in Subsection 5.5.1). In the following, we shall show that the operator $A_\sigma(k)$ is either selfadjoint or dissipative. This implies that there are no associated vectors in $\mathfrak{L}_{A_\sigma(k)}(0)$ (see Assertion 4° in Subsection 5.2.6), and so $\operatorname{Ker} A_\sigma(k) = \mathfrak{L}_{A_\sigma(k)}(0)$.

Proposition 5.7.3 *The number $\lambda = 0$ is an eigenvalue of the operator $A(k)$ if an only if k is a (real) number such that the homogeneous interior Dirichlet problem*

$$\Delta u(x) + k^2 u(x) = 0 \ \ in \ \ V^+, \qquad u^+(x) = 0 \ \ on \ \ S \tag{5.160}$$

has nontrivial solutions. Moreover,[12]

$$\operatorname{Ker} A(k) = \{\partial_\nu u^+ \ : \ \Delta u + k^2 u = 0 \ \ in \ \ V^+, \ u^+ = 0\}, \tag{5.161}$$

and the space of solutions of problem (5.160) *and the kernel* $\operatorname{Ker} A(k)$ *have the same dimension.*

Proof. Let φ be a nontrivial solution of the equation $A(k)\varphi = 0$. Set $u = -\mathcal{A}(k)\varphi$. Then $u^\pm = 0$, and $u = 0$ in V^- by virtue of the uniqueness for the exterior Dirichlet problem (e.g., see [31, Chapter 3]), so that $\partial_\nu u^- = 0$. In view of the properties of the operator $\mathcal{A}(k)$, the function u is a solution of problem (5.160); moreover, this solution is nontrivial, since $\partial_\nu u^+ = \varphi$ by virtue of (5.144).

Conversely, let u be a nontrivial solution of problem (5.160). Then $\partial_\nu u^+$ is not zero identically (by virtue of (5.150)), and $A(k)\partial_\nu u^+ = 0$ by (5.152). □

Obviously, *the operator $A(k)$ has a zero eigenvalue if and only if so does the operator $A_\sigma(k)$.*

The kernel $\operatorname{Ker} A(k)$ is nontrivial only for some isolated real values $k = \pm k_j$, where $k_j \to +\infty$ as $j \to \infty$. We shall discuss this in more detail in Subsection 5.8.1. Since $A(k)$ has a real principal symbol, it follows that the index of this elliptic operator is zero. Consequently, if $\operatorname{Ker} A(k) = \{0\}$, then there exists an inverse $A^{-1}(k)$, which is an elliptic pseudodifferential operator of order $+1$. The same is true of the pseudodifferential operator $A_\sigma(k)$.

2. Properties of solutions and estimates for their norms. We recall that the ball E_R and the domain V_R^- are defined by formulas (5.40) and (5.41).

Theorem 5.7.4 *Suppose that $\lambda \notin \Sigma(A_\sigma(k))$ and $\lambda \neq 0$. Then for each $g \in H^s(S)$, $s \geq 1/2$, problem* (5.134)–(5.137) *has a unique solution belonging to $H^{s+3/2}$ in V^+ and in V_R^- for any R.*[13] *Furthermore,*

$$\|u\|_{V^+, s+3/2} + \|u\|_{V_R^-, s+3/2} \leq C_s \|g\|_{S,s}, \tag{5.162}$$

where C_s is independent of u. If $\operatorname{Im} k < 0$, *then the above assertions hold for V^- instead of V_R^-.*

---

[12]Relation (5.161) must be understood as follows: $\operatorname{Ker} A(k)$ consists of functions v such that there exists a solution u of the Helmholtz equation in V^+ with $u^+ = 0$ and $v = \partial_\nu u^+$. A similar meaning must be assigned to relations (5.195), (5.196), and (5.313) below.

[13]If $n = 3$, then $u(x)$ is a classical solution of the problem (continuous together with the derivatives of order ≤ 2 in $\overline{V^\pm}$) for $s > 2$. If $n = 2$, then the same is true for $s > 3/2$. (See Assertion 11° in Subsection 5.3.2.)

This solution can be represented by formula (5.155), *where* φ *is the* (*unique*) *solution of Eq.* (5.157) *in* $H^s(S)$. *The solution* u *is infinitely smooth outside* S (*in* $\overline{V}^+$ *and* $\overline{V}^-$) *provided that* $g \in C^\infty(S)$. *Outside an arbitrary neighborhood of this surface, one has*

$$\sum_{|\alpha|\le m} |D^\alpha u(x)| \le C'_m(1+|x|)^{-(n-1)/2} e^{k_2|x|} \|g\|_{S,0} \tag{5.163}$$

for each positive integer m. *Furthermore,*

$$|\partial_r u(x) + iku(x)| \le \eta(r) r^{-(n-1)/2} e^{k_2 r} \|g\|_{S,0}, \tag{5.164}$$

where $\eta(r) \to 0$ *as* $r = |x| \to \infty$.

One can add $\|u\|_{S,s+1}$ and $\|\partial_\nu u^\pm\|_{S,s}$ to the left-hand side of (5.162).

The meaning of the above estimates is as follows. If a sequence g_q approximates g in $H^s(S)$, that is, $\|g - g_q\|_{S,s} \to 0$ as $q \to \infty$, and u_q is the solution of the problem with g replaced by g_q, then u_q approximates u in the sense that the left-hand sides in (5.162)–(5.164) tend to zero after the substitution of $u - u_q$ for u. If we find an approximate solution of Eq. (5.157) (e.g., by the spectral method suggested in this book), then, by substituting it into (5.155), we obtain an approximate solution of the original problem.

Proof of Theorem 5.7.4. We can assume (possibly, for a slightly larger value of R) that the homogeneous interior Dirichlet problem in the ball E_R for a given k does not have any nontrivial solutions. *Let us transform problem* (5.134)–(5.137) *into an equivalent problem in the ball* E_R. To this end, we use formula (5.151) with $S_R = \{x : |x| = R\}$ instead of S. By passing to the limit as $|x| \searrow R$ in this formula, we obtain the relation

$$A_1(k)\partial_r u^- = B_1(k)u^- + \frac{1}{2}u^- \quad \text{on } S_R, \tag{5.165}$$

where $A_1(k)$ and $B_1(k)$ are analogs of the operators $A(k)$ and $B(k)$ on S_R. Here, by Theorem 5.7.1, $A_1(k)$ is an elliptic pseudodifferential operator of order -1 with real principal symbol. It follows from a formula similar to (5.161) for $\operatorname{Ker} A_1(k)$ that $\operatorname{Ker} A_1(k)$ contains only zero; hence there exists a pseudodifferential operator $A_1^{-1}(k)$ of order $+1$. We shall show in Subsection 5.8.1 that $B_1(k)$ is a pseudodifferential operator of order -1. Using Assertion 8° from Subsection 5.3.2, we replace u^- by u^+ and $\partial_r u^-$ by $\partial_r u^+$ in (5.165). Now let us act on both sides by the operator $A_1^{-1}(k)$. We arrive at the relation

$$\partial_r u^+ - \frac{1}{2}A_1^{-1}(k)u^+ - A_1^{-1}(k)B_1(k)u^+ = 0 \quad \text{on } S_R. \tag{5.166}$$

Now, *instead of the original problem in* $\mathbf{R}^n$, *we can consider the problem in* E_R *for Eq.* (5.134) *outside* S *with conditions* (5.135) *and* (5.136) *on* S *and condition* (5.166) *on* S_R. (The solution of the new problem can readily be continued to the solution of the original problem; for this, one must find u^+ and $\partial_r u^+$ on S_R, replace these functions by u^- and $\partial_r u^-$, respectively, and substitute them into Eq. (5.151) with S_R instead of S.)

Let us verify that the new problem is elliptic. The ellipticity of the conditions on S (for $\lambda \neq 0$) was essentially verified in Subsection 5.5.5 (see Eq. (5.101)). Consider condition (5.166). Using the propositions from Section 5.4 and Theorem 5.7.1, we conclude that $A_1^{-1}B_1$ is a pseudodifferential operator of order zero, and the principal symbol of the operator $-\frac{1}{2}A_1^{-1}$ is equal to $|\xi|$ for $n = 3$ at any fixed point of S_R if $h_1 = h_2 = 1$ at this point. By substituting $|\xi'|$ for $a(0, \xi')$ into (5.91), we see that the ellipticity condition is satisfied on S_R as well. For $n = 2$, this is also true.

Let us show that the index of the problem is zero. First, condition (5.136) on S for nonzero λ can be continuously deformed, with the ellipticity preserved, to the condition $\partial_\nu u^+ = \partial_\nu u^-$. Under this condition and condition (5.135), the boundary is "erased" (cf. the end of Subsection 5.5.5), and we are left with a problem in E_R with the boundary condition only on S_R. Second, this boundary condition can be deformed to the Neumann condition, and the Neumann problem for the Helmholtz equation has index zero (this was mentioned in Subsection 5.5.5).

Thus, the problem in E_R is elliptic and has index zero, and so we can apply the main results stated in Subsection 5.5.5. It is easily seen that for $\lambda \notin \Sigma(A_\sigma(k))$ the solution of the problem is unique: should the problem have a nontrivial solution u for $g = 0$, this solution would be infinitely smooth in $\overline{V}^+$ and $\overline{V}^-$ (cf. 4° in Subsection 5.5.5), and formula (5.156) would give a nontrivial solution of Eq. (5.157) with $g = 0$.

This readily gives all the assertions in the first paragraph of the statement of Theorem 5.7.4 (for $\operatorname{Im} k < 0$, in the estimate (5.162) we can let $R \to \infty$). The remaining assertions are obvious. In particular, the left-hand sides in (5.163) and (5.164) can readily be estimated via $\|\varphi\|_{S,0}$ with the help of (5.155) and the Schwartz inequality, and $\|\varphi\|_{S,0}$ can be estimated via $\|g\|_{S,0}$ (see (5.33)). □

One can eliminate the condition $s \geq 1/2$ by using the results from [66, Chapter II]; we do not discuss this possibility.

3. The summability theorem. Let us equip $L_2(S)$ with the inner product

$$\langle \varphi, \psi \rangle_S = \int_S \sigma^{-1} \varphi \overline{\psi}\, dS. \tag{5.167}$$

The corresponding norm $\langle \varphi, \varphi \rangle_S^{1/2}$ is equivalent to the usual norm $\|\varphi\|_{S,0}$ in $L_2(S)$. Let $A_\sigma^*(k)$ be the adjoint of the operator $A_\sigma(k)$ with respect to the new inner product. Obviously, $A_\sigma^*(k) = \overline{A}_\sigma(k)$; hence we obtain the corollaries indicated in Subsection 5.2.6 (namely, 8° and 9°).

Remark 5.7.5 For $k_1 = \operatorname{Re} k = 0$, the operator $A_\sigma(k)$ is selfadjoint in $L_2(S)$ with respect to the inner product (5.167).

This can be seen from the structure of the kernel of the operator $A(k)$ (see (5.138)) and Eqs. (5.154) and (5.167). Obviously, in this case $A_\sigma(k)$ has an orthonormal basis of infinitely smooth eigenfunctions, which can be used for constructing expansions in $H^s(S)$ as well. In this case formula (5.168), given below, provides the asymptotics of the characteristic numbers of the operator $A_\sigma(k)$. Essentially, nothing can be added to this information. In the following, we mainly consider the other values of k, for which our operator is not selfadjoint. *Unless otherwise specified, we assume that* $k_1 \neq 0$.

By (5.80), the characteristic numbers ν_j of the selfadjoint operator $\operatorname{Re} A_\sigma(k)$ have the asymptotics

$$\nu_j = \begin{cases} -c_2 j + O(1) & \text{for } n = 2, \\ -c_3 j^{1/2} + O(1) & \text{for } n = 3, \end{cases} \tag{5.168}$$

where

$$c_2 = 2\pi\Big(\int_S \sigma\, dS\Big)^{-1} \quad \text{and} \quad c_3 = 4\sqrt{\pi}\Big(\int_S \sigma^2\, dS\Big)^{-1/2}. \tag{5.169}$$

Proposition 5.7.6 *The operator* $\operatorname{Im} A_\sigma(k)$ *has the order* $-\infty$ *if* $k_2 = 0$*; the order of this operator is* -3 *if* $k_2 < 0$ *and* $k_1 \neq 0$.[14]

Proof. First, let us consider the case $\sigma \equiv 1$. Since $A^*(k) = \overline{A}(k)$, it follows that $\operatorname{Im} A(k) = (A(k) - \overline{A(k)})/(2i)$ (see Assertion 9° in Subsection 5.2.6). The operator $\overline{A}(k)$ has the form (5.141) with $\overline{\Phi}$ instead of Φ. If $k_2 = 0$, then, as can be seen from (5.138), $\Phi - \overline{\Phi}$ is an infinitely smooth function, so that $\operatorname{Im} A(k)$ is an infinitely smoothing operator. Let $k_2 < 0$ and $k_1 \neq 0$. Then for $n = 3$ we have

$$\operatorname{Im}(e^{-ik|x|}|x|^{-1}) = -k_1 - 2k_1k_2|x| + \cdots \tag{5.170}$$

Since the singularity is here determined by the term containing $|x|$, it follows that $\operatorname{Im} A(k)$ is an operator of order -3 (by virtue of Proposition 5.4.1). One can readily verify that the last assertion is also valid for $n = 2$. The result can be extended to $A_\sigma(k)$ by using the inner product (5.167). □

Now we can apply the summability tests given in Subsections 5.6.1–5.6.2 to the operator $A_\sigma(k)$. See, in particular, our explanations after the formulation of Theorem 5.6.4. Set $\mathfrak{H}^s = H^s(S)$.

Theorem 5.7.7 $A_\sigma(k) \in \mathbf{B}(H^s(S))$ *for all* s *if either* $k_2 = \operatorname{Im} k = 0$ *or* $n = 2$ *and* $k_2 < 0$. *If* $n = 3$ *and* $k_2 < 0$, *then* $A_\sigma(k) \in \mathbf{R}(H^s(S))$ *for all* s.

The assertions of Theorem 5.6.3 *are valid for the operator* $A_\sigma(k)$ *with* $p = 1/2$ *for* $n = 3$, $p = 1$ *for* $n = 2$, $q = -\infty$ *for the case* $k_2 = 0$, *and* $q = -1$ *for the case* $n = 2$, $k_2 < 0$.

If $n = 3$ *and* $k_2 < 0$, *then* $A_\sigma(k)$ *satisfies the assertions of Theorem* 5.6.4 *with* $p = 1/2$ *and* $q = -1$.

---

[14] In the latter case, the operator is "farther" from being selfadjoint than in the case of real k; physically, this apparently corresponds to greater energy losses. A similar situation occurs for the operators considered below in Sections 5.8 and 5.11.

To clarify this statement, recall that $\sum P_l f$ and $\sum Q_l f$ are the Fourier series with parentheses with respect to the root functions of the operator $A_\sigma(k)$ and the eigenfunctions of the operator $\operatorname{Re} A_\sigma(k)$, respectively, on S. It follows from (5.168) that the parentheses can be arranged as was described in Subsection 5.6.2 (see (5.124)). Furthermore, the more differentiable the function f is, the more rapidly the series $\sum P_l f$ converges. In particular, if $f \in C^\infty(S)$, then the absolute values of the terms of this series decay faster than l^{-N} for any positive integer N, and the rate of convergence remains the same if we (locally) differentiate the series term by term arbitrarily many times. Moreover, except in the case $n = 3$, $k_2 < 0$, the series $\sum P_l f$ and $\sum Q_l f$ are equiconvergent for nonsmooth (or not very smooth) f; the convergence is especially rapid for $k_2 = 0$. For example, if $k_2 = 0$ and $f \in C^{(2)}(S)$, then the absolute values of the functions $P_l f - Q_l f$ decay as $l \to \infty$ faster than l^{-N} for an arbitrary positive integer N. By Theorem 5.7.4, the convergence of the corresponding series outside S can be characterized in a similar way.

4. Localization and asymptotics of the eigenvalues of $A_\sigma(k)$

Proposition 5.7.8 *Let* $k_1 = \operatorname{Re} k > 0$. *Then* $\operatorname{Im}\langle A_\sigma(k)\varphi, \varphi\rangle_S > 0$ *for* $A_\sigma(k)\varphi \neq 0$. *In particular, the operator* $A_\sigma(k)$ *is dissipative, and its nonzero eigenvalues* λ_j *lie in the upper half-plane* $\operatorname{Im}\lambda > 0$.[15] *For* $k_1 < 0$, *similar assertions are valid for the operator* $-A_\sigma(k)$. *For* $|k_1| \leq |k_2|$, *the real parts of the eigenvalues of the operator* $A_\sigma(k)$ *are negative.*

Proof. First, let $\sigma \equiv 1$ and $\varphi \in C^\infty(S)$. We define a function $u(x)$ by the formula $u(x) = \mathcal{A}(k)\varphi(x)$ and apply Green's formula (5.148) in the domains V^+ and V_R^- (see (5.41)) with sufficiently large R to $u(x)$ and $\overline{u(x)}$. By summing the results, we obtain

$$\int_{E_R} \Delta\overline{u} \cdot u\, dx = -\int_{E_R} |\nabla u|^2 dx + \int_S (\partial_\nu \overline{u}^+ - \partial_\nu \overline{u}^-) u\, dS + \int_{S_R} \partial_r \overline{u} \cdot u\, dS,$$

or

$$(A(k)\varphi, \varphi)_S = \overline{k}^2 \int_{E_R} |u|^2 dx - \int_{E_R} |\nabla u|^2 dx + \int_{S_R} \partial_r \overline{u} \cdot u\, dS. \tag{5.171}$$

By passing to the limit, we can extend this relation to $\varphi \in H^0(S)$. Let us use the radiation condition and separate the real and imaginary parts; we obtain

$$\operatorname{Re}(A(k)\varphi, \varphi)_S = (k_1^2 - k_2^2) \int_{E_R} |u|^2 dx - \int_{E_R} |\nabla u|^2 dx + k_2 \int_{S_R} |u|^2 dS + o(1), \tag{5.172a}$$

$$\operatorname{Im}(A(k)\varphi, \varphi)_S = -2k_1 k_2 \int_{E_R} |u|^2 dx + k_1 \int_{S_R} |u|^2 dS + o(1). \tag{5.172b}$$

[15]The latter assertion was already indicated in Chapter 2 for $\operatorname{Im} k = 0$ and $\sigma \equiv 1$.

If $k_2 = \operatorname{Im} k < 0$, then the passage to the limit as $R \to \infty$ gives

$$\operatorname{Re}(A(k)\varphi, \varphi)_S = (k_1^2 - k_2^2) \int_{\mathbf{R}^n} |u|^2 dx - \int_{\mathbf{R}^n} |\nabla u|^2 dx, \tag{5.173a}$$

$$\operatorname{Im}(A(k)\varphi, \varphi)_S = -2k_1 k_2 \int_{\mathbf{R}^n} |u|^2 dx, \tag{5.173b}$$

since the integral over S_R decays exponentially. The passage to the limit as $R \to \infty$ in (5.172b) is possible in the case $k_2 = 0$ as well, since the left-hand side is independent of R:

$$(\operatorname{Im} A(k)\varphi, \varphi)_S = k_1 \lim_{R \to \infty} \int_{S_R} |u|^2 dS. \tag{5.174}$$

If $(\operatorname{Im} A(k)\varphi, \varphi)_S = 0$, then the limit is zero. In this case, $u = 0$ in V^- (e.g., see [31, Lemma 3.11] (due to Rellich)). The same follows from (5.173b) if $k_2 < 0$. Consequently, $u^+ = 0$, which is impossible if $A(k)\varphi \neq 0$.

The above formulas imply all assertions in Proposition 5.7.8 for $\sigma \equiv 1$.

Now let $\sigma \not\equiv 1$. Then the above formulas remain valid if we replace $(A\varphi, \varphi)_S$ by $\langle A_\sigma \varphi, \varphi \rangle_S$ (see (5.167)). Thus, the results hold in this case as well. □

From Proposition 5.7.8 it follows, in particular, that $\operatorname{Ker} \operatorname{Im} A_\sigma(k)$ coincides with $\operatorname{Ker} A_\sigma(k)$ and $\operatorname{Ker} A(k)$ for $k_1 > 0$ (or $k_1 < 0$). See (5.161).

Proposition 5.7.9 *Formulas* (5.168) *remain valid for the characteristic numbers $\nu_j(A_\sigma(k))$ of the operator $A_\sigma(k)$. All $\nu_j(A_\sigma(k))$ lie in the union of some neighborhood of the origin with a domain of the form*

$$\{\mu : |\operatorname{Im} \mu| \leq C_q |\operatorname{Re} \mu|^q\},$$

where $q = -1$ if $k_1 \neq 0$ and $k_2 < 0$, and q is a negative number with arbitrarily large absolute value if $k_1 \neq 0$ and $k_2 = 0$.

This follows from what was said in Subsections 5.6.5 and 5.6.1.

5. Additional results in the two-dimensional case. Let $n = 2$ and, for simplicity, $\sigma(x) \equiv 1$. The operator $A(k)$ is elliptic; it has the order -1, and its symbol is real. Moreover, if we take the normalized arc length as the coordinate along S (see (5.47)), then the Lamé coefficient h is constant and the principal symbol is independent of x. The results outlined in Subsection 5.5.4 imply that the eigenvalues and the corresponding root functions of the operator $A(k)$ have complete asymptotic expansions in decreasing powers of the eigenvalue number. Prior to giving initial terms of these expansions, let us comment on some details of the calculation.

First, we must represent the operator $A(k)$ in local coordinates in a more detailed form than the one given in Subsection 1. By virtue of the well-known formulas for the Bessel and Neumann functions $J_0(z)$ and $N_0(z)$ of order zero (e.g., see [1]), the kernel

$$\Phi(x - y, k) = -(4i)^{-1} H_0^{(2)}(k|x - y|) \tag{5.175}$$

of the operator $A(k)$ is the sum of an analytic function of $|x-y|^2$ and a series in the functions $|x-y|^{2l}\log|x-y|$, $l=0,1,\ldots$. The initial terms of this series have the form

$$(2\pi)^{-1}\log|x-y| - k^2(8\pi)^{-1}|x-y|^2\log|x-y| + \cdots \tag{5.176}$$

On S we introduce the parametrization $x = x(t) = (x_1(t), x_2(t))$, where t is defined by formula (5.47); let $y = x(\tilde{t})$ and $\tau = t - \tilde{t}$. Then the operator takes the form

$$A(k)\varphi(t) = \int_0^{2\pi} K_A(t, t-\tilde{t})\varphi(\tilde{t})\,d\tilde{t}, \tag{5.177}$$

where

$$K_A(t,\tau) = \frac{i}{8\pi}\,|S|H_0^2(k|x(t) - x(t-\tau)|). \tag{5.178}$$

Now we must use the Taylor expansions of the functions $x_j(t-\tau)$ in τ and the relations $|x'(t)|^2 = (2\pi)^{-2}|S|^2$ and $x'(t)\cdot x''(t) = 0$. We find that the kernel $K_A(t,\tau)$ is the sum of an infinitely smooth function and an asymptotic series of the form $\sum a_l(t)\tau^{2l}\log|\tau|$ as $\tau \to 0$, whose initial part has the form

$$(2\pi)^{-2}|S|\log|\tau| - 2^{-2}(2\pi)^{-4}k^2|S|^2\tau^2\log|\tau| + \cdots \tag{5.179}$$

The coefficient in the first omitted term depends on t. We omit the subsequent calculations and only give the result. The initial part of the expansion for the complete symbol of the operator (5.177) has the form

$$a(x,\xi) = \alpha_0|\xi|^{-1} + \alpha_2|\xi|^{-3} + \alpha_4(x)|\xi|^{-5} + O(|\xi|^{-6}), \tag{5.180}$$

where

$$\alpha_0 = -\frac{|S|}{4\pi}, \quad \alpha_2 = -\frac{k^2|S|^3}{4(2\pi)^3}, \quad \alpha_4(x) = -\frac{k^2|S|^5}{4(2\pi)^5}\Big(\frac{3}{4}\,k^2 + q^2(x)\Big) \tag{5.181}$$

and $q(x)$ is the curvature (that is, the absolute value of the second derivative of the position vector of a point with respect to the arc length). This implies the formula

$$\lambda_m(A) = \alpha_0|m|^{-1} + \alpha_2|m|^{-3} + \alpha_4|m|^{-5} + O(|m|^{-6}), \qquad m \to \pm\infty, \tag{5.182}$$

for the eigenvalue, where

$$\alpha_4 = \frac{1}{2\pi}\int_{-\pi}^{\pi}\alpha_4(y)\,dy. \tag{5.183}$$

Of course, the main term in (5.182) is in concordance with that in (5.168) for $n=2$. The asymptotic expansion of the corresponding root functions[16] has the form

$$u_{\pm m}(x) = e^{\pm imx}[1 \pm \gamma_3(x)|m|^{-3} + O(|m|^{-4})], \qquad m \to \pm\infty, \tag{5.184}$$

[16] See the footnote in Subsection 5.5.4.

where

$$\gamma_3(x) = i\alpha_0^{-1} \int_{-\pi}^{x} (\alpha_4(y) - \alpha_4)\, dy. \tag{5.185}$$

See also the remark at the end of Subsection 5.5.4. The proofs, as well as the complete calculations, can be found in [10].

6. The Dirichlet problem.[17] Suppose that zero is not an eigenvalue of the operator $A(k)$, and consider problem (5.1), (5.2) with the radiation condition at infinity for $\lambda = 0$. This problem splits into the interior and exterior Dirichlet problems with $u^+ = u^- = g$ on S. The interior problem is uniquely solvable by virtue of Proposition 5.7.3, and the exterior problem is always uniquely solvable (see [31, Chapter 3]). For this "double" Dirichlet problem, we can readily obtain a theorem similar to Theorem 5.7.2. Here we only indicate that the estimate (5.162) will be replaced by the estimate

$$\|u\|_{V^+,s+1/2} + \|u\|_{V_R^-,s+1/2} \leq C_s \|g\|_{S,s}. \tag{5.186}$$

This problem is equivalent to the Fredholm equation of the *first* kind $A(k)\varphi = g$. To solve this equation, we can again use series expansions in the root functions φ_j of the operator $A(k)$. However, one must take into account the fact that for $g \in H^s(S)$ the solution φ lies in $H_{s-1}(S)$ and satisfies the estimate $\|\varphi\|_{s-1} \leq C_s \|g\|_s$ (since A^{-1} is an operator of the first order), so that the Fourier series with parentheses of the function φ with respect to the system $\{\varphi_j\}$ will be convergent to φ only in $H_{s-1}(S)$ in general.

If $\lambda = 0$ is an eigenvalue of the operator $A(k)$, then the equation $A(k)\varphi = g$ is solvable if and only if the function g is orthogonal to $\operatorname{Ker} A^*(k) = \operatorname{Ker} A(k)$ (these kernels coincide since either $A(k)$ or $-A(k)$ is a dissipative operator; see Assertion 4° in Subsection 5.2.6), that is, to finitely many functions forming a basis in this subspace. To get rid of this restriction for the solution of the exterior Dirichlet problem, one can use the well-known method of "improvement" of the kernel Φ in the single layer potential. We shall touch upon this method in Subsection 5.8.3.

7. Cases in which the operator $A(k)$ is normal. Let $\sigma \equiv 1$, and let S be a circle (if $n = 2$) or a sphere (if $n = 3$). In this case, one can readily verify (see [82]) that A is a normal operator: $AA^* = A^*A$. As we mentioned in Subsection 5.2.4, this is equivalent to the existence of an orthonormal basis of eigenfunctions of $A(k)$ in $L_2(S)$; there are no associated functions. One can readily find the eigenfunctions by solving problem (5.1), (5.2) by separation of variables; namely, the eigenfunctions are sines and cosines for $n = 2$ and spherical functions for $n = 3$. Since the completeness of these orthogonal systems is well known, we again find that A is a normal operator. The eigenvalues are not real; essentially, they are known (e.g., see Eqs. (7.2.51) and (11.3.44) in [75] and also the end of Subsection 2.2.1 of the present book).

If S is an ellipse or an ellipsoid, then the operator A will be normal for a special choice of $\sigma(x)$ (see [6]).

[17] Cf. Subsection 2.2.1. See also the remarks concerning the Dirichlet and Neumann problems below in Subsection 5.8.1.

8. An example of a problem with associated functions. Consider the one-dimensional analog of problem (5.1), (5.2) with the radiation condition at infinity. In this case, S consists of two points, a and b, so that $L_2(S)$ is just the two-dimensional complex space $\mathbf{C}^2$. The analog of the operator A has the form

$$\frac{i}{2k}\begin{pmatrix} \sigma(a) & e^{-ik(b-a)}\sigma(a) \\ e^{-ik(b-a)}\sigma(b) & \sigma(b) \end{pmatrix}\begin{pmatrix} \varphi(a) \\ \varphi(b) \end{pmatrix}. \tag{5.187}$$

If $\sigma(a) = A$, $\sigma(b) = 1$, $B = e^{-ik(b-a)}$, we have the matrix

$$\begin{pmatrix} A & AB \\ B & 1 \end{pmatrix}.$$

Its eigenvalues coincide if $B^2 = -(A-1)^2/4A$, and in this case the matrix has an associated vector. For example, we can set $A = 3 \pm 2\sqrt{2}$, $B = \pm i$, $b - a = 1$, $k = (1+2l)\pi/2$, $l = \pm 1, \pm 2, \ldots$.

9. The case of a complex-valued function $\boldsymbol{\sigma(x)}$. So far, we have assumed that $\sigma(x)$ is positive. Now we reject this assumption and suppose that $\sigma(x)$ is a nowhere vanishing complex function belonging to $C^\infty(S)$ with values in the sector $\Psi_0 = \{\lambda : |\arg\lambda| \le \beta\}$. In this case, we cannot treat $A_\sigma(k)$ as an operator close to a selfadjoint one, and so Theorem 5.7.7 is no longer valid. For simplicity, we assume that the interior homogeneous Dirichlet problem for the given value of k has no nontrivial solutions. Then the inverse $L_\sigma(k) = A_\sigma^{-1}(k)$ of $A_\sigma(k)$ exists and is a first-order pseudodifferential operator. Let us reduce Eq. (5.157) to the form

$$\mu\varphi - L_\sigma(k)\varphi = h, \tag{5.188}$$

where $\mu = \lambda^{-1}$ and $h = \mu L_\sigma(k)g$. Let Ψ_1 be the sector symmetric to Ψ_0 with respect to the origin and Ψ a closed sector with vertex at the origin having no common points with Ψ_1 except the origin.

Theorem 5.7.10 *Let $s \in \mathbf{R}$ be given. There exist $r > 0$ and $C_s > 0$ such that for any $\mu \in \Psi$ with $|\mu| > r$ and any $h \in H^s(S)$, Eq. (5.188) has a unique solution $\varphi \in H^{s+1}(S)$, and moreover, this solution satisfies the estimate*

$$\|\varphi\|_{S,s+1} + |\mu| \cdot \|\varphi\|_{S,s} \le C_s \|h\|_{S,s}. \tag{5.189}$$

If $2\beta < \pi/(n-1)$ and $\alpha \in (n-1, \pi/2\beta)$, then $L \in \mathbf{A}(\alpha, H^s(S))$.

Proof. To be definite, let $n = 3$. We assign the weight 1 to the parameter μ. Then the principal symbol with parameter of the left-hand side of (5.188) will be

$$\mu + 2(h_1^{-2}(x)\xi_1^2 + h_2^{-2}(x)\xi_2^2)^{1/2}\sigma^{-1}(x).$$

Obviously, this expression does not vanish if ξ_1 and ξ_2 are real, $\mu \in \Psi$, and $(\xi_1, \xi_2, \mu) \ne 0$. Hence we can apply the theorem about pseudodifferential operators elliptic with parameter (2° in Subsection 5.5.6), thus obtaining all but the last assertion of Theorem 5.7.10.

The last assertion follows from Theorem 5.6.5: since $L_\sigma(k)$ has the order 1 in the Sobolev scale, it follows that $s_j(L_\sigma^{-1}(k)) \leq Cj^{-1/(n-1)}$ (see 4° in Subsection 5.6.1), and the decay of the resolvent at a rate of $|\mu|^{-1}$ follows from (5.189). □

Note also that Theorem 5.7.4 remains valid and that Assumption 2° from Subsection 5.6.5 can be applied. Namely, *the eigenvalues* μ_j *of* $L_\sigma(k)$ *satisfy the relation* $|\mu_j| \asymp j^{1/(n-1)}$ *if* $2\beta < \pi/(n-1)$.

5.8 Other Scalar Problems with Spectral Parameter in Boundary or Transmission Conditions

1. Statement of the problems and reduction to equations on S

In this section we consider the following problems.[18]

Problem II (the interior problem with impedance condition). We seek a solution of the Helmholtz equation (5.1) in V^+ under the condition

$$\sigma(x)u^+(x) + \lambda\partial_\nu u^+(x) = g(x) \quad \text{on } S. \tag{5.190}$$

Problem III (the exterior problem with impedance condition). We seek a solution of Eq. (5.1) in V^- under the condition

$$\sigma(x)u^-(x) - \lambda\partial_\nu u^-(x) = g(x) \quad \text{on } S \tag{5.191}$$

and the radiation condition (5.137) at infinity.

Problem IV (the second problem with spectral parameter in transmission conditions). We seek a solution of Eq. (5.1) outside S under the conditions

$$\partial_\nu u^+(x) = \partial_\nu u^-(x) \tag{5.192}$$

and

$$\sigma(x)[u^-(x) - u^+(x)] - \lambda\partial_\nu u^\pm(x) = g(x) \tag{5.193}$$

on S and the radiation condition at infinity.

Just as in Section 5.7, k satisfies conditions (5.3), and $\sigma(x)$ is an infinitely smooth function on S; we assume that it is positive.

To reduce these problems to pseudodifferential equations on S, we shall use relations (5.152) and (5.153) on S. To this end, we need several auxiliary propositions.

[18]These problems are related to the w-method and the second version of the ρ-method (see Sections 2.1 and 2.2).

Proposition 5.8.1 *$B(k)$ is a pseudodifferential operator of order ≤ -1.*

Proof. To be definite, assume that $n = 3$. We shall use the notation introduced in the proof of Theorem 5.7.1. If the supports of functions $\omega_1, \omega_2 \in C^\infty(S)$ are disjoint, then $\omega_1 B(k)(\omega_2\varphi)$ is an infinitely smoothing operator. Let the supports of the functions ω_1 and ω_2 lie in the same coordinate neighborhood with local orthogonal coordinates t. Then $\omega_1 B(k)(\omega_2\varphi)$ is an integral operator with kernel

$$-(4\pi)^{-1}\omega_1(t)\Phi'(|x-y|,k)\partial_\nu|x-y|\,\omega_2(\tilde{t})h_1(\tilde{t})h_2(\tilde{t}),$$

where $x = x(t)$ and $y = x(\tilde{t})$. The vector product $(\partial x/\partial t_1) \times (\partial x/\partial t_2)$ is directed along the normal to S, and one can readily calculate that for an appropriate orientation of the t_1- and t_2-axes, we shall have

$$\partial_\nu|x-y| = |x-y|^{-1}h_1^{-1}(\tilde{t})h_2^{-1}(\tilde{t})D(t,\tilde{t}),$$

where

$$D(t,\tilde{t}) = \begin{vmatrix} x_1(\tilde{t}) - x_1(t) & x_2(\tilde{t}) - x_2(t) & x_3(\tilde{t}) - x_3(t) \\ \partial x_1(\tilde{t})/\partial\tilde{t}_1 & \partial x_2(\tilde{t})/\partial\tilde{t}_1 & \partial x_3(\tilde{t})/\partial\tilde{t}_1 \\ \partial x_1(\tilde{t})/\partial\tilde{t}_2 & \partial x_2(\tilde{t})/\partial\tilde{t}_2 & \partial x_3(\tilde{t})/\partial\tilde{t}_2 \end{vmatrix}. \tag{5.194}$$

Let us rewrite this function in the form $D(t,t+\tau) = D_1(t,\tau)$, where $\tau = \tilde{t}-t$. It belongs to C^∞, and its Taylor expansion in powers of τ does not contain terms of order < 2. Thus the operator $B(k)$ satisfies the assumptions of Proposition 5.4.1 with $p \geq 2$, $q = 3$, and $n = 3$; this provides the desired result. For $n = 2$, one can use Proposition 5.4.3. □

For $n = 2$, more accurate calculations show that $B(k)$ is of the order -3. We shall verify this at the end of Subsection 2.

Corollary 5.8.2 *$\frac{1}{2}I + B(k)$ and $\frac{1}{2}I - B(k)$ are elliptic operators of order zero. Their kernels $\operatorname{Ker}\left(\frac{1}{2}I + B(k)\right)$ and $\operatorname{Ker}\left(\frac{1}{2}I - B(k)\right)$ are finite-dimensional and consist of infinitely smooth functions.*

Proposition 5.8.3 *One has*

$$\operatorname{Ker}\left(\frac{1}{2}I + B'(k)\right) = \{\partial_\nu u^+ : \Delta u + k^2 u = 0 \;\; in \;\; V^+,\; u^+ = 0\}, \tag{5.195}$$

$$\operatorname{Ker}\left(\frac{1}{2}I - B(k)\right) = \{u^+ : \Delta u + k^2 u = 0 \;\; in \;\; V^+,\; \partial_\nu u^+ = 0\}. \tag{5.196}$$

The dimensions of these kernels coincide with the dimensions of the spaces of solutions of the corresponding homogeneous problems.

The proof of these well-known relations can be found in [31, Chapter 3]. It follows from (5.195) and (5.161) that $\operatorname{Ker} A$ and $\operatorname{Ker}\left(\frac{1}{2}I + B'(k)\right)$ coincide.

Let us now state two conditions; they will be used until Subsection 3.

Condition I *The interior homogeneous Dirichlet problem* (5.160) *has no nontrivial solution for a given k.*

Condition II *The interior homogeneous Neumann problem*

$$\Delta u + k^2 u = 0 \quad \textit{in } V^+, \qquad \partial_\nu u^+ = 0 \quad \textit{on } S \tag{5.197}$$

has no nontrivial solution for a given k.

The violation of Condition I means that k^2 is an eigenvalue of the operator $-\Delta_D$, the Laplacian (with the opposite sign) with homogeneous Dirichlet condition. It is well known that the eigenvalues k^2 are positive and form a sequence that tends to $+\infty$. If we number these eigenvalues taking their multiplicities into account, we obtain a sequence $\{k_j^2\}$ with the asymptotics $C'j^{2/n}$, where $C' = C'(n) > 0$ is indicated after formula (5.97). Likewise, the violation of Condition II means that k^2 is an eigenvalue of the operator $-\Delta_N$, the Laplacian (with the opposite sign) with homogeneous Neumann condition. The eigenvalues k^2 are nonnegative and also form a sequence tending to $+\infty$. If we number these eigenvalues taking their multiplicities into account, we obtain a sequence $\{\tilde{k}_j^2\}$ with the same asymptotics $C'j^{2/n}$ and with the same coefficients $C' = C'(n)$.

Let us also recall that the interior Dirichlet problem for Eq. (5.1) with condition $u^+ = g \in H^{3/2}(S)$ has a unique solution $u \in H^2(V^+)$ for all k satisfying Condition I. Likewise, the interior Neumann problem for (5.1) with condition $\partial_\nu u^+ = g \in H^{1/2}(S)$ has a unique solution $u \in H^2(V^+)$ for all k satisfying Condition II. The exterior Dirichlet and Neumann problems with the radiation condition at infinity are always uniquely solvable, and moreover, the inclusion $u^+ \in H^{3/2}(S)$ or $\partial_\nu u^+ \in H^{1/2}(S)$ implies $u \in H^2(V_R^-)$ for arbitrarily large R if k is real and $u \in H^2(V^-)$ if $k_2 < 0$.

Proposition 5.8.4 *Condition* I *is equivalent to the invertibility of the operators* $\frac{1}{2}I + B(k)$ *and* $\frac{1}{2}I + B'(k)$. *Condition* II *is equivalent to the invertibility of the operators* $\frac{1}{2}I - B(k)$ *and* $\frac{1}{2}I - B'(k)$.

Indeed, the pseudodifferential operators $\frac{1}{2}I + B(k)$, $\frac{1}{2}I + B^*(k)$, and $\frac{1}{2}I + B'(k)$ have the same real principal symbol $1/2$, and hence the index of each of these operators is zero. The first two of these operators are invertible or not invertible simultaneously (see Assertions 5° and 6° in Subsection 5.5.1); the same is true of the last two operators, since $B^*(k) = \overline{B'(k)}$. Thus, the kernel (5.195) is trivial if and only if all the three operators are invertible.

The second assertion of Proposition 5.8.4 is verified in a similar way. □

In particular, *all operators in the statement of Proposition* 5.8.4 *are invertible for nonreal* k.

Now we introduce the operators

$$T^+(k) = \left(\frac{1}{2}I - B(k)\right)^{-1} A(k) \quad \text{and} \quad T^-(k) = \left(\frac{1}{2}I + B(k)\right)^{-1} A(k) \tag{5.198}$$

under Conditions II and I, respectively. From (5.152) and (5.153) we obtain

$$u^+ = -T^+(k)\partial_\nu u^+ \quad \text{and} \quad u^- = T^-(k)\partial_\nu u^- \tag{5.199}$$

for solutions of the Helmholtz equation in V^+ and V^-, respectively, smooth up to S, with the radiation condition at infinity in the second case. Thus, the role of these operators is as follows: *they map the Neumann data for the Helmholtz equation into the Dirichlet data in the interior and exterior domains, respectively, with the sign being changed in the first case.*

Proposition 5.8.5 $B(k)A(k) = A(k)B'(k)$.

This relation can be found in the literature. To verify it, first we assume that k is not real and derive the second expression, say, for u^+ via $\partial_\nu u^+$. To this end, in V^+ we consider the single layer potential (5.139) with smooth density φ. According to (5.143) and (5.144), we have

$$u^+ = A(k)\varphi, \qquad \partial_\nu u^+ = -\left(\frac{1}{2}I - B'(k)\right)\varphi.$$

Hence

$$u^+ = -A(k)\left(\frac{1}{2}I - B'(k)\right)^{-1} \partial_\nu u^+.$$

By comparing this formula with the first formula in (5.199), we obtain (with regard to the first formula in (5.198))

$$A(k)\left(\frac{1}{2}I - B'(k)\right)^{-1} = \left(\frac{1}{2}I - B(k)\right)^{-1} A(k)$$

on smooth functions and hence on all functions from $H^0(k)$. It follows that $B(k)A(k) = A(k)B'(k)$. By passing to the limit, we obtain this relation for real k. □

Corollary 5.8.6 *One has*

$$T^{+*}(k) = \overline{T}^+(k), \qquad T^{-*}(k) = \overline{T}^-(k). \tag{5.200}$$

Indeed, for example, we have

$$T^{+*}(k) = \Big[A(k)\Big(\frac{1}{2}I - B'(k)\Big)^{-1}\Big]^*$$
$$= \Big(\frac{1}{2}I - \overline{B}(k)\Big)^{-1}A^*(k) = \Big(\frac{1}{2}I - \overline{B}(k)\Big)^{-1}\overline{A}(k) = \overline{T}^*(k).$$

Under Conditions I and II, we set

$$T(k) = T^+(k) + T^-(k) = \Big(\frac{1}{4}I - B^2(k)\Big)^{-1}A. \tag{5.201}$$

It follows from (5.200) that

$$T^*(k) = \overline{T}(k). \tag{5.202}$$

Proposition 5.8.7 (cf. [98]**)** *Under Conditions* I *and* II, *for* $\psi \in H^1(S)$ *we have*

$$\partial_\nu \mathcal{B}(k)\psi^\pm = -T^{-1}(k)\psi. \tag{5.203}$$

Proof. Let Condition I be satisfied. Consider the functions

$$u = \mathcal{B}(k)\varphi \quad \text{and } v = -\mathcal{A}(k)\psi, \quad \text{where } \psi = A^{-1}(k)\Big(\frac{1}{2}I - B(k)\Big)\varphi$$

and $\varphi \in H^1(S)$. By virtue of (5.145) and (5.143), u and v are solutions of the same exterior Dirichlet problem for Eq. (5.1). Hence $u = v$ in V^- and $\partial_\nu u^- = \partial_\nu v^-$, whence, with regard to (5.144), we obtain

$$\partial_\nu \mathcal{B}(k)\varphi^\pm = -\Big(\frac{1}{2}I + B'(k)\Big)A^{-1}(k)\Big(\frac{1}{2}I - B(k)\Big)\varphi.$$

It remains to apply Proposition 5.8.5. □

Now we set

$$T_\sigma^\pm(k) = \sigma T^\pm(k) \quad \text{and} \quad T_\sigma(k) = T_\sigma^+(k) + T_\sigma^-(k). \tag{5.204}$$

One can readily verify that the substitutions $\varphi^\pm = \partial_\nu u^\pm$ reduce Problems II and III to the equations

$$T_\sigma^+(k)\varphi^+ - \lambda\varphi^+ = -g \quad \text{and} \quad T_\sigma^-(k)\varphi^- - \lambda\varphi^- = g \tag{5.205}$$

on S, respectively, whereas Problem IV is reduced by the substitution $\varphi = \partial_\nu u^\pm$ to the equation

$$T_\sigma(k)\varphi - \lambda\varphi = g \tag{5.206}$$

on S. Here we assume that Conditions I and/or II hold, which guarantee the existence of the corresponding operators. These equations on S, for λ lying outside the spectrum of the corresponding operator, are equivalent to the original problems. For example, if φ is a solution of Eq. (5.206), then the corresponding solution u of Problem IV can be reconstructed as the solution of the interior and exterior Neumann problems with $\partial_\nu u^\pm = \varphi$. For the case $g = 0$, if λ belongs to the spectrum of the corresponding operator, the equivalence takes place only for the eigenfunctions. If the operators thus constructed are nonselfadjoint, they may have associated functions.

For $\sigma \not\equiv 1$, we shall use the inner product (5.167). It follows from Corollary 6 (which pertains to the case $\sigma \equiv 1$) that

$$T_\sigma^{+*}(k) = \overline{T}_\sigma^{+}(k), \quad T_\sigma^{-*}(k) = \overline{T}_\sigma^{-}(k), \quad T_\sigma^{*}(k) = \overline{T}_\sigma(k). \tag{5.207}$$

Theorem 5.8.8 *$T_\sigma^+(k)$, $T_\sigma^-(k)$, and $T_\sigma(k)$, as well as the real parts of these operators, are elliptic pseudodifferential operators of order -1. Their principal symbols are equal to $2a$, $2a$, and $4a$, respectively, where a is the principal symbol of the pseudodifferential operator $A_\sigma(k)$ (see (5.158)).*

This follows from the definitions of our operators and the composition theorem (Assertion 3° in Subsection 5.4.1).

Corollary 5.8.9 *The root functions of the operators $T_\sigma^\pm(k)$ and $T(k)$ belong to $C^\infty(S)$.*

For Problems II, III, and IV, theorems similar to Theorem 5.7.4 are valid. We do not dwell on these theorems.

2. Spectral properties of the operators $T_\sigma^\pm(k)$ and $T_\sigma(k)$

Proposition 5.8.10 *The operators $T_\sigma^\pm(k)$ and $T(k)$ are selfadjoint for pure imaginary k, and the operator $T_\sigma^+(k)$ is selfadjoint also for real k.*

Proof. It suffices to carry out the proof for $\sigma \equiv 1$ for the operators $T^\pm(k)$. For real k, the selfadjointness of $T^+(k)$ can be observed from Green's formula (5.149), which implies that for any two solutions of the Helmholtz equation one has

$$\int_S \partial_\nu u^+ \cdot \overline{v}^+ dS = \int_S u^+ \cdot \partial_\nu \overline{v}^+ dS.$$

For pure imaginary k, the operators $A(k)$ and $B(k)$ have real kernels; hence $T^\pm(k) = \overline{T}^\pm(k)$, and it suffices to use Corollary 6. □

Proposition 5.8.11 *If $k_2 = 0$, then $\operatorname{Im} T_\sigma^-(k)$ and $\operatorname{Im} T_\sigma(k)$ are operators of order $-\infty$. If $k_2 < 0$ and $k_1 \neq 0$, then $\operatorname{Im} T_\sigma^\pm(k)$ and $\operatorname{Im} T_\sigma(k)$ are operators of order -3.*

Proof. It again suffices to consider $T^\pm(k)$ for $\sigma \equiv 1$.

We set

$$B_1(k) = \frac{1}{2}(B(k) + \overline{B}(k)), \qquad B_2(k) = \frac{1}{2i}(B(k) - \overline{B}(k)) \tag{5.208}$$

and define $A_1(k)$ and $A_2(k)$ in a similar way (these are the real and imaginary parts of $A(k)$). We know that $A_2(k)$ is an operator of order $-\infty$ for $k_2 = 0$, $k_1 \neq 0$ and of order -3 for $k_2 < 0$, $k_1 \neq 0$ (Proposition 5.7.6 for $\sigma \equiv 1$). The same is true of $B_2(k)$. Indeed, this is an integral operator with kernel $\partial_{\nu_y} \operatorname{Im} \Phi(x-y,k)$. For $k_2 = 0$, $k_1 \neq 0$ this kernel is infinitely smooth together with $\operatorname{Im} \Phi(x-y,k)$. For $k_2 < 0$, $k_1 \neq 0$, and $n = 3$, the singularity of this kernel is determined by the term $\partial_{\nu_y}|x-y|$. We see from the proof of Proposition 5.8.1 that this is a kernel of a pseudodifferential operator of order -3. Now consider the case in which $k_2 < 0$, $k_1 \neq 0$, and $n = 2$. In the following, we shall show that for $n = 2$ even the pseudodifferential operator $B(k)$ has the order -3 for every k. It follows that the same is true of $\overline{B}(k)$.

To be definite, let us now consider the operator $T^-(k)$. In view of Corollary 6, we have

$$\begin{aligned}\operatorname{Im} T^-(k) &= \frac{1}{2i}\Big[\Big(\frac{1}{2}I + B(k)\Big)^{-1} A(k) - \Big(\frac{1}{2}I + \overline{B}(k)\Big)^{-1} \overline{A}(k)\Big] \\ &= \Big(\frac{1}{2}I + B(k)\Big)^{-1} A_2(k) - \Big(\frac{1}{2}I + B(k)\Big)^{-1} B_2(k)\Big(\frac{1}{2}I + \overline{B}(k)\Big)^{-1} \overline{A}(k).\end{aligned}$$

The first term on the right-hand side has the order -3, and the second term is even of the order -4.

In a similar way, one can consider $T^+(k)$. □

Theorem 5.8.12 *For the characteristic numbers of the operators $T_\sigma^\pm(k)$ and $\frac{1}{2}T_\sigma(k)$ and the real parts of these operators, formulas* (5.168) *hold with c_2 replaced by $c_2/2$ and c_3 replaced by $c_3/2$.*

The proof is based on the results formulated in Subsections 5.5.2 and 5.6.5 and on Theorem 5.8.8.

Theorem 5.8.13 *The assertions of Theorem* 5.7.7 *concerning the summability remain valid for the operators $T_\sigma^-(k)$ and $T_\sigma(k)$, as well as (if $k_2 \neq 0$) for $T_\sigma^+(k)$ instead of $A_\sigma(k)$.*

Proof. The proof is similar to that of Theorem 5.7.7. □

Proposition 5.8.14 *For $k_1 > 0$ and $k_2 < 0$, the operator $T_\sigma^+(k)$ is dissipative, and its eigenvalues lie in the open upper half-plane. For $k_1 < 0$ and $k_2 < 0$, the same is true of the operator $-T_\sigma^+(k)$. For the operators $\pm T_\sigma^-(k)$ and $\pm T_\sigma(k)$, similar assertions hold for $\pm k_1 > 0$, $k_2 \leq 0$. For $|k_1| \leq |k_2|$, the real parts of the eigenvalues of the operators $T_\sigma^\pm(k)$ and $T_\sigma(k)$ are negative.*

Proof. Just as in the proof of Proposition 5.7.8, we reduce the problem to considering the quadratic forms of the operators $T^\pm(k)$ for $\sigma \equiv 1$ with the aid of Green's formula (5.148) with $v = \overline{u}$.

For the operator $T^+(k)$, we obtain

$$(T^+(k)\varphi,\varphi)_S = \overline{k}^2 \int_{V+} |u|^2 dx - \int_{V+} |\nabla u|^2 dx, \tag{5.209}$$

where u is the solution of the Neumann problem for the Helmholtz equation (5.1) with $\partial_\nu u^+ = \varphi$. Hence

$$\mathrm{Re}(T^+(k)\varphi,\varphi)_S = (k_1^2 - k_2^2) \int_{V+} |u|^2 dx - \int_{V+} |\nabla u|^2 dx \tag{5.210}$$

and

$$\mathrm{Im}(T^+(k)\varphi,\varphi)_S = -2k_1k_2 \int_{V_+} |u|^2 dx. \tag{5.211}$$

For the operator $T^-(k)$, we apply Green's formula (5.148) to u and $v = \overline{u}$ in V_R^- for sufficiently large R, where u is the solution of the exterior Neumann problem with $\partial_\nu u^- = \varphi$. We obtain

$$\begin{aligned} -k^2 \int_{V_R^-} |u|^2 dx = &- \int_{V_R^-} |\nabla u|^2 dx \\ &- \int_{S_R} \partial_\nu u^- \cdot \overline{u}^- dx - ik \int_{S_R} |u|^2 dx + o(1). \end{aligned} \tag{5.212}$$

Hence for $k_2 < 0$, after passing to the limit as $R \to \infty$, by separating the real and imaginary parts, we obtain

$$\mathrm{Re}(T^-(k)\varphi,\varphi)_S = (k_1^2 - k_2^2) \int_{V-} |u|^2 dx - \int_{V-} |\nabla u|^2 dx, \tag{5.213}$$

$$\mathrm{Im}(T^-(k)\varphi,\varphi)_S = -2k_1k_2 \int_{V-} |u|^2 dx, \tag{5.214}$$

whereas for $k_2 = 0$ and $k_1 \neq 0$ we obtain

$$\mathrm{Im}(T^-(k)\varphi,\varphi)_S = k_1 \lim_{R\to\infty} \int_{S_R} |u|^2 dx. \tag{5.215}$$

Here the limit exists, since the left-hand side is independent of R; if the limit is zero, then $u = 0$ in V^- and $\varphi = \partial_\nu u^- = 0$.

Relations (5.211), (5.214), and (5.215) imply the dissipativity of the operators $\pm T^+(k)$ and $\pm T^-(k)$ for $\pm k_1 > 0$. Relations (5.210) and (5.213) imply the assertions about the signs of the real parts of the eigenvalues of the operators $T^\pm(k)$.

It remains to note that the quadratic form of the operator $T(k)$ is the sum of the quadratic forms of the operators $T^+(k)$ and $T^-(k)$ and that the formulas that we used remain valid for the quadratic forms $\langle T_\sigma^\pm(k)\varphi,\varphi\rangle_S$ of the operators $T_\sigma^\pm(k)$. □

Proposition 5.8.15 *For $k_1 \neq 0$, the characteristic numbers of the operators $T_\sigma^-(k)$ and $T_\sigma(k)$ lie in a domain of the same form as the characteristic numbers of the operator $A_\sigma(k)$ (see Proposition 5.7.9). For $T_\sigma^+(k)$ this is also true if $k_1 \neq 0$ and $k_2 < 0$.*

The proof is the same as that of Proposition 5.7.9.

For $k_2 = 0$, as we have already noted in Proposition 5.8.10, $T_\sigma^+(k)$ is a selfadjoint operator. Thus its characteristic numbers are real, and we know their asymptotics (see Theorem 5.8.12). The number 0 can be an eigenvalue of the operator $T_\sigma^+(k)$; if this is the case, then $\operatorname{Ker} T_\sigma^+(k) = \operatorname{Ker} A_\sigma(k)$.

Remark 5.8.16 If S is a circle or a sphere, then $T^\pm(k)$ and $T(k)$ are normal operators with the same eigenfunctions as $A(k)$ but with different eigenvalues; both can be found by separation of variables. Cf. Subsection 5.7.7 and also Subsection 0.1.2 and Subsection 2.2.1.

Remark 5.8.17 Now let us describe some *strengthening of the above results in the two-dimensional case*, where S is a closed curve on the plane. In this case, for $T^\pm(k)$ and $T(k)$ (as well as for $A(k)$, see Subsection 5.7.5) one can obtain complete asymptotic expansions of the eigenvalues and the corresponding root functions in negative powers of the eigenvalue number.

First, let us verify that in this case the pseudodifferential operator $B(k)$ *has the order* -3. By using the formula

$$H_0^{(2)\prime}(z) = -H_1^{(2)}(z) = -J_1(z) + iN_1(z),$$

we can write (e.g., see [1])

$$\begin{aligned}\partial_{\nu_y}\Phi(x-y,k) &= k(4i)^{-1}H_0^{(2)\prime}(k|x-y|)\partial_{\nu_y}|x-y| \\ &= c_1 H_1^{(2)}(k|x-y|)|x-y|^{-1}D(x-y) \\ &= [c_2\log|x-y| + c_3|x-y|^{-2} + \ldots]D(t,\tilde{t}),\end{aligned}$$

where the dots in square brackets stand for the sum of an entire analytic function of $|x-y|^2$ and a series in $|x-y|^{2l}\log|x-y|$, $l \geq 1$, and

$$D(t,\tilde{t}) = \begin{vmatrix} x_1(\tilde{t}) - x_1(t) & x_2(\tilde{t}) - x_2(t) \\ \partial x_1(\tilde{t})/\partial\tilde{t} & \partial x_2(\tilde{t})/\partial\tilde{t} \end{vmatrix} = \frac{\tau^2}{2}\begin{vmatrix} x_1''(t) & x_2''(t) \\ x_1'(t) & x_2'(t) \end{vmatrix} + O(\tau^3)$$

up to the sign. Here $x = x(t)$, $y = x(\tilde{t})$, $\tau = \tilde{t} - t$, and we use the notation from Subsection 5.7.5. In these local coordinates, we have

$$B(k)\varphi(t) = \int_{-\pi}^{\pi} L_B(t, t-\tilde{t})\varphi(\tilde{t})\,d\tilde{t},$$

where the kernel is the sum of an infinitely smooth function and an asymptotic series in $\tau^k\log|\tau|$, which starts from $\tau^2\log|\tau|$. (The coefficient of the last function is proportional to the curvature of S at $x(t)$.) Proposition 5.4.3 implies the desired assertion.

Now let us note that

$$\left[\frac{1}{2}I \pm B(k)\right]^{-1} A(k) = 2[I \mp 2B(k) + 2B^2(k)(I \mp 2B(k))^{-1}]A(k).$$

We see that the pseudodifferential operators $T^{\pm}(k)$ and $\frac{1}{2}T(k)$ differ from $2A(k)$ by a term of the order -4. Hence, without any additional computations, for the eigenvalues of these operators we obtain the formula

$$\lambda_m = 2\alpha_0|m|^{-1} + 2\alpha_2|m|^{-3} + O(|m|^{-4}), \qquad m \to \pm\infty \tag{5.216}$$

(see (5.181)), and for the root functions we obtain the formula

$$u_{\pm}(t) = e^{\pm imt} + O(m^{-2}), \qquad m \to \pm\infty. \tag{5.217}$$

Just as in Section 5.7, we can also consider the case in which the function $\sigma(x)$ is complex-valued; we do not dwell on this subject.

3. Cases in which Condition I or Condition II is violated. Recall that Conditions I and II pertain to real k and were not imposed in the study of Problem I. For simplicity, let us assume that $\sigma(x) \equiv 1$.

Problem II. So far, this problem was considered for k satisfying Condition II. If this condition is violated, then Problem II has the "eigenvalue" ∞ in the following sense: if we set $\lambda = 1/\mu$ and rewrite the homogeneous boundary condition in the form $\partial_\nu u^+ + \mu u^+ = 0$, then $\mu = 0$ is an eigenvalue. First, assume that in this case Condition I is satisfied, so that zero is not an eigenvalue of Problem II. Then this problem can be reduced to the equation

$$\Theta^+(k)\psi = \mu\psi, \tag{5.218}$$

where

$$\Theta^+(k) = A^{-1}(k)\Big[\frac{1}{2}I - B(k)\Big], \qquad \psi = u^+,\ \mu = \lambda^{-1}. \tag{5.219}$$

This is an unbounded operator in $H^0(S)$ with domain $H^1(S)$. The space $H^1(S)$ is compactly embedded in $H^0(S)$. For $k_2 = 0$, it follows from Green's formula that this is a selfadjoint operator. Consequently, for each k this is an operator with discrete spectrum. We can consider it as a weak perturbation of the operator $\operatorname{Re}\Theta^+(k)$ and obtain results similar to those given in Subsection 2.

However, we must also consider the case in which k satisfies neither Condition I nor Condition II. Then we have to transform Problem II. To this end, we again rewrite the homogeneous boundary condition in the form $\partial_\nu u^+ + \mu u^+ = 0$ and set $\mu = \tilde{\mu} + h$ and $\tilde{\lambda} = 1/\tilde{\mu}$. Now the boundary condition has the form

$$\tilde{\lambda}(\partial_\nu u^+ + hu^+) + u^+ = 0, \quad \text{where } \tilde{\lambda} = \lambda[1 - \lambda h]^{-1}.$$

The value $\lambda = 0$ corresponds to $\tilde{\lambda} = 0$, but the value $\lambda = \infty$ now corresponds to the finite value $\tilde{\lambda} = -h^{-1}$. From now on, we omit the tildes and write λ instead of $\tilde{\lambda}$, so that the new boundary condition has the form

$$\lambda(\partial_\nu u^+ + hu^+) + u^+ = 0. \tag{5.220}$$

Essentially, we have only shifted the spectral parameter. Let us rewrite relation (5.152) in the form

$$\left[\frac{1}{2}I - B(k) - hA(k)\right]u^+ = -A(k)(\partial_\nu u^+ + hu^+). \tag{5.221}$$

Now we must choose an h such that the operator on the left-hand side in (5.221) annihilates only the zero function. Since it is a Fredholm operator with index zero, it follows that it is invertible for such h. By setting

$$\widetilde{T}^+(k) = \widetilde{T}^+(k,h) = -\left[\frac{1}{2}I - B(k) - hA(k)\right]^{-1} A(k), \qquad \varphi = \partial_\nu u^+ + hu^+, \tag{5.222}$$

we reduce the transformed homogeneous Problem II to the equation

$$\widetilde{T}^+(k)\varphi = \lambda\varphi. \tag{5.223}$$

It is known that the desired h exists if S is smooth (see [31, Chapter 3]). Indeed, suppose that

$$\left[-\frac{1}{2}I + B(k) + hA(k)\right]\varphi = 0.$$

Since the operator on the left-hand side is elliptic (its principal symbol is equal to $-1/2$), it follows that φ is a smooth function. Consider the function $u = [\mathcal{B}(k) + h\mathcal{A}(k)]\varphi$. By (5.143) and (5.145), we have $u^- = 0$, so that $u = 0$ in V^- and $\partial_\nu u^- = 0$. Furthermore, by (5.143)–(5.146), we have $u^+ = \varphi$ and $\partial_\nu u^+ = -h\varphi$. Now we write out Green's first formula (5.148) with $v = \overline{u}$ and separate the imaginary parts. We obtain $0 = \operatorname{Im} h \int_S |\varphi|^2 dS$, and so we see that if h is not real, then $\varphi = 0$.

Thus we can set, say, $h = i$. Note, however, that if h_1 is not a characteristic number of the corresponding operator $\widetilde{T}^+(k,i)$, then it follows from the equation

$$-h_1 A(k)\varphi = \left[\frac{1}{2}I - B(k) - iA(k)\right]\varphi$$

that $\varphi = 0$. Then the operator $\dfrac{1}{2}I - B(k) - (i - h_1)A(k)$ is also invertible. Hence h can be chosen to be real, which guarantees the selfadjointness of the operator $\widetilde{T}^+(k)$. Essentially, we can choose an arbitrary h such that the problem

$$\Delta u + k^2 u = 0 \quad \text{in } V^+, \qquad \partial_\nu u^+ + hu^+ = 0 \quad \text{on } S \tag{5.224}$$

has only the trivial solution and replace $\partial_\nu u^+$ in the statement of Problem II by $\partial_\nu u^+ + hu^+$.

Problem III. So far, this problem was considered under the assumption that Condition I is satisfied. Now suppose that this condition is violated; to be definite, let k be *positive*. The operator $T^-(k)$, which maps the exterior Neumann data into the exterior Dirichlet data, exists for all k, but now the potentials used so far do not allow us to obtain an explicit expression for $T^-(k)$.

Let us use the well-known method of modification of potential kernels for the solution of exterior problems for exceptional values of k (see [31, Subsection 3.6], and the references therein). In this method, one replaces the fundamental solution (5.138) by Green's function of the exterior problem, which is constructed as follows.

Let E_r^+ be a ball of small radius r lying in V^+ together with its boundary S_r. If r is sufficiently small, then the given value of k satisfies Condition I in E_r^+. Indeed, without loss of generality, we can assume that $0 \in V^+$ and 0 is the center of this ball. Now if $v(x)$ is a solution of Eq. (5.1) in E_r^+, then $w(x) = v(rx)$ is a solution of the equation $\Delta w + (kr)^2 w = 0$ in the unit ball $E_1^+ = \{x : |x| < 1\}$, and it suffices to choose r so that $(kr)^2$ be less than the first eigenvalue of the operator $-\Delta_D$ in the unit ball.

We denote the complement of the closure of E_r^+ by E_r^-. Consider the problem

$$\Delta u + k^2 u = f \quad \text{in } E_r^-, \qquad \beta \partial_\nu u - u = 0 \quad \text{on } S_r \tag{5.225}$$

with compactly supported f and with the radiation condition at infinity, where ν is the normal pointing into the complement of the ball. For $f = 0$, this is a problem of the type III with spectral parameter $\lambda = \beta$, and we already know that for $\operatorname{Im}\beta < 0$ the uniqueness holds (since the signs of the imaginary parts of the eigenvalues coincide with the sign of k), and hence we have the unique solvability. Green's function of this problem has the form

$$\Psi(x,y,k) = \Phi(x-y,k) + \psi(x,y,k), \tag{5.226}$$

where the function ψ is chosen so that $\Psi(x,y,k)$, considered as a function of x, satisfies the Helmholtz equation (5.1) in E_r^- and the boundary condition in (5.225) for $y \neq x$ and $y \in E_r^-$. For $x, y \notin S_r$, the function ψ is infinitely smooth.

Let us define the potentials $\mathcal{A}_r(k)$ and $\mathcal{B}_r(k)$ by formulas of the form (5.139) and (5.140) and define the operators $A_r(k)$ and $B_r(k)$ by formulas of the form (5.141) and (5.142) with the function $\Psi(x,y,k)$ instead of $\Phi(x-y,k)$. The solutions in V^- satisfy the following relation similar to (5.153):

$$\left[\frac{1}{2} I + B_r(k)\right] u^- = A_r(k) \partial_\nu u^-. \tag{5.227}$$

However, here the operators on both sides are invertible, since the analog of the interior Dirichlet problem

$$\Delta u + k^2 u = 0 \quad \text{in } V^+ \setminus \overline{E_r^+}, \qquad u^+ = 0 \quad \text{on } S, \qquad \beta \partial_\nu u - u = 0 \quad \text{on } S_r \tag{5.228}$$

has no nontrivial solutions. This can be verified with the help of Green's first formula in $V^+ \setminus \overline{E_r^+}$ with $v = \overline{u}$. Indeed, by separating the imaginary parts in this formula, we find that $\partial_\nu u = 0$ on S_r, but then it follows from the boundary condition that $u = 0$ on S_r as well. Now we can use, say, the Holmgren uniqueness theorem (e.g., see [49, Chapter 5]).

Thus, Problem III is again reduced to the equation $T^{-}(k)\varphi^{-} - \lambda\varphi^{-} = g$; however, for the considered value of k we have

$$T^{-}(k) = \left[\frac{1}{2}I + B_r(k)\right]^{-1} A_r(k). \tag{5.229}$$

Now the operator $T^{-}(k)$ can be studied along the same lines.

Problem IV. Here we restrict ourselves to the following remark. For the values of k satisfying Conditions I and II, instead of $T(k)$ we can use the inverse operator

$$T(k)^{-1} = \Theta(k) = \left[\frac{1}{4}I - B'(k)^2\right] A^{-1}(k) \tag{5.230}$$

and, by setting

$$\psi = u^{-} - u^{+}, \qquad \mu = 1/\lambda, \tag{5.231}$$

reduce Problem IV to the equation

$$\Theta(k)\psi = \mu\psi. \tag{5.232}$$

This possibility can be used if k does not satisfy Condition II alone.

There are also other possibilities of avoiding Conditions I and II; cf. Subsection 5.11.9 below.

5.9 Scalar Problem with Spectral Parameter in the Equation

1. Statement of the problem and solvability theorems. It will be convenient to represent the problem as follows.[19] We seek the solution of the equation

$$\Delta u(x) + (k^2 - \mu\sigma(x))u(x) = f(x) \tag{5.233}$$

in $\mathbf{R}^n \setminus S$ (where $n = 1, 2$, or 3) under the radiation condition (5.137) at infinity and the conditions

$$u^{+} = u^{-}, \qquad \partial_\nu u^{+} = \partial_\nu u^{-} \quad \text{on } S. \tag{5.234}$$

We assume that the function $\sigma(x)$ is at least continuous in $\overline{V^+}$ and is equal to zero outside $\overline{V^+}$. Until Subsection 4, we assume that $\sigma(x) > 0$ in $\overline{V^+}$; however, μ may be complex. Just as in the preceding, k is a given number satisfying inequalities (5.3). By μ we denote the spectral parameter.

[19]The notation here is different from that used in Section 1.3. Namely, $k^2 - \mu\sigma(x)$ was earlier denoted by $k^2\varepsilon(x)\mu$.

We assume that the function $f(x)$ is zero outside V^+. This can readily be achieved by subtracting the integral $\int \Phi(x-y,k)f(y)\,dy$ from the solution, where Φ is the fundamental solution (5.138) of the Helmholtz equation (that is, by subtracting the field of the same sources in vacuum).

The values of u^- and $\partial_\nu u^-$ on S are related by (5.153). By substituting u^+ for u^- and $\partial_\nu u^+$ for $\partial_\nu u^-$ into this equation (see (5.234)), we obtain a relationship between u^+ and $\partial_\nu u^+$ on S (cf. the proof of Theorem 5.7.4). Suppose that Condition 1 of Subsection 5.8.1 is satisfied. (If this is not the case, than one has to modify the kernels of the operators $A(k)$ and $B(k)$, following the lines of Subsection 5.8.3, that is, to replace these operators by $A_r(k)$ and $B_r(k)$ with an appropriate r.) Then the operator $A(k)$ is invertible, and Eq. (5.153) can be written in the form

$$\partial_\nu u^+ - A^{-1}(k)\Big(\frac{1}{2}I + B(k)\Big)u^+ = 0. \tag{5.235}$$

We have obtained a boundary value problem for Eq. (5.233) in V^+ with pseudodifferential boundary condition (5.235). *This problem is equivalent to the original problem.* Indeed, if $u(x)$ is a solution of the new problem, then it can be continued to a solution of the original problem by formula (5.151), where u^- and $\partial_\nu u^-$ are taken from (5.234). Moreover, condition (5.153) will be satisfied, which guarantees that the values of u^- and $\partial_\nu u^-$ are consistent.

Now we can consider problem (5.233), (5.235) in V^+.

Let Θ_δ be the sector $\{\mu : |\arg\mu| \leq \pi - \delta\}$ on the complex plane, where δ is a small positive number.

Theorem 5.9.1 *Problem* (5.233), (5.235) *is elliptic with parameter in* Θ_δ *for arbitrarily small* $\delta > 0$. *Hence there exists an* $R = R_\delta > 0$ *such that for* $\mu \in \Theta_\delta$, $|\mu| \geq R_\delta$, *this problem has a unique solution* $u \in H^2(V^+)$ *for each* $f \in H^0(V^+)$, *and moreover, this solution satisfies the a priori estimate*

$$|\mu| \cdot \|u\|_{V^+,0} + \|u\|_{V^+,2} \leq C_\delta \|f\|_{V^+,0} \tag{5.236}$$

with a constant C_δ *independent of* f.

Proof. Let us verify the conditions of ellipticity with parameter (see Subsection 5.5.6). The principal symbol of the left-hand side in (5.233),

$$-|\xi|^2 - \mu\sigma(x), \tag{5.237}$$

does not vanish for $(\xi,\mu) \neq 0$, $\mu \notin \mathbf{R}_-$. The boundary value problem on a ray, corresponding to an arbitrary point of the boundary, has the form

$$v''(t) - (|\xi'|^2 + \mu\sigma(0))v(t) = 0, \qquad t > 0, \tag{5.238}$$

$$-v'(t) + |\xi'|v(t) = h, \qquad t = 0. \tag{5.239}$$

Here we use the same coordinate system as in Subsection 5.5.5 and assume that the principal symbol of the pseudodifferential operator $-\frac{1}{2}A^{-1}(k)$ is equal to $|\xi'|$ at the given point. The solutions of Eq. (5.238) decaying as $t \to +\infty$ have the form $Ce^{-(|\xi'|^2+\mu\sigma(0))^{1/2}t}$, where $(\cdots)^{1/2}$ is the value of the root with positive real part. One can readily see that C is uniquely determined by (5.239) for $(\xi', \mu) \neq 0$, $\mu \notin \mathbf{R}_-$.

It remains to use Assertion 1° of Subsection 5.5.6. □

To the spectral problem (5.233), (5.235) we can assign an *unbounded* operator $L_B(k)$ in $H^0(V^+)$ by setting

$$L_B(k)u = \sigma^{-1}[\Delta u + k^2 u]; \tag{5.240}$$

the domain $\mathfrak{D}(L_B(k)) \subset H^2(V^+)$ is specified by the boundary condition (5.235). The same operator can be interpreted as a *bounded* operator acting from the subspace $\mathfrak{D}(L_B(k)) \subset H^2(V^+)$ into $H^0(V^+)$. The index of this operator is zero, since the resolvent set is nonempty. Moreover, this operator is invertible, and we can readily write out the *inverse operator*:

$$K(k)f_1(x) = K_\sigma(k)f_1(x) = \int_{V^+} \Phi(x-y,k)\sigma(y)f_1(y)\,dy, \qquad x \in V^+. \tag{5.241}$$

Indeed, the substitution of $u = K(k)f_1$ into (5.240) yields $L_B(k)u = f_1$. The boundary condition is satisfied, since the right-hand side of (5.241), treated as a function on $\mathbf{R}^n$, belongs to $H^2(\mathbf{R}^n)$, so that conditions (5.234) are satisfied, and moreover, $\Delta u + k^2 u = 0$ in V^-. *The original problem is equivalent to the integral equation*

$$u(x) - \mu K(k)u(x) = f_2(x) \quad \text{in } V^+, \tag{5.242}$$

where $f_2 = K(k)(\sigma^{-1}f)$.

The following theorem supplements Theorem 5.9.1 and can be proved in the same way as Theorem 5.7.4.

Theorem 5.9.2 *Let $\mu \notin \Sigma(L_B(k))$. Then for each $f \in H^0(V^+)$, problem (5.233), (5.234), (5.137) has a unique solution, which belongs to $H^2(E_R)$ in the ball E_R containing $\overline{V}^+$ for an arbitrarily large R. Moreover,*

$$\|u\|_{E_R,2} \leq C_R \|f\|_{V^+,0}. \tag{5.243}$$

If $k_2 = \operatorname{Im} k < 0$, then in (5.243) we can replace E_R by $\mathbf{R}^n$. Furthermore, $u(x)$ is a C^∞ function in V^-; outside an arbitrary neighborhood of the set $\overline{V}^+$, we have

$$\sum_{|\alpha|\leq m} |D^\alpha u(x)| \leq C_m r^{-\frac{n-1}{2}} e^{k_2 r} \|f\|_{V^+,0} \tag{5.244}$$

for any m, and

$$\left|\frac{\partial u}{\partial r} + iku\right| \leq \gamma(r) r^{-\frac{n-1}{2}} e^{k_2 r} \|f\|_{V^+,0}, \tag{5.245}$$

where C_R, C_m, and $\gamma(r)$ are independent of f and $\gamma(r) \to 0$ as $r = |x| \to \infty$.

We can include the terms $\|u\|_{S,3/2}$ and $\|\partial_\nu u\|_{S,1/2}$ in the left-hand side of the estimate (5.243).

2. Spectral properties of the operators $L_B(k)$ and $K(k)$. Let us equip $H^0(V^+)$ with the inner product

$$\langle u, v\rangle_{V^+} = \int_{V^+} \sigma u \overline{v}\, dx. \tag{5.246}$$

The norm $\langle u, u\rangle_{V^+}^{1/2}$ is equivalent to the usual norm $\|u\|_{V^+,0}$.

Proposition 5.9.3 *The operator $K(k)$ is self-adjoint with respect to the inner product* (5.246) *for pure imaginary k.*

The verification is straightforward.

Proposition 5.9.4 *For $k_1 > 0$, the operator $K(k)$ is dissipative, and its eigenvalues lie in the open upper half-plane.*[20] *For $k_1 < 0$, similar assertions are valid for $-K(k)$. If $|k_1| \le |k_2|$, then the eigenvalues of the operator $K(k)$ lie in the left half-plane.*

Proof. Just as in the proof of similar assertions in Sections 5.7 and 5.8, everything is based on Green's formula (5.148). By applying this formula in a ball E_R of large radius R to the volume potential u with density $\sigma\varphi$ and to $v = \overline{u}$, we obtain, after separating the real and imaginary parts,

$$\operatorname{Im}\langle K(k)\varphi, \varphi\rangle_{V^+} = -2k_1k_2 \int_{E_R} |u|^2 dx + k_1 \int_{S_R} |u|^2 dS + o(1), \tag{5.247a}$$

$$\operatorname{Re}\langle K(k)\varphi, \varphi\rangle_{V^+} = (k_1^2 - k_2^2) \int_{E_R} |u|^2 dx - \int_{E_R} |\nabla u|^2 dx + k_2 \int_{S_R} |u|^2 dS + o(1). \tag{5.247b}$$

If $k_2 < 0$, we readily obtain the desired assertions by passing to the limit as $R \to \infty$, since if $u = 0$ everywhere, then $\sigma\varphi = \Delta u + k^2 u = 0$. If $k_2 = 0$, then we pass to the limit in (5.247a) and find that $\operatorname{Im}\langle K(k)\varphi, \varphi\rangle_{V^+} \ge 0$. Suppose that φ is an eigenfunction of the operator $K(k)$, $K(k)\varphi = \lambda\varphi$, and $\operatorname{Im}\lambda = 0$. (Note that $\lambda \ne 0$ since the operator $K(k)$ has an unbounded inverse.) Then $\operatorname{Im}\langle K(k)\varphi, \varphi\rangle_{V^+} = 0$ and $\int_{S_R} |u|^2 dS \to 0$, $R \to \infty$. It follows that $u = 0$ in V^- (by the Rellich lemma), so that $u^- = \partial_\nu u^- = 0$ on S. By (5.234), we have $u^+ = \partial_\nu u^+ = 0$ on S, and u is a solution of a homogeneous Cauchy problem for a second-order elliptic equation. It follows that $u = 0$ in V^+. However, $u = K(k)\varphi = \lambda\varphi$, and hence $\varphi = 0$ in V^+, which contradicts our assumption: by definition, an eigenfunction cannot vanish identically. Hence we see that $\operatorname{Im}\lambda > 0$. □

[20] Cf. inequalities (1.39).

Remark 5.9.5 From (5.247a) we see that $\mathrm{Im}\langle K(k)\varphi, \varphi\rangle_{V^+} > 0$ for nonzero φ if $k_2 < 0$ and, say, $k_1 > 0$.

However, we cannot claim that $\mathrm{Im}\langle K(k)\varphi, \varphi\rangle_{V^+} > 0$ for $\varphi \neq 0$ if $k_2 = 0$ and $k_1 > 0$ (cf. Proposition 5.7.8). Indeed, let $u(x)$ be an arbitrary function from $H^2(V^+)$ with zero Cauchy data: $u^+ = \partial_\nu u^+ = 0$. We extend $u(x)$ by zero to V^-. Clearly, $\Delta u + k^2 u = 0$ in V^-. Set $\varphi = \sigma^{-1}(\Delta u + k^2 u)$ in V^+. Then $K(k)\varphi = u$, and it follows from (5.247a) that

$$\mathrm{Im}\langle K(k)\varphi, \varphi\rangle_{V^+} = 0. \tag{5.248}$$

Conversely, if condition (5.248) is satisfied, then it follows for the volume potential $u = K(k)\varphi$ from (5.247a) that $u = 0$ in V^-, so that in V^+ the function u has zero Cauchy data on S.

Since the imaginary part $K_I(k)$ of the operator $K(k)$ is nonnegative, it follows that relation (5.248) is equivalent to $K_I(k)\varphi = 0$. Essentially, we have obtained a description of the kernel $\mathrm{Ker}\, K_I(k)$ for $k_2 = 0$, $k_1 > 0$: *this kernel consists of functions $\varphi \in H^0(V^+)$ of the form $\varphi = \sigma^{-1}(\Delta u + k^2 u)$, where u is a function from $H^2(V^+)$ with zero Cauchy data.* In particular, the kernel is infinite-dimensional (if $n > 1$). For $k_2 < 0$, $k_1 > 0$ the kernel is trivial.

Theorem 5.9.6 *For $k_1 \neq 0$, the operator $L_B(k)$ belongs to $\mathbf{A}(\alpha, H^0(V^+))$ if $\alpha > n/2$. Its eigenvalues μ_j have the following asymptotics for $n = 1, 2, 3$:*[21]

$$\mu_j = C_1 j^2 + o(j^2), \tag{5.249a}$$
$$\mu_j = C_2 j + o(j), \tag{5.249b}$$
$$\mu_j = C_3 j^{2/3} + o(j^{2/3}), \tag{5.249c}$$

respectively ($j \to \infty$), where

$$C_1 = -\pi^2\Big(\int_{V^+} \sigma^{1/2} dx\Big)^{-2}, \qquad C_2 = -4\pi\Big(\int_{V^+} \sigma\, dx\Big)^{-1},$$
$$C_3 = -(6\pi^2)^{2/3}\Big(\int_{V^+} \sigma^{3/2} dx\Big)^{-2/3}. \tag{5.250}$$

For arbitrarily small $\delta > 0$, all eigenvalues of the operator $L_B(k)$, starting from some number, lie in the sectorial neighborhood[22] *$\mathbf{C} \setminus \Theta_\delta$ of the ray $\mathbf{R}_-$.*

Proof. Since the inverse operator $K(k)$ acts boundedly from $H^0(V^+)$ to $H^2(V^+)$, it follows that $s_j(K(k)) \leq Cj^{-2/n}$ (see the Remark in Subsection 5.6.1). Now it suffices to apply Theorem 5.6.5 and Assertion 2° from Subsection 5.6.5. □

---

[21] For $k_1 = 0$, we have the same formulas with the error estimates $O(j)$, $O(j^{1/2})$, and $O(j^{1/3})$, respectively. See [52].

[22] In fact, even in a half-neighborhood; see Proposition 5.9.4.

Corollary 5.9.7 *For $\mu \notin \Sigma(L_B(k))$, the Fourier series of the solution $u(x)$ of the boundary value problem (5.233), (5.235) with respect to the root functions of the operator $L_B(k)$ is summable by the Abel–Lidskii method of order $\alpha > n/2$ in $H^2(V^+)$ and hence in $C^{(\beta)}(V^+)$ for $\beta < 2 - (n/2)$.*

Here we use Assertion 11° from Subsection 5.3.2.

Let us compare these results with those obtained in Sections 5.7 and 5.8. The summability theorems were obtained there in the scale $H^s(S)$, whereas here we prove the Abel–Lidskii summability only in the space $H^0(V^+)$ and in the subspace of $H^2(V^+)$ consisting of functions satisfying the boundary conditions (5.235). This is because the root functions are subjected to the boundary conditions (5.235) and so *cannot form a complete system in $H^s(V^+)$ for $s > 3/2$*. This fact is well-known for elliptic differential problems with spectral parameter in the elliptic equation.

The second difference is that instead of the results of Subsections 5.6.1 and 5.6.2, here we are only able to use the results of Subsection 5.6.3, even though, say, for $\operatorname{Im} k = 0$, the operator $\operatorname{Im} K(k)$ maps functions belonging to $L_2(V^+)$ to functions belonging to $C^\infty(\overline{V^+})$. Here the reason is that we cannot single out a leading selfadjoint part of the operator L, since the boundary conditions (5.235) are "nonselfadjoint." Let us discuss this in more detail.

The adjoint $K^*(k)$ of $K(k)$ with respect to the inner product (5.246) coincides with $\overline{K}(k)$. Let $\mathcal{R}$ be the range of $K(k)$. Then the range of $\overline{K}(k)$ is $\overline{\mathcal{R}}$. For simplicity, assume that $\sigma(x) \equiv 1$, and let us find the intersection of the two ranges for $k_1 \neq 0$.

Proposition 5.9.8 *For $k_1 \neq 0$, the intersection $\mathcal{R} \cap \overline{\mathcal{R}}$ consists of functions $u \in H^2(S)$ with zero Cauchy data u^+, $\partial_\nu u^+$ on S.*

Proof. Suppose that $u(x) \in H^2(V^+)$ has zero Cauchy data on S. Let us continue $u(x)$ by zero to V^- and set $\Delta u + k^2 u = \varphi$ and $\Delta u + \overline{k}^2 u = \psi$. Then

$$u(x) = \int_{V^+} \Phi(x-y,k)\varphi(y)\,dy = \int_{V^+} \overline{\Phi(x-y,k)}\psi(y)\,dy \tag{5.251}$$

in $\mathbf{R}^n$ and, in particular, in V^+, so that $u \in \mathcal{R} \cap \overline{\mathcal{R}}$.

Conversely, let $u(x) \in \mathcal{R} \cap \overline{\mathcal{R}}$. Then the function $u(x)$ belongs to $H^2(V^+)$ and can be represented by formulas (5.251) in V^+. It follows from these formulas that $\overline{u(x)} \in \mathcal{R} \cap \overline{\mathcal{R}}$; consequently, $\operatorname{Re} u(x)$ and $\operatorname{Im} u(x)$ belong to $\mathcal{R} \cap \overline{\mathcal{R}}$. Hence we shall assume that the function $u(x)$ is real-valued. Since $\Delta u + k^2 u = \varphi$ in V^+, it follows that $2k_1k_2 u = \operatorname{Im}\varphi$. Let us continue $u(x)$ into V^- using the first formula in (5.251). Formula (5.247a) shows that

$$0 = -2k_1k_2 \int_{V_R^-} |u|^2 dx + k_1 \int_{S_R} |u|^2 dS + o(1),$$

since

$$\operatorname{Im}\langle K(k)\varphi, \varphi\rangle = -2k_1k_2 \int_{V^+} |u|^2 dx.$$

It follows that $u(x) = 0$ in V^- (see the proof of Proposition 5.9.4), so that $u(x)$ has zero Cauchy data in V^+ on S. □

Note that $\mathcal{R} \cap \overline{\mathcal{R}}$ has an infinite codimension in both $\mathcal{R}$ and $\overline{\mathcal{R}}$ for $n = 2$ and $n = 3$ (u^+ can be prescribed arbitrarily, and then $\partial_\nu u^+$ can be computed from (5.235)). Thus, in these cases the ranges of the operators $K(k)$ and $K^*(k)$ differ substantially. The situation was different in Sections 5.7 and 5.8. For example, the ranges of $A(k)$ and $A^*(k)$ either coincide with some $H^s(S)$ or differ from it by finite-dimensional subspaces.

See, however, a remark about the results for classical boundary problems in [72] at the end of this section.

3. Special cases. For $n = 1$, Theorem 5.9.6 can be strengthened. In this case, the boundary value problem (5.233), (5.235) is differential:

$$u''(x) + k^2 u(x) = \mu\sigma(x)u(x) + f(x) \quad \text{on } (a, b), \tag{5.252}$$

$$u'(a) - iku(a) = 0, \qquad u'(b) + iku(b) = 0. \tag{5.253}$$

This problem belongs to the class of *regular problems for ordinary differential equations*, whose spectral properties are well studied. General theorems on regular problems (e.g., see [77, Chapter II]) imply the following assertion.

Proposition 5.9.9 *For $n = 1$, the eigenvalues μ_j are simple, at least, starting from some number. Moreover, for the eigenvalues and the eigenfunctions we have*

$$\mu_j(L_B(k)) = C_1 j^2 + O(1), \tag{5.254}$$

$$u_j(x) = \cos\left[\pi j \Big(\int_a^b \sqrt{\sigma}\, dx\Big)^{-1} \int_a^x \sqrt{\sigma}\, dt\right](1 + O(j^{-1})) \tag{5.255}$$

uniformly with respect to x for $j \geq j_0$.

It is known that if all the eigenvalues of a dissipative operator are simple starting from some number and if the system of its root functions is complete, then the following condition is sufficient for the normalized root functions to form a Bari basis (see [43, Section 6.4][23]):

$$\sum_{j \neq k} \frac{\operatorname{Im} \mu_j \operatorname{Im} \mu_k}{|\mu_j - \overline{\mu}_k|^2} < \infty. \tag{5.256}$$

One can readily verify that this condition is satisfied if the asymptotics (5.254) holds. Thus, the following theorem is valid.

Theorem 5.9.10 *For $n = 1$, the system of normalized root functions of the operator $L_B(k)$ is a Bari basis in $H^0(V^+)$. If $\mu \notin \Sigma(L_B(k))$, then the Fourier series of the solution $u(x)$ of problem (5.252), (5.253) with respect to this system converges to $u(x)$ in $H^2(V^+)$ and hence in $C^{(\alpha)}(V^+)$ for $\alpha < 3/2$.*

---

[23] Only bounded operators are considered there, but our assertion for an operator with discrete spectrum follows from the same assertion for the compact inverse operator.

For the case in which $\sigma(x) \equiv 1$ on $[a, b]$ and, say, $b - a = 1$, we set $l^2 = k^2 - \mu$. By substituting the general solution $C_1 e^{ilx} + C_2 e^{-ilx}$ of the equation with $f = 0$ into (5.253), we arrive at a transcendental equation for l, which is readily seen to split into the two equations

$$(l + k)e^{il} \pm (l - k) = 0$$

without common roots. To each simple root of this equation, there corresponds a simple eigenvalue μ. One can show that $k_2 > 0$ if there is a multiple root l (cf. [67]). Since in our case $k_2 \leq 0$, it follows that all eigenvalues are simple, and hence *there are no associated functions.*

One can also show that *there are no associated functions if $\sigma \equiv 1$ and V^+ is a disk* ($n = 2$) *or a ball* ($n = 3$). The eigenfunctions are calculated by separation of variables, and arbitrary functions from $H^0(V^+)$ can be expanded in double or triple series in these eigenfunctions (cf. Subsection 1.2.3). For details, see [5].

4. Generalizations and modifications. 1. In the preceding, we assumed $\sigma(x)$ to be positive in $\overline{V}^+$. Now we assume that it is different from 0 and takes *complex* values in the sector $\Psi_\beta = \Theta_{\pi-\beta} = \{\mu : |\arg \mu| \leq \beta\}$, $\beta < \pi$.

Theorem 5.9.1 remains valid if we assume that $\delta > \beta$. Indeed, one can readily verify that in this case the conditions of ellipticity with parameter in Θ_δ are preserved. Theorem 5.9.2 remains valid. The expression (5.246) no longer possesses the properties of an inner product, and Proposition 5.9.3 fails to be valid. The assertions about Abel–Lidskii summability hold for $\beta < \pi/n$, $\alpha = n/2 + \varepsilon$ with small $\varepsilon > 0$. Under the same assumption $\beta < \pi/n$, for the eigenvalues μ_j of $L_B(k)$ we have $|\mu_j| \asymp j^{2/n}$ by virtue of Assertion 2° in Subsection 5.6.5.

2. In this section, as well as in Sections 5.7 and 5.8, we can replace $\mathbf{R}^n$ by a bounded domain V_0 containing $\overline{V}^+$ and impose a boundary condition of the form (5.92) or the Dirichlet condition on the boundary $S_0 = \partial V_0$. The resulting problem can be studied in a similar way.

3. Let us briefly discuss the problems considered in Section 0.2. Their properties are similar to those of the problems considered in this section. It is convenient for us to represent these problems in the following form. We consider the equation

$$\Delta u(x) + \mu\sigma(x)u(x) = f(x) \tag{5.257}$$

in a bounded domain V^+ with the homogeneous ($g = 0$) boundary condition (5.92) or the Dirichlet condition on S. These are classical elliptic boundary value problems, very close to those considered in Subsections 5.5.5 and 5.5.6, and we can apply the assertions stated there as well as the theorems from Section 5.6. To the problem we assign the operator L_B in $H^0(V^+)$ acting as $-\sigma^{-1}(x)\Delta$ with domain $\mathcal{D} \subset H^2(V^+)$ specified by the boundary condition $Bu = 0$ on S. If $\sigma(x) > 0$ in $\overline{V}^+$, then we obtain an operator that is selfadjoint with respect to the inner product (5.246) and bounded below, under the Dirichlet condition or a condition of the form (5.92) with a real-valued function $b(x)$. (The boundedness below can be verified using Assertions 9° and

16° from Subsection 5.3.2 and Green's formula.) In this case, we know the asymptotics of the eigenvalues. For complex-valued $\sigma(x)$ with $|\arg\sigma(x)| \leq \beta$ we have ellipticity with parameter in the complement of a slightly wider sector. If $\beta < \pi/n$, we obtain the summability of the Fourier series in the root functions by the Abel–Lidskii method of order $n/2 + \varepsilon$ with small ε, and it is possible to indicate the rough asymptotics of the moduli of the eigenvalues (see Assertion 2° in Subsection 5.6.5). Finally, in the case of positive $\sigma(x)$ and under a condition of the form (5.92) with a complex-valued function $b(x)$, we have ellipticity with parameter outside an arbitrarily small sectorial neighborhood of the ray $\mathbf{R}_+$. This permits one to indicate the asymptotics of the eigenvalues, but without an additional remainder estimate. However, for these cases, when the problem is classical, there are similarity transformations, constructed in [72], which transfer the deviation of the problem from a selfadjoint one from the boundary conditions into the equation in V^+. One obtains a weak perturbation of a selfadjoint operator. This permits one to conclude that the eigenvalue asymptotics holds with the same remainder estimate as in (5.96) and (5.97).

5.10 Spectral Problems Related to the s-Method

1. Statement of spectral problems. The main goal of this section is to study the equations

$$I + A_{\sigma,R}(k) + hA_{\sigma,I}(k) = \mu A_{\sigma,I}(k) \tag{5.258}$$

and

$$I + K_R(k) + hK_I(k) = \mu K_I(k). \tag{5.259}$$

Here $A_{\sigma,R}(k)$ and $A_{\sigma,I}(k)$ are the real and imaginary parts of the operator $A_\sigma(k)$ on S considered in Section 5.7, and $K_R(k)$ and $K_I(k)$ are the real and imaginary parts of the operator $K(k) = K_\sigma(k)$ in V^+ considered in Section 5.9. By μ we denote the spectral parameter. The first of these equations appears in applications of the s-method (the introduction of a spectral parameter into the conditions at infinity) to problems of diffraction on a semitransparent surface or curve S, and the second equation arises in applications of this method to problems of diffraction on a dielectric body. If $\sigma = 1$ on S or in V^+, then these equations differ only in notation from Eqs. (2.147) and (2.161). By S we denote the *closed* infinitely smooth boundary of the domain $V^+ \subset \mathbf{R}^n$, $n = 2, 3$. Although positive values of k are of main interest, we shall assume that $k = k_1 + ik_2$, $k_1 > 0$, and $k_2 \leq 0$. The function $\sigma(x)$ will be assumed positive on S or in $\overline{V^+}$, respectively.

2. Abstract equations of the type (5.258) and (5.259). Let $T_0 = T_1 + iT_2$ be a compact operator in a separable Hilbert space H, and let T_1 and T_2 be the real and imaginary parts of this operator. Suppose that the equation

$$(I + T_0)f = g \tag{5.260}$$

is uniquely solvable, that is, -1 is not an eigenvalue of the operator T_0. Then the operator $I + T_1 + hT_2$ is invertible for all h except for some discrete set of values (see [43, Section 5.6]). Let us choose such a value of h and assume that it is real. Now set

$$J = I + T_1 + hT_2, \qquad T = J^{-1}T_2. \tag{5.261}$$

The operator J is selfadjoint, and T is a compact operator. Equation (5.260) is equivalent to the equation

$$(T - (h - i)^{-1}I)f = g_1, \tag{5.262}$$

where $g_1 = (i - h)^{-1}J^{-1}g$. Now we intend to find out whether it is possible to use expansions in root vectors of the operator T for solving Eq. (5.260) or (5.262). In other words, we wish to study the properties of the system of root vectors of the operator T.

To this end, following, say, [43, Section 5.8], we introduce the bilinear form

$$[f, g] = (Jf, g) \tag{5.263}$$

and note that T is selfadjoint with respect to this form:

$$[Tf, g] = (T_2 f, g) = (f, T_2 g) = [f, Tg].$$

The simplest case is the one in which all eigenvalues γ_j of the operator J are positive. Then there exists a minimal eigenvalue (indeed, $\gamma_j \to 1$, since they are the eigenvalues of the sum of the identity operator and a compact operator); let this eigenvalue be γ_1. We have $\gamma_1 \|f\|^2 \leq [f, f] \leq \|J\| \cdot \|f\|^2$, whence we see that the form (5.263) possesses all properties of an inner product, and moreover, the norms $\|f\|$ and $[f, f]^{1/2}$ are equivalent. Since T is an operator selfadjoint with respect to the form (5.263) and compact, it follows that in H there exists a basis of eigenvectors of this operator, orthonormal with respect to (5.263). In the original metric of H, this basis is a Riesz basis. Thus, the following assertion holds.

1° *If all eigenvalues of the operator J are positive, then in H there exists a Riesz basis consisting of eigenvectors of T and orthonormal with respect to the inner product* (5.263).

Sometimes one can ensure the positivity of the eigenvalues of J by an appropriate choice of h. However, this is impossible if there exists a vector f such that $T_2 f = 0$ and $((I + T_1)f, f) \leq 0$. On the other hand, if T_0 is a normal operator, then $I + T_1$ and T_2 commute and hence have a common orthonormal basis of eigenvectors. If in this case we assume that $\operatorname{Im}(T_0\varphi, \varphi) = (T_2\varphi, \varphi) > 0$ for $\varphi \neq 0$ (that is, all eigenvalues of the operator T_2 are positive), then all eigenvalues of J will be positive for sufficiently large $h > 0$. Indeed, these eigenvalues have the form $\alpha_j + \beta_j h$, where the eigenvalues α_j of $I + T_1$ tend to 1 and the eigenvalues β_j of T_2 are positive.

Now suppose that J has some negative eigenvalues. Then the quadratic form $[f, f]$ is not positive definite. However, the bilinear form (5.263) can be considered as an *indefinite inner product* in H. Let us indicate some properties of this inner product.

Let $\{f_j\}$, $j = 1, 2, \ldots$, be an orthonormal basis of eigenvectors of J in H, and let $Jf_j = \gamma_j f_j$. We assume that the eigenvalues are numbered so that $\gamma_j < 0$ for $j \leq \kappa$ and $\gamma_j > 0$ for $j > \kappa$. This is possible, since $\gamma_j \to 1$. Let $\sum \alpha_j f_j$ and $\sum \beta_j f_j$ be the Fourier series of vectors f and g in the system $\{f_j\}$. Then

$$[f, g] = \sum_{j=1}^{\infty} \gamma_j \alpha_j \overline{\beta}_j. \tag{5.264}$$

Here the first κ coefficients γ_j are negative, and the other coefficients are positive. The formula

$$\{f, g\} = \sum_{j=1}^{\kappa} (-\gamma_j) \alpha_j \overline{\beta}_j + \sum_{j=\kappa+1}^{\infty} \gamma_j \alpha_j \overline{\beta}_j \tag{5.265}$$

defines an inner product with the usual properties in H, and moreover, the norms $\{f, f\}^{1/2}$ and $\|f\|$ are equivalent, since $\gamma_j \to 1$.

In the conventional terminology, the space H equipped with the indefinite inner product (5.263) is the *Pontrjagin space* Π_κ. For the axiomatic definition of the Pontrjagin space, see [28] or [51].

If $[f, g] = 0$, then one says that the vectors f and g are *J-orthogonal.* An operator selfadjoint with respect to the form (5.263) is said to be *J-selfadjoint.*

Obviously, the operator T in (5.261) is J-selfadjoint and compact.

A compact J-selfadjoint operator in the Pontrjagin space can have nonreal eigenvalues and associated vectors corresponding to real and nonreal eigenvalues. However, the following three assertions are true of such an operator (see [81], where additional information can be found).

2° *The nonreal spectrum of T is symmetric with respect to the real axis taking multiplicities into account. The sum κ_1 of the multiplicities corresponding to the eigenvalues in the open upper half-plane is not greater than κ.*

3° *The number κ_2 of real eigenvalues such that there are associated vectors in the corresponding root space is not greater than κ. Moreover, $\kappa_1 + \kappa_2 \leq \kappa$.*

4° *If zero is an eigenvalue, then the root space $\mathfrak{L}_T(0)$ is the sum of two J-orthogonal T-invariant subspaces. The first of them is finite-dimensional, with dimension not greater than κ, and $[f, f] = 0$ on it. In the second, $[f, f] > 0$ for $f \neq 0$, and T is a usual selfadjoint operator without associated vectors.*

The following assertion was obtained in [23].

5° *Let T be a compact J-selfadjoint operator. The system of root vectors of T is complete in H if and only if $L_T(0)$ does not contain a vector $f \neq 0$ that is J-orthogonal to $L_T(0)$. Under this condition, in H there exists a Riesz basis of root vectors of the operator T.*

Let us additionally assume that the operator T is *J-nonnegative*: $[Tf, f] \geq 0$ for all f (or J-nonpositive: $[Tf, f] \leq 0$ for $f \neq 0$). This is equivalent to the assumption that T_2 is nonnegative: $(T_2 f, f) = \operatorname{Im}(T_0 f, f) \geq 0$ (respectively, nonpositive: $(T_2 f, f) \leq 0$). As was noted in the end of Subsection 5.2.6, in this case we have $(T_2 f, f) > 0$ (respectively, < 0) if $T_2 f \neq 0$.

6° *Let T be a compact J-selfadjoint J-nonnegative (or J-nonpositive) operator. Then all its eigenvalues λ are real, and $L_T(\lambda)$ does not contain associated vectors for $\lambda \neq 0$. Moreover, there are no associated vectors in $L_T(0)$ provided that* $\operatorname{Ker} T$ *does not contain a vector $f \neq 0$ that is J-orthogonal to* $\operatorname{Ker} T$;[24] *in this case, in H there exists a Riesz basis of eigenvectors of the operator T.*

Indeed, let $Tf = \lambda f$, $f \neq 0$, $\lambda \neq 0$. Then $T_2 f \neq 0$ and $\lambda[f,f] = (T_2 f, f) > 0$, so that $[f,f] \neq 0$ and λ is real. Suppose that $(T-\lambda)f_1 = f$; then $[f,f] = [f_1,(T-\lambda)f] = 0$, which is impossible, as we have just seen (cf. [28, p. 147]).

Suppose that $\operatorname{Ker} T$ does not contain a vector $f \neq 0$ that is J-orthogonal to $\operatorname{Ker} T$. If $Tf = 0$, $f \neq 0$, and $Tf_1 = f$, then for each vector $g \in \operatorname{Ker} T$ we have $[f,g] = [f_1, Tg] = 0$, which contradicts our assumption. Thus $\operatorname{Ker} T = L_T(0)$. It remains to refer to 5°. □

The equation for eigenvectors and eigenvalues of T can be rewritten in the form

$$[T_2 - \lambda(I + T_1 + hT_2)]f = 0. \tag{5.266}$$

3. The s-method in problems of diffraction on a semitransparent surface. Consider the equation

$$[I + A_\sigma(k)]\varphi = \psi. \tag{5.267}$$

It is uniquely solvable, since for the values of k that we consider according to the convention at the beginning of this section, $A_\sigma(k)$ cannot have real eigenvalues. Hence, the left-hand side of (5.258) is an invertible operator for all h except for a discrete set. Let us choose a real h for which this operator is invertible and set, just as in (5.261),

$$J = I + A_{\sigma,R}(k) + hA_{\sigma,I}(k), \qquad T = J^{-1}T_2, \quad \text{where } T_2 = A_{\sigma,I}(k). \tag{5.268}$$

According to Proposition 5.7.8, the operator T_2 is nonnegative and $\operatorname{Ker} T_2 = \operatorname{Ker} A_\sigma(k)$ is nontrivial only for some $k > 0$ (which do not satisfy Condition I). Moreover, $\operatorname{Ker} T_2$ is finite-dimensional, and if $0 \neq \varphi \in \operatorname{Ker} T_2$, then $J\varphi = \varphi$ and $[\varphi,\varphi] = \langle \varphi, \varphi \rangle_S \neq 0$. This permits us to use Assertion 6°.

Theorem 5.10.1 *Let h be a real number such that the operator $J = I + A_{\sigma,R}(k) + hA_{\sigma,I}(k)$ is invertible. Then in $H^0(S)$ there exists a Riesz basis $\{u_j(x)\}$, $j = 1, 2, \ldots$, of eigenfunctions of the operator $T = J^{-1}A_{\sigma,I}(k)$. The corresponding eigenvalues are real.*

Using the system of eigenfunctions $\{u_j\}$, we can apply the spectral method to Eq. (5.267), that is, construct its solutions in the form of series in the u_j. Indeed, this equation is equivalent to the Fredholm equation

$$I - (h - i)T = J^{-1}g. \tag{5.269}$$

[24] This is true without the assumption that T is J-nonnegative or J-nonpositive.

In (5.267) we can replace the sign "+" before $A_\sigma(k)$ by "−"; then the corresponding operator T_2 will be nonpositive.

Equation (5.267), as well as the similar equation with the minus sign before $A_\sigma(k)$, was written out in slightly different notation in Section 5.7 (see Eq. (5.157) with $\lambda = 1$ or $\lambda = -1$). As we know, it is equivalent to problem (5.134)–(5.137) with the same $\lambda = \pm 1$. The expansion of the solution φ on S in a series in the functions u_j corresponds to the expansion of the solution u of problem (5.134)–(5.137) in a series in the functions $u_j^S = \mathcal{A}_\sigma(k)u_j$. In Section 2.5, these are the functions (2.148) for $\sigma = 1$ in $V^\pm$.

4. The s-method in problems of diffraction on a dielectric body. Consider the equation

$$(I + K(k))f = g. \tag{5.270}$$

This equation is uniquely solvable, since $K(k)$ does not have real eigenvalues (by virtue of Proposition 5.9.4). Hence the left-hand side of (5.259) is an invertible operator for all h except for a discrete set. Let us choose a real h for which this operator is invertible and set

$$J = I + K_R(k) + hK_I(k), \qquad T = J^{-1}T_2, \quad \text{where } T_2 = K_I(k). \tag{5.271}$$

Since $\operatorname{Im}\langle K(k)\varphi, \varphi\rangle_{V^+} = \langle K_I(k)\varphi, \varphi\rangle_{V^+}$, it follows from Proposition 5.9.4 that the operator $K_I(k)$ is nonnegative. As we mentioned in Remark 5.9.5, for $k_2 < 0$ the kernel of this operator is trivial, and for $k_2 = 0$

$$\operatorname{Ker} K_I(k) = \{\varphi : \varphi = \sigma^{-1}(\Delta w + k^2 w),\ w \in H^2(V^+),\ w^+ = \partial_\nu w^+ = 0\}. \tag{5.272}$$

This is an *infinite-dimensional* subspace of $H^0(V^+)$ if $n > 1$. Let us clarify the meaning of the condition occurring in Assertion 6° of Subsection 1. This condition essentially says that if for some function $u \in \operatorname{Ker} K_I(k)$ and every function $v \in \operatorname{Ker} K_I(k)$ one has

$$\langle (I + K(k))u, v\rangle_{V^+} = 0, \tag{5.273}$$

then $u = 0$ in V^+. Let $u = \sigma^{-1}(\Delta w + k^2 w)$ and $v = \sigma^{-1}(\Delta w_1 + k^2 w_1)$, where $w, w_1 \in H^2(V^+)$ are functions with zero Cauchy data on S. We obtain

$$\int \sigma^{-1}(\Delta w + (k^2 + \sigma)w) \cdot (\Delta \overline{w}_1 + \overline{k}^2 \overline{w}_1)\, dx = 0. \tag{5.274}$$

Now let us assume that $\sigma(x)$ is a smooth function. Relation (5.274) means that $w(x)$ is a *generalized solution* of the following boundary value problem:

$$(\Delta + k^2)\sigma^{-1}(\Delta + k^2 + \sigma)w = 0 \quad \text{in } V^+, \qquad w^+ = \partial_\nu w^+ = 0 \quad \text{on } S. \tag{5.275}$$

Proposition 5.10.2 *The values of k for which problem* (5.275) *has nontrivial generalized solutions form at most countable set without finite accumulation points.*

Proof. To verify this statement, we outline a simple study of problem (5.275). However, this study is beyond the framework of Subsections 5.5.5 and 5.5.6, where only second-order equations were considered.

The equation in (5.275) is of the fourth order. It is elliptic, since the principal symbol of the left-hand side is $\sigma^{-1}(x)|\xi|^4$. Problem (5.275) is the Dirichlet problem for this equation. The Dirichlet problem for a scalar elliptic equation of an arbitrary order is always elliptic (e.g., see [4]). It follows from the fact that the homogeneous problem (5.275) is elliptic that its generalized solutions (our solution a priori belongs to $H^2(V^+)$) are usual smooth solutions (see [66]). Thus, *the condition in question is satisfied if problem* (5.275) *does not have nontrivial classical solutions.*

Next, problem (5.275) is elliptic with parameter $\lambda = k^2$ outside an arbitrarily narrow sectorial neighborhood Λ of the ray $\mathbf{R}_+$. This follows from the fact that the principal symbol with parameter of the left-hand side in (5.275) is equal to

$$\sigma^{-1}(x)(-|\xi|^2 + \lambda)^2; \tag{5.276}$$

it has zeros only for positive λ, and the Dirichlet problem for a scalar equation elliptic with parameter in some sector is elliptic with parameter in this sector (see [20]). For $\lambda \notin \Lambda$ sufficiently large in modulus, problem (5.275) has only trivial solutions.

Actually we can verify in a more elementary way that there are even no nonreal k such that problem (5.275) has nontrivial solutions. For this, set $w = w_1$ and integrate by parts the term $w \cdot \Delta\overline{w}$ in (5.274). Now let us separate the imaginary parts. We see that

$$2k_1k_2 \int_{V+} |w|^2 dx = 0.$$

Hence $k_2 = 0$ if w is not identically zero.

Finally, problem (5.275) can be linearized in λ by setting $u_1 = w$ and $u_2 = \lambda w$; then we obtain the equivalent Dirichlet problem for the system

$$\begin{pmatrix} 0 & -1 \\ \sigma\Delta[\sigma^{-1}(\Delta + \sigma)\cdot] & \sigma\Delta[\sigma^{-1}\cdot] + \Delta + \sigma \end{pmatrix} U + \lambda U = 0, \tag{5.277}$$

where U is the column vector with components u_1 and u_2. Here we have made yet another step beyond the framework of Section 5.5. This system is *elliptic in the sense of Douglis–Nirenberg* (e.g., see [49] or [14]): in the matrix of principal symbols

$$\begin{pmatrix} 0 & -1 \\ |\xi|^4 & -2|\xi|^2 \end{pmatrix},$$

the entries a_{ij}, $i,j = 1,2$, have the orders $l_i + m_j$, where in our case $l_1 = 0$, $l_2 = 2$, $m_1 = 2$, and $m_2 = 0$, and the determinant $|\xi|^4$ of this matrix is nonzero for $\xi \neq 0$, which means ellipticity. The vector-valued function $U(x)$ is treated as an element of the space $H^4(V^+) \times H^2(V^+)$. The ellipticity with parameter outside Λ is also preserved. In particular, the determinant of the principal symbol with parameter of system (5.277),

$$\begin{pmatrix} \lambda & -1 \\ |\xi|^4 & -2|\xi|^2 + \lambda \end{pmatrix},$$

is equal to $(-|\xi|^2 + \lambda)^2$.

To the last problem, we assign an operator which has a discrete spectrum. (It acts as the matrix in (5.277) in the subspace of $H^2(V^+) \times H^0(V^+)$ specified by the boundary conditions imposed on u_1. The domain of this operator lies in $H^4(V^+) \times H^2(V^+)$ and is specified by the boundary conditions imposed on u_1 and u_2). It follows that the values of k for which a nontrivial solution exists form at most countable set without finite limit points. □

Theorem 5.10.3 *Let h be a real number such that the operator $J = I + K_R(k) + hK_I(k)$ is invertible. If either $k_2 < 0$, or $k_2 = 0$ and problem (5.275) has no nontrivial solutions, then in $H^0(V^+)$ there exists a Riesz basis $\{v_j(x)\}$, $j = 1, 2, \ldots$, of eigenfunctions of the operator $T = J^{-1}K_I$.*

Note that we can always take $h = 0$ and use Assertion 1° in Subsection 1 provided that the norm of the operator $K_R(k)$ is less than 1. From the well-known estimates of the norm of an integral operator in L_2 (see, in particular, [61, Chapter 9, Section 2]) it follows that it suffices to require that one of the inequalities

$$\int_{V^+}\int_{V^+} \sigma(x)\sigma(y)[\operatorname{Re}\Phi(x-y,k)]^2\,dx\,dy < 1 \tag{5.278}$$

or

$$\max_x \int [\operatorname{Re}\Phi(x-y,k)]\sigma(y)\,dy < 1 \tag{5.279}$$

be satisfied (the similar inequality with x and y interchanged is equivalent to the one we have written out).

The system of functions $\{v_j(x)\}$ can be used to solve Eq. (5.270) by the spectral method, since this equation can be rewritten in the form

$$(I + K_R(k) + hK_I(k))f - (h-i)f = g,$$

or

$$I - (h-i)T = J^{-1}g. \tag{5.280}$$

The sign "+" in (5.270) can be replaced by "−"; then the corresponding operator T_2 will be nonpositive.

Set

$$v_j^S = K(k)v_j \tag{5.281}$$

(cf. Eq. (2.162)) These functions lie in the domain of the operator $L_B(k)$. By Theorem 5.9.2, they can be used to obtain an expansion of the solution of problem (5.233), (5.234), (5.137) with $\mu = \pm 1$, and moreover, the convergence in V^+ will take place in the sense of the norm in $H^2(V^+)$.

5. An example with associated functions. Theorem 5.10.3 does not say whether the system of root functions is complete for the exceptional k, for which problem (5.277) has nontrivial solutions. We do not consider this question but note that if for such k the system is complete, then it necessarily contains associated functions corresponding to the zero eigenvalue (see Assertion 5° in Subsection 2). Let us consider the following example.

Consider the problem

$$u''(x) + 4u(x) = f(x) \quad \text{on } (-\pi, \pi), \tag{5.282}$$

$$u'(-\pi) - iu(-\pi) = 0, \qquad u'(\pi) + iu(\pi) = 0. \tag{5.283}$$

This is the problem (5.252), (5.253) with $a = -\pi$, $b = \pi$, $k = 1$, $\mu = -1$, and $\sigma = 3$. (One can easily verify that the value $k = 1$ is exceptional.) By setting

$$\begin{aligned} T_1\varphi(x) &= \frac{3}{2}\int_{-\pi}^{\pi} \sin|x-y|\varphi(y)\,dy, \\ T_2\varphi(x) &= \frac{3}{2}\int_{-\pi}^{\pi} \cos(x-y)\varphi(y)\,dy, \end{aligned} \tag{5.284}$$

we reduce the problem to the equation

$$(I + T_1 + iT_2)\varphi(x) = f_1(x). \tag{5.285}$$

Here $T_1 + iT_2$ is the operator $K(k)$ from Section 5.9 with $n = 1$, $k = 1$, $\sigma = 3$, and $f_1 = K(k)f/3$. The operator $I + T_1$ is invertible. Indeed, if $(I + T_1)\varphi = 0$, then, by applying the operator $(d^2/dx^2) + 1$ to this relation, we obtain $\varphi'' + 4\varphi = 0$, whence $\varphi(x) = C_1 \cos 2x + C_2 \sin 2x$. But the functions $\cos 2x$ and $\sin 2x$ are not eigenfunctions of the operator T_1; one can verify that

$$T_1 \cos 2x = -\cos 2x - \cos x, \qquad T_1 \sin 2x = -\sin 2x - \sin x. \tag{5.286}$$

Below we shall again use these formulas. Let us reduce Eq. (5.285) to the form

$$(T - i)\varphi(x) = f_2(x), \quad \text{where } T = (I + T_1)^{-1}T_2. \tag{5.287}$$

We claim that the *trigonometric system*

$$\{1, \cos x, \sin x, \cos 2x, \sin 2x, \ldots\} \tag{5.288}$$

is a system of root functions of the operator T and that all these functions correspond to the eigenvalue $\lambda = 0$; moreover, $1, \cos 2x, \ldots$ are eigenfunctions, and $\cos x$ and $\sin x$ are associated functions.

Indeed, one readily sees that T_2 annihilates all functions $1, \cos 2x, \ldots$; hence, the same is true of T. The functions $\cos x$ and $\sin x$ are the eigenfunctions of T_2 corresponding to the eigenvalue $3\pi/2$. Hence it follows from (5.286) that

$$T \cos x = -\frac{3\pi}{2}\cos 2x, \qquad T \sin x = -\frac{3\pi}{2}\sin 2x. \tag{5.289}$$

Consequently, $\cos x$ and $\sin x$ are associated functions for T.

5.11 Vector Problems with Spectral Parameter in Boundary or Transmission Conditions

The problems that we intend to consider here will be stated below in Subsection 5. Before doing so, we must obtain some results similar to those given in Subsections 5.7.1 and 5.8.1.

1. Relations on S for solutions of the Maxwell equations. We consider the Maxwell system

$$\operatorname{rot}\mathbf{H} - ik\mathbf{E} = 0, \qquad \operatorname{rot}\mathbf{E} + ik\mathbf{H} = 0 \tag{5.290}$$

in $\mathbf{R}^3 \setminus S$ for $k = k_1 + ik_2 \neq 0$, $k_2 \leq 0$. The vector-valued functions $\mathbf{E}$ and $\mathbf{H}$ satisfy the radiation conditions at infinity. For $k_2 = 0$, these conditions can be represented in the form

$$\left.\begin{aligned} r^{-1}\mathbf{x} \times \mathbf{E} - \mathbf{H} = o(r^{-1}), \quad r^{-1}\mathbf{x} \times \mathbf{H} + \mathbf{E} = o(r^{-1}),\\ \mathbf{E} = O(r^{-1}), \qquad \mathbf{H} = O(r^{-1}), \end{aligned}\right\} \tag{5.291}$$

where $\mathbf{E} = \mathbf{E}(x)$, $\mathbf{H} = \mathbf{H}(x)$, $\mathbf{x}$ is the position vector of the point x, $r = |\mathbf{x}| \to \infty$, and $\times$ stands for the vector product of 3-vectors. For $k_2 < 0$, these conditions say that $\mathbf{E}$ and $\mathbf{H}$ must exponentially decay at infinity. Actually, conditions (5.291) are consequences of one of them, namely, of the first or the second condition in the first row.

We shall use some facts from the theory of system (5.290); they will be given with references mainly to [76]; also, see [31] and [48].

Let us introduce the vector fields

$$\boldsymbol{\varphi}^{\pm} = -\boldsymbol{\nu} \times \mathbf{H}^{\pm}, \qquad \boldsymbol{\psi}^{\pm} = \boldsymbol{\nu} \times \mathbf{E}^{\pm} \tag{5.292}$$

on S, where $\boldsymbol{\nu}$ is the unit outward normal vector. The solutions of system (5.290) in V^+ or V^-, respectively, smooth up to the boundary and satisfying the radiation conditions in the second case are given by the following formulas outside S:

$$\begin{aligned} \chi_{\pm}(x)\mathbf{E}(x) = \pm\Big[ik\int_S \Phi(x-y,k)\boldsymbol{\varphi}^{\pm}(y)\,dS_y + \int_S \boldsymbol{\psi}^{\pm}(y) \times \operatorname{grad}_y \Phi(x-y,k)\,dS_y \\ -ik^{-1}\int_S \operatorname{div}_0 \boldsymbol{\varphi}^{\pm}(y) \cdot \operatorname{grad}_y \Phi(x-y,k)\,dS_y\Big], \end{aligned} \tag{5.293a}$$

$$\begin{aligned} \chi_{\pm}(x)\mathbf{H}(x) = \pm\Big[ik\int_S \Phi(x-y,k)\boldsymbol{\psi}^{\pm}(y)\,dS_y - \int_S \boldsymbol{\varphi}^{\pm}(y) \times \operatorname{grad}_y \Phi(x-y,k)\,dS_y \\ -ik^{-1}\int_S \operatorname{div}_0 \boldsymbol{\psi}^{\pm}(y) \cdot \operatorname{grad}_y \Phi(x-y,k)\,dS_y\Big]. \end{aligned} \tag{5.293b}$$

These formulas were first given by Stratton and Chu (1939). We use the form of these formulas from [76], see Theorems 32, 33 and 36, 37 in this book. A somewhat different but equivalent form is contained in [31, Section 4.2]. Here $\chi_{\pm}(x) = 1$ in $V^{\pm}$ and

$\chi_\pm(x) = 0$ in $V^\mp$, and Φ is defined by the third row in (5.138). By div_0 we denote the *surface divergence*. In local orthogonal coordinates (t_1, t_2) on S, let $\boldsymbol{\varphi} = v^1\mathbf{e}_1 + v^2\mathbf{e}_2$ be the expansion of a vector field $\boldsymbol{\varphi}(t)$ in the corresponding unit vectors $\mathbf{e}_1(t)$ and $\mathbf{e}_2(t)$.[25] Then

$$\mathrm{div}_0\,\boldsymbol{\varphi} = \frac{1}{h_1h_2}\Big[\frac{\partial(h_2v^1)}{\partial t_1} + \frac{\partial(h_1v^2)}{\partial t_2}\Big], \tag{5.294}$$

where h_1 and h_2 are the Lamé coefficients. As we noted in Subsection 5.7.1, we can assume that $h_1 = h_2 = 1$ at a given point.

Note that

$$\mathrm{div}_0\,\boldsymbol{\varphi}^\pm = ik\mathbf{E}^\pm\cdot\boldsymbol{\nu} \quad\text{and}\quad \mathrm{div}_0\,\boldsymbol{\psi}^\pm = ik\mathbf{H}^\pm\cdot\boldsymbol{\nu}$$

(e.g., see [76, 158]).

In formulas (5.293) we can pass to the limit as $x \to S$. Obviously, the first integrals on the right-hand sides in (5.293) tend continuously to their values at $x \in S$ (cf. (5.143)). The second and third integrals undergo a jump (cf. (5.145)) of the tangent and normal components, respectively, and some terms outside the integrals arise. In particular, for the second integrals we have a formula of the form

$$\begin{aligned}\lim_{\substack{x\to x_0\in S\\ x\in V^\pm}} \boldsymbol{\nu}_x \times \int_S \mathbf{j}(y)\times \mathrm{grad}_y\,\Phi(x-y,k)\,dS_y \\ = \pm\frac{1}{2}\mathbf{j}(x_0) + \boldsymbol{\nu}_{x_0}\times\int_S \mathbf{j}(y)\times\mathrm{grad}_y\,\Phi(x-y,k)\,dS_y,\end{aligned} \tag{5.295}$$

where $\mathbf{j} = \boldsymbol{\psi}^\pm$ or $\boldsymbol{\varphi}^\pm$ (e.g., see [76, Theorem 46]). Note that

$$\mathrm{grad}_y\,\Phi(x-y,k) = -\,\mathrm{grad}_x\,\Phi(x-y,k)$$

and

$$\mathrm{grad}_x = \mathrm{grad}_{0x} + \boldsymbol{\nu}\partial_{\boldsymbol{\nu}_x}, \tag{5.296}$$

where the *surface gradient* grad_0 is defined by the formula

$$\mathrm{grad}_0\,U = \frac{1}{h_1}\frac{\partial U}{\partial t_1}\mathbf{e}_1 + \frac{1}{h_2}\frac{\partial U}{\partial t_2}\mathbf{e}_2. \tag{5.297}$$

Now let us introduce the operators

$$\begin{aligned}A\boldsymbol{\varphi}(x) &= 2k\boldsymbol{\nu}_x\times\int_S \Phi(x-y,k)\boldsymbol{\varphi}(y)\,dS_y \\ &\quad + 2k^{-1}\boldsymbol{\nu}_x\times\mathrm{grad}_{0x}\int_S \Phi(x-y,k)\,\mathrm{div}_0\,\boldsymbol{\varphi}(y)\,dS_y,\end{aligned} \tag{5.298}$$

$$\begin{aligned}B\boldsymbol{\varphi}(x) &= 2\boldsymbol{\nu}_x\times\int_S \boldsymbol{\varphi}(y)\times\mathrm{grad}_y\,\Phi(x-y,k)\,dS_y \\ &= -2\boldsymbol{\nu}_x\times\int_S \partial_{\boldsymbol{\nu}_y}\Phi(x-y,k)[\boldsymbol{\nu}_y\times\boldsymbol{\varphi}(y)]\,dS_y \\ &\quad + 2\boldsymbol{\nu}_x\times\int_S \boldsymbol{\nu}_y(\mathrm{grad}_{0y}\,\Phi(x-y,k)\cdot[\boldsymbol{\nu}_y\times\boldsymbol{\varphi}(y)])\,dS_y\end{aligned} \tag{5.299}$$

[25] We assume that $(\mathbf{e}_1, \mathbf{e}_2, \boldsymbol{\nu})$ is a right-handed trihedral, that is, $\boldsymbol{\nu} = \mathbf{e}_1\times\mathbf{e}_2$.

$(x \in S)$. In (5.298), we apply grad_{0x} to the integral. When differentiating this integral, we have to differentiate the kernel $\Phi(x-y,k)$, that is, grad_{0x} again enters the integrand. Thus a singular integral appears, which must be understood in the sense of Cauchy's principal value, that is, as the limit of the integral over $S \cap \{y : |x-y| \geq \varepsilon\}$ as $\varepsilon \to 0$.

One can readily verify that the following relations hold on S (cf. [48, Section 13]):

$$(I-B)\boldsymbol{\psi}^+ = iA\boldsymbol{\varphi}^+, \qquad (I-B)\boldsymbol{\varphi}^+ = -iA\boldsymbol{\psi}^+, \tag{5.300a}$$

$$(I+B)\boldsymbol{\psi}^- = -iA\boldsymbol{\varphi}^-, \qquad (I+B)\boldsymbol{\varphi}^- = iA\boldsymbol{\psi}^-. \tag{5.300b}$$

2. Pseudodifferential operators in Sobolev spaces of vector fields on S. Now we need to consider Sobolev spaces of vector-valued functions on S and pseudodifferential operators acting in these spaces. More precisely, these functions will be vector fields on S, that is, sections of the tangent bundle TS (see Subsection 5.4.3). We retain the notation $H^s(S)$ for the Sobolev spaces of vector fields. In the space $L_2(S) = H^0(S)$ of vector fields, the inner product of two fields $\boldsymbol{\varphi}$ and $\widetilde{\boldsymbol{\varphi}}$, locally represented in the form $\boldsymbol{\varphi} = v^1\mathbf{e}_1 + v^2\mathbf{e}_2$ and $\widetilde{\boldsymbol{\varphi}} = w^1\mathbf{e}_1 + w^2\mathbf{e}_2$, is defined by the formula

$$(\boldsymbol{\varphi}, \widetilde{\boldsymbol{\varphi}})_S = \int_S (v^1\overline{w}^1 + v^2\overline{w}^2)\, dS, \tag{5.301}$$

which is independent of the choice of orthogonal local coordinates. If the support of a vector field $\boldsymbol{\varphi}$ lies in a single coordinate neighborhood, then, by definition, $\|\boldsymbol{\varphi}\|^2_{S,s} = \|v^1\|^2_{S,s} + \|v^2\|^2_{S,s}$. *All assertions of Section* 5.3 *concerning the spaces H^s of scalar functions on a surface remain valid for the spaces H^s of vector fields.*

Pseudodifferential operators are first defined on the space $C^\infty(S)$ of infinitely smooth vector fields. If A is a pseudodifferential operator[26] and $\omega_1, \omega_2 \in C^\infty(S)$ are two scalar functions with supports in the same coordinate neighborhood, then the operator $\omega_1 A(\omega_2\boldsymbol{\varphi}) = \boldsymbol{\psi} = w^1\mathbf{e}_1 + w^2\mathbf{e}_2$ can be represented by the formula

$$\begin{pmatrix} w^1 \\ w^2 \end{pmatrix} = \omega_1 \begin{pmatrix} A_{11} & A_{12} \\ A_{21} & A_{22} \end{pmatrix} \left[\omega_2 \begin{pmatrix} v^1 \\ v^2 \end{pmatrix}\right], \tag{5.302}$$

where $\boldsymbol{\varphi} = v^1\mathbf{e}_1 + v^2\mathbf{e}_2$ and A_{ij} are scalar pseudodifferential operators in $\mathbf{R}^2$. The principal symbol of A is defined on nonzero cotangent vectors to S and is locally represented by a 2×2 *matrix* $a(t,\xi)$ that is infinitely smooth in (t,ξ) for $\xi \neq 0$ and positively homogeneous in $\xi = (\xi_1, \xi_2)$. *All assertions from Section* 5.4 *remain valid for such pseudodifferential operators* with the following natural modifications.

1. Let A and B be two pseudodifferential operators with principal symbols a and b, and let γ_1 and γ_2 be their orders. Then $AB - BA$ is an operator of order $\gamma_1 + \gamma_2 - 1$ if $ab = ba$.

2. The adjoint A^* of a pseudodifferential operator A with principal symbol a with respect to the inner product (5.301) is a pseudodifferential operator with principal symbol $a^* = \overline{a}'$, which is the Hermitian conjugate of a.

---

[26] The letters A and B had a different meaning in Subsection 1.

The ellipticity condition now reads $\det a(x,\xi) \neq 0$ for $\xi \neq 0$. *The assertions of Subsection* 5.5.1 *also remain valid.*

Finally, we can define pseudodifferential operators acting from scalar functions to vector fields on S, as well as pseudodifferential operators acting from vector fields to scalar functions. Examples are given by the differential operators grad_0 and div_0, respectively. Their principal symbols in orthogonal local coordinates have the form

$$i\begin{pmatrix} h_1^{-1}\xi_1 \\ h_2^{-1}\xi_2 \end{pmatrix} \quad \text{and} \quad i\,(h_1^{-1}\xi_1, h_2^{-1}\xi_2). \tag{5.303}$$

The composition theorem (Assertion 3° in Subsection 5.4.1) remains valid for any two pseudodifferential operators from these classes provided that the product is well defined.

3. The operators *A* and *B*. Here we shall show that the operators (5.298) and (5.299) are pseudodifferential operators of orders 1 and -1, respectively, and calculate the principal symbol of the pseudodifferential operator (5.298).

Proposition 5.11.1 *The operator* $\boldsymbol{\nu}_x\times$ *is a zero-order pseudodifferential operator with symbol*

$$J = \begin{pmatrix} 0 & -1 \\ 1 & 0 \end{pmatrix} \tag{5.304}$$

in orthogonal local coordinates.[27]

Indeed, this operator transforms each vector field $v^1\mathbf{e}_1 + v^2\mathbf{e}_2$ into the vector field $-v^2\mathbf{e}_1 + v^1\mathbf{e}_2$.

Obviously, $\boldsymbol{\nu}\times\boldsymbol{\nu}\times\ = -I$.

By the subscript $_{\mathrm{tg}}$ we shall indicate the tangent component of a vector at a given point on S.

Proposition 5.11.2 *The operators*

$$\Big[\int_S G(x-y,k)\varphi(y)\,dS_y\Big]_{\mathrm{tg}}, \qquad \Big[\int_S \partial_{\nu_y} G(x-y,k)\varphi(y)\,dS_y\Big]_{\mathrm{tg}} \tag{5.305}$$

are pseudodifferential operators of order -1. *The first of them has the principal symbol*

$$-\frac{1}{2}[h_1^{-2}\xi_1^2 + h_2^{-2}\xi_2^2]^{-1/2}E, \tag{5.306}$$

where E *is the* 2×2 *identity matrix.*

[27] In what follows we use only orthogonal local coordinates without mentioning this explicitly any more.

Proof. Let us consider the operators obtained by the multiplication of the operators (5.305) by ω_1 on the left and by the multiplication of φ in the integrand in (5.305) by ω_2, where ω_1 and ω_2 are functions from $C^\infty(S)$ with supports in a given coordinate neighborhood. Let t and $\tilde{t}$ be the local coordinates of points x and y from this neighborhood. Then

$$\mathbf{e}_1(\tilde{t}) = \alpha_{11}(t,\tilde{t})\mathbf{e}_1(t) + \alpha_{12}(t,\tilde{t})\mathbf{e}_2(t) + \alpha_{13}(t,\tilde{t})\boldsymbol{\nu}_x,$$
$$\mathbf{e}_2(\tilde{t}) = \alpha_{21}(t,\tilde{t})\mathbf{e}_1(t) + \alpha_{22}(t,\tilde{t})\mathbf{e}_2(t) + \alpha_{23}(t,\tilde{t})\boldsymbol{\nu}_x,$$

where $\alpha_{ij}(t,\tilde{t})$ are some infinitely smooth functions that are equal to δ_{ij} for $\tilde{t} = t$. For example, consider the first operator in (5.305). It takes each vector field $\boldsymbol{\varphi}(\tilde{t}) = v^1(\tilde{t})\mathbf{e}_1(\tilde{t}) + v^2(\tilde{t})\mathbf{e}_2(\tilde{t})$ to the vector field $\boldsymbol{\psi}(t) = w^1(t)\mathbf{e}_1(t) + w^2(t)\mathbf{e}_2(t)$ according to the formula

$$\begin{pmatrix} w_1(t) \\ w_2(t) \end{pmatrix} = \omega_1(t)\int_{\mathbf{R}^2} \Phi(x(t) - x(\tilde{t}), k)\begin{pmatrix} \alpha_{11}(t,\tilde{t}) & \alpha_{21}(t,\tilde{t}) \\ \alpha_{12}(t,\tilde{t}) & \alpha_{22}(t,\tilde{t}) \end{pmatrix}\begin{pmatrix} v_1(\tilde{t}) \\ v_2(\tilde{t}) \end{pmatrix}\omega_2(\tilde{t})h_1(\tilde{t})h_2(\tilde{t})\,dt.$$

The subsequent argument is the same as in the proof of Theorem 5.7.1. The second operator can be treated in a similar way (see the proof of Proposition 5.8.1). □

Proposition 5.11.3 *The operator*

$$\left[\int_S \boldsymbol{\nu}_y(\operatorname{grad}_{0y} \Phi(x-y,k)\cdot\boldsymbol{\varphi}(y))\,dS_y\right]_{\mathrm{tg}} \tag{5.307}$$

is a pseudodifferential operator of order -1.

Proof. Using the same notation as in the preceding proof, we write

$$\boldsymbol{\nu}_y = \beta_1(t,\tilde{t})\mathbf{e}_1(t) + \beta_2(t,\tilde{t})\mathbf{e}_2(t) + \beta_3(t,\tilde{t})\boldsymbol{\nu}_x,$$

where the $\beta_i(t,\tilde{t})$ are infinitely smooth functions and $\beta_1 = \beta_2 = 0$, $\beta_3 = 1$ for $t = \tilde{t}$. We must consider the operator

$$\begin{pmatrix} w^1(t) \\ w^2(t) \end{pmatrix} = \omega_1(t)\int_{\mathbf{R}^2}\begin{pmatrix} \beta_1(t,\tilde{t}) \\ \beta_2(t,\tilde{t}) \end{pmatrix}\Psi'(|x-y|)\Big[\frac{1}{h_1(\tilde{t})}\frac{\partial|x-y|}{\partial\tilde{t}_1}v^1(\tilde{t}) + \frac{1}{h_2(\tilde{t})}\frac{\partial|x-y|}{\partial\tilde{t}_2}v^2(\tilde{t})\Big]\omega_2(\tilde{t})h_1(\tilde{t})h_2(\tilde{t})\,d\tilde{t},$$

where $\Psi(t) = -(4\pi)^{-1}e^{-ikt}t^{-1}$ and $x = x(t)$, $y = x(\tilde{t})$. Here, say,

$$\frac{\partial|x-y|}{\partial\tilde{t}_1} = \frac{1}{|x-y|}\sum_{j=1}^{3}[x_j(\tilde{t}) - x_j(t)]\frac{\partial x_j(\tilde{t})}{\partial\tilde{t}_1}.$$

Now we can apply Proposition 5.4.1 with $p = 2$, $q = 3$, and $n = 3$. □

To simplify the representation of symbols, we set

$$\eta_1 = h_1^{-1}(t)\xi_1, \qquad \eta_2 = h_2^{-1}(t)\xi_2. \tag{5.308}$$

In this notation, the symbol (5.306) is equal to $-\dfrac{1}{2}|\eta|^{-1}E$.

Proposition 5.11.4 *The operators A and B (see Eqs. (5.298) and (5.299)) are pseudodifferential operators of order 1 and -1, respectively. The principal symbol $a(t,\xi)$ of A is given by the formula*

$$a(t,\xi) = k^{-1}|\eta|^{-1}J\Xi, \quad \textit{where } \Xi = \begin{pmatrix} \eta_1^2 & \eta_1\eta_2 \\ \eta_1\eta_2 & \eta_2^2 \end{pmatrix}. \tag{5.309}$$

Proof. It suffices to use the composition theorem (Assertion 3° in Subsection 5.4.1) and Propositions 5.11.1–5.11.3. Formula (5.309) follows with regard to the expressions (5.303), (5.304), and (5.306). □

4. The expressions for $\psi^\pm$ via $\varphi^\pm$ and for $\varphi^\pm$ via $\psi^\pm$. Until Subsection 9 we shall assume that the following condition is satisfied:

(a) *The problem in V^+ for system (5.290) with boundary condition $\boldsymbol{\nu} \times \mathbf{E}^+ = 0$ on S has no nontrivial solution smooth up to the boundary.*

Here $\mathbf{E}^+$ can be replaced by $\mathbf{H}^+$, since system (5.290) is invariant with respect to the replacement $\mathbf{E} \to \mathbf{H}$, $\mathbf{H} \to -\mathbf{E}$.

This condition is satisfied for all k except for $k = \pm k_j$, where $\{k_j\}_1^\infty$ is a sequence of positive numbers tending to infinity. This sequence has the asymptotics $k_j \sim \text{const}\, j^{1/3}$ (see [105]). The problem for system (5.1) in V^+ with the boundary condition

$$\boldsymbol{\nu} \times \mathbf{E}^+ = \mathbf{f} \quad \text{on } S, \tag{5.310}$$

where $\mathbf{f}$ is a given vector function, is known as the *interior Maxwell problem.* This problem is uniquely solvable for all other values of k in suitably chosen spaces. The similar *exterior Maxwell problem* for system (5.290) with boundary condition

$$\boldsymbol{\nu} \times \mathbf{E}^- = \mathbf{f} \quad \text{on } S \tag{5.311}$$

and the radiation conditions at infinity is uniquely solvable for all our k.

Concerning the study of these problems in Hölder function spaces, e.g., see [76] and [31]. See also [63] and [47] for especially convenient function spaces, and cf. Subsection 8 below. Here we restrict ourselves to the following remarks. It follows from the Maxwell equations that $\mathbf{E}$ and $\mathbf{H}$ have zero divergence and satisfy the vector Helmholtz equation. Hence, say, the solution of the interior Maxwell problem, under some a priori smoothness assumptions, is a solution of the *interior electric problem*

$$\Delta\mathbf{E} + k^2\mathbf{E} = 0 \quad \text{in } V^+, \quad \operatorname{div}\mathbf{E} = 0 \quad \text{and} \quad \boldsymbol{\nu} \times \mathbf{E} = \mathbf{f} \quad \text{on } S. \tag{5.312}$$

This is well known. One can readily verify that the latter problem is elliptic; hence, for smooth $\mathbf{f}$, its solutions, even in the generalized sense, are smooth up to S (cf. [66]). In particular, the solutions of the homogeneous interior Maxwell problem (for the exceptional values $k = \pm k_j$) are infinitely smooth in $\overline{V}^+$.

The situation is similar for problems obtained from the above problems by the substitution $\mathbf{E} \to \mathbf{H}$, $\mathbf{H} \to -\mathbf{E}$.

The following assertion is similar to Proposition 5.8.3.

Proposition 5.11.5

$$\operatorname{Ker}(I - B) = \{\varphi^+ : \mathbf{E}, \mathbf{H} \textit{ satisfy } (5.290) \textit{ in } V^+ \textit{ and } \psi^+ = 0\}. \tag{5.313a}$$

The dimension of this kernel coincides with the dimension of the space of solutions of the homogeneous Maxwell problem.

Since the Maxwell system is invariant with respect to the substitution $\mathbf{E} \to \mathbf{H}$, $\mathbf{H} \to -\mathbf{E}$, it follows that in (5.313) we can replace φ by ψ and ψ by $-\varphi$:

$$\operatorname{Ker}(I - B) = \{\psi^+ : \mathbf{E}, \mathbf{H} \text{ satisfy } (5.290) \text{ in } V^+ \text{ and } \varphi^+ = 0\}. \tag{5.314}$$

Proposition 5.11.5 is essentially contained in [31, Chapter 4, Theorem 4.23]. We give a proof using formulas from [76]. Let $\mathbf{E}$, $\mathbf{H}$ be a nontrivial solution of the Maxwell system with $\psi^+ = 0$. Then $\mathbf{E}$ is a solution of the homogeneous interior electric problem, and hence $\mathbf{E}$, $\mathbf{H} \in C^\infty(\overline{V}^+)$. We see from the second formula in (5.300a) that $(I - B)\varphi^+ = 0$. Moreover, φ^+ is not identically zero, since otherwise we would have $\mathbf{E} \equiv 0$ and $\mathbf{H} \equiv 0$ (see (5.293)). Conversely, let φ be a nontrivial solution of the equation $(I - B)\varphi = 0$. Since the operator $I - B$ is elliptic, it follows that $\varphi \in C^\infty(S)$. Consider the field

$$\begin{aligned} \mathbf{E} &= \int_S [ik\Phi(x-y,k)\varphi(y) - ik^{-1}\operatorname{div}_0 \varphi(y) \cdot \operatorname{grad}_y \Phi(x-y,k)]\, dS_y, \\ \mathbf{H} &= -\int_S \varphi(y) \times \operatorname{grad}_y \Phi(x-y,k)\, dS_y. \end{aligned} \tag{5.315}$$

This is a solution of the Maxwell system outside S (see Theorem 48 in [76]; also, cf. (5.293)). By virtue of what was said after Eq. (5.294) about the behavior of the potentials as $x \to S$, for this field we have

$$\psi^+ = \psi^-, \quad \varphi^+ = \frac{1}{2}(I + B)\varphi, \quad \varphi^- = \frac{1}{2}(-I + B)\varphi = 0,$$

whence $\varphi^+ = \varphi$. Consequently, $\mathbf{E} = \mathbf{H} = 0$ in V^- (by virtue of the uniqueness for the exterior problem with given φ^-), so that $\psi^- = 0$ and $\psi^+ = 0$ on S. Since φ is not zero identically, we have constructed a nontrivial solution of the homogeneous interior Maxwell problem. □

Proposition 5.11.6 *The dimensions of* $\operatorname{Ker}(I - B)$ *and* $\operatorname{Ker}(I + B)$ *coincide.*

This was proved in [76, Section 25]. One sees from the proof that the result follows from the easily verifiable relation

$$B^*\varphi = \boldsymbol{\nu} \times \overline{B}(\boldsymbol{\nu} \times \varphi), \tag{5.316}$$

from which we find, with regard to the fact that $\mathrm{Ker}(I - B)$ and $\mathrm{Ker}(I - \overline{B})$ coincide for real k (Lemma 112 in [76]), that

$$\boldsymbol{\nu} \times \mathrm{Ker}(I - B) = \mathrm{Ker}(I + B^*), \qquad \boldsymbol{\nu} \times \mathrm{Ker}(I + B^*) = \mathrm{Ker}(I - B). \tag{5.317}$$

Propositions 5.11.5 and 5.11.6 imply the following analog of Proposition 5.8.4.

Corollary 5.11.7 *Condition* (a) *is equivalent to the existence of the operators* $(I-B)^{-1}$ *and* $(I + B)^{-1}$.

Now, under condition (a), we can rewrite (5.300) in the form

$$\boldsymbol{\psi}^+ = i(I - B)^{-1}A\varphi^+, \qquad \varphi^+ = -i(I - B)^{-1}A\boldsymbol{\psi}^+, \tag{5.318a}$$
$$\boldsymbol{\psi}^- = -i(I + B)^{-1}A\varphi^-, \qquad \varphi^- = i(I + B)^{-1}A\boldsymbol{\psi}^-. \tag{5.318b}$$

Note that here $\boldsymbol{\psi}^-$ and $\boldsymbol{\psi}^+$ can be arbitrary smooth vector fields on S. This follows from the fact that the exterior Maxwell problem is always solvable, whereas the interior problem is solvable under condition (a). We find from (5.318) that $[(I \pm B)^{-1}A]^2\boldsymbol{\psi} = \boldsymbol{\psi}$ for smooth $\boldsymbol{\psi}$ and hence for arbitrary $\boldsymbol{\psi}$ belonging to $H^s(S)$ for some s:

$$[(I + B)^{-1}A]^2 = I, \qquad [(I - B)^{-1}A]^2 = I. \tag{5.319}$$

Thus, the first-order operator $(I \pm B)^{-1}A$ is the inverse of itself. (In particular, the principal symbol of the left-hand sides in (5.319) considered as second-order operators is zero. This can be verified directly: $a^2(t, \xi)$ is the zero matrix.)

Now note that for an arbitrary smooth vector field φ, for the solution (5.315) of the Maxwell system we have

$$\boldsymbol{\psi}^+ = \frac{i}{2}A\varphi, \qquad \varphi^+ = \frac{1}{2}(I + B)\varphi,$$

and from the second of these equations we can express φ via φ^+ (under condition (a)). We obtain the formula

$$\boldsymbol{\psi}^+ = iA(I + B)^{-1}\varphi^+.$$

By comparing it with the first formula in (5.318a), we see that $A(I+B)^{-1} = (I-B)^{-1}A$. We arrive at the following assertion (cf. Proposition 5.8.5).

Proposition 5.11.8 *One has* $AB = -BA$ *and* $A(I \pm B)^{-1} = (I \mp B)^{-1}A$.

By passing to the limit, we can extend the first of these relations to the values of k prohibited by condition (a).

5. Statement of spectral problems and their reduction to pseudodifferential equations on S. In a parallel way, we consider the following three problems,[28] assuming first that condition (a) is satisfied.

Problem I Find a solution of system (5.290) in V^+ under the condition

$$\lambda(\boldsymbol{\nu} \times \mathbf{H}^+) - i\mathbf{E}^+_{\mathrm{tg}} = \mathbf{g} \quad \text{on } S. \tag{5.320}$$

Problem II Find a solution of system (5.290) in V^- under the condition

$$\lambda(\boldsymbol{\nu} \times \mathbf{H}^-) + i\mathbf{E}^-_{\mathrm{tg}} = \mathbf{g} \quad \text{on } S \tag{5.321}$$

and the radiation conditions at infinity.

Problem III Find a solution of system (5.290) outside S under the conditions

$$\mathbf{H}^+_{\mathrm{tg}} = \mathbf{H}^-_{\mathrm{tg}}, \quad \lambda(\boldsymbol{\nu} \times \mathbf{H}) + \frac{i}{2}(\mathbf{E}^-_{\mathrm{tg}} - \mathbf{E}^+_{\mathrm{tg}}) = \mathbf{g} \quad \text{on } S \tag{5.322}$$

and the radiation conditions.

One can also consider the problems obtained from Problems I–III by the substitution $\mathbf{E} \to \mathbf{H}$ and $\mathbf{H} \to -\mathbf{E}$, but these problems are of no independent interest from the mathematical viewpoint. Note however that one of them was considered by Baum [24, 25] independently of Katsenelenbaum, Sivov, and Voitovich. Namely, Baum proposed to consider expansions in eigenfunctions of the problem actually equivalent to that obtained from Problem III by the substitution just indicated. See [13].

Let us introduce the operators

$$\begin{aligned} \mathrm{A_I} &= -\boldsymbol{\nu} \times (I - B)^{-1}A, \qquad \mathrm{A_{II}} = -\boldsymbol{\nu} \times (I + B)^{-1}A, \\ \mathrm{A_{III}} &= \frac{1}{2}(\mathrm{A_I} + \mathrm{A_{II}}) = -\boldsymbol{\nu} \times (I - B^2)^{-1}A. \end{aligned} \tag{5.323}$$

We see from (5.318) that *Problems* I–III *can be reduced to the equations*

$$(\mathrm{A_I} - \lambda)\varphi^+ = \mathbf{g}, \quad (\mathrm{A_{II}} - \lambda)\varphi^- = \mathbf{g}, \quad (\mathrm{A_{III}} - \lambda)\varphi = \mathbf{g}, \tag{5.324}$$

respectively, where $\varphi = \varphi^{\pm}$. More precisely, if g is a smooth vector field and $\mathbf{E}$, $\mathbf{H}$ is a solution of Problem I, II, or III, then φ^+, φ^-, or $\varphi = \varphi^{\pm}$ satisfies the respective equation in (5.324). Conversely, if φ^+, φ^-, or φ is a solution of the first, second, or third equation in (5.324), respectively, then the field $\mathbf{E}$, $\mathbf{H}$ giving the solution of the corresponding problem I, II, or III can be reconstructed as the solution of the Maxwell system in V^+, in V^-, or outside S with given $\boldsymbol{\nu} \times \mathbf{H}^+ = -\varphi^+$, $\boldsymbol{\nu} \times \mathbf{H}^- = -\varphi^-$, or $\boldsymbol{\nu} \times \mathbf{H}^{\pm} = -\varphi$ (and

[28] See the impedance method and the method of surface magnetic current in Section 2.6.

with the radiation conditions in the second and the third cases). Here we assume that λ does not belong to the spectrum of the corresponding operator A. If λ belongs to the spectrum, then the equivalence remains valid for the (vector-valued) eigenfunctions. We postpone the consideration of the problems and equations in Sobolev spaces until Subsection 8.

Proposition 5.11.9 *The operators* $\mathrm{A_I}$, $\mathrm{A_{II}}$, *and* $\mathrm{A_{III}}$ *are first-order pseudodifferential operators with the same principal symbol* $-J\cdot a(t,\xi) = k^{-1}|\eta|^{-1}\Xi$. *The inverse operators exist and are first-order pseudodifferential operators*

$$\mathrm{A_I^{-1}} = (I-B)^{-1}A\boldsymbol{\nu}\times, \quad \mathrm{A_{II}^{-1}} = (I+B)^{-1}A\boldsymbol{\nu}\times, \quad \mathrm{A_{III}^{-1}} = A\boldsymbol{\nu}\times \tag{5.325}$$

with principal symbol $a(t,\xi)J = k^{-1}|\eta|^{-1}J\Xi J$.

This follows from Propositions 5.11.1, 5.11.4, and 5.11.8 and relations (5.319). Since $\det\Xi \equiv 0$, we see that the operators $\mathrm{A_I}$, $\mathrm{A_{II}}$, and $\mathrm{A_{III}}$ are by no means elliptic.

6. The transformation of Eqs. (5.324) into elliptic equations. Let us introduce the new spectral parameter $\mu = \lambda - \lambda^{-1}$ (cf. [22]). To Eqs. (5.324) we apply, respectively, the operators $(\lambda\mathrm{A_I})^{-1}$, $(\lambda\mathrm{A_{II}})^{-1}$, and $(\lambda\mathrm{A_{III}})^{-1}$. Next, we add the resultant equations to the corresponding equations in (5.324). Thus we arrive at the equations

$$(\mathrm{B_I}-\mu)\varphi^+ = \mathbf{h}_\mathrm{I}, \quad (\mathrm{B_{II}}-\mu)\varphi^- = \mathbf{h}_\mathrm{II}, \quad (\mathrm{B_{III}}-\mu)\varphi = \mathbf{h}_\mathrm{III}, \tag{5.326}$$

which play the main role in the subsequent exposition. Here[29]

$$\mathrm{B} = \mathrm{A} - \mathrm{A}^{-1}, \qquad \mathbf{h} = \mathbf{g} + (\lambda\mathrm{A})^{-1}\mathbf{g}. \tag{5.327}$$

Theorem 5.11.10 *The operators* $\mathrm{B_I}$, $\mathrm{B_{II}}$, *and* $\mathrm{B_{III}}$ *are elliptic first-order pseudodifferential operators with the same principal symbol* $k^{-1}|\eta|E$.

This readily follows from Proposition 5.11.9.

It follows from Theorem 5.11.10 that the root vectors of the operators B belong to $C^\infty(S)$.

7. The spectral properties of the operators A and B.

Proposition 5.11.11 *Let* $\mathcal{T}$ *be any of the operators* A *and* B. *Then* $\mathcal{T}^* = \overline{\mathcal{T}}$. *If* $k_2 = 0$, *then* $\mathrm{A_I^*} = \mathrm{A_I}$ *and* $\mathrm{B_I^*} = \mathrm{B_I}$. *If* $k_1 = 0$, *then* $\mathcal{T}^* = -\mathcal{T}$.

Proof. All these relations follow from formulas similar to the Lorentz formula. More precisely, let $\mathbf{E}_1$ and $\mathbf{H}$ be smooth vector functions in $\overline{V}^+$. It follows from the formula $\operatorname{div}(\mathbf{H}\times\mathbf{E}_1) = \mathbf{E}_1\cdot\operatorname{rot}\mathbf{H} - \mathbf{H}\cdot\operatorname{rot}\mathbf{E}_1$ and the Gauss–Ostrogradskii formula that

$$\int_{V+}(\operatorname{rot}\mathbf{H}\cdot\mathbf{E}_1 - \mathbf{H}\cdot\operatorname{rot}\mathbf{E}_1)\,dx = \int_S(\boldsymbol{\nu}\times\mathbf{H}^+)\cdot\mathbf{E}_1^+\,dS. \tag{5.328}$$

[29]We hope that the reader has noticed the difference between the letters A and A, B and B.

Let $\{\mathbf{E}, \mathbf{H}\}$ and $\{\mathbf{E}_1, \mathbf{H}_1\}$ be two solutions of the Maxwell system (5.290) in V^+, smooth in $\overline{V^+}$. Then it follows from (5.328) and a similar formula with $\mathbf{H}_1$ instead of $\mathbf{H}$ and $\mathbf{E}$ instead of $\mathbf{E}_1$ that

$$\int_S \boldsymbol{\varphi}^+ \cdot (\boldsymbol{\nu} \times \boldsymbol{\psi}_1^+)\, dS = \int_S (\boldsymbol{\nu} \times \boldsymbol{\psi}^+) \cdot \boldsymbol{\varphi}_1^+\, dS \tag{5.329}$$

(the *Lorentz formula*). Now in (5.328) we replace $\mathbf{E}_1$ by $\overline{\mathbf{E}}_1$ or $\mathbf{H}$ by $\overline{\mathbf{H}}$, thus obtaining two similar formulas

$$\int_{V+} (\operatorname{rot} \mathbf{H} \cdot \overline{\mathbf{E}}_1 - \mathbf{H} \cdot \operatorname{rot} \overline{\mathbf{E}}_1)\, dx = \int_S (\boldsymbol{\nu} \times \mathbf{H}^+) \cdot \overline{\mathbf{E}}_1^+\, dS,$$

$$\int_{V+} (\operatorname{rot} \overline{\mathbf{H}} \cdot \mathbf{E}_1 - \overline{\mathbf{H}} \cdot \operatorname{rot} \mathbf{E}_1)\, dx = \int_S (\boldsymbol{\nu} \times \overline{\mathbf{H}}^+) \cdot \mathbf{E}_1^+\, dS.$$

These formulas yield

$$\int_S \boldsymbol{\varphi}^+ \cdot \overline{(\boldsymbol{\nu} \times \boldsymbol{\psi}_1^+)}\, dS = -\int_S (\boldsymbol{\nu} \times \boldsymbol{\psi}^+) \cdot \overline{\boldsymbol{\varphi}_1^+}\, dS \tag{5.330}$$

for $k_2 = 0$ and

$$\int_S \boldsymbol{\varphi}^+ \cdot \overline{(\boldsymbol{\nu} \times \boldsymbol{\psi}_1^+)}\, dS = \int_S (\boldsymbol{\nu} \times \boldsymbol{\psi}^+) \cdot \overline{\boldsymbol{\varphi}_1^+}\, dS \tag{5.331}$$

for $k_1 = 0$. (We have excluded the case $k = 0$, in which both sides of (5.330) and (5.331) are zero.) Since

$$\boldsymbol{\nu} \times \boldsymbol{\psi}^+ = -i\mathrm{A}_\mathrm{I}\boldsymbol{\varphi}^+ \quad \text{and} \quad \boldsymbol{\nu} \times \boldsymbol{\psi}_1^+ = -i\mathrm{A}_\mathrm{I}\boldsymbol{\varphi}_1^+$$

(see the first formulas in (5.318a) and (5.323)), we see that Eqs. (5.329), (5.330) and (5.331) prove our assertions about A_I. The assertions concerning A_II can be obtained in a similar way; instead of V^+ one takes V_R^- and then passes to the limit as $R \to \infty$ in (5.329) (for each k) and in (5.331) (for $k_1 = 0$ and $k_2 < 0$) with regard to the radiation conditions. As a consequence, we obtain the assertions concerning A_III and B. □

Proposition 5.11.11 shows that B_I for $k_2 = 0$ and all $i\mathrm{B}$ for $k_1 = 0$ ($k \neq 0$) are selfadjoint operators in $H^0(S)$ (and in $H^s(S)$ for an appropriate choice of the inner product in $H^s(S)$, see Subsection 5.5.2). In these cases, B has an orthonormal basis of eigenfunctions in $H^0(S)$ (which remains an orthogonal basis in any $H^s(S)$). In the following, in Proposition 5.11.12 and Theorem 5.11.13, we exclude these cases from our consideration.

Proposition 5.11.12 *The operators* $\operatorname{Im} \mathrm{B}_\mathrm{II}$ *and* $\operatorname{Im} \mathrm{B}_\mathrm{III}$ *have the order* $-\infty$ *if* $k_2 = 0$. *The operators* $\operatorname{Im}(k\mathrm{B})$ *have the order* -1 *if* $k_2 < 0$.

Proof. The proof is similar to that of Proposition 5.8.11. For $k_2 = 0$, it suffices to notice that the operators $A - \overline{A}$ and $B - \overline{B}$ have the form (5.298) and (5.299) with $\Phi - \overline{\Phi}$ instead of Φ. For $k_2 < 0$ we use the fact that $(k\mathrm{B})^* = \overline{k}\mathrm{B}^* = \overline{k}\,\overline{\mathrm{B}} = \overline{k\mathrm{B}}$. In this case,

$kA - \overline{kA}$ is an operator of order -1, as can be seen from (5.298), and $B - \overline{B}$, as can be seen from (5.299), is an operator of order -3. □

The principal symbol of the operators $\mathrm{Re}(k\mathrm{B})$ is a *scalar* matrix. It follows from [52] or [50] that the eigenvalues $\nu_j = \nu_j(\mathrm{Re}(k\mathrm{B}))$ of the operators $\mathrm{Re}(k\mathrm{B})$ have the asymptotics

$$\nu_j = \sqrt{2\pi|S|^{-1}}\, j^{1/2} + O(1), \qquad j \to \infty, \tag{5.332}$$

where $|S|$ is the surface area of S (cf. Eqs. (5.80) and (5.168)–(5.169) for scalar operators).

Now let us use the statements (and the notation) from Subsections 5.6.1 and 5.6.2 with $\mathfrak{H}^s = H^s(S)$.

Theorem 5.11.13 $\mathrm{B}_{\mathrm{II}}, \mathrm{B}_{\mathrm{III}} \in \mathbf{B}(H^s(S))$ *for* $k_2 = 0$, *and* $\mathrm{B} \in \mathbf{R}(H^s(S))$ *for* $k_2 < 0$ *and for all* s.

If $k_2 = 0$, *then the operators* B_{II} *and* $\mathrm{B}_{\mathrm{III}}$ *satisfy the assertions of Theorem* 5.6.3 *with* $p = 1/2$, $r = 0$, *and* $\inf q = -\infty$.

If $k_2 < 0$, *then the operators* B *satisfy the assertions of Theorem* 5.6.4 *with* $p = 1/2$, $r = 0$, *and* $q = -1$.

Here in the general case we treat $k\mathrm{B}$ as a weak perturbation of the selfadjoint operator $\mathrm{Re}(k\mathrm{B})$. For real k, the operator B is a very weak perturbation of the selfadjoint operator $\mathrm{Re}\,\mathrm{B}$. If k is not real, then B is a weak perturbation of the normal operator $k^{-1}\,\mathrm{Re}(k\mathrm{B})$, which is obtained by multiplying the selfadjoint operator $\mathrm{Re}(k\mathrm{B})$ by the complex number k^{-1}.

The eigenvalues of the operator B tend to infinity along the ray issuing from the origin and passing through the point k^{-1}. Namely, the following assertion holds.

Proposition 5.11.14 *The eigenvalues* $\mu_j = \mu_j(\mathrm{B})$ *of each of the operators* B *satisfy the asymptotic formula*

$$\mu_j = k^{-1}\sqrt{2\pi|S|^{-1}}\, j^{1/2} + O(1), \qquad j \to \infty. \tag{5.333}$$

All eigenvalues μ_j *of the operator* B *lie in the union of disks centered at the eigenvalues* ν_l *of the operator* $k^{-1}\,\mathrm{Re}(k\mathrm{B})$ *with radii* $C_h|\nu_l|^{-h}$, *where* h *is arbitrarily large if* $k_2 = 0$, *and* $h = 1$ *if* $k_2 < 0$.

Of course, the eigenvalues of B_{I} for $k_2 = 0$ lie on the real axis, and the eigenvalues of all B for $k_1 = 0$ lie on the imaginary axis.

Let us return to the operators A.

Proposition 5.11.15 *A number* λ *is an eigenvalue of the operator* A_{I} *if and only if* $\mu = \lambda - \lambda^{-1}$ *is an eigenvalue of the operator* B_{I}. *If, moreover, the equation* $\mu = \lambda - \lambda^{-1}$ *has distinct roots* λ_1 *and* λ_2, *then the root subspace* $L_{\mathrm{B}_{\mathrm{I}}}(\mu)$ *is the direct sum of the root subspaces* $L_{\mathrm{A}_{\mathrm{I}}}(\lambda_1)$ *and* $L_{\mathrm{A}_{\mathrm{I}}}(\lambda_2)$; *furthermore, the last two subspaces are isomorphic. If the roots coincide,* $\lambda_1 = \lambda_2$, *then* $L_{\mathrm{A}_{\mathrm{I}}}(\lambda_i) = L_{\mathrm{B}_{\mathrm{I}}}(\mu)$. *In particular, the subspaces* $L_{\mathrm{A}_{\mathrm{I}}}(\lambda)$ *are finite-dimensional and consist of infinitely smooth vector fields. Similar assertions are valid for the operators* A_{II} *and* B_{II}.

Proof. Let A, B stand for either $\mathrm{A_I}$, $\mathrm{B_I}$ or $\mathrm{A_{II}}$, $\mathrm{B_{II}}$. If $(\mathrm{A}-\lambda I)\varphi = 0$, then, as we know, $(\mathrm{B}-\mu I)\varphi = 0$, where $\mu = \lambda - \lambda^{-1}$. More generally, let $(\mathrm{A}-\lambda I)^m\varphi = 0$ for some positive integer m. Then

$$0 = (-\lambda^{-1}\mathrm{A}^{-1})^k(\mathrm{A}-\lambda I)^m\varphi = (\mathrm{A}^{-1}-\lambda^{-1}I)^k(\mathrm{A}-\lambda I)^{m-k}\varphi$$

for $k = 0,\dots,m$, whence it follows that $(\mathrm{B}-\mu I)^m\varphi = 0$. Consequently, $L_{\mathrm{A}(\lambda)} \subset L_{\mathrm{B}(\mu)}$.

The subspace $M = L_{\mathrm{B}}(\mu)$ is finite-dimensional, since B is an operator with discrete spectrum, and invariant with respect to A, since A commutes with $(\mathrm{B}-\mu I)^m$. In some basis of the subspace M, the matrix of the operator A in M has a Jordan form, and moreover, by virtue of the preceding its eigenvalues are the roots of the quadratic equation $\mu = \lambda - \lambda^{-1}$. Hence $M = L_{\mathrm{A}}(\lambda_1) \dot{+} L_{\mathrm{A}}(\lambda_2)$ if this equation has distinct roots λ_1 and λ_2, and $M = L_{\mathrm{A}}(\lambda_i)$ if the roots coincide.

If the roots are distinct, then

$$(\mathrm{A}-\lambda_1 I)^m\varphi = 0 \implies (\mathrm{A}-\lambda_2 I)^m(\boldsymbol{\nu}\times\varphi) = 0. \tag{5.334}$$

This can be verified as follows. The operator A (see (5.323)) is invertible (see (5.325)). If $(\mathrm{A}-\lambda_1 I)^m\varphi = 0$, then $(\lambda_1\mathrm{A})^m(\lambda_1^{-1}I - \mathrm{A}^{-1})^m\varphi = 0$; hence

$$(\mathrm{A}^{-1}+\lambda_2 I)^m\varphi = 0, \tag{5.335}$$

since the product $\lambda_1\lambda_2$ of the roots of the equation $\mu = \lambda - \lambda^{-1}$ is equal to -1. The operator A^{-1} is similar to $-\mathrm{A}$:

$$-\boldsymbol{\nu}\times\mathrm{A}^{-1}\boldsymbol{\nu}\times\; = -\mathrm{A}. \tag{5.336}$$

We set $\varphi = \boldsymbol{\nu}\times\varphi_1$, apply the operator $-\boldsymbol{\nu}\times$ on the left in (5.335), and insert $\boldsymbol{\nu}\times(-\boldsymbol{\nu}\times) = I$ between the factors $(\mathrm{A}^{-1}+\lambda_2 I)$, thus obtaining $(-\mathrm{A}+\lambda_2 I)^m\varphi_1 = 0$, which proves (5.334).

The mapping $\varphi \mapsto \boldsymbol{\nu}\times\varphi$ establishes an isomorphism between $L_{\mathrm{A}}(\lambda_1)$ and $L_{\mathrm{A}}(\lambda_2)$. □

Note that from (5.290) and (5.320) or (5.321) we also can conclude that $-1/\lambda$ is an eigenvalue along with λ (it suffices to use the invariance of the Maxwell system with respect to the replacement of **E**, **H** by **H**, −**E**).

Corollary 5.11.16 *The eigenvalues of each of the operators* $\mathrm{A_I}$ *and* $\mathrm{A_{II}}$ *have two accumulation points, namely,* 0 *and* ∞.

Proposition 5.11.17 *A number* μ *is an eigenvalue of the operator* $\mathrm{B_{III}}$ *if and only if at least one of the roots* λ_1,λ_2 *of the equation* $\mu = \lambda - \lambda^{-1}$ *is an eigenvalue of* $\mathrm{A_{III}}$. *Furthermore,* $L_{\mathrm{B_{III}}}(\mu)$ *is the direct sum of* $L_{\mathrm{A_{III}}}(\lambda_1)$ *and* $L_{\mathrm{A_{III}}}(\lambda_2)$. *In particular, the subspaces* $L_{\mathrm{A_{III}}}(\lambda)$ *are finite-dimensional and consist of infinitely smooth vector fields.*

The proof coincides with that of similar assertions in Proposition 5.11.15. There is no relation like (5.336) for the operator $\mathrm{A_{III}}$.

By virtue of Propositions 5.11.15 and 5.11.17, the system of root functions of the operator A has essentially the same basis properties as the system of root functions of the operator B (see Theorem 5.11.13).

Note also that for the operators A we still have Assertions 2°, 3°, 5°, and 6° of Subsection 5.2.6. If $\varphi_p, \ldots, \varphi_q$ is a basis in $L_{\mathrm{A}}(\lambda)$, then the matrix $\{(\varphi_j, \overline{\varphi}_k)_S\}$, $j, k = p, \ldots, q$, is nondegenerate; this follows from the similar assertion for $L_{\mathrm{B}}(\mu) \supset L_{\mathrm{A}}(\lambda)$. Hence Assertion 8° from Subsection 5.2.6 is also true for A.

The roots of the equation $\mu = \lambda - \lambda^{-1}$ may coincide only if $\lambda^{-1} = -\lambda$, that is, $\lambda = \pm i$. The following proposition implies that at least one of these cases, namely, $\lambda = -i$, is impossible.

Proposition 5.11.18 *The eigenvalues of the operators* $\mathrm{A_{II}}$ *and* $\mathrm{A_{III}}$ *lie in the half-plane* $\operatorname{Im}\lambda > 0$. *The same is true of* $\mathrm{A_I}$ *for* $k_2 = \operatorname{Im} k < 0$. *If* $k_2 < 0$, *then the eigenvalues satisfy also the inequality* $|\operatorname{Re}\lambda| / \operatorname{Im}\lambda \leq |k_1|/|k_2|$.

Proof. We carry out the proof for $\mathrm{A_{II}}$. It follows from Eq. (5.328) with $\overline{\mathbf{E}}$ instead of $\mathbf{E}_1$ and V_R^- instead of V^+ that

$$\begin{aligned} i \int_{V_R^-} (k|\mathbf{E}|^2 - \overline{k}|\mathbf{H}|^2)\, dx &= -\int_S (\boldsymbol{\nu} \times \mathbf{H}) \cdot \overline{\mathbf{E}}\, dS + \int_{S_R} (r^{-1}\mathbf{x} \times \mathbf{H}) \cdot \overline{\mathbf{E}}\, dS \\ &= i\overline{\lambda} \int_S |\boldsymbol{\nu} \times \mathbf{H}|^2\, dS - \int_{S_R} |\mathbf{E}|^2\, dS + o(1) \end{aligned}$$

with regard to (5.291) and (5.321) with $\mathbf{g} = 0$. We separate the real and imaginary parts:

$$-k_2 \int_{V_R^-} (|\mathbf{E}|^2 + |\mathbf{H}|^2)\, dx = \operatorname{Im}\lambda \int_S |\boldsymbol{\nu} \times \mathbf{H}|^2\, dS - \int_{S_R} |\mathbf{E}|^2\, dS + o(1), \tag{5.337a}$$

$$k_1 \int_{V_R^-} (|\mathbf{E}|^2 - |\mathbf{H}|^2)\, dx = \operatorname{Re}\lambda \int_S |\boldsymbol{\nu} \times \mathbf{H}|^2\, dS + o(1). \tag{5.337b}$$

Relation (5.337a) yields $\operatorname{Im}\lambda > 0$. Indeed, we readily see that $\operatorname{Im}\lambda \geq 0$ and that $\operatorname{Im}\lambda > 0$ for $k_2 < 0$. Suppose that $k_2 = 0$ and $\operatorname{Im}\lambda = 0$. Then the integral over S_R tends to zero as $R \to \infty$. It follows that $\mathbf{E} \equiv 0$, since the Cartesian components of the field $\mathbf{E}$ satisfy the Helmholtz equation and the corresponding radiation condition (the latter follows from the integral representation (5.293) of solutions of the Maxwell system).

Furthermore, for $k_2 < 0$ the passage to the limit in (5.337a) and (5.337b) as $R \to \infty$ implies $-k_2\alpha = \operatorname{Im}\lambda \cdot \beta$ and $-|k_1|\alpha < \operatorname{Re}\lambda \cdot \beta < |k_1|\alpha$, where α is the integral of $|\mathbf{E}|^2 + |\mathbf{H}|^2$ over V^- and β is the integral of $|\boldsymbol{\nu} \times \mathbf{H}|^2$ over S. Hence we obtain the desired inequality for $|\operatorname{Re}\lambda| / \operatorname{Im}\lambda$.[30]

[30]Some inequalities for $\operatorname{Re}\lambda / \operatorname{Im}\lambda$ could also be obtained in Subsections 5.7.4 and 5.8.2.

The statements concerning $\mathrm{A_I}$ and $\mathrm{A_{III}}$ can be proved in a similar way. For $k_2 = 0$, the eigenvalues of the operator $\mathrm{A_I}$ are real and distinct from zero. □

8. Uniquely solvable Problems I–III. Here we present a theorem pertaining to Problem II; similar theorems are valid for Problems I and III. Moreover, we restrict ourselves to the case $k_2 = 0$; for $k_2 < 0$, the inequalities given below can be strengthened.

Theorem 5.11.19 *Suppose that $k_2 = 0$, $\lambda \neq 0$, and λ is not an eigenvalue of the operator $\mathrm{A_{II}}$. Then the equation $(\mathrm{A_{II}} - \lambda I)\varphi^- = \mathbf{g}$ with a right-hand side from $H^s(S)$ has a unique solution in $H^s(S)$, and moreover, $\|\varphi^-\|_{S,s} \leq C_s\|\mathbf{g}\|_{S,s}$. If $s > 2$,[31] then Problem II has a unique classical solution. For this solution,*

$$\|\mathbf{E}\|_{V_R^-,s+1/2} + \|\mathbf{H}\|_{V_R^-,s+1/2} \leq C_s'\|\mathbf{g}\|_{S,s}; \tag{5.338}$$

moreover, for $|x| > R$ $(S \subset E_R)$ one has

$$|x| \sum_{|\alpha|\leq m} (|D^\alpha \mathbf{E}(x)| + |D^\alpha \mathbf{H}(x)|) \leq C_m'\|\mathbf{g}\|_{S,1}, \tag{5.339}$$

$$|x|^2[|\,|x|^{-1}\mathbf{x} \times \mathbf{E}(x) - \mathbf{H}(x)| + |\,|x|^{-1}\mathbf{x} \times \mathbf{H}(x) + \mathbf{E}(x)|] \leq C\|\mathbf{g}\|_{S,1}. \tag{5.340}$$

Proof. We omit the subscript $_{\mathrm{II}}$. Let $\mathbf{g} \in H^s(S)$. The equation $(\mathrm{A} - \lambda I)\varphi^- = \mathbf{g}$ cannot have two distinct solutions in $H^s(S)$, since λ is not an eigenvalue of the operator A. Let us transform this equation to the form $(\mathrm{B} - \mu I)\varphi^- = \mathbf{h}$. The latter equation has a unique solution in $H^s(S)$, since $\mathbf{h} = \mathbf{g} + \lambda^{-1}\mathrm{A}^{-1}\mathbf{g} \in H^{s-1}(S)$, B is a first-order elliptic operator, and $\mu = \lambda - \lambda^{-1}$ is not an eigenvalue of this operator. Using the fact that $\mathrm{A} - \lambda I$ commutes with $\mathrm{B} - \mu I$ and $(\mathrm{A} - \lambda I)\mathbf{h} = (\mathrm{B} - \mu I)\mathbf{g}$, for this solution we obtain $(\mathrm{B} - \mu I)[(\mathrm{A} - \lambda I)\varphi^- - \mathbf{g}] = 0$, so that φ^- is a solution of the original equation $(\mathrm{A} - \lambda I)\varphi^- = \mathbf{g}$. Furthermore,

$$\|\varphi^-\|_{S,s} \leq \widetilde{C}_s\|\mathbf{h}\|_{S,s-1} \leq C_s\|\mathbf{g}\|_{S,s}. \tag{5.341}$$

Let $s > 2$. Since $\varphi^- \in C^{(1)}(S)$ (see Assertion 11° in Section 5.3), it follows that there exists a unique classical solution of the exterior problem for system (5.290) with the condition $-\boldsymbol{\nu} \times \mathbf{H}^- = \varphi^-$ ([76, Section 25]). For this solution, formulas (5.318b) hold; in conjunction with the relation $(\mathrm{A} - \lambda I)\varphi^- = \mathbf{g}$, they imply (5.321).

Problem II cannot have two distinct classical solutions, since then we would have the corresponding two distinct solutions φ^- of the equation $(\mathrm{A} - \lambda I)\varphi^- = \mathbf{g}$.

Using Eqs. (5.293), we can obtain the estimates (5.339) and (5.340).

It follows from (5.321) that $\boldsymbol{\psi}^- \in H^s(S)$ and

$$\|\boldsymbol{\psi}^-\|_{S,s} \leq C_s''\|\mathbf{g}\|_{S,s}. \tag{5.342}$$

[31] We consider only these s to remain within the framework of the theory constructed in [76]; see also [31].

Finally, the estimate (5.338), say, for $\mathbf{E}$ can be obtained by using the ellipticity of the boundary conditions $\operatorname{div}\mathbf{E} = 0$, $\boldsymbol{\nu} \times \mathbf{E}^- = \boldsymbol{\psi}^-$ on S for the vector Helmholtz equation $\Delta\mathbf{E} + k^2\mathbf{E} = 0$ in conjunction with the above-mentioned uniqueness. □

Remark The properties of the operators A look somewhat paradoxical. Let us try to clarify them using the following simple model. Let $\{e_j\}_1^\infty$ be an orthonormal basis in a Hilbert space $\mathfrak{H}$. Define a selfadjoint operator L in $\mathfrak{H}$ by the relations $Le_j = \lambda_j e_j$, $j = 1, 2, \dots$, where $\lambda_{2l} = l$ and $\lambda_{2l+1} = l^{-1}$. Obviously the eigenvalues of L have two accumulation points 0 and ∞. The same is true of the eigenvalues of the inverse operator L^{-1}. Both operators can be considered as having the "order 1". The operator $L + L^{-1}$ is an "elliptic operator of order 1", its eigenvalues $\lambda + \lambda^{-1}$ have the asymptotics $j/2$. If λ is distinct from zero and all λ_j, then the resolvent $(L - \lambda_j I)^{-1}$ has bounded eigenvalues $(\lambda_j - \lambda)^{-1}$, this operator is "of zero order".

9. The case in which condition (a) is violated. Here we restrict ourselves to problems I and II. These cases are similar to those considered in Subsection 5.8.3, but here we use other possibilities. First, note that the interior Maxwell problem (5.310) is solvable for exceptional k if and only if

$$\mathbf{f} \perp \operatorname{Ker}(I + B^*) \tag{5.343}$$

(see [76, Section 25]). Moreover, if the solution exists, then it is not unique (it is defined modulo solutions of the homogeneous problem). For the solutions of the homogeneous problem, $(I - B)\boldsymbol{\varphi}^+ = 0$ by virtue of the second formula in (5.300a), and $\boldsymbol{\varphi}^+$ may be an arbitrary element of $\operatorname{Ker}(I - B)$ by virtue of (5.313a). All these assertions remain valid if we replace $\mathbf{E}$ by $\mathbf{H}$ and $\mathbf{H}$ by $-\mathbf{E}$.

Now let us consider Problem I. For the exceptional k, it has the eigenvalue 0, which corresponds to the fields $\mathbf{E}$, $\mathbf{H}$ with $\boldsymbol{\psi}^+ = 0$ and $\boldsymbol{\varphi}^+ \in \operatorname{Ker}(I - B)$, and the "eigenvalue" ∞, which corresponds to the fields with $\boldsymbol{\varphi}^+ = 0$ and $\boldsymbol{\psi}^+ \in \operatorname{Ker}(I - B)$, that is, with $\mathbf{E}^+_{\mathrm{tg}} = -\boldsymbol{\nu} \times \boldsymbol{\psi}^+ \in \operatorname{Ker}(I + B^*)$ (see (5.317)). Note that the subspaces $\operatorname{Ker}(I - B)$ and $\operatorname{Ker}(I + B^*)$ are orthogonal (see Assertion 5° in Subsection 5.2.6) and are contained in $C^\infty(S)$. Let $H^{(1)}$ be the orthogonal complement of the sum of these spaces, so that

$$\operatorname{Ker}(I - B) \oplus \operatorname{Ker}(I + B^*) = [H^{(1)}]^\perp,$$

and let P be the orthogonal projection on $H^{(1)}$ in $L_2(S)$. If $\mathbf{g} \in [H^{(1)}]^\perp$ and λ is finite and distinct from 0, then Problem I has a solution, which is the sum of the field $\mathbf{E}_1$, $\mathbf{H}_1$ with $\boldsymbol{\psi}^+ = 0$ and the field $\mathbf{E}_2$, $\mathbf{H}_2$ with $\boldsymbol{\varphi}^+ = 0$. Now let us assume that $\mathbf{g} \in H^{(1)}$ and reduce the problem to an equation in $H^{(1)}$.

Let $\boldsymbol{\varphi}^+$ be a smooth vector field on S, and let $\boldsymbol{\varphi}^+ = P\boldsymbol{\varphi}^+$; then $\boldsymbol{\varphi}^+$ satisfies condition (5.343), and, by virtue of the first equation in (5.300a), $A\boldsymbol{\varphi}^+$ belongs to the range of $I - B$. This permits us to write $\boldsymbol{\psi}^+ = i(I - B)^{-1}A\boldsymbol{\varphi}^+$; here $\boldsymbol{\psi}^+$ also satisfies condition (5.343) and is defined modulo an arbitrary element of $\operatorname{Ker}(I - B)$. Let us require that $\boldsymbol{\psi}^+$ be orthogonal to $\operatorname{Ker}(I - B)$, that is, $\boldsymbol{\psi}^+ = P\boldsymbol{\psi}^+$; this requirement determines $\boldsymbol{\psi}^+$ uniquely. Thus, instead of (5.318a), in $H^{(1)}$ we have the formulas

$$\boldsymbol{\psi}^+ = iP(I - B)^{-1}AP\boldsymbol{\varphi}^+, \qquad \boldsymbol{\varphi}^+ = -iP(I - B)^{-1}AP\boldsymbol{\psi}^+, \tag{5.344}$$

of which the second is a consequence of the first, since the Maxwell equations are invariant with respect to the substitution $\mathbf{E} \to \mathbf{H}$, $\mathbf{H} \to -\mathbf{E}$. These formulas show that in $H^{(1)}$ we have

$$[P(I - B)^{-1}AP]^2 = I. \tag{5.345}$$

Note that $[H^{(1)}]^{\perp}$ is invariant with respect to the operator $\boldsymbol{\nu}\times$ by virtue of (5.317); it follows that so is $H^{(1)}$. Now Problem I with $\mathbf{g} \in H^{(1)}$ can be reduced to the following equation in $H^{(1)}$:

$$(\widetilde{\mathrm{A}}_{\mathrm{I}} - \lambda)\boldsymbol{\varphi}^+ = \mathbf{g}, \quad \text{where } \widetilde{\mathrm{A}}_{\mathrm{I}} = -\boldsymbol{\nu} \times P(I - B)^{-1}AP. \tag{5.346}$$

In the same way as in Subsection 6, this equation can be transformed into the equation $(\widetilde{\mathrm{B}}_{\mathrm{I}} - \mu)\boldsymbol{\varphi}^+ = \mathbf{h}$, where $\widetilde{\mathrm{B}}_{\mathrm{I}} = \widetilde{\mathrm{A}}_{\mathrm{I}} - \widetilde{\mathrm{A}}_{\mathrm{I}}^{-1}$. This operator can be extended by setting, say, $\widetilde{\mathrm{B}}_{\mathrm{I}}\boldsymbol{\chi} = \boldsymbol{\chi}$ for $\boldsymbol{\chi} \in [H^{(1)}]^{\perp}$. One can readily verify that thus we obtain a first-order elliptic selfadjoint operator.

A similar approach could be applied to Problem II in Subsection 5.8.3.

Prior to proceeding to Problem II, let us derive some formulas replacing (5.318b). By P_1 and P_2 we denote the orthogonal projections on the ranges of the operators $I + B^*$ and $I + B$, respectively, that is, on the orthogonal complements of $\mathrm{Ker}(I + B)$ and $\mathrm{Ker}(I + B^*)$, respectively. The operator $(I + B)^{-1}$ does not exist for exceptional k, but $P_1(I+B)^{-1}P_2$ is well defined. The exterior problem with the condition $-\boldsymbol{\nu}\times\mathbf{H}^- = \boldsymbol{\varphi}^-$ is always uniquely solvable; moreover (see the first equation in (5.300b)), $A\boldsymbol{\varphi}^- = P_2A\boldsymbol{\varphi}^-$ and

$$\boldsymbol{\psi}^- = -iP_1(I + B)^{-1}P_2A\boldsymbol{\varphi}^- + \alpha_1\boldsymbol{\chi}_1 + \ldots + \alpha_m\boldsymbol{\chi}_m,$$

where $\boldsymbol{\chi}_1, \ldots, \boldsymbol{\chi}_m$ is an orthonormal basis in $\mathrm{Ker}(I + B)$. The coefficients α_j are determined as follows. Let $\boldsymbol{\varphi}_j^- = \boldsymbol{\nu}\times\overline{\boldsymbol{\chi}}_j$, $\mathbf{E}_j$, $\mathbf{H}_j$ be the solution of the exterior problem with $-\boldsymbol{\nu}\times\mathbf{H}_j^- = \boldsymbol{\varphi}_j^-$, and let $\boldsymbol{\psi}_j^- = \boldsymbol{\nu}\times\mathbf{E}_j^-$. Then, by using the Lorentz formula (Eq. (5.329) with $^+$ replaced by $^-$; it can be derived with the help of the radiation conditions), we obtain

$$\alpha_j = \int_S \boldsymbol{\psi}^- \cdot \overline{\boldsymbol{\chi}}_j \, dS = -\int_S \boldsymbol{\psi}^- \cdot (\boldsymbol{\nu} \times \boldsymbol{\varphi}_j^-)\, dS = \int_S \boldsymbol{\varphi}^- \cdot (\boldsymbol{\nu} \times \boldsymbol{\psi}_j^-)\, dS.$$

Thus

$$\boldsymbol{\psi}^- = -i\,[P_1(I + B)^{-1}P_2A + K]\boldsymbol{\varphi}^-, \tag{5.347}$$

where

$$K\boldsymbol{\varphi}^- = i\sum_{j=1}^{m}\boldsymbol{\chi}_j \int_S \boldsymbol{\varphi}^- \cdot (\boldsymbol{\nu} \times \boldsymbol{\psi}_j^-)\, dS \tag{5.348}$$

is a finite-dimensional infinitely smoothing operator. It follows that

$$\boldsymbol{\varphi}^- = i\,[P_1(I + B)^{-1}P_2A + K]\boldsymbol{\psi}^-, \tag{5.349}$$

since the Maxwell system is invariant with respect to the substitution $\mathbf{E} \to \mathbf{H}$, $\mathbf{H} \to -\mathbf{E}$. It follows from (5.347) and (5.349) that

$$[P_1(I+B)^{-1}P_2A+K]^2 = I. \tag{5.350}$$

After these preliminaries, we can consider Problem II and reproduce the argument used for the case of nonexceptional k with slight modifications. First, the problem is reduced to the equation

$$(\tilde{\mathrm{A}}_{\mathrm{II}} - \lambda)\boldsymbol{\varphi}^- = \mathbf{g}, \quad \text{where } \tilde{\mathrm{A}}_{\mathrm{II}} = -\boldsymbol{\nu} \times [P_1(I+B)^{-1}P_2A+K],$$

and then to the elliptic equation

$$(\tilde{\mathrm{B}}_{\mathrm{II}} - \mu)\boldsymbol{\varphi}^- = \mathbf{h}, \quad \text{where } \tilde{\mathrm{B}}_{\mathrm{II}} = \tilde{\mathrm{A}}_{\mathrm{II}} - \tilde{\mathrm{A}}_{\mathrm{II}}^{-1}.$$

A similar approach could be applied to Problem III in Subsection 5.8.3.

10. Two-dimensional problems. We can consider two-dimensional analogs of Problems I–III, in which $\mathbf{E}$, $\mathbf{H}$ is independent of x_3 and space $\mathbf{R}^3$ is divided into two parts by a cylindrical surface with generators parallel to the x_3-axis. Let the plane $x_3 = 0$ intersect this surface by an infinitely smooth closed curve S, which divides the plane into the interior and exterior parts V^+ and V^-. Let u and v be the projections of the vectors $\mathbf{E}$ and $\mathbf{H}$ on the x_3-axis. It is well known (and readily verifiable) that the Maxwell system is reduced in this case to the Helmholtz equations $\Delta u + k^2 u = 0$ and $\Delta v + k^2 v = 0$ (e.g., see [48, Section 4]). The analogs of Problems I, II, and III are reduced to problems for these equations in V^+, in V^-, and outside S, respectively, and in the last two cases the solution satisfies the radiation condition at infinity. The conditions on S have the following form: in Problem I,

$$\lambda k^{-1}\partial_\nu u^+ + u^+ = g_1, \qquad \lambda k v^+ - \partial_\nu v^+ = g_2;$$

in Problem II,

$$\lambda k^{-1}\partial_\nu u^- - u^- = g_1, \qquad \lambda k v^- + \partial_\nu v^- = g_2;$$

in Problem III,

$$\partial_\nu u^+ = \partial_\nu u^-, \qquad \lambda k^{-1}\partial_\nu u - \frac{1}{2}(u^- - u^+) = g_1,$$

$$v^+ = v^-, \qquad \lambda k v + \frac{1}{2}(\partial_\nu v^- - \partial_\nu v^+) = g_2.$$

These problems, in different notation, were considered in Sections 5.7 and 5.8. In particular, we see that each of problems I, II, and III splits into two scalar problems, in one of which the eigenvalues have the unique accumulation point at infinity, and in the other, at zero.

5.12 Scalar Spectral Problems in Lipschitz Domains

1. Lipschitz surfaces. Let S be a closed (hyper)surface in $\mathbf{R}^n$ (a curve for $n = 2$). It is called a *Lipschitz surface* if each point of S has a neighborhood whose intersection with S is the graph of a Lipschitz function. This means that after an appropriate rotation of the coordinate system, the surface S in this neighborhood is specified by the equation

$$x_n = f(x') = f(x_1, \dots, x_{n-1}), \tag{5.351}$$

where the function $f(x')$ is defined in some domain in $\mathbf{R}^{n-1}$ and satisfies the Lipschitz condition

$$|f(x') - f(y')| \leq C|x' - y'|. \tag{5.352}$$

We can assume that the Lipschitz constants C are bounded above, since, by virtue of the fact that S is compact, it can be covered by finitely many such neighborhoods.

Lipschitz curves in $\mathbf{R}^2$ may have corners, and Lipschitz surfaces in $\mathbf{R}^3$ may have edges and conical points. The reader is undoubtedly aware how important such cases are in diffraction problems.

In the subsequent exposition, S is either the common boundary of the interior and exterior domains V^+ and V^-, or the boundary of a domain V, bounded or unbounded.

Any function f satisfying the Lipschitz condition is differentiable almost everywhere in the sense of the Lebesgue measure in $\mathbf{R}^{n-1}$ (e.g., see [96, Chapter 8]); hence, on S there is a well-defined surface area element, which can be represented locally in the form

$$dS = \sqrt{1 + |\operatorname{grad} f(x')|^2}\, dx'. \tag{5.353}$$

It follows that the Lebesgue measure is defined on a Lipschitz surface; it is essentially induced by the Lebesgue measure in the ambient space $\mathbf{R}^n$.

Next, the normal direction is well defined almost everywhere on S; by ν_x we shall denote the unit outward normal vector at a point $x \in S$.

In the class of Lipschitz surfaces, we single out the subclass of *almost smooth* surfaces. The corresponding definition was introduced in [15]. A surface S is said to be almost smooth if it is C^∞ outside a closed subset S_0 of zero Lebesgue measure, referred to as the *singular subset.* In other words, each function (5.351) is assumed to belong to C^∞ outside a closed subset of zero measure in $\mathbf{R}^{n-1}$. An almost smooth surface in $\mathbf{R}^3$ may have edges and conical points; an almost smooth curve may have corners. Thus, it is the most interesting subclass of the class of Lipschitz surfaces.

A domain in $\mathbf{R}^n$ is called a *Lipschitz domain* if its boundary is a Lipschitz surface.

2. Function spaces. One can consider the Sobolev spaces $H^s(V)$ in a Lipschitz domain V for arbitrary $s \geq 0$ (we do not use negative s and L_p-spaces). The properties of these spaces are mainly the same as in the case of a smooth boundary. In particular, there exists a bounded extension operator from $H^s(V)$ to $H^s(\mathbf{R}^n)$ (see [96, Chapter 6]).

However, the spaces $H^s(S)$ can be defined in a natural way only for $|s| \leq 1$. The operator taking each function in V to its trace on S is a bounded operator from $H^{s+1/2}(V)$ to $H^s(S)$ only for $0 < s < 1$ in the general case. Here the inequality $s > 0$ is not new (this restriction also holds for smooth surfaces S), whereas the inequality $s < 1$ is specific to Lipschitz surfaces. However, the trace operator is bounded from $H^{s+1/2}(V)$ to $H^1(S)$ for $s > 1$ (see [53, 35]).

Furthermore, we shall see that being restricted to solutions of the homogeneous Helmholtz and Laplace equations (as well as some other equations), the trace operator is bounded from $H^{s+1/2}(V)$ to $H^s(S)$ for $s = 0$ and 1 as well. However, this is not easy to prove. For smooth surfaces S, the possibility of eliminating the restriction $s > 0$ for solutions of elliptic equations is known (see [66, 84]), but the restriction $s < 1$ is of course lacking.

So far, there is no calculus of pseudodifferential operators and even no notion of a pseudodifferential operator on Lipschitz surfaces. However, in the last two decades, the theory of the Dirichlet and Neumann problems was developed first for the Laplace equation (see the surveys [59, 60]), and then for the Helmholtz equation and more general equations. The initial success for the Laplace equation was achieved by Dahlberg [32, 33, 34], who used methods of harmonic analysis. This approach was later almost completely replaced by the analysis and application of surface potentials like $\mathcal{A}$ and $\mathcal{B}$ in Section 5.7, first for the Laplace equation [102, 103], and then for the Helmholtz equation [98, 74] and other equations. (Concerning close results for the Maxwell system, see [73] and references therein.) Before describing the corresponding results, we must specify the notion of passage to the boundary values for the case of a Lipschitz surface S and introduce the function spaces that are actually used in this theory.

Lipschitz surfaces can be characterized as surfaces satisfying the *uniform cone condition.* This condition means that each point $x \in S$ is the vertex of two closed cones $\Gamma^+(x)$ and $\Gamma^-(x)$ that lie in V^+ and V^-, respectively (excluding the point x itself), and are congruent to a given cone $\Gamma_0 = \{x : 0 \leq x_n \leq h, |x'| \leq \alpha x_n\}$, where $\alpha > 0$ and $h > 0$. The family of these cones can be subjected to some *regularity condition*, whose formal description is rather lengthy (e.g., see [102, 103]). We restrict ourselves to the following details: h and α are reduced by a sufficiently small factor from their maximum possible values, and the common axis of the cones $V^\pm(x)$ changes its direction continuously as x moves along S; the angle between the axis and the normal ν_x, where the latter exists, is bounded above by a constant that is less than $\pi/2$. In the following, we assume that some regular family of cones $\Gamma^\pm(x)$, $x \in S$, is fixed. The union of these cones covers some boundary strip, a neighborhood of the boundary S.

The Dirichlet boundary condition

$$u^\pm(x) = f(x) \tag{5.354}$$

is now interpreted as follows:

$$u(y) \to f(x) \quad \text{as } \Gamma^{\pm}(x) \ni y \to x \tag{5.355}$$

for almost all x on S. This is the so-called *nontangential convergence.*

The Neumann boundary condition

$$\partial_\nu u^{\pm}(x) = g(x) \tag{5.356}$$

is now treated as follows:

$$\partial_{\nu_x} u(y) = \operatorname{grad} u(y) \cdot \nu_x \to g(x) \quad \text{as } \Gamma^{\pm}(x) \ni y \to x \tag{5.357}$$

for almost all x on S. This is also nontangential convergence.

We see that now the boundary conditions in Problem I in Section 5.7 and in Problems II–IV in Section 5.8 must be understood in the sense of nontangential convergence. We could assume that the function $\sigma(x)$ is the restriction to S of a positive smooth function originally defined in a neighborhood of S. However, for simplicity, we restrict ourselves to the case in which $\sigma(x) \equiv 1$.

For a function $u(x)$ defined in V^+ or V^- we define the *nontangential maximal functions*

$$u_*^+(x) = \sup_{y \in \Gamma^+(x)} |u(y)| \quad \text{and} \quad u_*^-(x) = \sup_{y \in \Gamma^-(x)} |u(y)|. \tag{5.358}$$

These are functions on S.

Now let us define four spaces that are actually used in a number of papers concerning the Laplace and Helmholtz equations in Lipschitz domains (see [19]; cf. [98]).

The *space* $\mathcal{V}^+(k)$ consists of infinitely smooth solutions $u(x)$ of the Helmholtz equation (5.1) in V^+ with square integrable maximal functions: $u_*^+ \in L_2(S)$. The infinite smoothness in V^+ is assumed here from the very beginning, since the Helmholtz equation is elliptic.

The *space* $\mathcal{V}^-(k)$ consists of infinitely smooth solutions of the Helmholtz equation in V^- satisfying the radiation condition (5.137) and such that $u_*^- \in L_2(S)$.

The *space* $\mathcal{W}^+(k)$ consists of infinitely smooth solutions of the Helmholtz equation in V^+ such that u_*^+, $(\operatorname{grad} u)_*^+ \in L_2(S)$.

The *space* $\mathcal{W}^-(k)$ consists of infinitely smooth solutions of the Helmholtz equation in V^- such that the radiation condition is satisfied and u_*^-, $(\operatorname{grad} u)_*^- \in L_2(S)$.

The expression $\|u_*^+\|_{s,S}$ can serve as the norm in $\mathcal{V}^+(k)$. In particular, if it is zero, then $u(x)$ is identically zero near S and hence everywhere in V^+. Likewise, $\|u_*^-\|_{0,S}$ can be used as the norm in $\mathcal{V}^-(k)$, while $\|u_*^+\|_{0,S} + \|(\operatorname{grad} u)_*^+\|_{0,S}$ and $\|u_*^-\|_{0,S} + \|(\operatorname{grad} u)_*^-\|_{0,S}$ can be used as the norms in $\mathcal{W}^+(k)$ and $\mathcal{W}^-(k)$, respectively.

1° *The solution $u(x)$ of the Helmholtz equation in V^+ belongs to $\mathcal{V}^+(k)$ or $\mathcal{W}^+(k)$ if and only if it belongs to $H^{1/2}(V^+)$ or $H^{3/2}(V^+)$, respectively. The solution $u(x)$ of the Helmholtz equation in V^- with the radiation condition at infinity belongs to $\mathcal{V}^-(k)$*

or $\mathcal{W}^-(k)$ *if and only if it belongs to* $H^{1/2}(V_R^-)$ *or* $H^{3/2}(V_R^-)$, *respectively, where* R *is sufficiently large.*

For $k_2 = \operatorname{Im} k < 0$, we can omit R here. The main assertion is that for $\mathcal{V}^+(k)$, and the remaining assertions can be derived from it. The assertion for $\mathcal{V}^+(k)$ in the case of the Laplace equation ($k = 0$) was proved in [53] with the use of results from [34]. This assertion was generalized to arbitrary k in [19]. For $n \leq 3$, this is fairly easy: one has only to subtract the volume potential

$$v(x) = k^2 \int_{V+} G(x - y, k) u(y) \, dy \tag{5.359}$$

from the solution. Indeed, for $u \in H^{1/2}(V^+)$, as well as for $u \in \mathcal{V}^+(k)$, this function belongs to $L_2(V^+)$. Then, first, $v(x)$ is a bounded function (which can be verified using the Schwartz inequality), so that $v_*^+ \in L_2(S)$. Second, $v(x)$ belongs to $H^2(V^+)$ and all the more to $H^{1/2}(V^+)$. See [19].

3. Potential type operators and integral formulas. Now let us consider the operators $A(k)$, $B(k)$, $\mathcal{A}(k)$, and $\mathcal{B}(k)$, defined in Subsection 5.7.1, with the same k (see (5.3)). The operator $A(k)$ still is an integral operator with a weak singularity and so is obviously bounded in $L_2(S)$. The consideration of the operator $B(k)$ on a Lipschitz surface is much more difficult. It is now a singular integral operator defined in the sense of the Cauchy principal value. The nontrivial theorem stating the boundedness of singular integral operators in $L_2(S)$ for a Lipschitz surface S was proved in [30], where a more special result of [29] was generalized. Fortunately, this theorem permits one to estimate the maximal functions for the corresponding potential. It applies not only to $B(k)$ and $\mathcal{B}(k)$, but also to grad $\mathcal{A}(k)$ and $\operatorname{grad}_0 A(k)$ (the surface gradient). We obtain the following results.

1° *The operator* $B(k)$ *is bounded in* $L_2(S)$. *The same is true of the operators* $B'(k)$ *and* $\operatorname{grad}_0 A(k)$.

The operator $A(k)$ *is bounded from* $H^0(S)$ *to* $H^1(S)$.

The operator $\mathcal{B}(k)$ *is bounded from* $H^0(S)$ *to* $\mathcal{V}^\pm(k)$, *and the operator* $\mathcal{A}(k)$ *is bounded from* $H^0(S)$ *to* $\mathcal{W}^\pm(k)$.

The differences $A(k_1) - A(k_2)$ etc. are integral operators with a weaker singularity.

2° *The differences* $A(k_1) - A(k_2)$, $B(k_1) - B(k_2)$, *and* $B'(k_1) - B'(k_2)$ *are compact operators from* $H^0(S)$ *to* $H^1(S)$. *The difference* $\mathcal{B}(k_1) - \mathcal{B}(k_2)$ *is a bounded operator from* $H^0(S)$ *to* $\mathcal{W}^\pm(k)$.

Now let us examine relations (5.143)–(5.145).

3° *Relations* (5.143) *for* $\varphi \in H^0(S)$ *can be generalized to the case of Lipschitz surfaces. In this case, the limits* $\mathcal{A}(k)\varphi^\pm$ *must be understood in the sense of nontangential convergence, and they belong to* $H^1(S)$. *Relations* (5.144) *and* (5.145) *also remain valid in the sense of nontangential convergence; both sides of each of these relations belong to* $H^0(S)$.

To generalize Green's formulas to a Lipschitz domain, the following lemma on the approximation of S by smooth surfaces is helpful (see [78, 102, 103]).

4° *There exists a sequence of infinitely smooth closed surfaces* $S_j \subset V^+$ *with the following properties.* (a) S_j *is the boundary of a domain* V_j^+ *with closure* $\overline{V_j^+} \subset V^+$. (b) *The surfaces* S *and* S_j *are related by a Lipschitz diffeomorphism* $\Lambda_j \colon S \to S_j$; *in local coordinates, the Lipschitz constants for* Λ_j *and* Λ_j^{-1} *are bounded uniformly with respect to* j. (c) $\Lambda_j(x) \in \Gamma^+(x)$ *for all* j *and all* $x \in S$, *and* $\Lambda_j(x) \to x$ *uniformly with respect to* $x \in S$. (d) *The unit outward normal vectors* $\nu_{\Lambda_j(x)}$ *on* S_j *tend to* ν_x *almost everywhere and in* $L_2(S)$. (e) *There exist functions* $\omega_j(x)$ *on* S *bounded above and below by positive constants uniformly with respect to* j *and* x *and such that for each summable function* $f(x)$ *on* S *one has*

$$\int_{S_j} f(\Lambda_j^{-1}(x))\, dS = \int_S f(x)\omega_j(x)\, dS,$$

and moreover, $\omega_j(x) \to 1$ *almost everywhere and in* $L_2(S)$.

The sequence $\{S_j\}$ approximates the surface S "from the interior." There exists a similar sequence of smooth surfaces approximating S "from the exterior."

Green's formulas can be derived under the assumptions indicated in the following assertion (cf. [79, 45]).

5° *Let* S *be a Lipschitz surface. Then Green's first formula* (5.148) *remains valid for* $u \in \mathcal{W}^+(k)$ *and* $v \in H^1(V^+)$ *with* $(\operatorname{grad} v)_*^+ \in L_2(S)$. *Green's second formula* (5.149) *is valid for* $u \in \mathcal{W}^+(k_1)$ *and* $v \in \mathcal{W}^+(k_2)$.

We obtain the following corollary.

6° *For* $u \in \mathcal{W}^\pm(k)$, *the integral representations* (5.150) *and* (5.151) *of solutions of the Helmholtz equation (with the radiation condition at infinity), as well as relations* (5.152) *and* (5.153) *on* S, *are valid.*

4. The Dirichlet and Neumann problems and the invertibility of potential type operators. In connection with the investigation of the Dirichlet and Neumann problems for the Helmholtz equation, the invertibility of the operators $A(k)$ and $\frac{1}{2}I \pm B(k)$ on Lipschitz surfaces was studied in Sobolev spaces with small index ([102, 103, 98, 74]). We shall first state the final results and then try to explain how they were obtained. The crucial case was that of the Laplace operator ($k = 0$).

1° *The operator* $A(k)\colon H^0(S) \to H^1(S)$ *is invertible for all nonreal* k *and also for real* k *such that the interior Dirichlet problem does not have nontrivial solutions. For the other* k, *the operator* $A(k)$ *is not invertible, and*

$$\operatorname{Ker} A(k) = \{\partial_\nu u^+ : u \in \mathcal{W}^+(k),\ u^+ = 0\}. \tag{5.360}$$

2° *The operator* $B(k)$ *is bounded in* $H^1(S)$ *for all* k.

3° *The operator* $\frac{1}{2}I + B'(k)$ *is invertible in* $H^0(S)$ *for the same* k *for which* $A(k)$ *is invertible as an operator from* $H^0(S)$ *to* $H^1(S)$. *For the other* k, *the kernel* $\operatorname{Ker}\left[\frac{1}{2}I + B'(k)\right]$ *coincides with* $\operatorname{Ker} A(k)$. *The operator* $\frac{1}{2}I + B(k)$ *is invertible in* $H^0(S)$ *si-*

multaneously with $\frac{1}{2}I + B'(k)$*; moreover, in this case* $\frac{1}{2}I + B(k)$ *is also invertible in* $H^1(S)$.

4° *The operator* $\frac{1}{2}I - B(k)$ *is invertible in* $H^0(S)$ *and* $H^1(S)$ *for nonreal* k *and also for real* k *such that the interior homogeneous Neumann problem does not have nontrivial solutions. For the other* k*, the operator is not invertible, and the kernel* $\operatorname{Ker}\left[\frac{1}{2}I - B(k)\right]$ *lies in* $H^1(S)$ *and is given by the formula*

$$\operatorname{Ker}\left[\frac{1}{2}I - B(k)\right] = \{u^+ : u \in \mathcal{W}^+(k),\, \partial_\nu u^+ = 0\}. \tag{5.361}$$

The operator $\frac{1}{2}I - B'(k)$ *is invertible in* $H^0(S)$ *simultaneously with* $\frac{1}{2}I - B(k)$.

The following two assertions deal with the Dirichlet and Neumann problems.

5° *The exterior Dirichlet problem with* $u^- \in H^0(S)$ *is uniquely solvable for all* k*, and the solution belongs to* $\mathcal{V}^-(k)$*. For* $u^- \in H^1(S)$*, the solution belongs to* $\mathcal{W}^-(k)$*. The interior Dirichlet problem with* $u^+ \in H^0(S)$ *is uniquely solvable for all nonreal* k *as well as for the nonexceptional real* k*. For* $u^+ \in H^1(S)$*, the solution belongs to* $\mathcal{W}^+(k)$.

By saying that k is nonexceptional, we here mean that $\operatorname{Ker} A(k)$ is trivial; this is Condition I of Section 5.8.

6° *The exterior Neumann problem with* $\partial_\nu u^- \in H^0(S)$ *is uniquely solvable for all* k*, and the solution belongs to* $\mathcal{W}^-(k)$*. The interior Neumann problem with* $\partial_\nu u^+ \in H^0(S)$ *is uniquely solvable for all nonreal* k *as well as for the nonexceptional real* k*. The solution belongs to* $\mathcal{W}^+(k)$.

By saying that k is nonexceptional, here we mean that $\operatorname{Ker}\left[\frac{1}{2}I - B'(k)\right]$ is trivial; this is Condition II of Section 5.8.

Let us give some explanations. As was already indicated, $k = 0$ is the key case. One must establish the invertibility of the operators $A(0)$ and $\frac{1}{2}I + B(0)$, the Fredholm property of the operator $\frac{1}{2}I - B(0)$, and the fact that the index of the latter operator is zero without using the ellipticity and the compactness of $B(0)$ (in the present case, these two properties are lacking). In [102, 103], this was done by verifying and using the *Rellich inequalities*

$$\left\|\left[\frac{1}{2}I \pm B'(0)\right]\varphi\right\|_{0,S} \le C\left\{\left\|\left[\frac{1}{2}I \mp B'(0)\right]\varphi\right\|_{0,S} + \left|\int A(0)\varphi\, dS\right|\right\}, \tag{5.362}$$
$$\varphi \in L_2(S).$$

See these papers for further details.

As soon as the case $k = 0$ has been considered, one can proceed to other k by using Assertion 2° of the preceding subsection applied to the differences $A(k) - A(0)$, $B(k) - B(0)$, and $B'(k) - B'(0)$. First one considers the nonreal values of k, for which it is possible to prove the invertibility of the desired operators. It is easy to prove the

Fredholm property and the fact that the index is zero; then, by using Green's formulas, one verifies that the kernels of these operators are trivial, which implies the invertibility. For real k, the triviality of the kernel is equivalent to invertibility.

The additional important details are as follows. We can verify, first for nonreal k, that the formula from Proposition 5.8.5 is valid:

$$B(k)A(k) = A(k)B'(k).$$

This formula extends to real k by passing to the limit. We see that the boundedness of $B'(k)$ in $H^0(S)$ implies the boundedness of $B(k)$ in $H^1(S)$. Furthermore, it follows that if the operator $A(k)\colon H^0(S) \to H^1(S)$ has an inverse, then

$$\left[\frac{1}{2}I \pm B(k)\right]^{-1} = A(k)\left[\frac{1}{2}I \pm B'(k)\right]^{-1} A^{-1}(k), \tag{5.363}$$

so that the invertibility of the operator $\frac{1}{2}I + B'(k)$ or $\frac{1}{2}I - B'(k)$ in $H^0(S)$ implies the invertibility of $\frac{1}{2}I + B(k)$ or $\frac{1}{2}I - B(k)$ in $H^1(S)$. However, in the latter assertion one can eliminate the assumption that $A(k)$ is invertible. To see this, we verify the following analog of the smoothness theorem for elliptic equations (Assertion 4° in Subsection 5.5.1):

7° *For every k, the relation*

$$\left[\frac{1}{2}I \pm B(k)\right]\varphi = \psi \tag{5.364}$$

and the inclusions $\varphi \in H^0(S)$ and $\psi \in H^1(S)$ imply that $\varphi \in H^1(S)$.

It follows that if the operator $\frac{1}{2}I \pm B(k)$ is invertible in $H^0(S)$, then it defines a one-to-one (and continuous) self-mapping of $H^1(S)$. Then the inverse operator is continuous by the Banach closed graph theorem. In turn, to verify Assertion 7°, it suffices to apply, say, the operator $\left[\frac{1}{2}I \pm B(i)\right]^{-1}$, which plays the role of a parametrix, to both sides of Eq. (5.364).

Yet another important assertion concerning smoothness pertains to the Dirichlet problems.

8° *Let u be a solution of the interior or exterior Dirichlet problem in $\mathcal{V}^{\pm}(k)$ with $u^{\pm} \in H^1(S)$. Then $u \in \mathcal{W}^{\pm}(k)$.*

For $k = 0$, this was proved in [32, 33]. The result can be extended to the case $k \neq 0$ by using the volume potential (5.359) with the help of Assertion 1° in Subsection 2. See [19].

The uniqueness of solutions in $\mathcal{W}^{\pm}(k)$ can be verified for the Dirichlet and Neumann problems for all k in the case of exterior problems and for nonreal k in the case of interior problems with the help of Green's formulas. By virtue of Assertion 8°, this implies the uniqueness for the Dirichlet problems in $\mathcal{V}^{\pm}(k)$ for the same k. As to the existence, for nonexceptional k the solution can be constructed in the same form as for smooth S. Namely, these are the formulas

$$u(x) = \mathcal{A}(k)A^{-1}(k)u^{\pm} \quad \text{and } u(x) = \mathcal{B}(k)\Big[B(k) \pm \frac{1}{2}I\Big]^{-1} u^{\pm} \tag{5.365}$$

for the solutions of the Dirichlet problems (in the first formula, we have $u^{\pm} \in H^1(S)$ and $u \in \mathcal{W}^{\pm}(k)$) and the formulas

$$u = \mathcal{A}(k)\Big[\mathcal{B}'(k) \mp \frac{1}{2}I\Big]^{-1} \partial_\nu u^{\pm} \tag{5.366}$$

for the solutions of the Neumann problems. We do not consider modifications of these formulas for the case of problems in V^- for exceptional k; neither do we study the solvability conditions and the construction of solutions for the case in which there is no uniqueness and the problem is only Fredholm of index zero. In these situations, one can use methods known for the case of smooth S (see [31, Chapter 3]). Similar reasoning was used in Subsections 5.8.3 and 5.11.9.

We point out that for the exceptional values of k, just as in the case of a smooth S, the corresponding k^2 are eigenvalues of the operator $-\Delta_D$ or $-\Delta_N$ in the domain V^+.

However, the difference from the case of smooth S is as follows: now these operators admit in general only the "variational" definition, and their domains consist in general of less smooth functions.

The operator $-\Delta_D$ is defined as follows. Consider the bilinear form

$$a_D[u, v] = \int_{V^+} \operatorname{grad} u(x) \cdot \operatorname{grad} \overline{v}(x)\, dx \tag{5.367}$$

on the subspace $\overset{\circ}{H}{}^1(V^+)$ of functions with zero Dirichlet data, that is, on the closure of the linear manifold $C_0^\infty(V^+)$ in $H^1(V^+)$. The corresponding quadratic form $a_D[u, u]$ satisfies the Gårding inequality

$$\varepsilon \|u\|_1^2 \le a_D[u, u] \tag{5.368}$$

($u \in \overset{\circ}{H}{}^1(S)$). This inequality is even valid in an arbitrary bounded domain without the assumption that the boundary is Lipschitz (e.g., see [3]). According to the general scheme, which can be found, say, in [54, Chapter 6], the form $a_D[u, v]$ defines an unbounded selfadjoint operator A_D in $L_2(V^+)$ with domain $D(A_D) \subset \overset{\circ}{H}{}^1(V^+)$ such that

$$\int_{V^+} A_D u \cdot \overline{v}\, dx = a_D[u, v] \quad \text{for } u \in D(A_D),\ v \in \overset{\circ}{H}{}^1(V^+). \tag{5.369}$$

This operator A_D is, by definition, $-\Delta_D$. The behavior of its eigenvalues was studied in [27]; they obey the formula $k_j^2 \sim C' j^{2/n}$ (in our case, $n = 2$ or $n = 3$) with the same $C' = C'(n)$ as for the case of smooth S (see Subsection 5.8.1).

To define the operator $-\Delta_N$, we introduce the form

$$a_N[u,v] = \int_{V+} (\operatorname{grad} u \cdot \operatorname{grad} \overline{v} + u \cdot \overline{v})\, dS \tag{5.370}$$

on the space $H^1(S)$. The corresponding quadratic form $a_N[u,u]$ satisfies an analog of the Gårding inequality (5.368) for $u \in H^1(V^+)$ in Lipschitz domains. The same scheme permits one to define an unbounded selfadjoint positive operator A_N in $L_2(V^+)$ with $D(A_N) \subset H^1(V^+)$ such that

$$\int_{V+} A_N u \cdot \overline{v}\, dx = a_N[u,v] \quad \text{for } u \in D(A_N),\ v \in H^1(V^+). \tag{5.371}$$

The operator A_N is $I - \Delta_N$. The operator $-\Delta_N$ is nonnegative, and zero is an eigenvalue of this operator. The same asymptotic formula $k_j^2 \sim C' j^{2/n}$ holds for its eigenvalues (see [27]).

8° *The set of squares k^2 of the numbers k for which Condition* I *is violated coincides with the set of eigenvalues of the operator $-\Delta_D$. Moreover, the multiplicity of an eigenvalue is equal to the dimension of* $\operatorname{Ker} A(\pm k)$.

Likewise, the set of squares k^2 of the numbers k for which Condition II *is violated coincides with the set of eigenvalues of the operator $-\Delta_N$. Moreover, the multiplicity of an eigenvalue is equal to the dimension of* $\operatorname{Ker}\left[\frac{1}{2}I - B'(k)\right]$.

The eigenfunctions of the operators $-\Delta_D$ and $-\Delta_N$ belong to $H^{3/2}(V^+)$.

The proof can be found in [19].

Note also that formulas (5.203) remain valid for $\psi \in H^1(S)$.

5. Problems I–IV and spectral properties of the corresponding operators. Problems I–IV for the case of a Lipschitz surface S can be reduced to the same integral equations on S as for smooth S. These are Eq. (5.157) for Problem I, Eqs. (5.205) for Problems II and III (under Conditions II and I, respectively), and Eq. (5.206) for Problem IV (under Conditions I and II). Recall that for simplicity we assume that $\sigma(x) \equiv 1$.

The operators $A(k)$, $T^{\pm}(k)$, and $T(k)$ in these equations are bounded operators from $H^0(S)$ to $H^1(S)$ and are invertible for all nonreal k as well as for real k satisfying Conditions I and II. (Condition II is unnecessary when one considers the operators $T^-(k)$ and $[T^-(k)]^{-1}$.) The inverses of $T^{\pm}(k)$ and $T(k)$ are given by the formulas

$$[T^{\pm}(k)]^{-1} = A^{-1}(k)\left[\frac{1}{2}I \mp B(k)\right] \quad \text{and} \quad T^{-1}(k) = A^{-1}(k)\left[\frac{1}{4}I - B^2(k)\right]. \tag{5.372}$$

1° *The operators $A(k)$, $T^{\pm}(k)$, and $T(k)$ are selfadjoint for pure imaginary k, and $T(k)$ is also selfadjoint for real k.*

This can be verified in the same way as for smooth S.

For the other values of k, we consider the operators $A(k)$, $T^{\pm}(k)$, and $T(k)$ as weak perturbations of selfadjoint operators. However, in contrast with the case of smooth S, the latter are now not the real parts of the former but their values for pure imaginary

k, say, for $k = i$. It is important that we can gain a large amount of information about the properties of the operators $A(i)$, $T^{\pm}(i)$, and $T(i)$. In particular, we know that they are invertible as operators from $H^0(S)$ to $H^1(S)$.

2° *$A(k)$, $T^{\pm}(k)$, and $T(k)$ are relatively compact perturbations of the operators $A(i)$, $T^{\pm}(k)$, and $T(i)$, respectively.*

This means that, for example,

$$A(k) = A(i)[I + A_1(k)], \tag{5.373}$$

where $A_1(k)$ is a compact operator in $L_2(S)$. Indeed,

$$A_1(k) = A^{-1}(i)[A(k) - A(i)],$$

and it suffices to use Assertion 2° from Subsection 3. Likewise,

$$T^{\pm}(k) = T^{\pm}(i)[I + T_1^{\pm}(k)] \quad \text{and } T(k) = T(i)[I + T_1(k)], \tag{5.374}$$

where $T_1^{\pm}(k)$ and $T_1(k)$ are compact operators in $L_2(S)$. Indeed, for example,

$$T_1^{+}(k) = [T^{+}(i)]^{-1}[T^{+}(k) - T^{+}(i)],$$

and here we have

$$T^{+}(k) - T^{+}(i) = A(k)\Big\{\Big[\frac{1}{2}I - B'(k)\Big]^{-1} - \Big[\frac{1}{2}I - B'(i)\Big]^{-1}\Big\} + [A(k) - A(i)]\Big[\frac{1}{2}I - B'(i)\Big]^{-1},$$

where the expression in braces can be represented in the form

$$\Big[\frac{1}{2}I - B'(k)\Big]^{-1}[B'(k) - B'(i)]\Big[\frac{1}{2}I - B'(i)\Big]^{-1}.$$

We see from these formulas that $T_1^{+}(k)$ is a compact operator.

If $T = T_0 + T_1$, where T_0 is a selfadjoint operator and T_1 is only known to be relatively compact with respect to T_0, then the subordination of the operator $T_1 = T - T_0$ to T_0 is very weak. However, even in this case one can make important conclusions about the properties of the operator T (using assertions from Subsection 5.6.6).

3° *Let S be a Lipschitz surface. Then Proposition* 5.7.8 *is valid for the operator $A(k)$, and Proposition* 5.8.14 *is valid for the operators $T^{\pm}(k)$ and $T(k)$.*

Indeed, Green's formulas used to prove these assertions are at hand.

In particular, we conclude that $A(i)$, $T^{\pm}(i)$, and $T(i)$ are negative operators, that is, operators with negative eigenvalues.

4° *The operators $A(k)$, $T^{\pm}(k)$, and $T(k)$ belong to $S^{(\frac{1}{n-1})}(L_2(S))$, that is, their s-numbers satisfy the inequality $s_j \leq Cj^{\frac{1}{n-1}}$.*

This follows from the fact that these operators are bounded from $H^0(S)$ into $H^1(S)$ (cf. Remark in Subsection 5.6.1). We can also apply the estimate of s-numbers given in [15] to $A(k)$ and use the inequalities in (5.114).

Theorem 5.12.1 *The system of root functions of each of the operators $A(k)$, $T^{\pm}(k)$, and $T(k)$ for k such that this operator is not selfadjoint is complete in $H^0(S)$, and the Fourier series with respect to this system of a function $f \in H^0(S)$ is summable to f by the Abel–Lidskii method of order $n-1+\varepsilon$, $\varepsilon > 0$. Outside an arbitrarily narrow sectorial neighborhood[32] of the ray $\mathbf{R}_-$, there may be at most finitely many eigenvalues of each of these operators. The root functions corresponding to nonzero eigenvalues belong to $H^1(S)$. The same is true of the eigenfunctions for k such that these operators are selfadjoint.*

Let us explain how this can be proved for the operator $A(k)$. This operator can be represented in the form (5.373), where $A(i)$ is a selfadjoint negative operator belonging to $S^{(\frac{1}{n-1})}(L_2(S))$ and $A_1(k)$ is a compact operator. Moreover, either $A(k)$ is the inverse of $A^{-1}(k)$, or $A(k)$ has a finite-dimensional kernel, on which it coincides with a selfadjoint operator. We can eliminate the kernel by adding an appropriate finite-dimensional operator to $A(k)$. (Cf. our explanations after the formulation of Theorem 5.6.4.) Hence the completeness of root functions and the inclusion $A(k) \in \mathbf{A}(\alpha, L_2(S))$, $\alpha = n-1+\varepsilon$, follow from Assertion 1° in Subsection 5.6.6.

Note that *the assertions concerning the completeness and the summability by the Abel–Lidskii method are also valid in $H^1(S)$ if the operator is invertible.* In the opposite case, the completeness and the Abel–Lidskii summability hold in the range of the operator in question in the sense of the norm in $H^1(S)$.

It is obvious that the root functions corresponding to nonzero eigenvalues belong to $H^1(S)$. Note that in [19] there is an example of an operator $A(k)$ with an eigenfunction that corresponds to the zero eigenvalue and does not belong to $H^1(S)$. The corresponding domain is obtained in the two-dimensional case from the disk by deleting a narrow sector with the same center, and in the three-dimensional case, from the right circular cylinder by deleting a narrow dihedral sector with edge on the axis of the cylinder.

For the case of an almost smooth surface (or curve), the eigenfunctions and associated functions are C^∞ in the complement of the singular set S_0 by virtue of the local version of the smoothness theorem for elliptic equations (mentioned after Assertion 4° in Subsection 5.5.1).

The asymptotics of the characteristic numbers of the operators $A(k)$, $T^{\pm}(k)$, and $T(k)$ can be obtained for the case in which the surface S is almost smooth. First, this is done for k such that the corresponding operators are selfadjoint. Here we use Theorem 3.2 in [15]. It can be applied directly to $A(k)$, whereas to $T^{\pm}(k)$ and $T(k)$ it can be applied with regard to the fact that these operators admit two representations; for example,

$$T^+(k) = \left[\frac{1}{2}I - B(k)\right]^{-1} A(k) = A(k)\left[\frac{1}{2}I - B'(k)\right]^{-1}.$$

Thereafter the results are extended to the other values of k with regard to the fact

[32] Even a half-neighborhood, since the operators $A(k)$ or $-A(k)$, $T^{\pm}(k)$ or $-T^{\pm}(k)$, and $T(k)$ or $-T(k)$ are dissipative.

that our operators can be treated as weak perturbations of their values at $k = i$ (see Assertion 2° in Subsection 5.6.6).

Theorem 5.12.2 *Let S be an almost smooth surface. Then the following formulas are valid for the characteristic numbers of each of the operators $A(k)$, $\frac{1}{2}T^{\pm}(k)$ and $\frac{1}{4}T(k)$:*

$$\mu_j = \begin{cases} -c_2 j + o(j) & \text{for } n = 2, \\ -c_3 j^{1/2} + o(j^{1/2}) & \text{for } n = 3, \end{cases} \tag{5.375}$$

where c_2 and c_3 are indicated in (5.169) $(\sigma \equiv 1)$.

As to the situations with exceptional values of k, where Condition I or II is violated, we only note that the considerations given in Subsection 5.8.3 (see also Subsection 5.11.9) still can be applied here.

5.13 Concluding Remarks

The considerations in this chapter demonstrate a diversity of mathematical techniques that prove to be necessary for the analysis of the nonselfadjoint problems posed in the preceding chapters. Essentially, these problems stimulated certain revision and further development of the mathematical tools that had earlier been used for the study of spectral properties of elliptic nonselfadjoint operators. In particular, the research activity of the author of this chapter was strongly influenced by these problems.

Until 1960–70s, abstract theorems on the summability of series in root vectors of nearly selfadjoint operators were mainly applied (as far as many-dimensional problems were concerned) to elliptic differential boundary value problems for equations with selfadjoint principal part and “selfadjoint” boundary conditions. The spectral parameter in these problems enters into the equation. (An example is given by the equation $\Delta u + \ldots = \lambda u$ in the domain V^+ with the condition $u^+ = 0$ on S; the dots stand for lower-order terms.) Such problems can be reduced to operators of the form $L = L_0 + L_1$, where L_0 is a selfadjoint operator of positive order and the order of the operator L_1 is less than that of L_0 but nonnegative; the aim is, in particular, to cover the case in which the difference of orders is small.

The case in which the difference of orders is large attracted little attention, even though it was considered in [69]. However, this is just the case needed in the problems considered in Chapter 2. These are problems for second-order elliptic equations with spectral parameter in a boundary or transmission condition on a compact surface S. By reducing these problems to equations on S, we arrive at elliptic pseudodifferential operators of the form $L = L_0 + L_1$ of order ± 1, which are very close to selfadjoint semibounded operators L_0 (see Sections 5.7 and 5.8). In particular, L_1 is an operator of order $-\infty$ if the nonselfadjointness is caused only by the radiation conditions (without damping) or by a boundary condition on a surface that has no common points with S.

It is natural to consider such an operator L in the scale of Sobolev spaces $H^s = W_2^{(s)}$ on S, since one can assume that the operator L_0 or L_0^{-1} generates this scale. Moreover, we can use the known asymptotic formula with a remainder estimate for the eigenvalues of L_0. Here one encounters some new phenomena. First, the eigenvalues of L tend to those of L_0 very rapidly provided that the order of L_1 is very low. Second, if the order of L_1 is so low that the theorems in [69, 56, 55, 71] on the basis property with parentheses of the system of root functions of L can be applied, then we can improve these theorems as follows. We can give an estimate of how frequently the parentheses can be inserted into the series and how rapidly the resultant series with parentheses converges in $H^s(S)$. The more smooth is the function to be expanded, the more rapidly the corresponding series converges. For nonsmooth or not very smooth functions, the convergence rate of the difference between this series and the corresponding series with parentheses in the eigenfunctions of L_0 is the higher, the lower is the order of L_1. These facts can be stated in an abstract form (see Subsection 5.6.2). These convergence properties are preserved for the corresponding series outside S, that is, in the domains V^+ and V^- (by virtue of the a priori estimates for elliptic problems in question).

Along with theorems on nearly selfadjoint operators, we use the theorem on the summability by the Abel–Lidskii method of the series in root functions of operators with rays of minimal growth (see Subsection 5.6.3). This theorem, which strengthens the summability theorem in [65] and the completeness theorem in [39, Chapter XI, Section 9] and in [2], proves to be useful when it is impossible to single out a leading selfadjoint part of the operator (see Subsection 5.7.9 and Section 5.9). Cf. [62]. The estimates needed for the application of this theorem are provided by theorems on equations elliptic with parameter [2, 20, 89, 46] formulated in Subsection 5.5.6.

The operators associated with most of the problems considered in the present book possess the important properties of dissipativity and complex symmetry. In this chapter, we have repeatedly used these properties starting from Subsection 5.2.6.

The mathematical tools used in our investigation are mainly collected in Sections 5.2–5.6.

The spectral problems posed in Chapters 1 and 2 are mainly not interior problems but either exterior or transmission problems. However, the spectral parameter does not occur in the equation in the infinite domain. It occurs either in the boundary condition or in the equation in the bounded domain, on one side of the surface in question. In Subsection 5.7.2 and Section 5.9, we reduced scalar problems of this form to problems in a bounded domain with pseudodifferential boundary conditions. This allowed us, in particular, to derive the desired assertions on the unique solvability as well as the a priori estimates from known theorems.

The theorems in Sections 5.7–5.9 can be generalized to dimensions $n > 3$; however, in this case the operators corresponding to the problems in Sections 5.7 and 5.8 for $\operatorname{Im} k < 0$ will only belong to $\mathbf{A}(\alpha, H^s(S))$ with some $\alpha > 0$. Needless to say, one can also consider more general equations.

Problems for the Maxwell system are always interesting in that they seldom fit in the framework of general mathematical theorems; in this sense, the problems considered in Section 2.6 are not an exception. In Section 5.11 we show that these problems have two accumulation points of the eigenvalues (the property rarely found in problems of mathematical physics), but their other spectral properties are similar to those of the scalar problems considered in Sections 5.7 and 5.8. The proofs are based on reducing these problems for the Maxwell system to pseudodifferential systems on the boundary with the subsequent transformation of them to elliptic systems.

The theorem on the basis property in Section 5.7 was preceded by a completeness theorem [82], where also the cases in which the operator $A(k)$ is normal were indicated.

The main results of Section 5.8 were obtained by Z. N. Golubeva [44], who was then a post-graduate student of the author of this chapter. The main results of Section 5.11 were obtained in the joint paper [17].

The s-method (Section 2.5), which considerably differs from the other methods in this book, is analyzed in Section 5.10 with the use of some theorems about operators in spaces with indefinite metric.

In Subsections 5.7.5 and 5.8.2 we use theorems on the representation of pseudodifferential operators on a closed curve by Fourier type series and the theorems from [85, 10] on asymptotic series for the eigenvalues of elliptic pseudodifferential operators with real principal symbol on a closed curve. These results are stated in Subsections 5.4.4 and 5.5.4.

In Subsections 5.7.4 and 5.8.2 we use the deep theorem from [72] stating that the eigenvalue asymptotics with a remainder estimate is preserved under a weak perturbation of a selfadjoint operator.

In Subsection 5.7.9 and 5.9.4 we use a theorem from [18] about the rough asymptotics of the moduli of eigenvalues of an elliptic operator without leading selfadjoint part. In Section 5.9 we use the asymptotic formula for the eigenvalues of an elliptic pseudodifferential operator with an isolated ray on which ellipticity with parameter fails (from [11]). Both results are formulated in Subsection 5.6.5.

The results of Section 5.12 concerning spectral problems in Lipschitz domains are quite new. The exposition in this section is based on the paper [19], which uses a series of results of many authors on boundary value problems for the Laplace and Helmholtz equations in Lipschitz domains published in the last two decades, and on the results from [15] about the eigenvalue asymptotics for integral operators on Lipschitz surfaces.

Problems similar to those studied in the present book for the Lamé system in linear elasticity were considered in [16].

We hope that further research will make it possible to achieve new results concerning the problems studied in this book, or similar problems, with possible influence on the theory of nonselfadjoint operators. The authors would be happy to learn that the book has drawn the attention of new mathematicians to these problems.

Bibliography for Chapter 5

[1] M. Abramovitz and I. A. Stegun. *Handbook of Mathematical Functions. Appl. Math. Series*, Vol. 55, National Bureau of Standards, USA 1964.

[2] S. Agmon. On the eigenfunctions and on the eigenvalues of general elliptic boundary value problems. *Comm. Pure Appl. Math.* **15**(1962), 119–147.

[3] S. Agmon. *Lectures on Elliptic Boundary Value Problems.* Van Nostrand, Princeton 1965.

[4] M. S. Agranovich. Elliptic singular integro-differential operators. *Uspekhi Mat. Nauk* **20**(5)(1965), 3–120. Engl. transl. in: *Russian Math. Surveys* **20**(5)(1965), 1–121.

[5] M. S. Agranovich. Nonselfadjoint operators in diffraction type problems for a dielectric body. *Radiotekhn. i Èlektron.* **19**(5)(1974), 970–979. Engl. transl. in: *Radio Engrg. Electron. Phys.* **19**(1974), 34–42.

[6] M. S. Agranovich. Nonselfadjoint operators in scalar diffraction problems on a smooth closed surface. *Radiotekhn. i Èlektron.* **20**(1)(1975), 39–48. Engl. transl. in: *Radio Engrg. Electron. Phys.* **20**(1)(1975), 11-19.

[7] M. S. Agranovich. On nonselfadjoint integral operators with kernels of the Green function type and corresponding diffraction problems. *Radiotekhn. i Èlektron.* **20**(7)(1975), 1370–1378. Engl. transl. in: *Radio Engrg. Electron. Phys.* **20**(7)(1975), 26–32.

[8] M. S. Agranovich. On summability of series in root vectors of non-selfadjoint elliptic operators. *Funktsional. Anal. i Prilozhen.* **10**(3)(1976), 1–12. Engl. transl. in: *Functional Anal. Appl.* **10**(1977), 165–174.

[9] M. S. Agranovich. On the convergence of series in root vectors of almost selfadjoint operators. *Trudy Moskov. Mat. Obshch.* **41**(1980), 163–180. Engl. transl. in: *Trans. Moscow Math. Soc.* **41**(1982), 167–183.

[10] M. S. Agranovich. Elliptic pseudodifferential operators on a closed curve. *Trudy Moskov. Mat. Obshch.* **47**(1984), 22–67. Engl. transl. in: *Trans. Moscow Math. Soc.* **47**(1985), 23–74.

[11] M. S. Agranovich. Some asymptotic formulas for elliptic pseudodifferential operators. *Funktsional. Anal. i Prilozhen.* **21**(1)(1987), 63–65. Engl. transl. in: *Functional Anal. Appl.* **21**(1987), 53–56.

[12] M. S. Agranovich. Elliptic operators on closed manifolds. In: *Itogi Nauki i Tekhniki, Sovrem. Probl. Matem. Fundament. Napravl.*, Vol. 63, VINITI, Moscow 1990, 5–129. Engl. transl. in: *Encycl. Math. Sci.*, Vol. 63, Springer-Verlag, Berlin 1994, 1–130.

[13] M. S. Agranovich. A nonselfadjoint boundary value problem for the Maxwell system. In: *Spectral and Evolution Problems, Proc. of the Fourth Crimean Autumn Mathematical School*, Vol. 4, Simferopol 1995, 208–209 (Russian).

[14] M. S. Agranovich. Elliptic boundary problems. In: *Encycl. Math. Sci.*, Vol. 79 Springer-Verlag, Berlin 1997, 3–144.

[15] M. S. Agranovich and B. A. Amosov. Estimates of s-numbers and spectral asymptotics for integral operators of potential type on nonsmooth surfaces. *Funktsional. Anal. i Prilozhen.* **30**(2)(1996), 1–18. Engl. transl. in: *Functional Anal. Appl.* **30**(2)(1996), 75–89.

[16] M. S. Agranovich, B. A. Amosov, and M. Levitin. Spectral problems for the Lamé system in smooth and nonsmooth domains with spectral parameter in boundary conditions. Preprint 1998.

[17] M. S. Agranovich and Z. N. Golubeva. Some problems for Maxwell's system with a spectral parameter in the boundary condition. *Dokl. Akad. Nauk SSSR* **231**(4)(1976), 777–780. Engl. transl. in: *Soviet Math. Dokl.* **17**(1976), 1614–1619.

[18] M. S. Agranovich and A. S. Markus. On spectral properties of elliptic pseudo-differential operators far from self-adjoint ones. *Z. Anal. Anwendungen* **8**(3)(1989), 237–260.

[19] M. S. Agranovich and R. Mennicken. Spectral problems for the Helmholtz equation with spectral parameter in boundary conditions on a nonsmooth surface. *Mat. Sb.* **190**(1)(1999), 29–68.

[20] M. S. Agranovich and M. I. Vishik. Elliptic problems with parameter and parabolic problems of general form. *Uspekhi Mat. Nauk* **19**(3)(1964), 53–161. Engl. transl. in: *Russian Math. Surveys* **19**(3)(1964), 53–157.

[21] N. I. Akhiezer and I. M. Glazman. *Theory of Linear Operators in Hilbert Space. Vols. 1, 2.* 3rd ed. Vysshaya Shkola, Moscow 1977 (Russian). Engl. transl.: Pitman, London 1981.

[22] N. G. Askerov, S. G. Krein, and G. I. Laptev. On a class of nonselfadjoint boundary value problems. *Dokl. Akad. Nauk SSSR* **155**(3)(1964), 499–502. Engl. transl. in: *Soviet Math. Dokl.* **5**(1964), 424–427.

[23] T. Ya. Azizov and I. S. Iohvidov. A test of whether the root vectors of a completely continuous J-selfadjoint operator in a Pontrjagin space Π_κ are complete and form a basis. *Mat. Issled.* **6**(1)(1971), 158–161.

[24] C. E. Baum. On the eigenmode expansion method for electromagnetic scattering and antenna problems. Pt. 1. *Interaction Notes* **229**(1975). Pt. 2. *Interaction Notes* **472**(1988).

[25] C. E. Baum. Representation of surface current density and far scattering in EEM and SEM with entire functions. *Interaction Notes* **487**(1992).

[26] Yu. M. Berezanskii. *Expansions in Eigenfunctions of Selfadjoint Operators.* Naukova Dumka, Kiev 1965 (Russian). Engl. transl.: Amer. Math. Soc., Providence 1968.

[27] M. S. Birman and M. Z. Solomyak. Spectral asymptotics of nonsmooth elliptic operators. Pt. 1. *Trudy Moskov. Mat. Obshch.* **27**(1972), 3–52. Pt. 2. *Trudy Moskov. Mat. Obshch.* **28**(1973), 3–34. Engl. transl. in: *Trans. Moscow Math. Soc.* **27**(1975), 1–52, **28**(1975), 1–34.

[28] J. Bognár. *Indefinite Inner Product Spaces.* Springer-Verlag, Berlin 1974.

[29] A. P. Calderón. On the Cauchy integral on Lipschitz curves, and related operators. *Proc. Nat. Acad. Sci. U.S.A.* **74**(1977), 1324–1327.

[30] R. Coifman, R. McIntosh, and Y. Meyer. L'intégrale de Cauchy définit un opérateur borné sur L^2 pour les courbes lipschitziennes. *Ann. of Math.* **116**(1982), 361–387.

[31] D. Colton and R. Kress. *Integral Equation Methods in Scattering Theory.* Wiley, New York 1983.

[32] B. Dahlberg. On estimates of harmonic measure. *Arch. Rational Mech. Anal.* **65**(1977), 272–288.

[33] B. Dahlberg. On the Poisson integral for Lipschitz and C^1 domains. *Studia Math.* **66**(1979), 13–24.

[34] B. Dahlberg. Weighted norm inequalities for the Lusin area integral and the nontangential maximal functions for functions harmonic in a Lipschitz domain. *Studia Math.* **67**(1980), 279–314.

[35] Z. Ding. A proof of the trace theorem for Sobolev spaces on Lipschitz domains. *Proc. Amer. Math. Soc.* **124**(2)(1996), 591–600.

[36] C. L. Dolph. Recent developments in some non-self-adjoint problems of mathematical physics. *Bull. Amer. Math. Soc.* **67**(1)(1961), 1–69.

[37] B. A. Dubrovin, A. P. Fomenko, and S. P. Novikov. *Modern Geometry. Methods and Applications.* 2nd ed. Nauka, Moscow 1986 (Russian). Engl. transl. (Pts. I and II) in: *Grad. Texts in Math.* Springer-Verlag, Berlin **93**(1984), **104**(1985).

[38] N. Dunford and J. T. Schwartz. *Linear Operators. Part I: General Theory.* Interscience Publishers, New York–London 1958.

[39] N. Dunford and J. T. Schwartz. *Linear Operators. Part II.* Interscience Publishers, New York–London 1963.

[40] I. M. Gel'fand and G. E. Shilov. *Generalized Functions. Properties and Operations.* Academic Press, New York 1964.

[41] G. Geymonat and P. Grisvard. Eigenfunction expansions for non-self-adjoint operators and separation of variables. *Lect. Notes in Math.* **1121**(1985), 123–136.

[42] G. Geymonat and P. Grisvard. Expansions in generalized eigenvectors of operators arising in the theory of elasticity. *Diff. Int. Equat.* **4**(1991), 459–481.

[43] I. Ts. Gohberg and M. G. Krein. *Introduction to the Theory of Linear Nonselfadjoint Operators in Hilbert Spaces.* Nauka, Moscow 1965 (Russian). Engl. transl.: Amer. Math. Soc., Providence 1969.

[44] Z. N. Golubeva. Some scalar diffraction problems and the corresponding nonselfadjoint operators. *Radiotekhn. i Èlektron.* **21**(2)(1976), 219–227. Engl. transl. in: *Radio Engrg. Electron. Phys.* **21**(7)(1976).

[45] P. Grisvard. *Elliptic Problems in Nonsmooth Domains.* Pitman, London 1985.

[46] G. Grubb. *Functional Calculus of Pseudo-Differential Boundary Problems.* 2nd ed. Birkhäuser, Boston 1996.

[47] P. Häner. An exterior boundary-value problem for the Maxwell equations with boundary data in a Sobolev space. *Proc. Roy. Soc. Edinburgh Sect. A* **109**(1988), 213–224.

[48] H. Hönl, A. W. Maue, and K. Westphal. *Theorie der Beugung.* Springer-Verlag, Berlin 1961.

[49] L. Hörmander. *Linear Partial Differential Operators.* Springer-Verlag, Berlin 1963.

[50] L. Hörmander. The spectral function of an elliptic operator. *Acta Math.* **121**(1968), 193–218.

[51] I. S. Iohvidov and M. G. Krein. Spectral theory of operators in spaces with an indefinite metric. *Trudy Moskov. Mat. Obshch.* **5**(1956), 367–432. Engl. transl. in: *Amer. Math. Soc. Transl.* (2) **13**(1960), 105–175.

[52] V. Ivrii. *Precise Spectral Asymptotics for Elliptic Operators. Lect. Notes in Math.* **1100**(1982).

[53] D. Jerison and C. Kenig. The inhomogeneous Dirichlet problem in Lipschitz domains. *J. Funct. Anal.* **113**(1995), 161–219.

[54] T. Kato. *Perturbation Theory for Linear Operators.* Springer-Verlag, Berlin 1966.

[55] V. E. Katsnel'son. Conditions under which systems of eigenfunctions form a basis. *Funktsional. Anal. i Prilozhen.* **1**(2)(1967), 39–51. Engl. transl. in: *Functional Anal. Appl.* **1**(1967), 122–131.

[56] V. E. Katsnel'son. *Convergence and summability of series in root vectors for some classes of nonselfadjoint operators.* PhD dissertation, Kharkov University 1967 (Russian).

[57] M. V. Keldysh. Eigenvalues and eigenfunctions of some classes of nonselfadjoint equations. *Dokl. Akad. Nauk SSSR* **77**(1)(1951), 11–14.

[58] M. V. Keldysh. On the completeness of the eigenfunctions of some classes of non-selfadjoint operators. *Uspekhi Mat. Nauk* **26**(4)(1971), 15–41. Engl. transl. in: *Russian Math. Surveys* **26**(4)(1971), 15–44.

[59] C. Kenig. Recent progress on boundary value problems on Lipschitz domains. *Proc. Sympos. Pure Math.* **43**(1985), 175–205.

[60] C. Kenig. Elliptic boundary value problems on Lipschitz domains. In: *Bejing Lectures on Harmonic Analysis, Ann. of Math. Studies*, Vol. 112, Princeton Univ. Press, Princeton 1986, 131–183.

[61] A. N. Kolmogorov and S. V. Fomin. *Elements of the Theory of Functions and Functional Analysis.* Nauka, Moscow 1989 (Russian). German transl.: *Reele Funktionen und Funktionalanalysis.* Deutscher Verlag der Wissenschaften, Berlin 1975.

[62] A. G. Kostyuchenko and G. V. Radzievskii. On the summability by the Abel method of n-component expansions. *Sibirsk. Mat. Zh.* **15**(4)(1974), 885–780.

[63] R. Kress. On the boundary operator in electromagnetic scattering. *Proc. Roy. Soc. Edinburgh Sect. A* **103**(1986), 91–98.

[64] V. D. Kupradze. *Randvertaufgaben der Schwingungstheorie und Integralgleichungen.* Deutscher Verlag der Wissenschaften, Berlin 1956.

[65] V. B. Lidskij. On summability of series in principal vectors of nonselfadjoint operators. *Trudy Moskov. Mat. Obshch.* **11**(1962), 3–35. Engl. transl. in: *Amer. Math. Soc. Transl. Ser.* (2) **40**(1964), 193–228.

[66] J.-L. Lions and E. Magenes. *Problèmes aux limites non homogènes et applications. Vol. I.* Dunod, Paris 1968.

[67] G. I. Makarov and V. V. Novikov. Some spectral problems arising in the theory of wave propagation. In: *VI Soviet Symposium on Diffraction and Wave Propagation*, Moscow–Erevan 1973, 9–17 (Russian).

[68] A. I. Mal'cev. *Foundations of Linear Algebra.* Moscow–Leningrad 1970 (Russian). Engl. edition: W. H. Freeman & Co., San Francisco–London 1963.

[69] A. S. Markus. Expansion in root vectors of a slightly perturbed selfadjoint operator. *Dokl. Akad. Nauk SSSR* **142**(3)(1962), 538–541. Engl. transl. in: *Soviet Math. Dokl.* **3**(1)(1962), 104–109.

[70] A. S. Markus. *Introduction to the Spectral theory of Polynomial Operator Pencils.* Stiintsa, Kishinev 1986 (Russian). Engl. transl. in: Transl. Math. Monographs **71**, Amer. Math. Soc., Providence 1988.

[71] A. S. Markus and V. I. Matsaev. Convergence of expansions in eigenvectors of an operator close to selfadjoint. *Mat. Issled.* **61**(1981), 104–129.

[72] A. S. Markus and V. I. Matsaev. Comparison theorems for spectra of linear operators and spectral asymptotics. *Trudy Moskov. Mat. Obshch.* **45**(1982), 133–181. Engl. transl. in: *Trans. Moscow Math. Soc.* **45**(1984), 139–187.

[73] D. Mitrea, M. Mitrea, and J. Pipher. Vector potential theory on nonsmooth domains in $\mathbf{R}^3$ and applications to electromagnetic scattering. *J. Fourier Anal. Appl.* **3**(2)(1997), 131–192.

[74] M. Mitrea. Boundary value problems and Hardy spaces associated to the Helmholtz equation in Lipschitz domains. *J. Math. Anal. Appl.* **202**(1996), 819–842.

[75] P. M. Morse and H. Feshbach. *Methods of Theoretical Physics. Pts. I, II.* McGraw-Hill 1953.

[76] Cl. Müller. *Foundations of the Mathematical theory of Electromagnetic Waves.* Springer-Verlag, Berlin 1969.

[77] M. A. Naimark. *Linear Differential Operators.* Nauka, Moscow 1969 (Russian). Engl. edition: Parts I and II, Frederic Ungar Publishing Co., New York 1967, 1968.

[78] J. Nečas. Sur les domaines du type N. *Czechoslovak Math. J.* **12**(1962), 27–48.

[79] J. Nečas. *Les méthodes directes en theéorie des équations elliptiques.* Academia, Prague 1967.

[80] V. I. Paraska. Asymptotics of eigenvalues and singular numbers of linear smoothing operators. *Mat. Sb.* **68**(4)(1965), 621–631. Engl. transl. in: *Amer. Math. Soc. Transl. Ser.* (2) **79**(1969), 87–95.

[81] L. S. Pontryagin. Hermitian operators in spaces with indefinite metric. *Izv. Akad. Nauk SSSR Ser. Mat.* **8**(1944), 175–274. Engl. summary: pp. 275–280.

[82] A. G. Ramm. Expansion in eigenfunctions of discrete spectrum in diffraction problems. *Radiotekhn. i Èlektron.* **18**(3)(1973), 496–501. Engl. transl. in: *Radio Engrg. Electron. Phys.* **18**(1973), 364–369.

[83] S. Rempel and B.-W. Schulze. *Index Theory of Elliptic Boundary Problems.* Akademie Verlag, Berlin 1982.

[84] Ya. A. Roitberg. *Elliptic Boundary Value Problems in Generalized Functions.* Kluwer, Dordrecht 1996.

[85] G. V. Rozenbljum. Almost-similarity of operators and spectral asymptotics of pseudodifferential operators on a circle. *Trudy Moskov. Mat. Obshch.* **36**(1978), 59–74. Engl. transl. in: *Trans. Moscow Math. Soc.* **36**(2)(1979), 57–82.

[86] G. V. Rozenbljum, M. A. Shubin, and M. Z. Solomyak. Spectral theory of differential equations. In: *Itogi Nauki i Tekhniki*, Sovrem. Probl. Matem. Fundament. Napravl., 1989. VINITI, Moscow. Engl. transl. in: *Encycl. Math. Sci.* **64**, Springer-Verlag, Berlin 1994.

[87] M. Schechter. A generalization of the problem of transmission. *Ann. Scuola Norm. Sup. Pisa* **14**(3)(1960), 207–236.

[88] R. T. Seeley. Refinement of the functional calculus of Calderón and Zygmund. *Konink. Nederl. Acad. Wetensch. Proc. Ser. A* **68**(3)(1965), 521–531.

[89] R. T. Seeley. Complex powers of an elliptic operator. In: *Proc. Sympos. Pure Math.*, Vol. 10, 1967, 288–307.

[90] R. T. Seeley. A sharp asymptotic remainder estimate for the eigenvalues of the Laplacian in a domain of $\mathbf{R}^3$. *Adv. in Math.* **29**(1978), 244–269.

[91] A. A. Shkalikov. Estimates of meromorphic functions and summability theorems. *Pacific J. Math.* **103**(1982), 569–582.

[92] A. A. Shkalikov. On estimates of meromorphic functions and summability of series in root vectors of nonselfadjoint operators. *Dokl. Akad. Nauk SSSR* **268**(6)(1983), 1310–1314. Engl. transl. in: *Soviet Math. Dokl.* **27**(1983), 259–263.

[93] M. A. Shubin. *Pseudodifferential Operators and Spectral Theory.* Nauka, Moscow 1978 (Russian). Engl. transl.: Springer-Verlag, Berlin 1985.

[94] L. N. Slobodetskii. Generalized Sobolev spaces and their applications to boundary problems for partial differential equations. *Uchen. Zap. Leningr. Gos. Ped. Inst.* **197**(1958), 54–112. Engl. transl. in: *Amer. Math. Soc. Transl. Ser.* (2) **57**(1966), 207–215.

[95] M. Spivak. *A Comprehensive Introduction to Differential Geometry*, Vol. 4. 2nd ed. Publish or Perish, Wilmington, Delaware 1979.

[96] E. M. Stein. *Singular Integrals and Differentiability Properties of Functions.* Princeton Univ. Press, Princeton 1970.

[97] M. E. Taylor. *Pseudodifferential Operators.* Princeton Univ. Press, Princeton 1981.

[98] R. H. Torres and G. V. Welland. The Helmholtz equation and transmission problems with Lipschitz interfaces. *Indiana Univ. Math. J.* **42**(1993), 1454–1485.

[99] H. Triebel. *Interpolation Theory, Function Spaces, Differential Operators.* Deutscher Verlag der Wissenschaften, Berlin 1978.

[100] F. Ursell. On the exterior problems of acoustics. *Math. Proc. Cambridge Philos. Soc.* **74**(1973), 117–125.

[101] I. N. Vekua. On metaharmonic functions. *Trudy Tbiliss. Mat. Inst. Razmadze Akad. Nauk Gruzin. SSR* **12**(1943), 105–174.

[102] G. Verchota. *Layer potentials and boundary value problems for Laplace's equation on Lipschitz domains.* PhD thesis, Univ. of Minnesota 1982.

[103] G. Verchota. Layer potentials and regularity for the Dirichlet problem in Lipschitz domains. *J. Funct. Anal.* **59**(1984), 572–611.

[104] V. N. Vizitej and A. S. Markus. Convergence of multiple expansions in eigenvectors and adjoint vectors of an operator pencil. *Mat. Sb.* **66**(2)(1965), 287–320. Engl. transl. in: *Amer. Math. Soc. Transl. Ser.* (2) **87**(1970), 187–227.

[105] H. Weyl. Über die Randwertaufgabe der Strahlungstheorie und asymptotische Spectralgesetze. *J. Reine Angew. Math.* **143**(1913), 177–202.

Index